For conditic

OXFORD SERIES ON ADVANCED
MANUFACTURING

SERIES EDITORS

J. R. CROOKALL
MILTON C. SHAW
NAM P. SUH

OXFORD SERIES ON ADVANCED MANUFACTURING

3. Milton C. Shaw: *Metal cutting principles*
4. Shiro Kobayashi, T. Altan, and S. Oh: *Metal forming and the finite element method*
5. Norio Taniguchi, Masayuki Ikeda, Toshiyuki Miyazaki, and Iwao Miyamoto: *Energy-beam processing of materials: advanced manufacturing using a wide variety of energy sources*
6. Nam P. Suh: *Axiomatic approach to manufacturing process and product optimization*
7. N. Logothetis and H. P. Wynn: *Quality through design: experimental design, off-line quality control, and Taguchi's contributions*
8. John L. Burbidge: *Production flow analysis for planning group technology*
9. J. Francis Reintjes: *Numerical control: making a new technology*
10. John Benbow and John Bridgewater: *Paste flow and extrusion*
11. Andrew J. Yule and John J. Dunkley: *Atomization of melts for powder production and spray deposition*
12. John L. Burbidge: *Period batch control*
13. Milton C. Shaw: *Principles of abrasive processing*

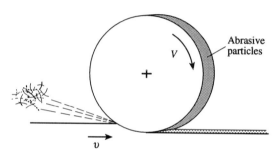

Schematic representation of a stock removal grinding operation showing what happens in the spark stream. The swarf is invisible at first, then only slightly visible as dull red particles are followed by white hot ones, and finally disintegration by explosion occurs as the temperature rises due to air resistance and exothermic reaction of chips with the air. The sequence of events resembles that of a meteor entering the earth's atmosphere, but on a much less grand scale.

Principles of
Abrasive Processing

MILTON C. SHAW
Professor Emeritus of Engineering
Arizona State University

CLARENDON PRESS • OXFORD
1996

Oxford University Press, Walton Street, Oxford, OX2 6DP
Oxford   New York
Athens   Auckland   Bangkok   Bombay
Calcutta   Cape Town   Dar es Salaam   Delhi
Florence   Hong Kong   Istanbul   Karachi
Kuala Lumpur   Madras   Madrid   Melbourne
Mexico City   Nairobi   Paris   Singapore
Taipei   Tokyo   Toronto
and associated companies in
Berlin   Ibadan

Oxford is a trade mark of Oxford University Press

Published in the United States
by Oxford University Press Inc., New York

© Milton C. Shaw, 1996

All rights reserved. No part of this publication may be
reproduced, stored in a retrieval system, or transmitted, in any
form or by any means, without the prior permission in writing of Oxford
University Press. Within the UK, exceptions are allowed in respect of any
fair dealing for the purpose of research or private study, or criticism or
review, as permitted under the Copyright, Designs and Patents Act, 1988, or
in the case of reprographic reproduction in accordance with the terms of
licences issued by the Copyright Licensing Agency. Enquiries concerning
reproduction outside those terms and in other countries should be sent to
the Rights Department, Oxford University Press, at the address above.

This book is sold subject to the condition that it shall not,
by way of trade or otherwise, be lent, re-sold, hired out, or otherwise
circulated without the publisher's prior consent in any form of binding
or cover other than that in which it is published and without a similar
condition including this condition being imposed
on the subsequent purchaser.

A catalogue record for this book is available from the British Library

Library of Congress Cataloging-in-Publication Data
Shaw, Milton Clayton, 1915–
Principles of abrasive processing / Milton C. Shaw.
(Oxford series on advanced manufacturing : 13)
Includes indexes.
1. Grinding and polishing.   I. Title.   II. Series.
TJ1280.S4461   1996   621.9′2–dc20   95-38013

ISBN 0 19 859021 0

Typeset by Colset Pte Ltd, Singapore
Printed in Great Britain by
Bookcraft (Bath) Ltd Midsomer Norton, Avon

*Dedicated to
my wife*

Mary Jane

The photograph on the cover shows 8620 steel being ground with a cubic boron nitride (CBN) wheel under the following conditions: wheel speed, 4000 f.p.m. (20 ms^{-1}); work speed, 8 f.p.m. (40.6 ms^{-1}); wheel depth of cut, 0.0015 in (0.04 mm); fluid, air (courtesy G. E. Superabrasives, Worthington, Ohio).

PREFACE

The removal of superfluous material plays a major role in materials processing, and abrasive processes as well as cutting operations are important in this connection.

While there is some similarity between cutting and grinding there are important differences. Grinding speeds are usually at least an order of magnitude greater than cutting speeds and undeformed chip thicknesses are about two orders of magnitude smaller in grinding than in cutting. The very small chip size in grinding is mainly responsible for the specific energy (energy required to remove a unit volume of material) being about an order of magnitude greater in grinding than in cutting. The high speeds normally involved in grinding together with high values of specific energy give unusually high tool-workpiece temperatures with important consequences. Thus, grinding is not just cutting with small chips.

Grinding, once considered primarily a finishing operation involving low rates of removal, has evolved as a major competitor to cutting, as the term 'abrasive machining' suggests. As near-net-shape production has progressed, grinding has assumed an ever more important role.

This monograph discusses a wide range of abrasive technology in fundamental terms. The emphasis is on why things happen as they do rather than a how-to-do-it approach. Topics covered extend from the production and evaluation of abrasive particles to the theory of comminution. First the mechanics of the simplest grinding process (surface grinding) is thoroughly covered. This is followed by simplified methods of extending principles thus developed to more complicated cases such as internal, external, creep feed, centerless, belt, and nontraditional grinding processes. Special attention is given to the following important variables: chip size, temperatures, surface integrity, wear, workpiece hardness, and the grinding environment.

The overall aim of this book is to present a unified approach to abrasive processing that will be useful in solving new grinding problems of the future. It should be of value to those doing research on abrasive processes in universities as well as to those involved in solving production problems in industry.

Where possible, theoretical aspects have been kept as simple as possible. In general, solutions of problems encountered in abrasive technology involve change of a procedure rather than prediction of a first time result. Therefore, solutions that identify variables of importance and their relative importance are emphasized. Adverse consequences of high temperature usually involve atomic diffusion and hence time and temperature are both important. There is little merit in seeking a specific value of temperature that may act for a microsecond or less since experiences that may be called on for interpretation are essentially those corresponding to equilibrium conditions. A more practical approach is to seek a proportionality that identifies variables of major significance and their relative importance. In this connection, dimensional analysis is a useful approach. Since emphasis is on principles involved in grinding behavior rather than on solutions to current problems, this monograph should be valuable far into the future.

As with many subjects it is convenient to divide abrasive behavior into two regimes. One of these is very coarse grinding that involves large chips, while fine grinding that involves chips of much smaller undeformed chip thickness is the other. These two regimes are referred to as stock removal grinding (SRG) ànd form and finish grinding (FFG). While chip formation in SRG approaches that of cutting, FFG involves an entirely different model leading to different specific energy, surface temperatures, and surface integrity. Much of the experimental and analytical interpretation presented here was done in conjunction with graduate students and research scholars in the author's laboratories over three extended periods of time:

(1) Massachusetts Institute of Technology, 1946-1961
(2) Carnegie Mellon University 1961-1978
(3) Arizona State University 1978-present

During the first period, the work was supported primarily by funds from industry. During the second period, most of the support was from two forward looking trade associations — The Abrasive Grain Association (AGA) for research on abrasive grits where no bonding material was involved and the Grinding Wheel Institute (GWI) for research involving bonded products. Research during the third period was supported primarily by the National Science Foundation. I am indebted to the many contributions and valuable discussion with coworkers that are included here.

In undertaking a work of this magnitude it is to be expected that some errors and inconsistencies will be uncovered by discerning readers and I should be grateful to be informed of these. My wife, Mary Jane, spent many hours editing the text, reading proofs and generally improving the manuscript, for which I am most grateful.

Tempe, AZ M.C.S.
May 1995

CONTENTS

List of Symbols   xiv

1. Introduction   1
Cutting Grinding Grinding Overview Important Considerations Special Topics *Absolute Versus Relative Solutions* Systems Approach *References*

2. Abrasive Tools   13
Introduction Common Abrasives Types Surface Morphology *Superabrasives* Surface Morphology *Boron Carbide Commercial Friability Tests Bonded Products Coated Products References*

3. Single Grit Performance   43
Introduction Grit Strength Static Tests Dynamic Tests Results and Discussion *Single-Grit Wear Tests* Introduction Literature Review Rubbing Wear Tests *Fly Milling Tests* Introduction Plain Fly Milling High-speed low-pressure Tests Overcut Fly Milling Face Fly Milling *Bench Tests Specific Energy* Introduction Experimental Results Sharpened Spherical Grits *References*

4. Form and Finish Grinding Mechanics   89
Introduction Hardness Model of Chip Formation Undeformed Chip Thickness Active Grit Density, C Ratio of Scratch Width to Scratch Depth Wheel-Work Contact Length Equivalent Chip Thickness References

5. Form and Finish Grinding Performance   107
Introduction Grinding Forces and Power Size Effect in FFG Mean Force per Grit Measurement of Wheel Grade Wheel Grading Tests Based on Elastic Modulus Wheel-Work Deflection Trueing, Dressing, and Conditioning The Interrupted Grinding Principle Grinding Swarf Grinding with Axial Feed References

6. Abrasive Cut-off   164
Introduction Economics Representative G Versus $\dot{d}$ curves Mechanics Wheel Speed General Observations Machine and Operating Considerations Optimization Oscillation and Rotation Competition for Abrasive cut-off Friction Sawing Precision Crack-off References

7. Conditioning of Slabs and Billets   189

*Introduction   Mechanics   Wear   Economics   Role of Abrasive Type
High-Speed – Constant-Power Conditioning   The Future   References*

8. Vertical Spindle Surface Grinding   201

*Introduction   Chip Geometry   Cost Optimization   Optimum Work Area
Wheel Wear   References*

9. Grinding Temperatures   213

Introduction   Form and Finish Grinding Temperatures   Introduction
Conductive Heat Transfer   Jaeger   Linear Jaeger Model
The Galileo Principle   Dimensional Analysis   FFG Temperatures
Grinding With Coolant   Representative Applications   Example
Surface Melting in Grinding   Sparks and Chips   Modeling and
Simulation   Thermoelectric Measurements   Radiation Measurements
Infrared Photographic Film   *Stock Removal Temperatures*
Introduction   Experimental Results   Analysis   *References*

10. Surface Integrity   261

Introduction   Surface Finish   Introduction   Factorial Experimental
Design   Surface Finish Experiments   Analytical Approach
Dynamic Active Grit Density, C Area Continuity Approach
Example   Transverse Profile Approach   Spark-out   Tracing Direction
Chatter   *Metallurgical Damage*   Introduction   Martensitic
Transformations   Surface Grinding   Analogy   Example 1   Annealed
Steel   Example 2   *Residual Stresses*   Introduction   Measurement
Examples   Origins of Residual Grinding Stresses   Ceramics   *References*

11. Wheel Life   314

Introduction   Cluster Overcut Fly Grinding (COFG)   Example 1
Example 2   Local Wheel Deflection   COFG Summary
Materials that are Difficult to Grind   Wear-resistant Tool Steel
Improved Grindability of T15 Tool Steel   *References*

12. The Grinding Environment   337

*Introduction   Gases   Liquids   High-speed Grinding   Inorganic Fluids
for Titanium Alloys*   Corrosion Inhibitors   Summary: Inorganic
Solutions for Titanium   *Belt Grinding of Titanium Alloys
Diamond Grinding of Titanium Alloys   Fluid Evaluation by COFG
Grinding Aids   Wheel Loading   References*

13. Special Processes   379

Introduction   Internal Grinding   Introduction   Constant-Force Grinding
Honing   *Creep Feed Grinding*   Introduction   Wheel Speed with Al_2O_3
Up versus Down Grinding   Cooling   Temperature Instability   Creep
Feed Machines   Surface Finish   Cubic Boron Nitride   High-efficiency
Deep Grinding   Creep Feed Cylindrical Grinding   Speed Stroke Grinding

CONTENTS xiii

Cylindrical Grinding Introduction CIRP Study Wheel Wear
Wheel Life Speed Ratio Axial Force in Profile Grinding
Centerless Grinding Introduction Applications Mechanics
Machines Optimization Example Production Rate Optimum
Belt Grinding Introduction Contact Wheels Belt Life Surface Finish
Special Operations The Future *Nontraditional Operations*
Introduction Electrochemical Grinding (ECG) Abrasive Jet Machining
(AJM) Abrasive Water Jet (AWJ) Cutting *Deburring, Lapping, and
Polishing* Burr Technology Abrasive Flow Machining (AFM)
Ultrasonic Cleaning Superfinishing Lapping Free Abrasive Grinding
Magnetic Field Assisted Finishing Polishing Precision Polishing of
Glass Elastic Emission Machining (EEM) Large Optical Components
References

14. Hard Work Materials 471

Introduction Rock and Concrete Diamond Wire Sawing Fine Wire
Sawing Grindability of Granite Diamond Circular Sawing of Granite
Power Analysis for large d/D Abrasive Cut-off Creep Feed Grinding
Summary: Performance when d/D is Large Factorial Experimental
Design for Power Chip Storage Wheels with a Single Layer of Grits
Basalt Comminution *Ceramics* Introduction Pendulum Grinding
Creep Feed Grinding Abrasive Cut-off Electrolytic in-Process Dressing
Ferrites Temperatures *Glasses* Introduction Edge Grinding
Slitting of Glass Machining of Glass Ultrasonic Grinding *References*

15. Precision Finishing 535

*Introduction Consequences of Small Chip Size Single-Point Diamond
Turning (SPDT) Ultraprecision Diamond Grinding (UPDG)
Continuous Electrolytic Dressing 'Ductile' Regime Removal
Subsurface Crack Detection Mean Radial Force per Grit Volume
Deformed in Fine Grinding The Possibility of Indentation Involving
Pulverization The Special Situation for Glass References*

Author Index 567

Subject Index

SYMBOLS

A	area
	apparent area of contact
A_R	real area of contact
B	mean volume of single chip
C	mass specific heat
	mean number of active grits per unit area of wheel face
	constant in equation 6.1
$\dot{C}$	number of chips per unit time
C_D	discharge coefficient (eqns 13.41 and 13.42)
D	wheel diameter
	diffusivity coefficient
D_S	wheel diameter
D_W	work diameter
E	Young's modulus of elasticity
F	total force (fundamental dimension in dimensional analysis)
F'	force per unit width ground
F''	force per grit
F_P	power (horizontal) component of force in surface grinding
F_Q	nonpower (vertical) component of force in surface grinding
F_T	tangential component of force (Chapter 14 for large d/D)
F_R	radial component of force (Chapter 14 for large d/D)
G	grinding ratio
G^*	cost optimum grinding ratio
H	hardness
K	constant
L	fundamental dimension of length in dimensional analysis
	sliding length
	length of cut
	length of cutting edge (eqn 4.16)
	nondimensional Peclet number (Vl/α)
M	volume removed
$\dot{M}$	volume removal rate
$\dot{M}'_W$	volume removal rate per width ground
$\dot{M}'_S$	volume rate of wheel loss per width ground
N	r.p.m.
	nondimensional wear number
	number
P	load (force)
	probability of grit fracture

LIST OF SYMBOLS

P'	power per unit width ground
Q	total heat flux
	total rate of flow
R	resultant force on wheel
	fraction of energy to work
R'	resultant force per grit
R_a	arithmetic average (centerline average) roughness
R_t	peak to valley roughness
S	screen size (openings/inch)
	volume per chip in abrasive cut-off (eqn 6.16)
	traverse rate in single point dressing
	wheel sharpness parameter (eqn 13.4)
S'	spacing of adjacent scratches
T	time
	fundamental dimension in dimensional analysis
	time between dressing operations
T_D	down time to dress wheel
T_P	time to grind one part
T_d	time point on workpiece at temperature θ_d
U	total energy
V	wheel speed
V_C	mean volume per chip
V_W	wear, volume
W	weight
b	width of cut or grind
b'	mean width of single abrasive scratch
d	wheel depth of cut
$\dot{d}$	downfeed rate
d_d	dressing depth
d'	maximum scratch depth
d^*	cost optimum downfeed rate
g	mean grit diameter
h	wear depth
hp	horsepower
k	coefficient of thermal conductivity
l	wheel–work contact length
	undeformed chip length
	length of cut
l_c	circumferential length of cluster in COFG
l_r	mean spacing of grits cutting in same groove
m	strain rate sensitivity
n	strain hardening index
	a number
	exponent in size effect (eqn 1.3)
	constant in eqn 6.1

LIST OF SYMBOLS

p	pressure
	pitch of feed marks on finished surface
	mean protrusion of active grits in wheel surface
q	chip equivalent thickness ($t_e = vd/V$), eqn 4.15
q'	speed ratio (V/v)
r	ratio of scratch width to scratch depth ($b'/\bar{t}$)
t	maximum undeformed chip thickness
$\bar{t}$	mean undeformed chip thickness
t_c	equivalent chip thickness ($= vd/V$), eqn 4.15
u	specific energy (energy per unit volume removed)
v	work speed
v_a	radial infeed per revolution of work in plunge cylindrical grinding
x	coordinate direction
	cost of labor, machine and overhead per unit time
y	coordinate direction
	cost of usable wheel per unit volume
$\hat{y}$	distance in Fig. 9.5(c)
y_f	distance to thermal front below surface
z	coordinate direction
	cost of work material per unit volume
α	rake angle in cutting
	thermal diffusivity ($k/\rho C$)
	coefficient of linear expansion
	angle
β	heat diffusivity, $(k\rho C)^{0.5}$
	friction angle (Fig. 4.2)
	angle
δ	depth of deflection
ε	log (true) strain
$\dot{\varepsilon}$	true strain rate
θ	mean grinding temperature
	fundamental dimension in dimensional analysis
θ_m	maximum grinding temperature
θ_H	nondimensional homologous temperature
Λ_W	work removal paramater (eqn 13.2)
Λ_S	wheel wear parameter (eqn 13.3)
λ	ratio of mean roughness (R_t) to downfeed per cut in VSSG
μ	coefficient of friction
ν	Poisson's ratio
ρ	mass density
	grit tip radius
ρC	volume specific heat
σ	normal stress
σ_f	plastic flow stress

LIST OF SYMBOLS

τ	shear stress
	time
ϕ	shear angle in cutting
	angle between grinding wheel and regulating wheel axes in centerless grinding
ψ	some function (in dimensional analysis)
¢	cost per part
¢$_D$	total cost of dressing
¢*	optimum cost per part

Abbreviations

AGA	Abrasive Grain Association
ASME	American Society of Mechanical Engineers
CIRP	Collège International pour l'Etude Scientifique des Techniques de Production Mécanique (International Institution for Production Engineering Research)
GWI	Grinding Wheel Institute
ICMTDR	International Conference for Machine Tool Design and Research
NAMRC	North American Metalworking Research Conference
JSGE	Japan Society of Grinding Engineers
JSPE	Japan Society of Precision Engineers
SME	Society of Manufacturing Engineers
VDI	Verein Deutsche Ingenieurs
CBN	Cubic boron nitride
COFG	Cluster Overcut Fly Grinding
FFG	Form and Finish Grinding
HAZ	Heat Affected Zone
OTM	Over Tempered Martensite
SEM	Scanning Electron Microscope
SPDT	Single Point Diamond Turning
SRG	Stock Removal Grinding
UPDG	Ultraprecision Diamond Grinding
UTM	Untempered Martensite
VSSG	Vertical Spindle Surface Grinding
w/o	Weight percent
v/o	Volume percent
AFM	Abrasive Flow Machining
AJM	Abrasive Jet Machining
AWJ	Abrasive Water Jet Machining
CHM	Chemical Milling
EBM	Electron Beam Machining
ECG	Electrochemical Grinding
ECM	Electrochemical Machining

EDM	Electrodischarge Machining
EEM	Elastic Emission Machining
ELID	Electrolytic-in-process Dressing
IBM	Ion Beam Machining
LBM	Laser Beam Machining
USG	Ultrasonic Grinding

Subscripts

D	down
H	homologous
P	power component
Q	radial component
S	wheel
U	UP
W	work
c	chip
	critical
	creep
e	effective
	equivalent
m	maximum
p	pendulum

1

INTRODUCTION

Grinding is one of the most important operations employed in Production Engineering to remove unwanted material and to introduce desired geometry and surface properties. Once considered only a secondary finishing operation, grinding is now more widely employed, as suggested by the term 'abrasive machining'. In grinding, material is removed by small extremely hard (>2000 kg mm^{-2}) brittle refractory particles. The chips produced are relatively small and cutting speeds are relatively high (usually 30 m s^{-1} or higher). The abrasive particles are bonded together to form a porous body of revolution (grinding wheel), coated on a moving substrate (coated product) or employed in a free state between surfaces in relative motion. The approach to grinding taken here is to consider the basic processes in fundamental terms rather than to attempt to present a large mass of empirical results. Thus, the emphasis is placed on why things occur as they do, rather than on specific solutions to problems ordinarily met. The rationale for this approach is that it better enables an engineer in the workshop to find solutions to new and unusual problems.

While it is often suggested that grinding is merely cutting with small chips, nothing could be further from the truth. Metal cutting is covered in detail in Shaw (1984). Only a few of the more important aspects are reviewed below for purposes of comparison with grinding.

Cutting

Metal cutting involves the removal of excess material that lies above the finished surface by a process of concentrated shear. Figure 1.1 shows a two-dimensional model of the cutting process, in which a sharp tool is producing a new surface in a planing mode. There is essentially no deformation until the advancing work material reaches shear plane A–B. The material is abruptly sheared as it crosses shear plane A–B and there is essentially no further deformation as it is convected away as a chip. The surface of the chip is subjected to high pressure and hence high frictional resistance. This results in secondary subsurface plastic flow, extending about halfway along the chip–tool contact length (i.e. from A to C), followed by sliding friction from C to D. About three-quarters of the total energy is associated with shear along A–B and the remaining quarter with shear and sliding along A–D. Essentially all of the energy consumed in cutting ends up as heat. Because of the plane strain character of most cutting operations (width of cut $b > t$) the resultant forces on the shear plane (R') and the tool face (R) will be equal in magnitude and

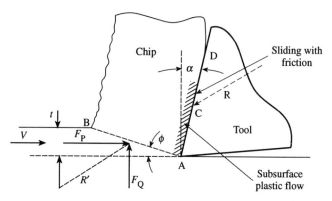

FIG. 1.1. A basic model of an orthogonal cutting operation (after Merchant 1945).

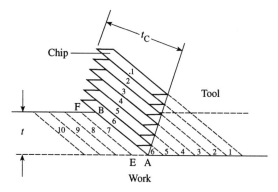

FIG. 1.2. A 'card' model for metal cutting (after Piispanen 1937).

oppositively directed. This results in a very close coupling between changes on the tool face and shear plane.

Figure 1.2 shows a remarkably accurate representation of cutting action considering its simplicity. According to this representation, not all atomic planes are active shear planes, but only those associated with a structural defect (second phase particles, missing atoms, impurities, grain boundaries, etc.). This gives rise to a process that resembles the sequential sliding of a deck of cards, having a card thickness that depends on the spacing of imperfections (Shaw 1950). This in turn results in inhomogeneous strain and a series of points on both surfaces of the chip. The points on the tool face side of the chip are removed by the secondary shear process occurring from A to C and by burnishing from C to D in Fig. 1.1. The points on the free surface of the chip remain unchanged and serve to indicate the degree of strain inhomogeneity pertaining (Shaw *et al.* 1991).

If structural defects are widely spaced, the free surface of the chip will have a saw-tooth apparance. For less intense, more closely spaced defects, the free surface of the chip will be relatively smooth and the 'teeth' will not extend continuously across the chip but will be staggered. Each of the 'cards' in Fig. 1.2 is produced by a tensile fracture extending from A to E and shear fracture from E to F. In the case of continuous chips, rewelding occurs along E–F when displacement ceases as the shear line shifts to the next position.

The specific energy u is a very useful concept for all material removal and deformation processes. For cutting and grinding, it is the energy per unit volume of material removed; while in forming it is the energy per unit volume of material deformed. In metal cutting (Fig. 1.1), the specific energy is

$$u = F_P V / Vbt = F_P / bt, \tag{1.1}$$

where F_P is the power component of force on the tool (horizontal in Fig. 1.1), b is the width of cut, and t is the undeformed chip thickness (Fig. 1.1).

Specific energy is an intensive quantity that characterizes the cutting resistance offered by a material just as tensile stress and hardness characterize the strength and plastic deformation resistance of a material respectively. The specific energy when cutting mild steel is about 300 000 in lb in^{-3} (2.07 J mm^{-3}).

In metal cutting, the specific energy is essentially independent of the cutting speed V, varies slightly with rake angle α (increased by about 1 percent per degree decrease in rake angle), but varies inversely with undeformed chip thickness t to an appreciable degree. For practical values of undeformed chip thicknesss in cutting, ($t > 0.001$ in $= 25\ \mu$m), u varies with t approximately as follows:

$$u \sim 1/t^{0.2}. \tag{1.2}$$

The range of rake angle α for most metal cutting tools is $0 \pm 10\,°$.

As previously mentioned, for a sharp tool, practically all of the energy is consumed along the shear plane and along the tool face. A worn tool will generally have a small radius at the tool tip, a region of zero clearance on the clearance surface (wear land), and a crater along the chip–tool contact area on the tool face. The wear land and tool tip radius will cause additional energy to be dissipated in the finished surface and slightly below the finished surface respectively, thus causing an increase in specific cutting energy as wear progresses. The crater will induce curl which, in turn, will decrease the length of chip–tool contact and hence result in a decrease in specific energy.

The bulk of the energy involved in metal cutting is consumed along the shear plane, and essentially all of this energy is convected away by the chip. The energy consumed along the tool face is divided between chip and tool, with most of it going to the extensive member (the chip). The energy consumed on the finished surface with a worn tool is divided between the tool and the work, with most of the energy again going to the extensive member (the work). The net result for a typical cutting operation is an energy distribution as follows:

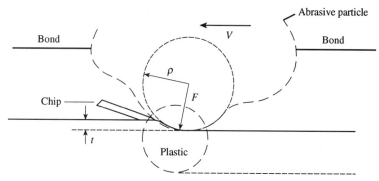

FIG. 1.3. The mechanism of chip formation at a single abrasive particle in fine grinding (after Shaw 1972).

> 90 percent of total energy to chip,
> 5 percent of total energy to tool,
> 5 percent of total energy to work.

Grinding

Grinding is performed by refractory abrasive particles of relatively uncontrolled geometry, producing many small chips at very high speed. The entire field of grinding may be divided into two regimes:

- stock removal grinding (SRG)
- form and finish grinding (FFG)

The first regime involves those processes in which the main objective is to remove unwanted material without regard for the quality of the resulting surface. The abrasive cut-off operation and the conditioning of slabs and billets in the steel industry (snagging) are typical processes of this type. In these cases, undeformed chip thickness is relatively large and wheel wear is so rapid that it is not necessary periodically to dress the wheel to remove wear flats and metal adhering to the tool face.

The second regime involves those operations in which form and finish are a major concern, and wheels must be periodically dressed to provide sharp cutting edges that are relatively free of adhering metal and wear flats.

The mean undeformed chip thickness in FFG is relatively small, and this gives rise to important differences in the metal removal mechanism compared to that in metal cutting.

In fine grinding (FFG), the undeformed chip thickness is so small relative to the radius of curvature at the tip of a cutting point that the cutting mechanism of Fig. 1.1 is inappropriate. The effective rake angle is so highly negative that metal removal by concentrated shear is replaced by a process more accurately described as extrusion. Figure 1.3 shows an active cutting point of radius ρ on the surface of an abrasive particle. The resultant force between

INTRODUCTION

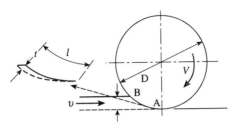

FIG. 1.4. Horizontal spindle surface grinding operation showing a single undeformed chip in the inset to the left.

the abrasive particle and the work (F) gives rise to a plastic zone that resembles that produced in a Brinell hardness test. The dashed curve is the elastic–plastic boundary involved, which is so positioned due to the inclination of resultant force F that it leaves a small region of the plastic zone in front of the grit unconstrained. The chip is extruded through this unconstrained region (Shaw 1972).

The mechanism of Fig. 1.3 for fine grinding gives rise to a size effect that is much greater than that involved in cutting, where the increase in u is due to a decrease in probability of encountering a structural defect with a decrease in size of the deformation zone. In fine grinding, a relatively large volume of material must be plastically deformed under conditions of very high specific energy in order that a relatively small volume can escape as a chip. This causes the specific energy in fine grinding to increase much more rapidly with a decrease in the undeformed chip thickness t than in cutting.

For fine grinding,

$$u \sim 1/t^n, \tag{1.3}$$

where n is 0.8–1.0, compared with about 0.2 for cutting.

In fine grinding very little of the total energy involved is convected away by the chips. Practically all of the energy ends up as heat, which is divided between the work and the wheel. An important quantity in grinding is the energy distribution coefficient R, which is the fraction of the total energy involved that ends up in the work, while $(1 - R)$ is the fraction of total energy going to the wheel.

When fine grinding with Al_2O_3 without a coolant, about 80 percent of the total energy ends up in the work (Sato 1961; Malkin 1968). This is because the bulk of the energy involved in chip formation is used to deform a layer of appreciable thickness beneath the finished surface. The bulk of the energy ends up in the work because it is not generated in the surface but beneath the surface.

Figure 1.4 depicts a fine surface grinding operation with the theoretical shape of a single undeformed chip shown to one side. The undeformed chip thickness is seen to increase as the chip is formed. In actual practice, forward chip extrusion begins only after the radial force component reaches a certain

TABLE 1.1 *Representative values of undeformed chip thickness (t), specific energy (u), and exponent n for cutting, stock removal grinding (SRG) and form and finish grinding (FFG)*

	t		u		n
	μin	μm	in lb in^{-3}	J m^{-3}	
Cutting	0.01	250	300 000	$2\,068 \times 10^6$	0.2
SRG	10^3	25	2×10^6	$13\,790 \times 10^6$	0.3
FFG	50×10^6	1.25	10×10^6	$68\,930 \times 10^6$	0.8–1.0

value. At first only rubbing occurs, followed by plowing of metal to the side and finally extrusion to the front (Hahn 1962). The picture is further complicated by the fact that all active grits in the surface of the wheel are not at the same elevation and hence the undeformed chip thickness has a range of values. It is therefore only feasible to deal with average values in fine grinding.

For the grinding operation shown in Fig. 1.4, the specific grinding energy is

$$u = (F_\mathrm{P} V)/(vbd). \tag{1.4}$$

In stock removal grinding (SRG), the undeformed chip thickness is much greater than in fine grinding and the removal mechanism of Fig. 1.1 will be approached in extreme cases. As a result, the specific energy u and exponent n in eqn (1.3) versus representative values of undeformed chip thickness t for mild steel are approximately as given in Table 1.1.

Grinding Overview

The abrasives in use today are hard refractory brittle particles that are crushed and sorted relative to size and shape. The chemical and physical properties of these materials are covered in Chapters 2 and 3, together with some performance tests for these materials. Before about 1900 grinding was done using natural stone discs that were rotated at relatively high speed so that hardened steel knives and other implements could be sharpened. At about the turn of the century, relatively pure synthetic abrasive grits (Al_2O_3 and SiC) were introduced. These have been gradually improved by alloying, control of structure, and other changes. In addition, materials of entirely different chemistry have been developed (synthetic diamond and cubic boron nitride). In the USA some companies market only abrasive particles, while others manufacture finished abrasive products (wheels and belts). Some larger companies do both. The companies concerned with abrasive particle manufacture in the USA have a trade association (The Abrasive Grain Association) that collectively considers a variety of nonproprietary problems (standards, safety, specifications, government regulations, etc.). A similar trade association (The Grinding Wheel Institute) is concerned with problems associated with abrasive products

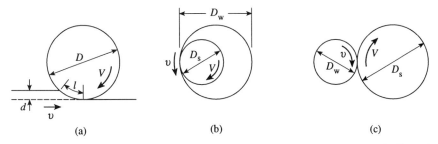

FIG. 1.5. Basic FFG grinding operations: (a) horizontal spindle surface grinding; (b) internal grinding; (c) external grinding.

incorporating a bonding system (wheels and belts). Both of these associations support nonproprietary research and, together, have had a major influence in increasing our understanding of grinding processes in fundamental terms.

Three important grinding processes employing grinding wheels are:

- horizontal spindle surface grinding (Fig. 1.5(a))
- internal grinding (Fig. 1.5(a))
- external grinding (Fig. 1.5(c))

In Figure 1.5, D_S is the diameter of the grinding wheel and D_W is the diameter of the work. In the case of surface grinding, the work diameter is infinite. A significant difference between the three operations is the conformity between wheel and work, which has an important influence on contact length (l) and influences grinding performance. Two surface velocities are shown in Fig. 1.5:

- velocity of work $= v$
- velocity of wheel $= V$

When V and v are in opposite directions, as in Figs 1.4 and 1.5, this is referred to as up grinding. Since $V \gg v$, the undeformed chip thickness (t) gradually increases from zero to its maximum value where the grit leaves the work. If V and v are in the same direction, this is referred to as down grinding, and the undeformed chip thickness goes progressively from the maximum value to zero.

All three of the operations shown in Fig. 1.5 are frequently employed with axial feed per pass (in surface grinding) and per revolution of the work for internal and external grinding. When there is no feed per pass (or per revolution of the work) and the wheel is fed radially into the work, this is called plunge grinding. When the periphery of the wheel has a pattern dressed into it, this is called form plunge grinding.

The nature and the amount of bonding material employed in bonded abrasive products plays an important role relative to grinding forces, temperatures,

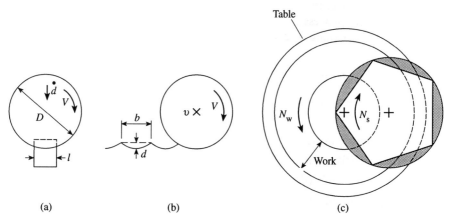

FIG. 1.6. Representative SRG grinding operations: (a) plunge type abrasive cut-off; (b) conditioning operation, wheel feeding into paper with velocity v; (c) Vertical spindle surface grinding (plan view showing segmental wheel grinding on face), N_S = wheel r.p.m,. N_W = work r.p.m.

wheel life, and surface integrity (finish, residual stress and subsurface damage). The undeformed chip thickness (t in Fig. 1.4) plays a very important role relative to the specific grinding energy, grinding forces, surface temperature, and surface integrity. Approximate methods of estimating the mean undeformed chip thickness ($\bar{t}$) are presented in Chapter 4 for FFG. The roles of grinding forces, energy, and wheel–work deflection are discussed in Chapter 5 for FFG.

Three important SRG operations are shown in Fig. 1.6:

- abrasive cut-off (Fig. 1.6(a), discussed in Chapter 6)
- conditioning of slabs and billets (Fig. 1.6(b), discussed in Chapter 7)
- vertical spindle surface grinding (Fig. 1.6(c), discussed in Chapter 8)

In the abrasive cut-off operation, a relatively thin wheel operating at high surface speed (V) is fed rapidly through the workpiece at a rate $\dot{d}$.

The conditioning of billets in the steel mill is an operation performed between hot and cold working operations in order to remove surface defects (scale, blisters, cracks, decarburized and burned material, etc.). In this operation, a large resin-bonded wheel operating at very high speed is fed rapidly across the work surface with a relatively large wheel depth of cut. This is known as snagging. An alternative process, in which unsound surface material is removed from a billet by burning using an oxyacetelene torch, is called scarfing.

The vertical spindle surface grinding operation employs a large rotary table on which the work is attached and a somewhat smaller diameter segmental wheel. Unlike most grinding operations, in this case grinding takes place on

INTRODUCTION

the face of the wheel instead of the periphery. All three of the operations in Fig. 1.6 are SRG processes having a rate of wheel wear that is so high that they are self-sharpening and periodic dressing is unnecessary.

Important Considerations

The following topics are very important to grinding and are discussed in the chapters indicated:

- grinding temperatures (Chapter 9)
- surface Integrity (Chapter 10)
- grinding wheel life (Chapter 11)
- grinding fluids (Chapter 12)

The mean surface temperature in grinding is the single most important variable influencing grinding performance. In many grinding operations, space to accommodate chips along the wheel–work contact length (l) is an important consideration. This is particularly the case in SRG where removal rates are high, as well as for operations involving grits of small size (diamond and CBN grinding). If the removal rate is too high or l is too great, this can lead to wheel loading, excessively high forces, temperatures, and wear rates.

Surface integrity involves all aspects of the quality of ground surfaces including: surface roughness, burn, subsurface transformation, overtempering, crack formation, oxidation, and residual surface stresses.

Grinding wheel wear is an important consideration not only because it influences the life of a wheel but also because it influences the frequency of dressing. Normally, when a wheel is being dressed it is not removing metal, and this down time results in added production cost. Continuous dressing is a way around this problem.

The cost of a grinding operation consists of:

(1) machine, labor, and overhead costs;
(2) abrasive costs.

As the removal rate increases, item (1) will decrease but item (2) will increase. There is usually a cost optimum removal rate as well as an optimum removal rate for maximum output.

Special Topics

Several special topics are considered in Chapter 13. These include the following:

- internal grinding
- constant force grinding
- honing

- creep feed grinding
- speed-stroke grinding
- cylindrical grinding
- centerless grinding
- form grinding
- belt grinding
- nontraditional grinding operations — electrochemical grinding (ECG), abrasive jet machining (AJW), and abrasive water jet machining (AWJM)
- deburring, lapping, and polishing

The grinding of unusually hard materials, including rock, concrete, ceramic, and glass, is discussed in Chapter 14.

The following two ultra-precision finishing operations are considered in Chapter 15:

- single point diamond turning (SPDT)
- ultra-precision diamond grinding (UPDG)

Absolute Versus Relative Solutions

Many of the problems associated with abrasive processing are extremely complex. Frequently, solutions are of limited range, and different approximations and simplified models are required for different operating conditions. Uncertainty of quantities to be used in evaluating a final solution is an important limitation of what may be achieved by a purely analytical approach. Another important problem is how an absolute analytical result (a specific temperature or stress, as opposed to a proportionality) should be interpreted. This is best illustrated by an example.

Surface temperature is one of the most important secondary variables involved in grinding. While approximate solutions may be obtained by complex analysis, the quantities to be substituted to obtain a specific temperature are elusive. For example, the ratio of real to apparent area of wheel–work contact (A_R/A) will normally range from near zero to about 2 percent. Unless a precise value of A_R/A is known, the estimated temperature will vary substantially. An alternative to obtaining a 'complete' (absolute) solution is to obtain a relative solution in the form of a proportionality. This may be achieved in the case of surface temperature in grinding by dimensional analysis, which avoids a great deal of complex analysis. While the end result does not give an absolute value, it does provide a relative proportionality. Since the time at temperature is so very short in grinding (μs) it is difficult to interpret the consequence of a high temperature, even though a precise value could be determined, since most experience pertains to equilibrium conditions.

In Chapter 9 temperatures are discussed primarily from the point of view of relative solutions, where the question to be answered is what variables

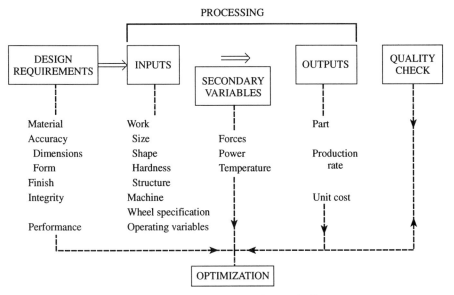

FIG. 1.7. A systems approach to grinding.

should be changed and by how much when a temperature is observed to be too high. For the reasons indicated in Chapter 9, this approach has been termed the 'Galileo Principle'.

Systems Approach

Figure 1.7 represents a systems approach to grinding, in which inputs and outputs to the process are identified and integrated with design requirements and quality control. This shows items which are self-explanatory that are set in the design office and checked against the finished product, together with processing details. The secondary variables are sometimes considered to be outputs, but in reality they are merely quantities that are useful in the solution of problems and in optimization.

In the past there has been relatively little interaction between the design and production aspects of the complete manufacturing system. The importance of a broader approach is now realized, which requires design engineers to have a basic understanding of processing technology and processing engineers of design concepts. This requires a team approach to all major decisions, ranging from materials and process selection to product quality and performance improvement.

A systems approach to grinding is discussed in detail by Peklenik (1985). Two approaches may be taken to the grinding process — a deterministic one and one in which the stochastic aspects are emphasized with the help of statistical methods. This monograph involves a deterministic approach. Peklenik (1988) outlines what is involved if a stochastic approach is taken.

References

Hahn, R. S. (1962). *Proc. 3rd ICMTDR*. Pergamon Press, Oxford, p. 129.
Merchant, M. E. (1945). *J. appl. Phys.* 16, 277 (a) and 318 (b).
Peklenik, J. (1985). *ASME PED 16*, p. 127.
Piispanen, V. (1937). *Eripaines Teknillisesla Aikakauslehdesli* 27, p. 315.
Shaw, M. C. (1950). *J. appl. Phys.* 21, 599.
Shaw, M. C. (1972). *Mech. Chem. Engng Trans., Inst. Engrs* **MC8**, 73.
Shaw, M. C. (1984). *Metal cutting principles*. Oxford. Clarendon Press.
Shaw, M. C., Janakiram, M., and Vyas, A. (1991). *Proc. NSF Grantees Conf.*, Dearborn, Michigan.

Volumes devoted to abrasive technology:
ASME Grinding Symposium (1985). *PED 16*. ASME, New York.
King, R. I. and Hahn, R. S. (1984). *Handbook of modern grinding technology*. Chapman and Hall, New York.
Malkin, S. (1989). *Grinding technology*. Ellis Horwood, Chichester, UK.
New developments in grinding (1972). Proc. Int. Grinding Conf., Pittsburgh. Carnegie Press, Carnegie-Mellon University.
Proceedings of SME International Grinding Conferences (1986, 1988, 1990, 1992, 1994). SME, Dearborn, Michigan.
Salmon, S. C. (1992). *Modern grinding technology*. McGraw-Hill, New York.

In addition, articles are to be found regularly in several journals:
Ann. CIRP (Hallweg, Berne)
Bull. Japan Soc. Precision *Engrs* (Tokyo)
J. Engng Ind. (*Trans ASME*) (New York)
Proc. Instn Mech. Engrs (London)
Proc. N. Am. Metal Working Res. Conf, (SME, Dearborn, Michigan)
Z. VDI (Berlin)

2

ABRASIVE TOOLS

Introduction

The abrasive grits employed in all grinding operations are hard brittle refractory particles that may be classified according to their hardness or chemistry. While all abrasives are hard (indentation hardness >2000 kg mm^{-2}), those that are unusually hard are often called superabrasives. There are two superabrasive types—diamond (D) and cubic boron nitride (CBN)—which have hardnesses of about 6000 and 4500 kg mm^{-2} respectively. All other abrasives (common abrasives) have a much lower hardness and wear resistance. Common abrasives comprise a wide variety of chemistry and structure that results in different degrees of fracture stress, toughness, and chemical stability.

Abrasive particles in bonded abrasive products (grinding wheels) are blocky crystals that roughly resemble a sphere with many sharp edges (Fig. 2.1). The size of an abrasive grit is expressed in terms of a screen number S, which corresponds to the number of openings per linear inch, or in terms of a mean grit diameter (g). The mean diameter of an abrasive grit (g) in inches is related to the screen number approximately as follows:

$$Sg = 0.7 \qquad (2.1)$$

Thus, a number 8 grit size has a diameter of about 0.09 in (2.3 mm) while a 46 grit size is about 0.015 in (0.38 mm) diameter. Standard grit sizes are covered by American National Standards Institute (ANSI) standard B74-20. The nondimensional product Sg would be 1 if the diameter of the wire were zero. The value 0.7 on the right side of eqn (2.1) infers that the wire diameter is about 30 percent of the wire spacing. In Europe a different grit size designation is sometimes used (FEPA, Federation of European Producers of Abrasive Products), which corresponds to the approximate mean grit diameter in micrometers.

Figure 2.2 shows the approximate relation between grit diameter (g) and screen number (S) over six orders of magnitude. While screens are normally not used to grade grits below about $S = 400$, much larger values of S are often referred to in the literature. Even though Fig. 2.2 is approximate, it is sufficiently accurate for estimating mean grit diameters for analytical purposes. This becomes evident when it is realized that S represents a range of values that may be as large as ± 50 percent of the mean value for very small grit sizes.

The depth of penetration of an abrasive grit into the work is a small percentage of its diameter. For example, in the conditioning of steel, which is one of the coarsest grinding operations, the grit size S is about 8 and the maximum depth of grit penetration (t) is about 0.004 in (0.1 mm). The maximum value

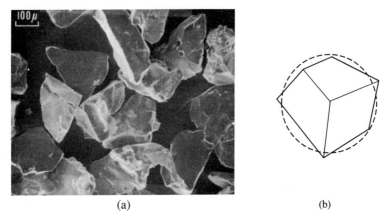

FIG. 2.1. (a) Regular aluminum oxide particles, showing the typical blocky shape. (b) The similarity between a blocky abrasive particle and a sphere.

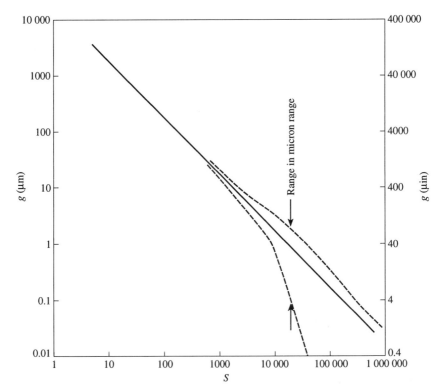

FIG. 2.2. The Approximate variation of the mean grit diameter (g) with screen number (S) over a wide range of grit sizes.

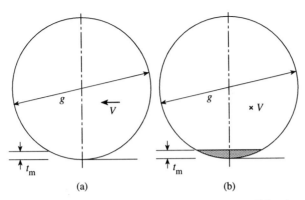

FIG. 2.3. Grit penetration (t_m) versus grit diameter (g): (a) Side elevation; (b) end view. For solid line $t_m/g = 5$ percent; for dashed line $t_m/g = 2$ percent.

of t/g is thus about $0.004/0.09 = 4.5$ percent. In horizontal spindle surface grinding a 46 grit size is common and the maximum grit penetration is about 0.0003 in (0.008 mm). The corresponding value of t/g in this case is $0.0003/0.015 = 2\%$. Figure 2.3 shows two views of an abrasive grit with a 5 percent depth of penetration, which represents the maximum value likely to be found. In practice, the depth of penetration will usually be a fraction of that shown in Fig. 2.3. It is thus evident that an extremely small percentage of the surface of a grit is operative at any one time.

Common Abrasives

Types

Before about 1890, abrasives were hard materials found in nature (typically emery, which is about 50 percent Al_2O_3 with other oxides, principally iron oxide). The first synthetic abrasive was silicon carbide (SiC). In 1891, Acheson (1893, 1896), a former employee of T. A. Edison, produced the first SiC by fusing layers of sand, coke, and sawdust with salt to remove impure iron as volatile iron chloride. In this process electric power is used to produce temperatures in excess of 3000 °F (1650°C), and the resulting mass of SiC at the center of the furnace is then crushed and screened to produce abrasive grits of different size. The overall reaction pertaining is

$$SiO_2 + 3C \Rightarrow SiC + 2Co. \tag{2.2}$$

There are two grades of SiC. The one which has fewer impurities is colored green, while that made using unconverted recycled charge is black. The green variety is somewhat more friable than the black but has about the same hardness. Most of the SiC used is black. Silicon carbide is the second most widely used type of common abrasive.

Shortly after SiC was produced, Jacobs (1900) fused bauxite to produce aluminum oxide, another class of synthetic abrasive. The composition of a typical bauxite is

- alumina, 58 w/o
- ferric oxide, 5 w/o
- silica, 5 w/o
- titania, 2.5 w/o
- water, 29.5 w/o

This is first calcined at 950 °C to remove water and then fused in an electric furnace to produce regular Al_2O_3. This contains up to 2.5 w/o titania as an impurity. By heat treating regular Al_2O_3 under oxidizing conditions, more highly oxidized and less soluble titanium oxides are formed, which precipitate, giving rise to dispersion hardening. At the same time heat treatment may cause the mechanical defects induced in crushing to be healed. Thus, the properties of regular Al_2O_3 abrasive grit may be altered by heat treatment. Aluminum oxide, the most widely used common abrasive, is the hardest oxide and there are several forms in use.

To produce a pure form of Al_2O_3, Bayer alumina must be used as the starting material instead of bauxite. Bayer alumina is produced by treating bauxite with hot caustic to produce relatively pure aluminum hydroxide which, when calcined, produces a relatively pure (99 percent) α alumina powder. When this is fused in an electric furnace, the result is white aluminum oxide. This is pure Al_2O_3 plus about 0.5 w/o Na_2O and soft β Al_2O_3 dispersed as very small voids. These voids increase the friability of the grit, resulting in a material with many sharp cutting points. This gives rise to an unusually sharp free cutting grit at the expense of a reduced tensile strength—a combination that is advantageous in fine tool grinding operations. Other varieties of Al_2O_3 are alloys of chromium oxide (<3 percent), which are pink in color, and alloys of vanadium oxide, which have a green color.

Tone (1916) produced microcrystalline Al_2O_3 of enhanced toughness by very rapidly cooling a thin stream of molten Al_2O_3 issuing from a furnace. This produced a grit of unusual toughness for use in SRG operations. Another method of producing a tough abrasive was to alloy Al_2O_3 and ZrO_2 (Saunders and White 1917). This produced a two-phase composite consisting of very hard α (hexagonal) alumina dendrites in a softer eutectic (40 w/o ZrO_2) matrix. This material is now produced by fusion of a mixture of calcined bauxite, zircon sand, petroleum coke, and iron borings in an electric arc furnace. This is poured in thin layers on to a water-cooled steel plate to ensure rapid solidification and a fine grain structure. While pure ZrO_2 is normally not hard enough to be used alone as an abrasive, it provides a tough and wear-resistant composite when alloyed with Al_2O_3.

Ridgway (1935) devised a process for producing a hydrolyzable ingot from which single crystal grits are released. These are grits that are not produced by crushing and hence possess fewer mechanical defects, which results in enhanced toughness. Still another method of producing abrasive grit that is free of mechanical defects due to crushing is to consolidate the material by sintering (~1500 °C) rather than melting. Ueltz (1963) produced the first commercial grain of this type by compacting finely divided bauxite paste under pressure, and granulating the resulting compacts to the desired grit size and shape, followed by sintering in a kiln at a temperature somewhat below the melting point. Impurities in the bauxite act as sintering aids. This approach also enables grits of controlled shape to be produced by extruding or pressing a paste.

More recently, Leithauser and Sowman (1982) and Cottringer et al. (1986) have produced Al_2O_3 abrasives by a sol-gel process in which the abrasive is not fused but sintered. By this technique, submicron α alumina particles are formed by hetergeneous nucleation (MgO seeding to produce seeds of alumina-magnesia spinel). The seed material is usually about 1 w/o. The resulting grits are essentially pore free, are not crushed, and have a submicron grain size, all of which results in a material of unusual toughness. The very fine grain size provides a self-sharpening action at an acceptable wear rate. There is a seed of the order of 0.05 μm in diameter for each resulting crystallite which has a size of about 0.2 μm. The production of extremely fine grain size (~0.1 μm) abrasive grits of α Al_2O_3 is an important development with a promising future. These abrasives are unusually tough and self-sharpening.

In addition to heat treatment to cause a change in structure or to heal mechanically induced defects, grits are sometimes coated. For example, a red iron oxide is sometimes applied to enhance vitreous bond strength. Also, a silane coating is sometimes applied to improve resin bond strength and to make it possible for resin-bonded wheels to be used with water-based grinding fluids without excessive loss of bond strength.

Fillers of cryolite, pyrite, and hollow Al_2O_3 spheres are sometimes used to improve grinding performance. Cryolyte (sodium hexafluoroaluminate) is used to lower grinding temperatures, pyrite (iron sulfide) serves as a solid lubricant to reduce grinding energy, while hollow Al_2O_3 spheres are employed to increase porosity without decreasing wheel strength.

While there is a large variety of alumina-based abrasive grits, there are only two SiC-based materials. This is because SiC does not form substitutional solid solutions as Al_2O_3 does, and also because the defect content cannot be so readily manipulated as for Al_2O_3. Additional details concerning the synthesis of common abrasives may be found in Coes (1971), Cichy (1972) and Ueltz (1972).

Surface Morphology

The surface structure of an abrasive grit, as revealed by the scanning electron microscope (SEM), plays an important role in abrasive performance. Microsopic details such as voids, micro- and macrocracks, and details of surface

TABLE 2.1 *Chemical analysis—common abrasives, w/o*

Abrasive type	Fe	Fe$_2$O$_3$	Si	SiO$_2$	Ti	TiO$_2$	Na	SO$_3$	C	S	Loss on ignition	ZrO$_2$	MnO	CaO	MgO
Sintered bauxite		5.72		3.21		3.31						0.06	0.01	0.18	0.26
40 percent ZrO$_2$		0.25		1.84		1.44						42.05			
10 percent ZrO$_2$		0.10		1.85		2.73						10.18			
Microcrystalline high TiO$_2$ (2.7 percent)															
Regular		0.26		0.80		2.43									
Monocrystalline		0.21		0.73		2.65									
White	0.03	0.10	0.02	0.11	<0.01	0.61	0.25	<0.01	0.07	0.17	0.07	0.11	<0.005	0.05	0.02
Black SiC			1.29	1.96					0.83		0.66			0.01	0.03
Green SiC			0.63	1.18					0.65		0.40			0.02	0.03

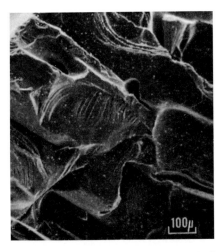

FIG. 2.4. The surface of green SiC grit (after Komanduri and Shaw 1974a).

structure are important since, as previously mentioned, the extent of abrasive engagement is so very shallow. In some instances, the size of the grit influences the detailed surface geometry, but in other cases it does not. A comprehensive study of the surface characteristics of the abrasives listed in Table 2.1 has been made (Komanduri and Shaw 1974a) and representative results are given below.

Figure 2.4 shows the surface of a 12 grit size green silicon carbide abrasive, which reveals surface features similar to that of fractured glass. As the grit size is reduced, weak spots are eliminated as a result of crushing to fine grit size. SiC ($S > 80$) is stronger than larger grit size material.

Figure 2.5 shows the surface of a regular Al_2O_3 grit at high magnification. This surface is relatively void free and lacks sharp corners. It appears to be a two-phase material, the softer phase being removed during a final milling operation that develops many microdepressions with rounded edges. This type of abrasive has low friability and is used primarily for relatively rough grinding operations.

Figure 2.6(a) shows a coarse grit size ($S = 12$) white Al_2O_3 grit. Many pores are evident, but there are few sharp cutting edges. Figure 2.6(b) shows the surface of a smaller ($S = 36$) grit size specimen, where many sharp fracture facets are evident. Apparently, the sharp edges are produced when the material fractures through the many pores present in the coarser grit. This type of grit is used primarily in FFG in relatively fine sizes ($S > 46$).

Figure 2.7 shows a rapidly solidified 40 w/o ZrO_2 - 60 w/o Al_2O_3 alloy. The white phase is ZrO_2, while the dark phase is α Al_2O_3. The dendrites are as small as 1 μm with rapid solidification. This material has a structure independent of grit size and no sharp edges. This suggests that it is best suited for rough SRG operations, which is where it gives best performance.

Figure 2.8 shows the surface of a monocrystalline Al_2O_3 grit. The same

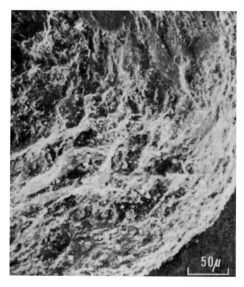

FIG. 2.5. The surface of regular Al_2O_3 grit $S = 12$) (after Komanduri and Shaw 1974a).

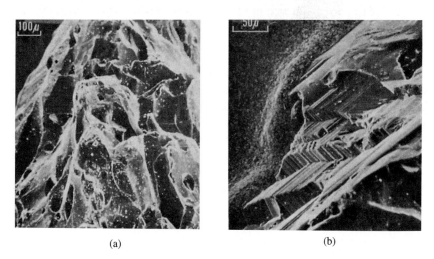

FIG. 2.6. (a) An SEM micrograph of white Al_2O_3 grit $(S = 12)$. (b) An SEM micrograph of white Al_2O_3 grit $(S = 36)$. (After Komanduri and Shaw 1974a.)

structure is found over a wide range of grit sizes. The presence of sharp facets makes this a good abrasive for FFG operations. In contrast, abrasives that are best suited for SRG operations do not have such a fine structure.

Figure 2.9 shows sintered bauxite, which reveals no sharp edges. This material is relatively tough, has a low friability, and is best suited for SSG operations such as snagging and abrasive cut-off. The very fine structure is due to the very fine powder size ($\sim 4\,\mu$m) employed.

FIG. 2.7. An SEM micrograph of 40 w/o ZrO_2 – 60 w/o Al_2O_3 grit showing the dentritic growth pattern (after Komanduri and Shaw 1974a).

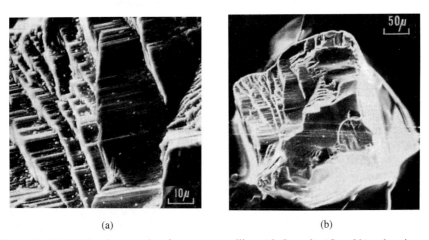

(a)   (b)

FIG. 2.8. (a) SEM micrograph of monocrystalline Al_2O_3 grit ($S = 80$), showing a series of sharp stepwise facets. (b) The same as (a), but at higher magnification. (After Komanduri and Shaw 1974a).

Figure 2.10 shows a chromium-modified abrasive grit, while Fig. 2.11 shows a vanadium-modified abrasive grit. It is reported (Coes 1971) that the purpose of these additions is partly esthetic (ruby color for Cr_2O_3 and green for V_2O_5) and partly since these materials in solid solution in α Al_2O_3 provide a hardening effect. There appear to be many sharp cutting edges in the surface of the chromium-modified abrasive which are not present on the vanadium-modified abrasive. Instead, the vanadium-modified grit has many microcavities on the surface. The presence of the micropyramids on the chromium-

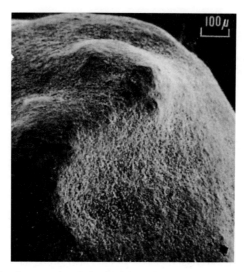

FIG. 2.9. An SEM micrograph of sintered bauxite grit, showing the rounded surface and the absence of voids and sharp cutting edges (after Komanduri and Shaw 1974a).

modified grit and the microcavities on the vanadium-modified grit could have a significant effect on chip–grit friction and on the adherence of bonding materials.

Superabrasives

Natural diamond has been in use for grinding very hard nonferrous materials, notably glass and ceramics, for a very long time (since about 1890 for saws and since about 1940 for cutting tools). Even before the end of the nineteenth century, many attempts were made to synthesize diamond, without success. It gradually became evident that extremely high temperatures and pressures would be required to convert graphite into diamond (Rossini and Jessops 1938).

Much of the pioneering experimental work that led to the first successful synthesis of diamond was performed by P. W. Bridgman (1947). While Bridgman was able to maintain temperatures near 3000 K at about 29 kbar pressure for short intervals of time, this was not sufficient to convert graphite directly to diamond (1 kbar = 1000 atmospheres). It remained for a team of innovative researchers at the General Electric Company to discover that certain catalysts made it possible to increase the rate of conversion to a practical level at relatively low temperatures and pressures (Hall 1960a, Suits 1964) Subsequently, Bundy (1963) discovered that the direct conversion of graphite to diamond without a catalyst is possible, but only at pressures above about 125 kbar and temperatures in the vicinity of 3000 K. Under these conditions, graphite spontaneously collapses into polycrystalline diamond.

The elements that are effective as catalysts include chromium, manganese,

FIG. 2.10. (a) An SEM micrograph of chromium-modified Al_2O_3. (b) The upper right-hand part of (a) at higher magnification. (After Komanduri and Shaw 1974a).

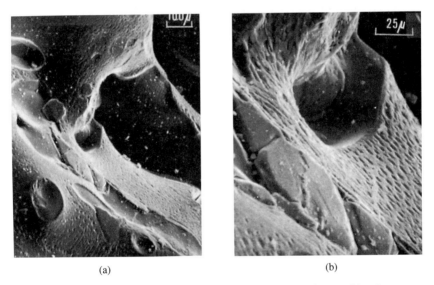

FIG. 2.11. (a) An SEM of vanadium-modified Al_2O_3, showing cavities from gasses formed during manufacture. (b) The central region of (a) at higher magnification. (After Komanduri and Shaw 1974a.)

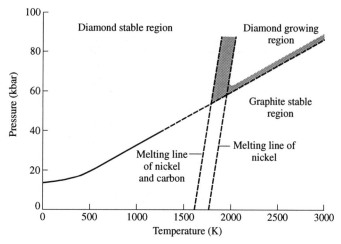

FIG. 2.12. A pressure–temperature diagram, showing the diamond growing region with nickel as catalyst (after Bovenkirk *et al.* 1959).

and tantalum, plus all elements of Group VIII of the periodic table. Carbides and compounds of these elements that decompose at or below diamond synthesis conditions may also be used. These elements are believed to play a dual role: (a) that of catalyst; and (b) as a good solvent for graphite, but a poor solvent for diamond. It appears that graphite first dissolves into the catalyst and is then converted into diamond at the appropriate conditions of pressure and temperature within the diamond stable region. Being relatively insoluble in the molten catalyst, diamond precipitates and thus allows more of the non-diamond form of carbon to go into solution.

Each catalyst has a different pressure–temperature region of effectiveness. Figure 2.12 is the pressure–temperature equilibrium diagram, according to Bovenkerk *et al.* (1959), when nickel is the catalytic solvent. This shows the diamond–graphite equilibrium line, as well as the melting lines of nickel and nickel–carbon eutectic. While graphite is a preferred starting material, since it is available in a very pure form, other types of carbonaceous material may be used.

The higher the pressure above the equilibrium line at a given temperature, the greater is the rate of diamond nucleation and growth. Diamonds formed at pressures substantially above the equilibrium line develop from many nuclei and have a skeletal structure. Such diamonds are very friable, and are designated RVG, since they are used in resin or vitrified grinding wheels. On the other hand, by subjecting the reaction mixture to pressures and temperatures closer to the equilibrium line for a longer time, fewer nucleation sites develop, and larger and more perfect single crystals of diamond are formed.

Temperature has a major influence on the crystal habit of the diamonds produced. At the lowest possible temperatures, the cubic form is predominant.

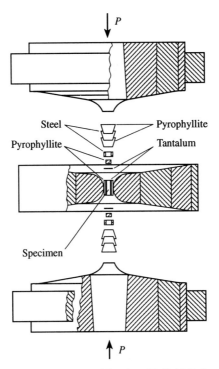

FIG. 2.13. The 'belt' (after Hall 1960a).

Cubo-octahedra are predominant at intermediate temperatures, while octahedra form at the highest temperatures.

The conversion of graphite to diamond is carried out in a high-temperature, high-pressure apparatus called a 'belt' (Fig. 2.13) by its designer, Hall (1960b). Nickel appears to be the most common metal used with graphite, although other catalytic solvents can be used. The common operating range, with nickel as catalytic solvent, appears to be 75–95 kbar and up to about 2000 °C. Since diamond size increases with time, the reaction conditions are normally maintained for several minutes when larger grits are desired. The crystal structures of hexagonal graphite and cubic diamond are shown in Fig. 2.14. The synthesis of diamond was announced by General Electric on February 16, 1955, with more details following in articles prepared by a research team (Bundy et al. 1955a,b).

Similarities in the layer-type crystal structure of hexagonal boron nitride and graphite prompted Wentdorf to investigate the possibility of a high-temperature, high-pressure stable cubic form of boron nitride similar to diamond. Early work using hexagonal boron nitride powder with the then already familiar diamond forming catalysts did not result in a cubic form, even at pressures as high as 100 kbar and temperatures up to 2000 °C, (Wentdorf 1960). The

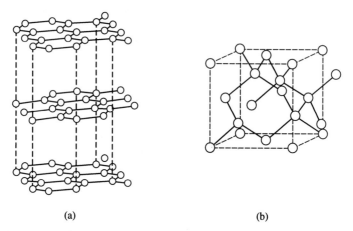

FIG. 2.14. The crystal structure of (a) graphite and (b) diamond.

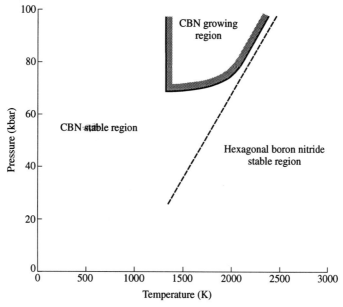

FIG. 2.15. A pressure-temperature equilibrium diagram showing CBN growing regions using magnesium as a catalyst (after Wentdorf 1960).

catalyst solvents useful for CBN production were found to be the alkali metals, the alkaline earth metals, their nitrides, antimony, tin, and lead, or combinations of all of these.

Larger size CBN crystallites (300 μm or larger) of commercial importance were found to form in the presence of a catalyst. Figure 2.15 is a pressure-temperature equilibrium diagram due to Wentdorf (1960), indicating the regions in which hexagonal boron nitride and CBN are stable in the presence of magnesium as catalyst. Larger crystals were produced when pressures and temperatures were close to the equilibrium line. The particular catalyst used was found to influence the pressure–temperature necessary for conversion. The higher the atomic weight of the catalyst, the higher was the pressure necessary to effect the transformation. The most effective catalysts were found to be nitrides of magnesium, calcium, or lithium. A mixture of magnesium nitride and sodium metal, when used as a catalyst at 80 kbar and 1700–1900 °C, was found to result in large crystals. The size of cubic boron nitride particles increases with time, and hence it is advantageous to maintain reaction conditions from 3 to 5 min, even though the reaction time is only about 0.5 minute.

CBN may be produced in monocrystalline form when a catalyst is used or in polycrystalline form in the absence of a catalyst. Monocrystalline CBN is tougher than its polycrystalline counterpart and gives better results when grinding hard M-2 tool steel or superalloys (Kumar 1993).

While diamond, the hardest known substance, has a hardness of about 6000 kg mm^{-2}, the hardness of CBN, the second hardest known substance, is 4500 kg mm^{-2}. The main reason why CBN is of interest as an abrasive is that it is much more chemically stable than diamond in the presence of hot iron. CBN is also more refractory than diamond, as the hardness temperature plots in Fig. 2.16 indicate. CBN is stable in air to about 1300 °C, but diamond is stable in air to only about 800°C.

Surface Morphology

As in the case of the regular abrasives, the surface morphology of superabrasives plays an important role in their performance. The superiority of synthetic diamonds over natural diamonds lies in the ability to control the structure, shape, and friability of the grains during their production. A comprehensive SEM study of a number of different types of superabrasives has been made (Komanduri and Shaw 1972b) and representative results are given below.

In the lower temperature region, but at pressures exceeding that of diamond–graphite equilibrium by a few kilobars, the rate of nucleation and growth of diamonds is very high. The growing crystals interfere and produce a skeletal structure with many defects. The resulting grits are relatively weak and friable, and are designated RVG by the principal US producer of synthetic diamonds. Figure 2.17(a) shows RVG diamonds of 50–60 grit size, while Fig. 2.17(b) is a single grit at higher magnification, showing the skeletal structure.

Polycrystalline diamonds of the RVG type are used in resin-bonded wheels for grinding tough materials such as cemented tungsten carbide. Diamonds of this type are self-sharpening and provide sharp cutting edges by microfracture instead of forming wear flats. Although microchipping wear may be considerable, it is frequently justified in terms of improved surface integrity. Polycrystalline forms of diamond are available in grit sizes as coarse as $S = 20$-30.

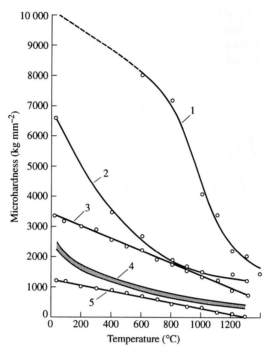

FIG. 2.16. The variation of Knoop hardness with temperature for several hard materials: 1, diamond; 2, CBN; 3, SiC; 4, varieties of Al_2O_3; 5, tungsten carbide (92 w/o WC, 8 w/o Co) (after Loladze and Bockuchava 1972).

The life of RVG grits is improved by coating them with a metal sheath corresponding to 55 percent of the weight of the diamond which corresponds to an increase in mean diameter of about 15 percent. The grit size indicated for coated superabrasive particles corresponds to the uncoated particle size. In addition to providing more area for bonding, the metal sheath tends to strengthen the grits and keep fragments from escaping and, at the same time, improves heat transfer from the surface of the wheel. However, there is an increase of power consumption of about 100 percent, and therefore an additional increase in heat flux when metal-coated diamonds are used. It is felt that these two opposing effects (improved bonding and higher power consumed with metal-coated diamonds) tend to cancel in wet grinding. However, in dry grinding, with the same type of metal coating, the increased power consumption is predominant, leading to higher temperatures and poorer wheel life.

Diamond has a very high thermal conductivity (about ten times that of nickel and twice that of copper) and a high heat capacity. Therefore, the coating should also act as a heat barrier to protect diamond from overheating during grinding, which may otherwise result in partial oxidation or graphitization. In wet grinding, the coolant extracts some heat, but this is absent in

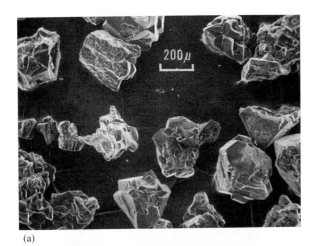

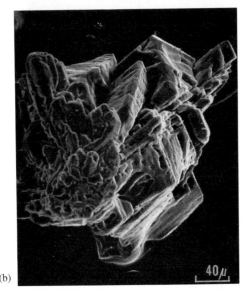

FIG. 2.17. (a) SEM micrograph of several RVG polycrystalline diamond 50–60 grits. (b) A single grit shown at higher magnification, where skeletal structure is evident. (After Komanduri and Shaw 1974b.)

dry grinding. Therefore, the grits used in dry grinding should be coated with a material to take heat away as fast as possible. Since copper is one of the best thermal conductors, it is generally used as the coating material for dry grinding. This is done despite the fact that copper is not as strong as nickel and gives a weaker bond. However, the additional bond strength with nickel is required only when a wheel is used with a water-based fluid, since the water lowers bond strength by interfacial penetration. RVG grits coated with nickel

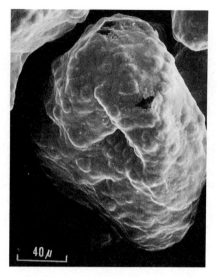

FIG. 2.18. An SEM micrograph of a single nickel-coated RVG-W diamond grit (after Komanduri and Shaw 1974b).

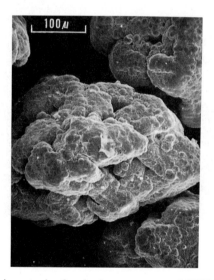

FIG. 2.19. An SEM micrograph of a single copper-coated RVG-D diamond grit (after Komanduri and Shaw 1974b).

and recommended for wet grinding are designated RVG-W, and those meant for dry grinding which are coated with copper are designated RVG-D. Unlike the copper coating, which conforms to the topography of polycrystalline diamond, the nickel coating completely encompasses the grit and the topographical features of the base crystal are barely discernible (Fig. 2.18). Figure 2.19 shows a copper-coated RVG grit (designation RVG-D).

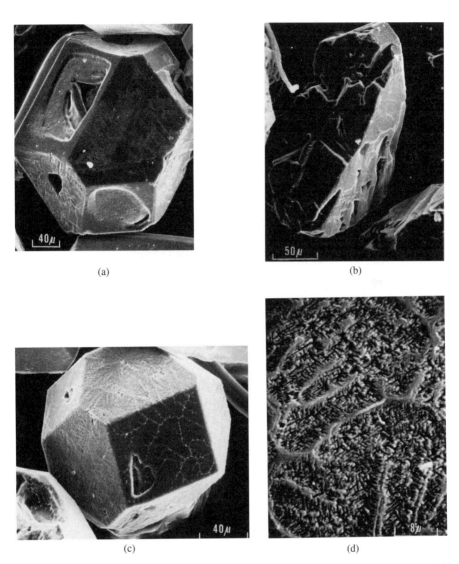

FIG. 2.20. MBG diamond grits: (a) MBG-II; (b) MGB-P; (c) MBG-T. (d) Same as (c) but at higher magnification. (After Komanduri and Shaw 1974b).

MBG grits are single crystals, are used with a metal bond, and are found to be stronger than the polycrystalline type (RVG). The principal crystal habit of growth of these crystals is cubo-octahedral, although many faces are not well formed. There are at least three varieties of MBG; namely, MBG-II, MBG-P, and MBG-T. Due to imperfections in the structure, these crystals have medium friability and strength and fall between the weak polycrystalline (RVG) type and the stronger, more fully developed, blocky MBS type.

Figure 2.20 shows examples of the three MBG types of grits. The MBG-II

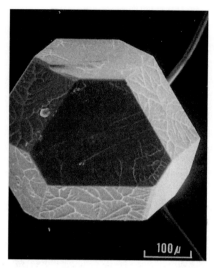

FIG. 2.21. An SEM micrograph of MBS diamond grit: note the network of streams on the cubic and octahedral faces (after Komanduri and Shaw 1974b).

variety is shown in Fig. 2.20(a), which reveals fairly well developed surfaces with some holes. Figure 2.20(b) illustrates an MBG-P crystal, which shows a characteristic elongated shape and rougher surfaces than the MBG-II type. Figures 2.20(c) and (d) are of an MBG-T type grit, showing fairly well developed cubo-octahedral faces and microrough surfaces as revealed by etching.

Figure 2.21 shows a crystal with fully developed cubo-octahedron faces. The strength of crystals of this type is the highest of the synthetic varieties (Brecker *et al.* 1973). Being well formed, with practically no imperfections, these crystals are the toughest and are used in metal-bonded saw blades for the sawing, grinding, and shaping of nonmetallic materials including stone and concrete. On both cube and octahedral faces, surface features resembling a network of streams can be seen. Such networks arise during the last stages of the diamond growth process, when some of the catalyst freezes epitaxially upon the diamond face, and diamond formation ceases where the catalyst has frozen. However, nondiamond carbon can migrate to regions where the catalyst is still molten and convert to diamond, thereby forming ridges on the diamond faces. This process continues until all the catalyst has frozen. This surface feature is observed on all types of synthetic diamonds except the polycrystalline variety.

Tarasov (1959) was among the first to report the successful use of polycrystalline diamond (RVG) for grinding tool and die steels of high vanadium, chromium, and carbon content. Tarasov showed that such steels could be ground more rapidly, economically, and with less surface damage with polycrystalline diamond wheels than with conventional aluminium oxide wheels. Further improvement in the economics of grinding these steels resulted when

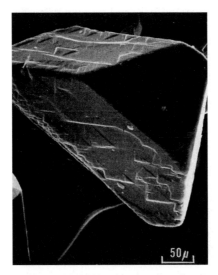

Fig. 2.22. An SEM micrograph of CBN grit, showing sharp edges and smooth surfaces on well developed planes (after Komanduri and Shaw 1974b).

polycrystalline diamonds were metal coated. Still further improvement resulted in 1969 when CBN became available (Navarro 1970).

CBN crystals are stronger than polycrystalline diamonds and some of the less perfectly formed MBG crystals, although considerably weaker than well formed single-crystal diamonds of the MBS type. Figure 2.22 is a single crystal of CBN, indicating well developed planes and sharp edges. Unlike the various types of diamond, fully developed surfaces of CBN are found to be relatively featureless and smooth. The network of streams found on both cubic and octahedral faces of nearly all single crystals of diamond (excepting the RVG type) is not found on CBN crystals. Since the manufacture of diamonds and CBN involves essentially the same process and use of a catalyst, the reason for the missing network structure is not clear at present, although it may be associated with the different types of catalytic elements used in the two cases.

The inherent smoothness of CBN faces makes bonding difficult. One way of overcoming this is to etch the crystals to provide surfaces similar to those on MBG-T crystals (Figs 2.20(c) and (d)). However, this will result in a reduction in crystal size, and since these crystals are already small, this may not be wise. Alternately, a rough surface may be created by coating the CBN with another material. This also results in an increase in apparent grit size which is advantageous. Figure 2.23 is a metal-coated cubic boron nitride crystal, showing details of the surface structure of the metal coating. Using an energy-dispersive X-ray analyzer, an elemental analysis was made and a strong nickel K_α peak was observed, thus verifying the metal coating on the CBN crystals to be nickel.

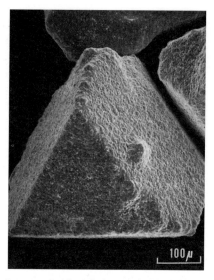

Fig. 2.23. An SEM micrograph of nickel-coated CBN-II grit (after Komanduri and Shaw 1974b).

The ability to control the strength and friability of diamonds by controlling the defect structure, shape, and surface morphology during manufacture has increased the extent of industrial application over that pertaining when only natural diamonds were available. Since the appearance of synthetic diamonds, natural diamonds have also become available in a variety of types made possible by special selection processes.

An important feature of synthetic diamonds and CBN is that little or no communition is involved in sizing. Common abrasives such as aluminum oxide and silicon carbide are usually crushed during manufacture, which introduces strength-reducing cracks into their structure. Since less dressing is involved with diamond and CBN wheels, this results in lower labor costs and a higher degree of accuracy. At the same time, better finish and surface integrity frequently result. The higher hardness of these two materials enables grits to remain sharper, while the high conductivity of diamond may result in lower surface temperatures, leading to compressive residual stresses provided that the right combination of abrasive and grinding conditions is chosen for a given work material. The use of higher strength grits for high metal-removal rates enables friability to be matched to requirements.

Boron Carbide

Boron carbide (B_4C), having a hardness of 2750 kg mm^{-2}, was developed because it was thought to have potential for use in grinding wheels. Rapid oxidation in air, a tendency to react with work materials, and a very high

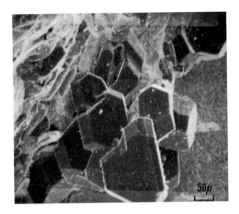

FIG. 2.24. An SEM micrograph of the surface of 12 grit size boron carbide (after Komanduri and Shaw 1974b).

friability are some of the reasons why this material has not been successful in ordinary grinding operations. It is used as a water slurry in the ultrasonic grinding process discussed in Chapter 14, and in applications in which high abrasive wear resistance at relatively low temperatures ($<1000\,°C$) and stresses is required, such as in rocket nozzles.

Boron carbide is manufactured (Ridgway 1933) by adding carbon in the form of high-purity petroleum coke to boric acid (B_2O_3) and heating to $2400\,°C$ in an electric furnace. SEM examination reveals a very porous structure. Figure 2.24 is an SEM of the surface of a 12 grit size particle of B_4C, which shows a number of sharp carbides that should be effective as microcutting tools. As the grit size is reduced by crushing, particles remain that are relatively free of pores but which contain many sharp cutting edges. This is the material that is used in ultrasonic grinding applications.

Commercial Friability Tests

Two widely used values to characterize the friability of abrasive grits are the friability index and the single blow friability exponent (r). The friability index test is performed by ball milling a 100 g abrasive sample ($S = -12$ to $+14$) in a jar rotating at 77 r.p.m. for 850 revolutions with a charge of 2000 g of 0.75 in (19 mm) steel balls. The friability index (F.I.) is

$$\text{F.I} = \frac{\text{weight of fines }(-16\text{ mesh})}{\text{total weight of abrasive recovered}} \times 100.$$

In the single-blow impact test a large number of grits (10 000) of size g_0 are dropped, one at a time, into a rubber-lined chamber in which two flat paddles rotate at 1500 r.p.m. The rate of dropping is such that each grit is struck one blow. The fracture debris is collected and passed through a set of standard

TABLE 2.2 *Properties of representative common abrasives*

Type	Friability index, F.I.	Single blow friability exponent, r	Specific gravity (g cm^3)	Knoop hardness (kg mm^{-2})†
Aluminum oxides				
Sintered bauxite	6.5	0.046	3.66	1372
40 percent ZrO$_2$	7.9	0.106	4.36	1462
10 percent ZrO$_2$	10.9	0.083	4.07	1965
Microcrystalline high TiO$_2$ (2.7 percent)	29.6	0.258	3.91	1951
Regular	35.6	0.274	3.92	2042
Monocrystalline	47.7	0.275	3.90	2276
White	56.6	0.505	3.72	2122
Ruby (3 percent Cr)	65.0	1.579	3.86	2258
Silicon carbides				
Black SiC	57.2	0.806	3.14	2679
Green SiC	62.5	1.075	3.12	2837

† Load = 100 g for all but the SiC tests, which employed a 200 g load.

screens (range of S values) and the weight proportion W of the total sample remaining on each screen size (screen opening = g) is determined. A plot of W versus g/g_0 is found to correspond approximately to the following equation:

$$W = (1 - g/g_0)^r \tag{2.3}$$

where r is the friability exponent.

When these two tests were performed on the grits of Table 2.1 ($S = 12$), the results given in Table 2.2 were obtained. These are arranged in order of increasing friability for each abrasive. The Knoop hardness values are average values obtained on diamond-polished grit surfaces (see Shaw 1984 for testing details).

While these two tests are relatively unscientific, they do serve to characterize friability in about the same way (Fig. 2.25). Figure 2.26 shows the relation between the friability index and Knoop hardness for several alumina grit types.

Bonded Products

Grinding wheels are bodies of revolution consisting of abrasive particles, bond bridges, and pores. The bond holds the entire body together, while the pores provide space for chips and fluid. Increasing the amount of bond increases the strength of the bond bridges but decreases the volume of pores. The force required to dislodge a grit increases with increase in bond content. The grade

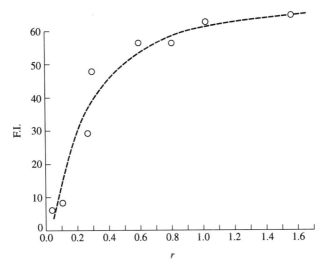

FIG. 2.25. The relation between the friability index (F.I.) and the friability exponent (*r*) for grits of Tables 2.1 and 2.2.

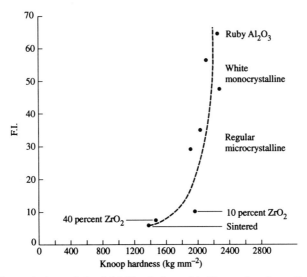

FIG. 2.26. The variation of the friability index with Knoop hardness for several grit types of alumina.

of a wheel (sometimes called the hardness) increases with bond content, and is designated by a letter of the alphabet as follows:

soft	medium	hard	very hard
E, F, G, H, I	J, K, L, M	N, O, P, Q, R	S, T, U, V, ...

There are several types of bonding materials, designated by a letter:

- V = vitrified bond material which is a brittle glass-ceramic
- B = resin bond material which is relatively soft and nonrefractory, but less brittle than a V bond
- R = natural or synthetic rubber bond
- E = shellac
- M = metal bond

The first grinding wheels were produced with a resin bond in 1909, and appeared soon after L. H. Baekeland received his patent on Bakelite (phenol-formaldehyde) in 1907.

The structure number is a measure of the volume and size of the pores as follows:

structure number = pore size

Small	Regular	Large	Very large
1, 2, 3, 4	5, 6, 7	8, 9, 10	11, 12, 13, 14

The openness of the structure plays an important role relative to chip accommodation, and is controlled by the amount of bonding material employed or by incorporating a volatile filler as an initial ingredient. Sawdust, which is converted to a gas when the wheel is fired, or a material which sublimes on firing, such as hexachlorethane (C_2Cl_6), is introduced to the mix before pressing. When these materials leave as a gas, voids are left behind.

The grit size is designated by the S number (screen number = number of openings per linear inch) as follows:

coarse	medium	fine	very fine
8, 12, 20, 24, 36	46, 54, 60, 80, 100, 120	150, 180, 240, 280	320, 400, 600

The grit type is designated by a letter, preceded by a symbol that indicates the particular type of generic material employed by the wheel manufacturer. The principal grit types are as follows:

- A = predominantly alumina
- C = predominantly SiC
- D = diamond
- B = CBN

A universal marking system is used to convey all of this information in a compact manner, as follows:

grit type	S	grade	structure	bond	manufacturer's symbol
WA	46	K	8	V	XX

This corresponds to a white aluminum oxide wheel of 46 grit size, a medium (K) grade, an open (8) structure and having a vitrified bond. Any special treatment such as impregnation with sulfur, wax, graphite, etc. would be incorporated in XX at the end of the designation.

The size of a grinding wheel is usually specified by giving the following dimensions in the order indicated:

diameter   thickness   bore

A typical vitrified grinding wheel such as that with the foregoing designation would correspond approximately to the following volume percentages:

50 percent abrasive
10 percent bond
40 percent pores

A typical hot pressed resin-bonded cut-off wheel would correspond to the following volume percentages before curing:

70 percent abrasive
12 percent resin
4 percent liquid
14 percent filler

The pore content of the resin-bonded wheel will be much smaller than that for a vitrified wheel. This is important for cut-off wheels, which operate at very high speeds and hence are subjected to unusually high centrifugal stresses. However, the pore volume in the surface of such wheels is much larger than in bulk due to the scouring action of the chips on the relatively soft non-refractory bond material. Cut-off wheels are subjected to considerable shock loading, and hence the relatively higher shock resistance of the resin bond also plays an important role.

Grinding wheels are trued and dressed by machining the active surface with a hard steel or diamond tool. Dressing details are given in Chapter 5. Some wheels are operated at such a high removal rate they are self-dressing in the sense that grit fracture and dislodgement (wheel wear) is so great that wear flats and metal adhesion are not a problem. This is normally the case in SRG operations. However, in FFG operations, wheel geometry usually must be periodically reestablished, the wheel surface cleaned of adhering metal, and the abrasive particles sharpened.

The abrasive content of superabrasive wheels is usually designated by a concentration number. A concentration of 100 corresponds to 4.2 carats of D or CBN per cm^3 (1 carat = 200 mg). Since the densities of D and CBN are about the same (3.52 and 3.48 $g\,cm^{-3}$ respectively), the volume percentage of abrasive corresponding to a concentration of 100 is about 25 percent in both cases. Also, the volume percentage of abrasive is proportional to the concentration number for other cases (i.e. 12.5 volume percent for a concentration of 50, etc.).

Superabrasives are used in a variety of grit sizes:

- For sawing, $S = 20\text{-}60$
- For grinding, $S = 40\text{-}325$
- For lapping and polishing, $S = 325\text{-}10^4$
- For precision grinding, $S =$ up to 5000

Superabrasive grinding wheels are frequently produced by anchoring a single layer of nickel-coated D or CBN particles on the periphery of a metal blank by a layer of electrodeposited nickel or by brazing. Such wheels are particularly useful for form grinding operations, since the desired form may be applied to the blank before coating rather than being provided by a very slow and costly truing operation.

Superabrasive grinding wheels are also used with resin (B), metal (M), and vitreous (V) bonding materials. Such wheels are usually produced with an abrasive containing region that is only about 2 mm (0.08 in) thick, attached to a metallic core. A standardized marking system indicates the core shape, location, and shape of the abrasive section and any special modification to the core. These details are specified in the U.S.A. by an American National Standards Institute (ANSI) standard (B74.3—1986) and in Europe by the corresponding FEPA standard.

The designation used to specify a CBN wheel is similar to that for a wheel with common abrasive:

grit type	S	grade	concentration	bond	mfg.
B	120	O	50	B	XX

This corresponds to a CBN wheel of 120 grit size, an O grade, a concentration of 50, and a resin bond.

When substituting a superabrasive wheel for a common abrasive wheel, it is advisable to employ a wheel of smaller width and a grit size that is two or three steps smaller than the conventional abrasive size. Brittle materials such as rock or glass are best ground with a wheel of low concentration (50 or lower), while more ductile materials are usually best ground with a wheel of higher concentration (150 or more) As the grit size of a superabrasive wheel is increased, the removal rate may be increased and wheel life increases, but the finish obtained is poorer. A wheel of higher concentration normally enables a higher removal rate with more power required and a longer wheel life.

Coated Products

Coated abrasive products consist of a single layer of abrasive particles attached to some sort of substrate (cloth, paper, polymer, etc.). The abrasive particles are farther apart than those in a grinding wheel, so there is ample chip accommodation space available for high removal rates. The substrate is usually in the form of a belt or disk.

An adhesive layer called the make coat is first applied to the backing material

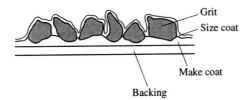

FIG. 2.27. A sectional view of a coated abrasive product.

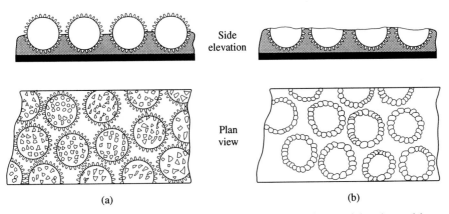

FIG. 2.28. Abrasive grits, consisting of small common abrasive particles sintered into the form of a hollow sphere: (a) two views before grinding; (b) two views after grinding (after Borkowski and Szymanski 1992).

before the abrasive particles are applied. As the backing material passes through a strong electrostatic field, the particles are oriented with their longest dimension vertical. This provides good grit retention and also orients the particles with their sharpest edges outward. A second layer of adhesive, called the size coat, is then applied over the entire assembly. Figure 2.27 shows a typical coated product in section.

An interesting type of abrasive grit used for coated products is shown in Fig. 2.28. The abrasive particles are hollow ceramic spheres, with smaller common abrasive particles in the peripheries. of the spheres. The size of the individual abrasive particles is about one-tenth that of the entire particle. The resulting particle size corresponds to that normally used in a belt or disc for the intended purpose. Figure 2.28(b) shows the grits after break-in. The space available to accomodate chips with this design is even greater than for an ordinary coated product, which is important at high removal rates. Solid agglomerate grits are also used in coated products (see Fig. 13.32). Applications of coated products of these types are discussed in chapter 13 under 'Belt grinding.'

References

Acheson, E. G. (1893). *U.S. Pat. 492,767.*
Acheson, E. G. (1896). *U.S. Pat. 560,291.*
American National Standards Institution (ANSI) *Standard B74.3* – 1986.
Borkowski, J. and Szymanski, A. (1992). *Uses of abrasives and abrasive Tools.* Ellis Horwood, New York, p. 46.
Bovenkirk, H. P., Bundy, F. P. Hall, H. T., Strong, H. M., and Wentdorf, R. H. (1959). *Nature* **184**, 1094.
Brecker, J. N., Komanduri, R., and Shaw, M. C., (1973). *Ann. CIRP* **22**(2), 189.
Bridgman, P. W. (1947). *J. chem. Phys.* **15**, 92.
Bundy, F. P. (1963). *J. chem. Phys.* **38**, 631.
Bundy, F. P., Hall, H. M., Strong, H. M., and Wentdorf, R. H. (1955a). *Chem. Engng News* **33**, 718.
Bundy, F. P., Hall, H. M., Strong, H. M., and Wentdorf, R. H. (1955b). *Nature* **176**, 55.
Cichy, P. (1972). *Met. Soc. AIME*, TMS paper EFC7.
Coes, L. (1971). *Abrasives.* Springer Verlag, New York.
Cottringer, T., van de Merwe, R. H., and Bauer, R. (1986). *U.S. Pat. 4,623,364.*
Hall, H. T. (1960a). *U.S. Pat. 2,947,608.*
Hall, H. T. (1960b). *U.S. Pat. 2,947,610.*
Jacobs, C. F. (1900). *U.S. Pat. 659,207.*
Komanduri, R. and Shaw, M. C. (1974a). *J. Engng. Mat. (Trans. ASME)* **96**, 145.
Komanduri, R. and Shaw, M. C. (1974b). *Int. J. Machine Tool Des. Res.* **14**, 63.
Kumar, K. V. (1993). *Proc. 5th Int. Grinding Conf.*, SME, Dearborn, Michigan.
Leitheiser, M. C. and Sowman, H. G. (1982). *U.S. Pat. 4,314,827.*
Loladze, T. N. and Bokuchava, G. V. (1972). *Proc. Int. Grinding Conf.* Carnegie Press, Pittsburgh, Pennsylvania, p. 432.
Navarro, N. P. (1970). *Soc. Man. Engrs, Tech. Paper MR-70-198.*
Ridgway, R. R. (1933). *U.S. Pat. 1,897,214.*
Ridgway, R. R. (1935). *U.S. Pat. 2,003,867.*
Rossini, F. P. and Jessops, R. S. (1938). *J. Res. Nat. Bur. Stds* **21**, 491.
Saunders, L. E. and White, R. H. (1917). *U.S. Pats. 1,240,490 and 1,240,491.*
Shaw, M. C. (1984). *Metal cutting principles.* Clarendon Press, Oxford, p. 76.
Suits, C. G. (1964). *Am. Scient.* **52**, 395.
Tarasov, L. P. (1959). *Tool Engr* **13**, 109.
Tone, F. J. (1916). *U.S. Pat. 1,192,709.*
Ueltz, H. F. G. (1963). *U.S. Pat. 3,079 243.*
Ueltz, H. F. G. (1972). *Proc. Int. Grinding Conf.*, Carnegie Press, Pittsburgh, Pennsylvania, p. 1.
Wentdorf, R. H. (1960). *U.S. Pat. 2,947,617.*

3

SINGLE GRIT PERFORMANCE

Introduction

The evaluation of abrasive grits for use in grinding is a complex problem due to a wide range of grinding conditions and the nonuniform geometry of grits. Since chip size is very small, grinding mechanics is easily affected by the variations in geometry and properties from grit to grit. The grit properties of primary concern are:

- hardness
- bulk strength (static and with impact)
- friability
- wear resistance (attritious and microchipping)
- chemical stability

Since the relative hardness of the contacting bodies is of prime importance in determining abrasive wear, abrasives of high hardness are desired. Static strength is required to withstand grinding forces, while impact strength determines chipping wear. Susceptibility to failure due to thermal crack formation is also of importance. Wear resistance is related to adhesion and brittle fracture. Chemical stability influences oxidation and diffusion wear.

The high temperatures, high pressures, and chemical effects present in actual grinding operations make evaluations based on room-temperature properties inadequate. It is necessary to devise laboratory bench tests to evaluate abrasive performance characteristics under conditions that more closely resemble those in actual grinding than do conventional room-temperture properties. Bench tests are useful in screening new abrasive candidates for FFG and SRG, since they provide valuable information based on a small fraction of the effort and sample size required to test a complete grinding wheel or belt. The performance of abrasive grits may be approached from two points of view:

- the behavior of individual grits
- the behavior of complete abrasive tools (principally bonded and coated products), including abrasive particles, bond, and pores

In this chapter, the behavior of single grits will be considered, while the performance of more complex composite abrasive tools is considered in Chapter 11.

Important characteristics that may be studied for isolated abrasive particles include the following:

- grit strength

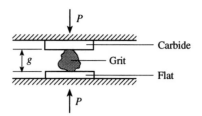

FIG. 3.1. The static tensile strength test.

- sliding wear rate
- single grit milling (fly milling)
- heat treatment of grit
- forces and energy
- dressability
- grit tip radius
- specific energy

Grit Strength

Static Tests

Fracture strength determines the grinding loads which a grit can withstand without failure. It can be measured by loading individual grits between hard platens (Fig. 3.1) until rupture occurs, and early work on this test has been presented by Yoshikawa and Sata (1960). Although the applied load is compressive, brittle fracture occurs due to a tensile stress which develops perpendicular to the loading axis. This loading is similar to the loads applied to grits in grinding, and provides a useful comparison of abrasive materials relative to their bulk strength.

Assuming that brittle failure occurs by the maximum tensile stress criterion (Takagi and Shaw 1983), the effective stress σ_e (psi) for a blocky abrasive particle will be (Brecker 1974)

$$\sigma_e = 1.37 \, P/g^2 \tag{3.1}$$

where P is the fracture load (in lb) and g is nominal grit diameter (in), or

$$\sigma_e = 0.962 \times 10^{-3} P/g^2$$

when P is in kg and g is in mm.

As in all failure testing, there is considerable variation in results and a statistical treatment of the data is necessary. The statistics that Weibull applied to ball bearing failures is a convenient method to use (see Shaw 1984 for details). At least 20 grits are tested to give reasonable accuracy. The grits are

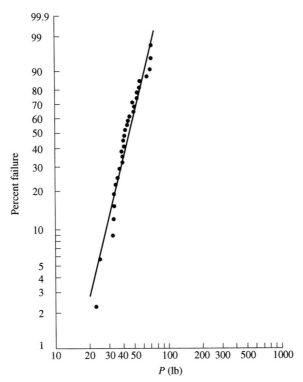

FIG. 3.2. A Weibull plot of tensile breaking loads for regular Al$_2$O$_3$ 12 grit size (after Brecker 1974).

ordered on the basis of increasing failure load and then the statistically expected percent failed is plotted against the failure load for the given grit sample. For example, Fig. 3.2 shows results obtained for a regular aluminum oxide sample. The mean breaking load P_{50} is about 45 lb and thus the mean tensile strength, according to eqn (3.1), is:

$$\sigma_e = 1.37[45/(0.060)^2] = 17\,100 \text{ p.s.i.} (118 \text{ MPa}).$$

The slope of the plot is indicative of the degree of brittleness of the material, with glass (brittle) having a slope of about 1, while steel (ductile) has a slope of about 40. Abrasives have a slope of 2–4, indicating brittle behavior.

This test can also be used to evaluate the bulk tensile strength of the abrasive material by grinding parallel flats on the grit so that stress concentrations due to irregular geometry are largely eliminated. This is representative of grit strength after a large wear land has developed on the grit. Abrasive materials with loading flats have a bulk strength that is several times their strength in the sharp condition. Since the strength of a grit in its 'as-crushed' condition is indicative of initial grit strength in grinding and is easier to obtain, abrasive

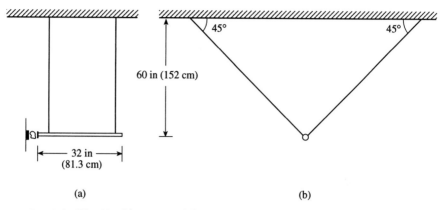

FIG. 3.3. The Hopkinson pendulum apparatus: (a) side view; (b) end view.

grits are usually tested in the 'as-manufactured' condition. When loading flats are used they are made to have a chordal extent of 20 percent of the nominal grit diameter.

Another static test for abrasive grit strength is an indentation test in which flats ground on abrasive particles are indented by a Vickers hardness indenter at increasing loads (depth of impression) until cracks are evident at the corners of the impression. These results correlate very well with the compression strength test but the tests are considerably more tedious to perform.

Dynamic Tests

The strength tests that have been considered thus far are static tests. In practice, abrasive grits are subjected to shock loading. One method employed to incorporate shock consists of a Hopkinson pendulum test, in which a long cylinder is impacted against a stationary grit to cause fracture (Fig. 3.3). The 60 in (1.5 m) pendulum length combined with a 0.25 lb (1.1 N) striker rod provides sufficient impact energy (~0.1 in lb–5.7 mm N) to fracture all types of number 12 grit. The striker is pulled back a distance corresponding to the desired energy level and then released. The energy is increased in increments of 0.016 in lb (0.09 mm N) until fracture occurs. It is necessary to test about 30 grits in order to obtain a reasonable Weibull plot of percent failure versus fracture energy. When multiple-impact tests were compared with single-blow tests, it was found that 75 percent of the failures occurred on the first two impacts, which suggests essentially no fatigue effect. Each grit type was therefore subjected to five impacts before advancing to the next energy level. Figure 3.4 shows the Weibull plot for percentage failure versus impact energy level at fracture for number 12 grit size regular Al_2O_3 in the as-received (no flats) condition. However, this dynamic test procedure proved to be very tedious to perform.

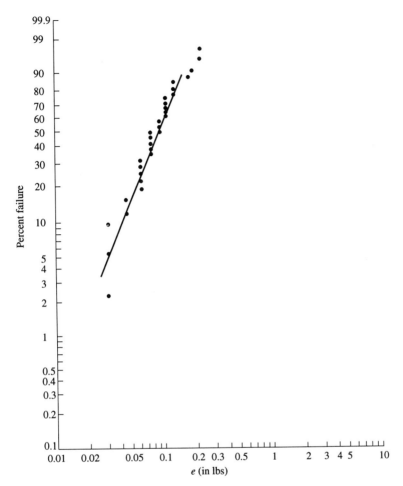

FIG. 3.4. A Weibull plot for percent failed versus fracture energy (*e*) for unground regular Al_2O_3.

A more convenient determination of dynamic strength, particularly for small grits, can be obtained with a roll crushing device (Brecker 1974). As shown in Fig. 3.5, grits are dropped one at a time between hardened steel rolls that have a gap equal to 60–80 percent of the nominal grit diameter. The force required to crush the grit is measured with a strain gage circuit and recorded for use in generating a Weibull plot. This test can also be used to evaluate the details of grit fracture. Typical force recordings made on an oscilloscope are shown in Fig. 3.6.

The desired mode of failure is one in which the particle splits into a few large pieces along the loaded diameter (Fig. 3.6(b)). The vibration indicated is the natural frequency of the test device after the sudden release of energy

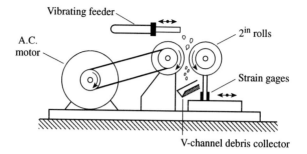

FIG. 3.5. The roll crushing tensile test (after Brecker 1974).

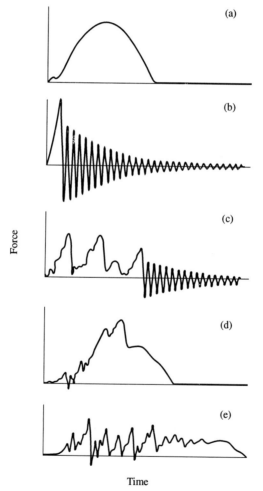

FIG. 3.6. Typical force traces obtained during roll crushing tensile strength tests (after Brecker 1974). (a) No fracture; (b) fracture into two parts; (c) partial failures before fracture; (d) fracture followed by nonfracture loading of fragment; (e) continuous crushing.

when the particle fails. The smaller grit sizes tend to crush gradually (Fig. 3.6(e)) rather than to fracture suddenly.

Results and Discussion

Table 3.1 gives mean strength values for a wide variety of commercial grits, all tested in the dry condition and in number 12 grit size. The quantity m is the slope of the Weibull plot. The 'Friability exponent' is that discussed in Chapter 2. Values of e_{50}, the median fracture energy in the pendulum test, are also given in Table 3.1.

Several interesting observations can be made from the tensile strength data. The tests on unground grits and the friability tests distinguish between FFG and SRG grits, but the ordering of abrasives within the categories varies between the tests. Both white and regular aluminum oxide show greater strength than expected, considering their friability. The roll crushing tests include some dynamic grit loading and should be more representative of strength in grinding. The roll crushing strengths agree closely with static strength values, except for the regular and microcrystalline grits which exhibit significantly greater strength when tested dynamically. This characteristic of increased strength under somewhat dynamic loading indicates why these abrasives are used for moderate to heavy grinding conditions.

The static tests with ground flats show that all of the abrasives can withstand much greater loads when their irregular shapes do not magnify stresses. Particularly noteworthy is the much improved strength of a microcrystalline grit when a wear flat develops on the grit. Thus, this grit type is more applicable to circumstances in which fracture is not wanted when a wear land develops on the grit.

Since grinding conditions cause grits and/or wheels to go from a sharp condition to a dull condition, the operations cover both areas in which flats do or do not exist. Thus, a compromise on the characteristics exhibited in both regions is necessary.

The low strength of silicon carbide is the principal reason for its use on materials that are difficult to process, where chipping or fracture is necessary to maintain sharp cutting edges.

The tensile strength can vary significantly as the abrasive diameter decreases. A typical result shown in Fig. 3.7 indicates a large increase in strength for regular aluminum oxide below 46 grit size ($g = 0.015$ in). The Weibull slope increases as grit size decreases, indicating less brittle behavior at small grit size. Since a more defect-free material is approached as grit size decreases, an increase in strength is not unexpected.

The majority of the 12 and 24 grit size particles failed in a brittle mode, as in Fig. 3.6(b), but only about 10 percent of the fine size particles failed. This difference in behavior could be due to a strain energy size effect. Glucklick (1970) has observed that the size effect noted in the strength and fracture of materials is not only a function of sharp crack initiation but also of sharp crack growth. The weakest-link theory, in which the defects in a brittle

TABLE 3.1 *Hardness, friability, tensile strength, and impact strength for common abrasives, 12 grit size*

Type	Knoop hardness ($kg\ mm^{-1}$)†	Friability index	Friability exponent	Static $\bar{\sigma}_e$, with ground flats		$\bar{\sigma}_e$, unground		$\bar{\sigma}_e$, roll crushing, unground		e_{50}, pendulum	
				p.s.i.	m	p.s.i.	m	p.s.i.	m	in lb	m
Aluminum oxide											
Modified (3 percent Cr)	2 258	65.0	1.60	57 000	3.1	11 400	3.3	9 600	1.9	0.064	1.6
White	2 122	56.6	0.50	53 400	3.0	17 900	3.5	17 000	3.2	0.076	2.4
Monocrystalline	2 276	47.7	0.28	71 000	3.1	13 500	2.8	14 400	2.1	0.065	2.8
Regular	2 042	35.6	0.27	63 300	2.6	19 100	4.0	24 000	2.7	0.090	2.5
Microcrystalline	1 951	29.6	0.26	142 000	1.9	17 100	2.8	26 600	3.4	0.107	1.9
10 percent ZrO_2	1 965	10.9	0.08	175 000	3.3	25 800	3.2	22 800	1.9	0.185	2.5
40 percent ZrO_2	1 462	7.9	0.11	172 500	3.0	22 800	3.1	24 700	2.7	0.240	1.8
Sintered	1 372	6.5	0.05	278 000	2.6	43 300	3.2	39 900	2.2	0.260	2.7
Silicon carbide											
Green	2 837	62.5	1.10	79 500	1.7	9 000	1.9	10 550	2.0	0.063	2.4
Black	2 679	57.2	0.81	106 000	2.0	15 700	2.3	16 500	1.9	0.075	2.0

† Load 100 g except for SiC, which was 200 g.
1 in lb = 11.3 cm N; 1000 p.s.i. = 6.895 MPa.

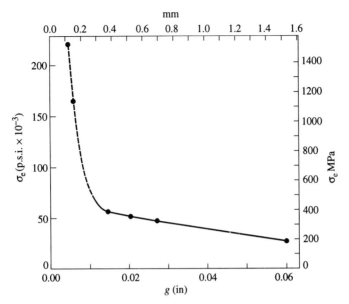

FIG. 3.7. The mean fracture stress of regular Al$_2$O$_3$ grit as a function of grit size (after Brecker 1974).

material determine its strength, is valid where sharp crack formation and growth coincide. However, stable crack growth has been found to exist in many materials thought to be brittle. For instance, plastic flow has been observed in small glass specimens. All materials should exhibit a brittle–ductile transition, such as illustrated in Fig. 3.8 for a few materials. The transition in strength for regular aluminum oxide occurs at a particle size of about 0.7 mm (24 grit size) and agrees closely with the approximate 1 mm transition noted in glass. The effective strength data presented for particle sizes below 24 grit size is therefore approximate due to the transition from brittle to ductile behavior.

The grit shape also can affect the results. Samples of several abrasive materials were separated according to shape by using a vibrating table with 12 pockets on the side. Shapes obtained run from blocky to plate-like, with the predominant portion usually being on the blocky end of the scale. Typical results are shown in Fig. 3.9, where larger forces are required to fracture the blocky particles. However, since particles tend to rest on their longest dimension and thus be loaded across the smallest dimension, the effective tensile strength is found to be roughly independent of loading direction as long as the appropriate loaded diameter is used. An exception to this occurs for plate-like grits loaded in the direction of the longest axis and which show a sharp

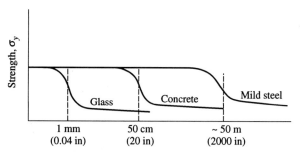

FIG. 3.8. The ductile–brittle transition size for different materials (after Glucklich 1970).

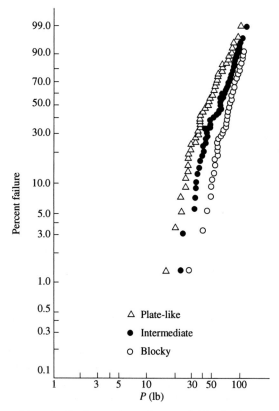

FIG. 3.9. Roll crushing results for various shapes of 12 grit size regular Al_2O_3 (after Brecker 1974).

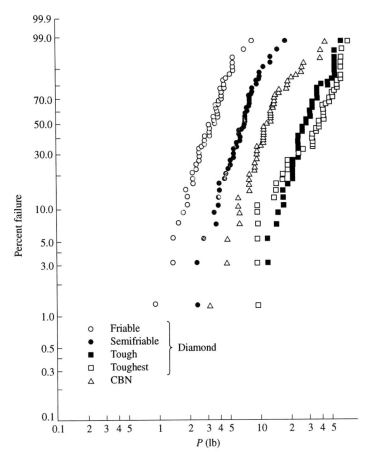

FIG. 3.10. Fracture loads for several types of 60 grit size diamond and CBN (after Brecker 1974).

decrease in strength, probably due to flaws being oriented parallel to the long axis of the grit and thus perpendicular to the tensile stress. When these plate-like grits are loaded parallel to their shortest axis they show greater strength, but this is misleading since the grit no longer fits the blocky shape assumed in the stress model. Thus, this test does not work on plate-like materials.

Several types of manufactured diamonds were also evaluated on the basis of tensile strength for comparison purposes. Figure 3.10 shows Weibull plots for several types of diamond and cubic boron nitride (CBN) grits of 60 size tested in the roll crushing test using carbide rolls of 2 in (50 mm) diameter. The slopes of the curves are the same as for the aluminum oxides and silicon carbides and indicate brittle behavior. The regular aluminum oxide of 60 grit size had a mean tensile strength of 85 000 p.s.i. (586 MPa). The mean tensile

TABLE 3.2 *Tensile strength for diamonds, CBN and Al_2O_3, 60 grit size, dry*

Abrasive	$\bar{\sigma}_e$	
	p.s.i.	MPa
Diamonds		
Friable	47 000	324
Semifriable	96 000	662
Tough	411 000	2 834
Very tough	480 000	3 310
CBN	164 000	1 131
Regular Al_2O_3	85 000	586

1000 p.s.i. = 6.895 MPa.

strengths for the diamonds and cubic boron nitride are given in Table 3.2. The friable diamond has a strength half that of the regular aluminum oxide, while the more perfectly formed diamond has a strength five times greater than regular aluminum oxide. The cubic boron nitride is about twice as strong as the aluminum oxide. The superabrasives offer very high hardness and high tensile strength. Also to be considered is the chemical reactivity of these materials, which limits their range of application. Despite their high cost, there are numerous applications in which they perform well and hence result in a lower cost per part than when common abrasive materials are used. The wide range of tensile strengths for synthetic diamonds enables the user to select the strength or friability best suited to the application.

The friable diamonds were found to fail primarily by crushing (80 percent of sample) whereas the tough, blocky, more perfect diamonds gave good tensile fractures (75 percent) or did not fail (~5 percent). The cubic boron nitride failed by crushing almost as frequently as it failed by fracture. Thus, the friability of these abrasives is inversely related to their tensile strength. Except for the sintered aluminum oxide, which had a number of nonfailing grits, all the common abrasives gave good tensile breaks.

Single-grit Wear Tests

Introduction

The wear of individual abrasive particles is mechanical (attritious), involving removal of very small particles and microchipping on the one hand and chemical wear on the other. There also are interactions between these two types of wear. In addition, there is abrasive pull-out (bond failure), which contributes to the wear of a complete grinding wheel or belt. The chemical aspects of wear involve affinity of the abrasive–workpiece pair, temperature, pressure, and time. This is best studied by exaggerating time by use of a continuous rubbing test. The mechanical aspects are best studied by a fly milling test, in which a large number of microchips are generated with the mechanical shock, thermal cycling, and generation of nascent surface involved in an actual grinding operation. Literature pertaining to the chemical aspects of abrasive wear will first be reviewed, followed by some rubbing results designed to extend our understanding of the chemical aspects of wear. This will be followed by considering several types of fly milling tests, in which a single abrasive particle attached to the periphery of a high-speed disc is fed into a workpiece to generate chips. A review of the more important principles of tribology pertaining to material removal operations is to be found in Shaw (1984).

Literature Review

Coes (1955) suggested that the wear of grinding wheels has a chemical side that is at least as important as the generally recognized physical side. He described work on single unbonded abrasive grits which were conical in shape and made to scratch a helical groove in a cylindrical specimen under a constant normal load that was insufficient to cause grit fracture. The rate at which a flat developed and the helical path length at which a scratch was no longer produced indicated the rate of total wear of the abrasive. While hardness and strength are important abrasive properties from the point of view of mechanical wear, the chemical interaction with the metal cut is equally important relative to chemical wear. The disappointing grinding results with extremely hard B_4C are due to its tendency to oxidize at grinding temperatures. Silicon carbide was found to wear more slowly on glass than did aluminum oxide, but aluminum oxide wore less rapidly on steel than did SiC. Water vapor was found to increase the wear rate of aluminum oxide on steel.

Coes (1955) also observed that oxygen normally found in air is beneficial to grinding, since it prevents chips from rewelding, an observation that had been independently made a few years earlier (Outwater and Shaw 1952). More slowly oxidizing steels were found by Coes to require an atmosphere reacting more quickly than oxygen (i.e. sulfur, chlorine, or phosphorus). It was suggested that since titanium reacts so slowly, even with chlorine, it should be ground at greatly reduced speeds to prevent chips from rewelding to the ground surface. Cryolite was found to weaken the abrasive but also to lubricate, the net effect being a beneficial one. Tellurium was found to be very effective as a wheel filler in grinding stainless steel, since it attacked the grain boundaries of the stainless steel at high speed. It was suggested that analog diffusion studies in which different metals are held at high temperatures (700 °C) in contact with abrasives might yield important data relative to the chemical characteristics of different grinding systems. This idea had been introduced into the study of carbide tool–workpiece combinations much earlier by Dawihl (1940) and was useful in understanding their behavior.

Diffusion studies have been performed on pure iron in static contact with diamond (D) and CBN abrasive particles at high temperature (1300 °C) by Loladze and Bockuchava (1972), to explain the very great difference in stability between D and CBN in contact with hot iron. Figure 3.11

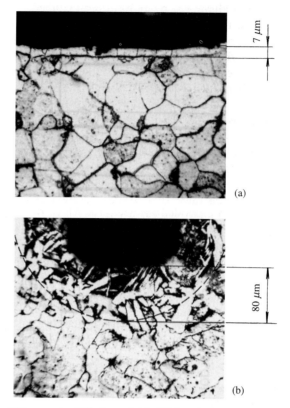

FIG. 3.11. The difference in diffusion of (a) CBN and (b) D into pure iron: polished and etched sections at tips of indenters held in contact with iron at 1300 °C for (a) 720 s for CBN and (b) 0.5 s for D (after Loladze and Bockuchava 1972).

dramatically shows the difference in behavior of D and CBN in contact with hot iron.

Geopfert and Williams (1959) studied the reactions of molten iron with silicon carbide and aluminum oxide. While no reaction was observed between aluminum oxide and iron in an inert atmosphere, iron was found to react readily with SiC. It was observed that as the Si and C content of the iron increased, the tendency for SiC to react decreased. These facts were offered in explanation for the superior behavior of Al_2O_3 on steel and the superior behavior of SiC on cast iron. This work has been repeated and made more quantitative by Bielawski (1966), who has also shown that a decrease in contact angle (signifying greater wetting) between the molten metal and the abrasive corresponds to greater reactivity. It was found that SiC gives a high contact angle relative to cast iron, but that Al_2O_3 gives a high contact angle relative to steel or 304 stainless steel. These results are in agreement with grinding performance.

Single crystals of α aluminum oxide have been studied fairly extensively. Coffin (1961) studied the wear characteristics of Al_2O_3 at low sliding speeds and found the atmosphere to play an important role. Steijn (1961) and Duwell (1962) observed that the orientation of Al_2O_3 crystals influences their wear rate. The fact that the maximum wear rate occurs on the plane and in the direction of easiest glide (Duwell 1962) suggests that plastic flow may contribute to the wear of Al_2O_3 crystals. While it might appear surprising that a normally brittle material such as Al_2O_3 shows evidence of slip and plastic flow in electron photomicrographs of a surface after rubbing, it should be kept in mind that the local surface temperatures and pressures pertaining are extremely high. The increased temperature will tend to lower the flow stress, while the pressure will tend to postpone fracture until plastic flow can occur.

It is expected that iron must first oxidize before reacting with Al_2O_3 to form a spinel ($FeAl_2O_4$) and therefore air is necessary for spinel formation. This explains why Goepfert and Williams (1959) found no spinel in their experiments between Al_2O_3 and molten iron (no air was present), and also why wear of a ceramic turning tool is observed to be much greater where the chip rubs under light pressure than where it is under heavy pressure. Where the contact pressure is high, oxygen from the air will be excluded, but not where rubbing pressure is light.

Duwell (1962) performed tests in which single aluminum oxide and silicon carbide grits were caused to rub on a rotating cylinder along a helical path. These tests revealed the following:

1. In general, wear was of both the attritious (gradual) and chipping (fragmentation) types.
2. Silicon carbide on steel did not chip but attritious wear was rapid.
3. Aluminum oxide on steel both chipped and showed attritious wear such that its total wear rate was greater than that for SiC.

When steel is ground with an abrasive belt, Al_2O_3 grit is capable of removing about 40 times as much metal as SiC, even though the total rate of abrasive loss is greater for Al_2O_3. The explanation for this is that, in the absence of chipping, flats form quickly on the SiC grits but chipping of the Al_2O_3 keeps this material sharp, and hence it continues to cut at a rapid rate for a much longer period of time when operating under constant normal force. While it is not clear why Al_2O_3 chips more than SiC when cutting steel, this could be due to the lower strength (hardness) of Al_2O_3 or due to its lower thermal shock resistance. When heated to a high temperature and dropped into water, Al_2O_3 shows many cracks, but SiC does not. This suggests that Al_2O_3 is more sensitive to thermal shock than SiC.

Duwell and McDonald (1961) ran sliding tests both with helical motion and where the specimen ran over the same track repeatedly. In the latter case the friction coefficient was appreciably higher than when a new surface was continuously encountered. This could be due to the greater tendency for oxidation

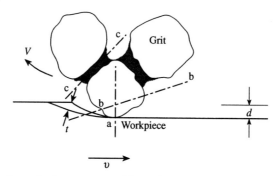

FIG. 3.12. Grinding wheel wear: a, attritious; b, partial fracture of grit; c, fracture of bond post.

when the same track is repeatedly traversed. The greater amount of iron oxide should cause a greater amount of the low melting eutectic spinel to form which, in turn, would cause an increase in the real area of contact and hence greater friction. Use of an active gas (HCl or HBr in the presence of water vapor) was found to reduce the friction and increase the cutting rate under constant pressure grinding (Duwell and McDonald 1961). In general, oxides (corundum, garnet, and flint) gave less wear on steel than carbides or borides. The fact that flint (820 Knoop hardness) gave a lower wear rate on steel than B_4C (2750 Knoop hardness) emphasizes the fact that something other than hardness (chemical stability) under grinding wear conditions is involved under grinding wear conditions.

In discussing the effective action of HCl and HBr gas when Al_2O_3 rubs steel, Duwell and McDonald (1961) suggest this is due to the lower melting points of the iron salts ($FeCl_3$, $FeBr_3$) than for steel. It seems reasonable that these materials might rather act by reducing the strength of the metal cut, as in the case of a grinding wheel containing tellurium when cutting stainless steel (Coes 1955).

An excellent review of all of the interesting work performed by Duwell and his associates is to be found in Duwell and McDonald (1965).

Rubbing Wear Tests

Grinding wheel wear can be broadly classified as either attritious or fracture wear. Attritious wear occurs at the grit–workpiece contact surface, as shown at 'a' in Fig. 3.12, and it is generally accepted that plastic flow, crumbling, and chemical reaction have a significant effect on this phenomenon. This results in the flattening or dulling of the abrasive grits and accounts for the glazed appearance of a used grinding wheel. Fracture wear, on the other hand, is due to removal of abrasive particles from the wheel either by partial fracture of grits (at 'b' in Fig. 3.12) or by fracturing away of the bond post (as at 'c').

A typical sequence of events for a grit comprises the dulling by attritious wear and then the fracturing away of a small porion (giving a sharper grit) until the bond posts break, resulting in the final release of the remainder of the grit.

Wear in grinding is both physical and chemical in nature. In fine grinding, the dulling of a grit is mainly due to the formation of a wear flat which grows in size during grinding in a manner which is analogous to the growth of a wear land on the flank of a cutting tool. In general, however, both cutting forces and temperatures increase with the increase in the wear land area. This fact makes the study of the effects of chemical reaction between the grit and the workpiece much more important, at least for fine grinding.

The rubbing test appears to be of greatest interest relative to the possible chemical reaction occurring between the grit and the disc material. By traversing the same track, a considerable amount of reaction product may be accumulated for identification. However, the purely physical aspect of grit wear appears to be better approximated by a fly milling test, in which a single grit is mounted on the periphery of a metal disc and used in place of a grinding wheel to produce chips. Such tests are discussed below. The basic difference between the two tests concerns the influence that time has upon the physical and chemical action. In rubbing tests, as mentioned earlier, the chemical action associated with grit wear is exaggerated by the long time available for reaction to occur. The contact time in fly milling is 10^{-4} s or less, depending upon the wheel speed. The amount of chemical product resulting from this short contact time is difficult to detect, despite the fact that these products play an important role in the chip-forming process. Thus, it would appear to be advantageous to think of the rubbing test as being primarily one which is suitable for the study of the chemical aspects of grit behavior and the fly milling test as being more suitable for studying the physical aspects.

Two types of rubbing tests on abrasive grits have been used:

- pin-on-disc tests, in which a stationary abrasive particle is loaded against a rotating disk, repeatedly encountering the same circular track on the disk;
- rubbing tests, in which an abrasive particle traverses a helical path on a rotating cylinder.

A pin-on-disc apparatus is shown schematically in Fig. 3.13(a), while a pin-on-cylinder apparatus is shown in Fig. 3.13(b).

In the pin-on-disc test the 6 in (150 mm) diameter circular plate A is driven by the spindle of an inverted drill press (B), while the stationary grit (C) is attached to a horizontal arm (D). This arm is supported by a shaft mounted between two pillow block bearings (E). The arm is attached to the shaft through a thin flexible spring member (F), which provides high vertical stiffness but low horizontal stiffness. The arm is prevented from moving horizontally under the action of the friction force by ring G, which is fitted with wire resistance strain gages used to measure the friction force. The arm and specimen-holder are counterbalanced by weight W so that the vertical load supported by the grit is equal to the applied load (P).

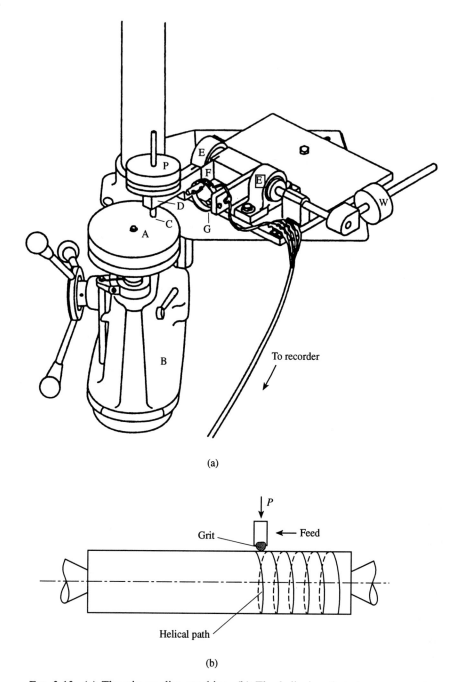

FIG. 3.13. (a) The pin-on-disc machine. (b) The helical path test arrangement.

In making a test, the surface of the plate is ground to a 4 μin (0.1 μm) AA finish and then mounted on the machine. The grit is then subjected to a run-in period under high load in order to develop an initial wear area. The grit is removed from the machine periodically and the wear height determined using a precision micrometer (±0.0001 in, ±2.5 μm). The wear profile is traced on paper using an optical projector at 50×, the wear area being obtained from the tracing with the aid of a planimeter. The spindle is driven by an infinitely variable speed drive in order to utilize a wide range of speeds.

The wear volume (V_W) is determined as follows from the change in height Δh and the initial and final areas, A_1 and A_2 respectively, of the flat worn on the grit:

$$V_W = [\tfrac{1}{2}(A_1 + A_2)]\Delta h. \qquad (3.2)$$

Since there is considerable test-to-test variation, it is necessary to use the average of six or more grits in making comparisons. Representative results are given in Fig. 3.14 for several common abrasives tested against A-6 tool steel ($R_C = 60$) at a speed of 1000 f.p.m. (5 m s^{-1}) and a load of 1 lb (4.45 N). The common abrasive materials tested in Fig. 3.14 were those the properties of which are given in Table 2.1. The volume worn away is seen to be proportional to the rubbing distance. Similar results, including the same relative wear ranking, were obtained for a wide range of materials, including AISI 1044 steel, AISI 4340 steel, 18-8 stainless steel, Waspaloy, and a titanium alloy (Ti 150A).

From classical adhesive wear theory the volume worn away (B) is related to applied load (P), the sliding distance (L) and the hardness of the wearing member (H) as follows:

$$(BH)/LP = N_W, \qquad (3.3)$$

where N_W is the wear number, a nondimensional constant for a given sliding system (see Shaw 1984, p. 227). For a given disc–abrasive combination and a given applied load, the wear volume should be proportional to L, as shown in Fig. 3.14.

In the helical path rubbing tests, the workpice is mounted on centers in a lathe and the abrasive grit loaded by gravity and fed axially across the workpiece to give a once-over helical path (Fig. 3.13(b)). When the grit has traversed the length of the cylinder, a light cut with a carbide tool is taken on the surface of the cylinder before the next test to remove the wear scar.

Table 3.3 gives a comparison of the wear corresponding to the same load (1 lb = 4.2 N), speed (1000 f.p.m. = 5 m s^{-1}), and sliding distance for several common grit materials for the disc (single track) and for the 4 in (100 mm) diameter cylinder (single-track and helical-track) tests on two steels. Here it is seen that the wear is considerably less when the grit rubs on a new surface than when the same path is repeatedly traversed. When the single-track test was made on the cylinder, a decrease in the wear rate relative to that for a

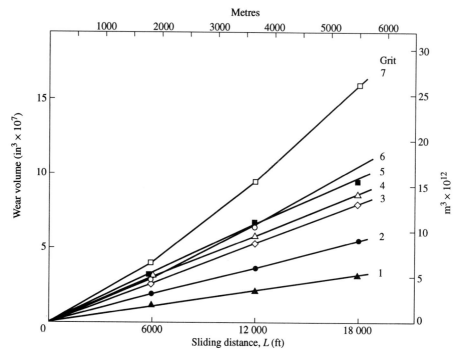

FIG. 3.14. The volumetric wear of several types of common grit types operating on A-6 tool steel ($R_C = 60$) at a 1 lb (4.45 N) load and a speed of 1000 f.p.m. (5 m s^{-1}) (after Lal *et al.* 1973). Grit materials: 1, black SiC; 2, regular Al_2O_3; 3, ruby Al_2O_3; 4, white Al_2O_3; 5, 40 percent ZrO_2-Al_2O_3; 6, sintered Al_2O_3; 7, 100 percent zirconia.

TABLE 3.3 *A comparison of results between the helical path and single track tests*

Grit type	Test piece (AISI steel)	Single track rubbing (disc) (in^3)	Single track rubbing (cylinder) (in^3)	Helical track (in^3)
Regular Al_2O_3	1045	0.76×10^{-6}	—	0.43×10^{-6}
Microcrystalline Al_2O_3	1045	0.91×10^{-6}	—	0.65×10^{-6}
Sintered Al_2O_3	4340	0.99×10^{-6}	0.55×10^{-6}	0.24×10^{-6}
Regular Al_2O_3	4340	0.66×10^{-6}	0.51×10^{-6}	0.35×10^{-6}
White Al_2O_3	4340	0.66×10^{-6}	0.50×10^{-6}	0.18×10^{-6}

disc was observed. This is believed to be due to the difference in conformity of the two geometries, the situation with the lower conformity (cylinder) giving the lower wear rate. The grit wear rates when operating on soft steels were found to be greater than for the corresponding hard steel, apparently because of the greater real area of contact for the soft steel.

The wear rate for a wide variety of grits was found to increase in the disc or cylinder tests in the following order:

$$\text{increasing wear rate} \downarrow \left| \begin{array}{l} \text{A-6 tool steel } (R_C = 60) \\ \text{1045 steel (soft)} \\ \text{4340 steel (soft)} \\ \text{Waspaloy} \\ \text{18–8 stainless steel} \\ \text{titanium alloy (150A)} \end{array} \right.$$

The wear rates for disc or cylinder materials that oxidize in air (A-6, 1045, or 4340 steels) were found to be less for the more friable grits than for the more ductile grits. The reverse was true for disk or cylinder materials that do not oxidize in air (Waspaloy, stainless steel, and titaniuim alloy). This is believed to be due to the oxide preventing large bonds from being established which, in turn, decreases microchipping and thus enables hard brittle abrasives to be more effective. The titanium alloy was found to give unusually high rates of wear when operated against a friable grit material.

The coefficient of sliding friction was found to vary between 0.3 and 0.5 for a wide variety of grit–metal combinations. The coefficient of friction increases with a decrease in sliding speed. This suggests that a greater number of stronger bonds are formed at lower speeds than at higher speeds.

When metal discs were replaced by a metal-supported bonded abrasive paper containing 320 mesh size diamond particles on the pin-on-disc apparatus, quite different results were obtained (Fig. 3.15). On this occasion the wear rate increased as the friability increased, instead of decreasing as the friability increased (compare Figs. 3.14 and 3.15). It appears that the one test procedure has a strong chemical component (tool steel), while the other (bonded diamond disc) is essentially a physical test for toughness. The tougher grits appear to be more reactive with the steel and hence subject to a greater amount of chemical wear than with diamond. This could also be due to a larger real area of contact between the steel and the softer grit materials which, in turn, could lead to a greater rate of chemical action with iron oxide that forms on the steel surface. Photomicrographs of wear tracks on steel and the bonded diamond disc appear to bear this out. The worn steel disc had an etched appearance, while the worn bonded diamond disc showed a large number of abrasive tracks on its surface.

A few tests on a sintered tungsten carbide disc gave results similar to those for the bonded diamond disc, suggesting that — as with diamond — there is less chemical action with WC than with steel.

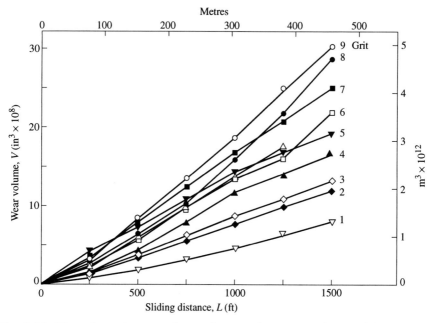

FIG. 3.15. The volumetric wear of several types of common grit materials operating on a 320 screen size diamond-coated abrasive paper at 0.25 lb (1.05 N) load and a sliding speed of 500 f.p.m. (2.5 m s^{-1}) (after Scrutton *et al.* 1973). Grit materials: 1, sintered Al_2O_3; 2, 10 percent ZrO_2-Al_2O_3; 3, regular Al_2O_3; 4, monocrystalline Al_2O_3; 5, 40 percent ZrO_2-Al_2O_3; 6, microcrystalline Al_2O_3; 7, green SiC; 8, black SiC; 9, white Al_2O_3.

Matsuo (1981) has run pin-on disc tests on number 6 Stellite (R_A 63) using several abrasive materials including diamond and CBN. Representative curves are given in Fig. 3.16, and these results are in general agreement with Fig. 3.15. The wear rate for Si_3N_4 (Knoop hardness = 1470 kg mm^2) and D are seen to be particularly low relative to the wear rates for the common abrasive types. The wear rate for CBN was only slightly higher than that for D. Figure 3.16 is in agreement with the previous observation that the wear rates of common abrasives increase as the friability of the abrasive increases when rubbing on a material that does not oxidize in air (Stellite). The coefficient of friction of D rubbing against Stellite was found to be an order of magnitude lower than that for common abrasives rubbing against stellite.

The conditions pertaining on the rubbing surfaces in the pin-on-disc test are less severe than those involved in grinding, and therefore other more severe tests, such as fly milling, must be employed on abrasive particles in order to rate their performance.

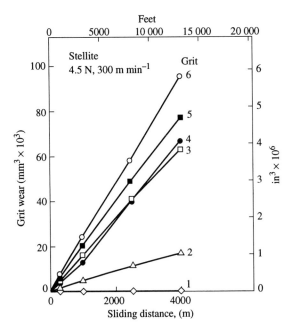

FIG. 3.16. The volumetric wear of several types of abrasives operating on a disc of stellite 6 on a pin-on-disc machine at a load of 1 lb (4.45 N) and a speed of 2950 f.p.m. (15 m s^{-1}) (after Matsuo 1981). Grit materials: 1, diamond; 2, Si_3N_4; 3, 25 percent ZrO_2-Al_2O_3; 4, black SiC; 5, regular Al_2O_3; 6, ruby Al_2O_3.

Fly Milling Tests

Introduction

Fly milling tests are performed by mounting a single grit in the head cavity of a small (number 10) Allen head screw with ceramic cement. This is screwed into the periphery of an aluminum disc, which is then used in place of a grinding wheel on a horizontal spindle surface grinder.

Brown (1957) has performed single-grit wear tests and Grisbrook et al. (1965) studied the extent of side flow in grinding scratch formation on soft aluminum. Takenaka (1964) conducted single-grit studies on hard steel and cast iron and found very little side flow or rubbing. Electron photomicrographs of flats worn on single abrasive grits revealed that they consisted of many small ridges, which cause the bottoms of cut grooves to have many parallel scratches.

Plain Fly Milling

The plain fly milling test (Fig. 3.17), in which a number of full crescent-shaped chips are made, gives contact times, contact lengths (l), shock, and newly

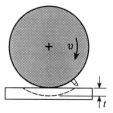

FIG. 3.17. The plain fly milling test (side view): the specimen is inclined in direction perpendicular to the paper.

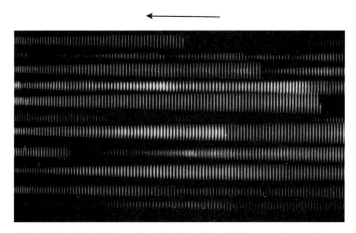

FIG. 3.18. Typical results of plain fly milling tests (top view).

generated surface areas similar to those in actual grinding. A slightly inclined specimen (0.004 in per 6 in = 0.1 mm in 152 mm) is used to provide a gradually increasing depth of cut (t) until fracture occurs. Typical specimens are shown in Figs 3.18 and 3.19. The depth of any cut (t) can be obtained from the length of that cut (l) by:

$$t = l^2/4D, \tag{3.4}$$

where D is the wheel diameter.

The size and frequency of fractures can be determined, as well as an approximate attritious wear rate.

The test is most useful as a means of quickly evaluating the area of applicability of a new grit or a treated grit. If many fractures occur at small depth of cut, the abrasive is self-sharpening and applicable to FFG if the wear rate is not too high. Grits which repeatedly cut to deep depths are applicable to SRG. The change in depth of cut compared to the pitch of the specimen can also be used quickly to assess attritious wear.

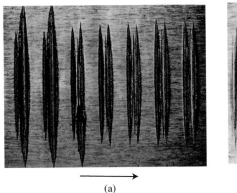
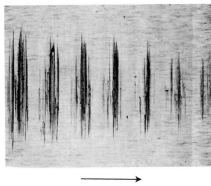

(a)   (b)

FIG. 3.19. Photomicrographs of fly milling cuts made with regular Al_2O_3 grit on AISI 52100 steel: (a) soft ($R_B = 63$); (b) hard ($R_C = 63$).

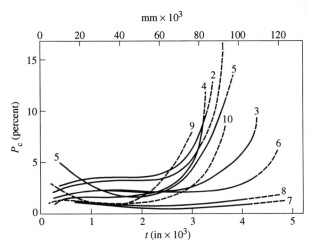

FIG. 3.20. The fracture probability (P_C) in plain fly milling for various abrasives used on 52100 steel ($R_B = 92$) as a function of depth of cut (t) (after Brecker et al. 1973). Grit types: 1, white aluminum oxide; 2, modified (3 percent Cr) Al_2O_3; 3, monocrystalline Al_2O_3; 4, regular Al_2O_3; 5, microcrystalline Al_2O_3; 6, 10 percent zirconia-Al_2O_3; 7, 40 percent zirconia-Al_2O_3; 8, sintered Al_2O_3; 9, green silicon carbide; 10, black SiC.

Values of the probability of fracture (P_C) for several grit types are given in Fig. 3.20 for cuts of different depth (t), while Fig. 3.21 gives similar values for attritious wear in terms of mean change in depth of cut per cut (Δt versus t). The tough aluminum oxides were found to give very small fracture wear even at very large depths of cut. Most of the other aluminum oxides fracture when cuts above 0.003 in, (0.076 mm) depth are attempted on the soft AISI

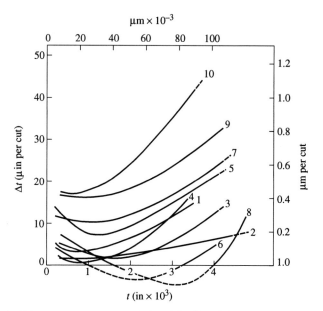

FIG. 3.21. Attritious wear per cut (Δt) in plain fly milling for various abrasives used on 52100 steel ($R_B = 92$) as a function of depth of cut (t) (after Brecker *et al.* 1973). Grit types are given in Fig. 3.20.

52100 steel workpiece used. At the lower depths of cut, the fracture rates correlate well with tensile strength results given in Table 3.1. The silicon carbides wear rapidly and thus develop wear flats which greatly improve their strength, as seen in the static tensile strength studies. The high-hardness, friable aluminum oxides show low attritious wear and thus do not develop wear flats rapidly enough to enhance their tensile strength as much.

The attritious wear data does not always vary inversely with hardness. For example, the silicon carbides (Fig. 3.21) show especially high wear, which can be explained by their chemical reaction with the iron (ferrite) matrix in steel (Geopfert and Williams 1959). The low wear rate, which at times was negative for the two-phase zirconia alloys (numbers 6 and 8) and for the sintered alumina, may be related to the ductility, weldability, and tendency to build-up of these materials.

In order to incorporate an element of shock in plain fly milling tests, the workpiece may be positioned with the center of cut over the entering or exiting edge of the specimen. The rate of wear is greater in these tests than in those without shock loading at entrance or exit. The greatest increase in wear rate is observed for the more friable grit types and when shock occurs on entering a cut than when exiting a cut.

Plain fly milling tests are useful in evaluating the improved performance of grits that have been heat treated to heal defects. In one series of tests, regular

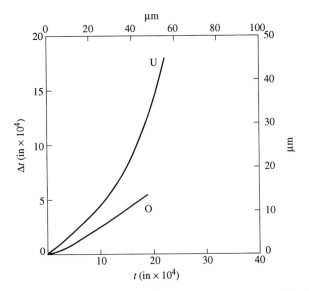

FIG. 3.22. The variation in the change in depth of cut (Δt) when grit chips with maximum grit depth of cut (t) in plain fly milling operation on regular Al_2O_3 with different heat treatments when cutting hard AISI 52100 steel at 6000 f.p.m. (30 m s^{-1}) with a lateral specimen inclination of 0.0048 in per 6 in (108 mm per 152 mm). Curve U is for untreated grit, while curve O is for the optimum heat treatment of 1300 °C for 15 minutes.

Al_2O_3 grits of 12 screen size were heated at different time–temperature combinations before testing at a speed of 6000 f.p.m. (30 m s^{-1}). Two specimens were used; soft ($R_B = 93$) and hard ($R_C = 63$) AISI 52100 steel. The workpieces were inclined at 0.0048 in in 6 in (1.08 mm in 152 mm).

Untreated grits not only fractured more frequently but gave a higher rate of attritious wear. For this particular type of grit a heat treatment of 1300 °C for 15 min gave the best results (less frequent fracture, smaller fracture size (Δt), and less attritious wear, where Δt is the change in maximum depth of cut at the midpoint of a cut). Figure 3.22 shows the change in Dt occurring with fracture versus the depth of cut (t) for several heat treatments Curve U is for untreated grit while curve O is for the optimum heat treatment (1300 °C for 15 min).

The ratio of scratch width to scratch depth (r) plays an important role in the analysis of complete grinding wheels, as will be seen in the next chapter. The plain fly milling test with single abrasive particles may be used to understand this variable better. When plain fly milling tests with specimen inclination are performed for different depths of cut, values of r may be obtained corresponding to different values of maximum depth of cut t. Figure 3.23

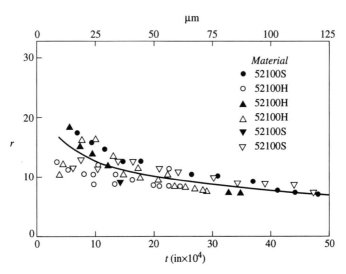

FIG. 3.23. The variation of the scratch width to scratch depth ratio (r) with maximum depth of cut (t) for several 40 percent ZrO_2-Al_2O_3 abrasive particles cutting hard (H) and soft (S) specimens of AISI 52100 steel in plain fly milling with specimen inclination.

shows such results for several sintered Al_2O_3 grits cutting hard and soft AISI 52100 steel. While there is considerable scatter in the data, it is evident that r increases as t decreases. Figure 3.24 shows mean curves corresponding to Fig. 3.23 for cutting hard and soft AISI 52100 steel with three different grit types. From Fig. 3.24, the value of r is seen to lie in the range 15–30 for fine grinding, while in the coarse grinding regime the value of r is closer to 10.

High-speed Low-pressure Tests

Komanduri and Shaw (1976a) have employed a special form of plain fly milling to study metal build-up and attritious wear at very high sliding speeds. This involves a large-diameter aluminum disc with an abrasive particle, mounted in the head of a hollow-headed screw attached as shown in Fig. 3.17. The stationary specimen is advanced into the rotating grit at a rate of a few angstrom units (Å) per encounter. This is accomplished by slowly heating a hollow aluminum column by d.c. resistance as shown in Fig. 3.25. The rotating specimen (grit) is capable of traversing the stationary specimen at speeds up to 20 000 f.p.m. (100 m s^{-1}). The rotating specimen may be periodically removed and examined for attritious wear. In this test essentially no material is removed from the stationary member even after several minutes of operation. The objective is to study build-up and attritious wear under conditions of very high speed and light pressure.

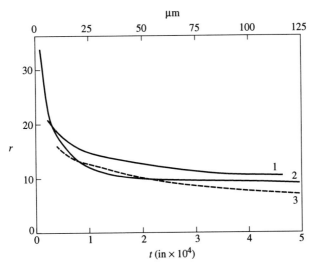

FIG. 3.24. Mean curves showing the variation of r with t for three different common abrasive materials: 1, regular Al_2O_3; 2, 40 percent ZrO_2-Al_2O_3; 3, sintered Al_2O_3.

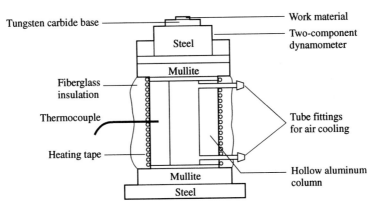

FIG. 3.25. A schematic of the microfeed device (after Komanduri and Shaw 1976a).

In Komanduri and Shaw (1976a) a polished workpiece of cobalt-based superalloy (5 w/o Co, 10 w/o Ni, 23.5 w/o Cr, and 7 w/o W) was tested with a green SiC abrasive particle. SEM and auger electron microscope (AES) studies failed to reveal any metal build-up on the SiC abrasive. AES and transmission electron microscope (TEM) investigation of wear debris suggested that attritious wear of SiC is due to one or more of the following mechanisms:

- preferential removal of suface atoms from the abrasive by oxidation at high temperature
- dissociation of SiC, diffusion of Si and C into the work material and formation of unstable carbides (Ni_3C and CO_3C) which decompose on cooling,
- cleavage fracture of SiC to micron-sized crystallites

At high temperatures SiC should react exothermically with O_2 in the atmosphere as follows:

$$SiC + O_2 \rightarrow SiO_2 + C \quad (-183.6\,\text{kcal}). \quad (3.4)$$

This may be considered as a chemical etching of the SiC surface. Some of the carbon may react with O_2, forming either CO or CO_2. Several other possible reactions that may account for the rapid attritious wear that takes place when SiC is used to grind a cobalt base alloy at high speeds and extremely low feed rates are discussed in Komanduri and Shaw (1976a). However, there was no evidence of metal build-up on the SiC abrasive or other evidence of direct reaction between the abrasive and the work material. Instead, the SiC appears to oxidize and then to be subsequently removed. The presence of appreciable amounts of amorphous carbon found in the wear debris is explained by the lack of affinity of cobalt for carbon and is consistent with the foregoing oxidation reaction (eqn 3.4).

Komanduri (1976) has used the test arrangment of Fig. 3.25 to study the behavior of Al_2O_3 when traversing the same cobalt-based alloy at very high speeds and low feeds. In this case there was considerable metal build-up. SEM, AES, and X-ray diffraction studies revealed that the work material first oxidizes and then undergoes a solid phase reaction with the Al_2O_3 abrasive. Metals that readily form stable oxides (i.e. metals having a high affinity for oxygen and a large negative free energy of oxidation) should form strong bonds with Al_2O_3 and hence give rise to metal build-up. Such metals include Zr, Al, Ti, Si, and Cr. The oxides of these metals also have about the same thermal expansion as Al_2O_3 and this adds to their tendency to build up on an Al_2O_3 abrasive.

Material that builds up cyclically on abrasive particles causes periodic chipping of the abrasive away from the point of contact, resulting in an increased wear rate and poor surface finish.

In the case of the previously considered cobalt-based alloy, build-up on Al_2O_3 is due to the oxidation of Cr at the surface to Cr_2O_3 and formation of a strong solid solution with Al_2O_3 (Al_2O_3–Cr_2O_3). Similarly, Ti_2O_3 is quite soluble in Al_2O_3 and this accounts for the tendency of titanium to build up on an Al_2O_3 abrasive.

Although the cobalt-based alloy is high in cobalt, no spinel formation corresponding to CoO–Al_2O_3 was identified by X-ray diffraction. Instead, Cr_2O–Al_2O_3 was found in the built-up layer. Most of the chromium in this alloy will be in the form of carbides, which oxidize to Cr_2O_3 and form a strong solid solution bond with Al_2O_3.

Stainless steel is also a material that is difficult to grind due to metal build-up. In this case chromium will diffuse to the suface, oxidize to Cr_2O_3 and form solid solution bonded built-up layers on Al_2O_3.

The foregoing discussion emphasizes the difference in the behavior of SiC and Al_2O_3 in grinding difficult-to-grind metals such as cobalt-based superalloys, titanium alloys, and stainless steels. In the case of SiC there is no build-up due to solid state bonding, as there is with Al_2O_3. Instead, SiC wears attritiously to give submicron wear debris by chemical decomposition of the SiC surface by oxidation. The high-speed–low-feed plain fly cutting test is useful for studying the mechanism of attritious wear and build-up associated with abrasive materials of different chemistries.

Komanduri and Shaw (1976b) have also used the device shown in Fig. 3.25 to study the chemical interaction between a single-cystal synthetic diamond and pure iron (99.999 w/o Fe). AES revealed an extremely thin (~200 Å) diffusion layer after high-speed–low feed grinding contact. Komanduri and Shaw (1975) have considered a mechanism of diamond wear that involves graphitization of diamond at the surface; this was first proposed by Ikawa and Tanaka (1971). This mechanism involves low contact pressure, high temperature, and rapid transport of a material capable of removing graphite (such as low carbon steel or pure iron). The fact that diamond is worn more rapidly by a soft low carbon steel than by a strong alloy steel, giving rise to higher surface temperatures, tends to rule out oxidation as the primary wear mechanism. As discussed in Komanduri and Shaw (1976b) the subsurface chemistry revealed by AES study after sputter etching supports the graphitization mechanism.

Ikawa and Tanaka (1971) and Loladze and Bokucheva (1972) have studied the chemical interaction between diamond and ferrous materials, including pure iron, under quasi-equilibrium conditions. This was done by loading a diamond against hot metal in vacuum for an appreciable period of time (thousands of seconds) compared with abrasive contact times (tenths of a millisecond). In lkawa and Tanaka's experiments, the couple was heated at 1000 °C for 30 min under a static load of 15 lb (66.4 N). A diffusion layer of about 0.2 mm (0.008 in) was identified. Lolanize and Bokuchava loaded a conical diamond into commercially pure iron at 1300 °C for 0.5 s and a diffusion layer of about 80 μm (1600 μin) was found. In both of these cases, the results are in reasonably good agreement with elementary atomic diffusion theory.

The use of the high-speed–low-feed technique employing the feed device of Fig. 3.25 provides a closer approximation to an actual grinding situation than the static heated couple approach, and is therefore better adapted to studies of the wear mechanisms involved. This is illustrated by the possibility of distinguishing between oxidation and graphitization as the primary action involved in the rapid wear of diamond when grinding a low-carbon steel or pure iron (Komanduri and Shaw 1976b).

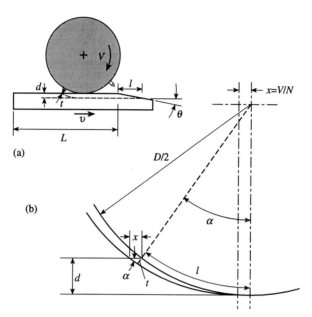

FIG. 3.26. The overcut fly milling test: (a) the method of groove generation (side view); (b) the undeformed chip shape.

Overcut Fly Milling

In overcut fly milling (OCFM) a single grit is used to cut a groove at a very low table speed, yielding a chip of similar geometry and thickness to that generated in FFG (Fig. 3.26). A gentle taper at the starting end of the specimen (Fig. 3.26(a)) minimizes shock and enables a rapid evaluation of wear characteristics. Accurate evaluation is made by tracing across the grooves with a stylus instrument and determining the wear or the decrease in the depth of cut (t) between successive grooves made by an abrasive grit. The attritious wear per cut, Δt, can be obtained by dividing the total wear between grooves, $\Sigma \Delta t$, by the number of cuts per groove:

$$\Delta t = \Sigma \Delta t \, v/NL, \tag{3.5}$$

where v is the table speed, N is the wheel r.p.m. and L is the length cut between measuring points. The value of undeformed chip thickness (t) pertaining will be

$$t = 2(vl)/(ND), \tag{3.6}$$

since the work feed per cut will be much smaller than l, and v is the work speed, l is the length of cut equal to $(dD)^{0.5}$, if grit deflection is negligible, N is the wheel r.p.m., and D is the wheel diameter.

In this test fracture is evidenced by an abrupt change in groove depth. This

SINGLE GRIT PERFORMANCE

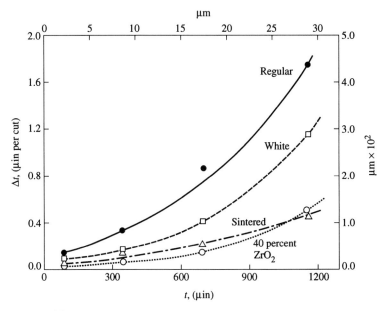

FIG. 3.27. Attritious wear per cut (Δt) as a function of chip thickness (t) for various abrasives used on AISI 1020 steel with a 20.5 in (521 mm) wheel diameter (D) at a speed (V) of 9000 f.p.m. (45 m s^{-1}) and a depth of cut of 0.0005 in (12.5 μm).

occurs very infrequently, but when it does it is not used in data reduction since what is desired is an accurate value of attritious wear in the absence of chipping or gross fracture.

At a moderate wheel speed (6000 f.p.m. = 30 m s^{-1}), no significant differences in wear between regular, white, 40 percent zirconia, and sintered alumina is observed when cutting soft AISI 1020 steel. Differences do become evident at a higher wheel speed, as shown in Fig. 3.27. The results do not correlate with abrasive hardness, since tough grits give the lowest wear. Factors other than Knoop hardness must be found to provide an interpretation of these results. More details concerning overcut fly milling results are available in Lal and Shaw (1972).

One possible explanation of why relatively soft grits such as 40 percent ZrO_2-Al_2O_3, which wear less rapidly in overcut fly milling than regular Al_2O_3 grits, are not used in FFG is related to differences in dressability. A study of differences in dressability between relatively soft and hard grit types has been made by Lal and Shaw (1973). Overcut fly milling grooves were produced on flat AISI 1020 steel specimens that were inclined in the longitudinal direction at 0.00035 in in^{-1}. The depth of cut (d) thus increased progressively with each cut until a value of 0.002 in (0.05 mm) was reached, or until gross grit fracture occurred. Tests were run on a horizontal spindle surface

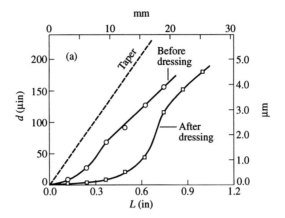

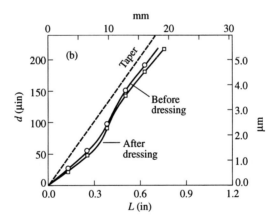

FIG. 3.28. The variation of the depth of cut (d) with distance (L) along flat inclined (0.00035 in/in) AISI 1020 steel specimen subjected to overcut fly milling: (a) 40 percent ZrO_2-Al_2O_3; (b) regular Al_2O_3 (after Lal and Shaw 1973).

grinder on grits of 12 and 36 mesh size, table speeds from 6 to 120 in min^{-1} (2.5–51 mm s^{-1}), a wheel speed of 6000 f.p.m. (30 m s^{-1}), and a wheel diameter of 8.25 in (210 mm). The grooves produced were traced transversely at several points along their length (L) so that plots of maximum depth d versus L could be drawn. Figure 3.28 shows representative results for a tough grit (40 percent ZrO_2) in part (a) and for a more friable grit (regular Al_2O_3) in part (b) before and after fine dressing with a single sharp diamond (two passes with 0.0005 in (12.5 μm) in-feed at 3 in min^{-1} (76 mm min^{-1}) cross-feed,

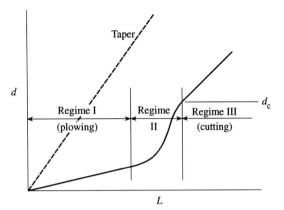

FIG. 3.29. A schematic of a d versus L curve identifying three regimes: I, plowing; II, transition; III, cutting (after Lal and Shaw 1973)

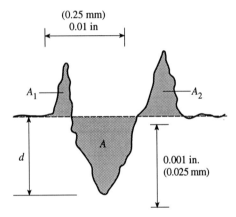

FIG. 3.30. The transverse trace of an OCFM groove in regime I, where $A = A_1 + A_2$ (after Lal and Shaw 1973)

followed by two passes at 0.0002 in (5 μm) in-feed at 3 in/min (76 mm min^{-1}) and then spark-out.

Figure 3.29 is a schematic of a typical d vs L curve, with three regimes identified. The first of these (I) is a region of rubbing and plowing with no chip formation. The third (III) is where chip formation occurs, and in this region the curve is roughly parallel to the longitudinal specimen taper, deviating only slightly due to gradual attritions and microchipping wear of the grit. Region II is a transition region. At the critical depth of cut for chip formation, d_c, a transverse trace ceases to appear, as in Fig. 3.30, where $A = A_1 + A_2$. Beyond this point the area removed begins to be greater than $A_1 + A_2$.

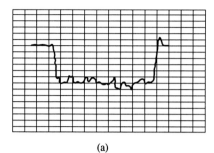 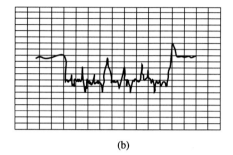

(a)   (b)

FIG. 3.31. Transverse traces of grooves produced in OCFM tests with dressed grits at $V = 6000$ f.p.m. (30 m s^{-1}), $v = 15 \text{ in min}^{-1}$ (6.35 mm s^{-1}) (a) 40 percent $ZrO_2 - Al_2O_3$; (b) regular Al_2O_3. Vertical magnification = $10 \times$ horizontal (after Lai and Shaw 1973).

The curves of Fig. 3.28 are consistent with Fig. 3.29. It is evident that the tough grit (a) is less sharp and has a larger d_c value after dressing than before, whereas the more friable grit (b) is at least as sharp after dressing as before and has about the same value of d_c. The difference in dressability is further evident in typical transverse traces in the cutting regime III, as shown in Fig. 3.31, where the bottom of the groove for the more friable grit in part (b) has many more sharp points than the bottom of the less friable grit in part (a). The same general results were obtained for different grit sizes, dressing rates, and table speeds. It would appear that there is a grit property that may be characterized as dressability, and that the more friable grits that are used in FFG give a grit surface with more sharp points with either self- or diamond dressing than do less friable grits.

The effective radius (ρ) of an abrasive grit at the point of action is considerably less than the nominal radius of the grit ($g/2$). The effective radii for grits of different type may be determined by making transverse traces of overcut fly milling grooves of different depths (d). The radius at the tip of the grit (ρ) may be calculated from

$$\rho = d/2 + b'^2/(8d), \qquad (3.7)$$

since the transverse trace of an OCFM groove is well approximated by a circular arc. In eqn (3.7)

d = depth of OCFM groove, b' = width of OCFM groove.

When OCFM tests were run on a number of work materials (AISI 1030, 303 SS, and AISI 52100 hard steel) under a wide range of cutting conditions (d, V, D, v, and fluids) using abrasive grits of several sizes and types in the as-crushed condition, it was found that the effective grit radius of curvature (ρ) was approximately independent of all variables except grit size (g). Figure 3.32 shows how ρ varies with g for a number of grit materials. The equation of the mean curve is

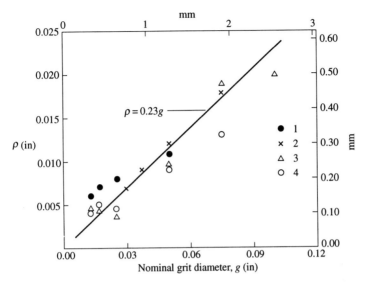

FIG. 3.32. The variation of the effective grit radius (ρ) with the nominal grit diameter (g) for several grit types: 1, sintered Al_2O_3; 2, 40 percent ZrO_2-Al_2O_3; 3, regular Al_2O_3.

$$\rho = 0.23\,g. \tag{3.8}$$

Thus, the only important variable influencing ρ is grit size g and, for all situations, the effective transverse grit radius is about one-quarter of the mean grit diameter (g).

Ikawa and Tanaka (1971) have used the OCFG technique to study the relative wear characteristics when different metals are ground with a diamond cone having an apex angle of 120°. Figure 3.33 shows new and worn grits, where l is the measure of wear employed. Figure 3.34 shows the variation of l with accumulated wheel-work contact length L for four different work materials. Since the nickel, iron, and brass employed had about the same hardness, this suggests that the differences in wear rates for these three materials are not mechanical but chemical. This is borne out by loss of diamond being much more rapid when heated in contact with iron than when heated in contact with titanium, nickel, or brass. This in turn cannot be explained by differences in the rate of diffusion of C in these various metals.

Instead, it appears that the rate of graphitization of diamond is responsible, which is different when in contact with different metals at elevated temperatures and is particularly high for iron. The idea that the rate of graphitization of diamond is the controlling mechanism, first suggested by Ikawa and Tanaka (1971), has been verified and extended by Komanduri and Shaw (1976b).

Tanaka et al. (1981) have found that Mn acts as a catalyst for graphitization of diamond at elevated temperatures just as iron does.

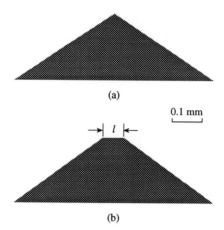

FIG. 3.33. A profile of a diamond cone (a) before and (b) after OCFG, showing the measure of wear, l (after Ikawa and Tanaka 1971).

Face Fly Milling

Chips in certain grinding operations tend to be long and of constant thickness (t). For example, in the abrasive cut-off operation (Fig. 3.35(a)) the wheel is fed downward into the work and individual chips are of length l. A single grit fly milling test well suited to this purpose is shown in Figure 3.35(b). In this case individual chips will resemble those produced in abrasive cut-off if the mean undeformed chip thickness is the same and the width of the specimen in the horizontal direction is equal to that of the workpiece in the corresponding cut-off operation.

Bench Tests

The laboratory bench tests described above are useful in screening new abrasive materials for FFG and SRG. The tensile strength as measured in the diametral compression test indicates the loads which a grain can withstand in grinding. Materials with high strength are used for SRG, while lower strength materials are generally used in FFG. Wear characteristics of abrasives are also important and are better evaluated by milling tests, which simulate grinding conditions better than pin-on-disc tests do.

The plain fly milling test with its gradually increasing depth of cut allows quick evaluation of the area of applicability of abrasive materials. Deep-cutting grits which exhibit few fractures and low attritious wear are applicable to SRG, while those which self-sharpen by fracturing at small depths of cut but have low attritious wear are applicable to FFG. Overcut fly milling yields more accurate attritious wear data for both FFG and SRG abrasive materials. The face fly milling test is useful in evaluating attritious wear when long, constant thickness chips are involved.

SINGLE GRIT PERFORMANCE

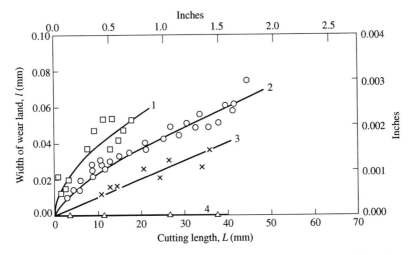

FIG. 3.34. The wear land (l) versus the cumulative distance cut (L) with a diamond cone for different metals: 1, sintered carbide; 2, pure iron; 3, nickel; 4, brass. Wheel speed, 1800 m min^{-1} (5900 f.p.m.); work speed, 4 m m^{-1} (13.1 f.p.m.); depth of cut, 2 μm (80 μin); coolant, dry (after Ikawa and Tanaka 1971).

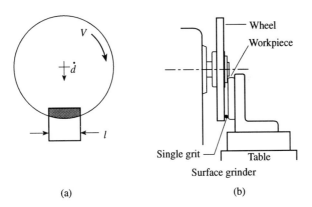

FIG. 3.35. (a) The abrasive cut-off operation. (b) A face fly milling test in which a specimen of width 1 in (25.4 mm) in the horizontal direction is fed vertically against a single rotating abrasive grit.

Since the performance of abrasive materials is determined by a combination of tensile strength, fracture characteristics, and attritious wear characteristics, there is no single test which permits accurate evaluation of abrasive performance for all situations. Tests under actual grinding conditions are the most useful, since they duplicate the environment in which the abrasive material will be used, including the role of bonding material. Complete wheel performance is discussed in Chapter 11.

Specific Energy

Introduction

Backer *et al.* (1952) observed a significant increase in the specific energy as chip thickness decreased when grinding SAE1112 steel with a 32A46-H8-VBE wheel at 3000 f.p.m. (30 m/s). Their results utilized a chip thickness calculation which assumed a constant number of cutting edges per unit wheel surface area. Nakayama and Shaw (1967) later showed that the number of active cutting edges decreases very dramatically at small chip thicknesses. Therefore, it was necessary to test single cutting edges in order to establish the size effect more accurately in the light of the variations in cutting edge density, and to show that the size effect was not attributable to increased rubbing effects on the wheel surface at small chip thickness.

The piezo-electric dynamometer developed by Crisp *et al.* (1968) made the measurement of forces on a single cutting edge during plain fly milling at 30 m s^{-1} possible. Using a $10 \times 5 \times 1.5$ mm workpiece results in horizontal and vertical natural frequencies much greater than 30 000 Hz, so that the dynamometer responds accurately to the approximate 5000 Hz excition of a single cutting edge (Brecker 1967). Using a tapered workpiece, forces for cuts of varying depths were recorded by an oscilloscope. The specific energy u is obtained by dividing the maximum horizontal force by the maximum cross-sectional area of the cut. Further details concerning the dynamometer employed for single grit force studies may be found in Shaw (1984).

The cross-sectional area at the deepest actual depth of cut is obtained by tracing the cut and than using a planimeter to measure the area.

Experimental Results

Because common friable abrasive grits have irregular shapes and fracture repeatedly, initial tests utilized relatively slowly wearing diamonds. The specific energies when cutting A-6 tool steel, as shown in Fig. 3.36, indicate a definite size effect. A similar result obtained by Tanaka and Ikawa (1964) when cutting molybdenum high-speed steel on a special continuous cutting apparatus is also shown.

When a wear land develops on the diamond cutting edge, plowing instead of cutting occurs at small depths of cut and at the beginning of deep cuts. When plowing gives way to cutting, the depth of cut and thus the area of cut increase sharply and forces drop.

The specific energy curve for a single grit in the surface of a white aluminum oxide wheel used to cut hard ($R_C = 63$) and soft ($R_C = 34$) AISI 52100 steel without significant fracture is given in Fig. 3.37. The observed size effect results in the specific energy being related to the maximum undeformed chip thickness t by

$$u \sim t^{-0.8}. \tag{3.9}$$

SINGLE GRIT PERFORMANCE

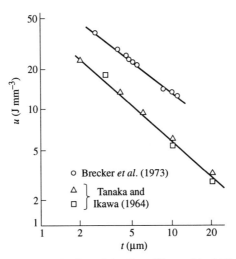

FIG. 3.36. Specific energy results for plain fly milling with 0.75 mm (0.03 in) radius diamond on A-6 tool steel. Also shown are similar results on tool steel by Tanaka and Ikawa (1964) (after Brecker and Shaw 1974).

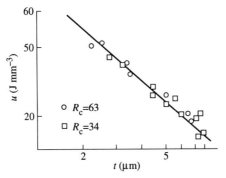

FIG. 3.37. The specific energy for a single abrasive grit on the surface of white Al_2O_3 (vitrified bond) wheel in plain fly milling of hard and soft AISI 52100 steel. Wheel speed, 6000 f.p.m. (30 m s^{-1}). All grits on the wheel surface but one were dressed away (after Brecker and Shaw 1974).

The hardness of the steel does not affect the results. Figure 3.38 shows the variability present when using single grits in the surface of several different wheels. The area measurements are sensitive to plowing and the triangles in Fig. 3.38 indicate area measurements which are low due to area measurement errors caused by plowed material in the cut. Actual specific energy values are lower than those shown for such cases.

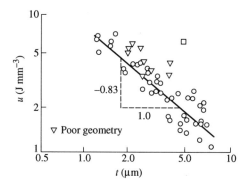

FIG. 3.38. The specific energy for single abrasive grits on surfaces of various white aluminum oxide abrasive wheels in plain fly milling of $R_C = 63$ AISI 52100 steel at 6000 f.p.m. (30 m s^{-1}).

Tests on hardened aluminum indicate a predominance of plowing instead of cutting. The size effect was stronger and the exponent in eqn (3.9) changed from -0.8 to -1.2.

Thus, the size effect in grinding has been verified using masurements on single cutting edges at typical grinding speeds. Dressing diamond and abrasive grains of comparable size yielded virtually identical specific energies. The results agree with those obtained by Backer et al. (1952) using complete abrasive wheels at large chip thicknesses. Very small depths of cut could not be obtained in these single grit tests to permit identification of an upper limit of specific energy.

The fact that specific grinding energy is appoximately the same for hard and soft steels is a paradoxical result that requires explanation. It is believed to be due to the fact that soft steel is more ductile than hard steel and hence will involve greater strain in the formation of chips. What the soft steel lacks in strength is compensated for by its greater ductility.

Single-grit fly milling tests have also been performed at the Aachen Technische Hochschule (Koenig et al. 1985). In these tests a single abrasive particle was mounted in a low-mass piezoelectric dynamometer which, in turn, was mounted on the periphery of a metal disc. Rotary mercury contacts were used to carry the signal to an oscilloscope. This apparatus was used to study subsurface structural changes, grinding forces with and without lubricants, the difference in wear rates for hard and soft steel, and the influence of the temperature of the workpiece on the ratio of forces F_P/F_Q, which was regarded as a grinding friction coefficient, all for single-grit grinding.

It was found that the wear rate increased inversely with the hardness of the work material. Annealed material gave the highest wear rate due to greater adhesion between grit and chips. While abrasive wear (microchipping) was greater for a hard work material, adhesive wear was dominant. In the tests

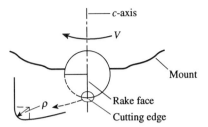

FIG. 3.39. A single-crystal sapphire sphere ground to provide a 0° rake angle and mounted in the surface of a 10 in (25.4 cm) wheel.

with different fluids, it was found that the depth of cut for the onset of chip formation increased with lubricity and that the lubricant had a far greater influence on F_P than on F_Q. F_P was substantially less with a lubricant except at very low values of depth of cut. The grinding coefficient of friction experiments were performed by radially plunging a single grit into the work (AISI 1045 steel) to a depth of 10 μm and then measuring F_P/F_Q without further radial infeed. This gave a ratio corresponding to zero depth of cut (0 removal) that had a minimum value of about 0.08 at 300 °C.

The dimensional units of specific energy in terms of force (F), length (L) and time (T) are $[FL^{-2}]$. The preferred units in the English system are in lb in^{-3}, but in the workshop the unusual combination of units h.p. in^{-3} min^{-1} is frequently used. In the metric system specific energy is usually expressed as J mm^{-3}. The following relationship may be used to convert from one set of units to another:

$$10^6 \text{ in lb in}^{-3} = 2.525 \text{ h.p. in}^{-3} \text{ min}^{-1} = 6.895 \text{ J mm}^{-3}.$$

Sharpened Spherical Grits

Single crystal sapphire (Al_2O_3) spheres were oriented with the C axis vertical and flats were ground to provide grits with predetermined rake angles of 0° and −20° (Fig. 3.39). These special grits were mounted in the periphery of a 10 in (25.4 cm) diameter disc which replaced the grinding wheel in a horizontal spindle surface grinder. Small (1/4 × 1/4 × 0.04 in = 6.4 × 6.4 × 1 mm) specimens were mounted in the piezoelectric dynamometer (Shaw 1984, p. 161) and plain fly milling cuts were taken on an inclined flat specimen, as shown in Fig. 3.40. The maximum depth (t) was obtained by tracing transversely to the cutting direction at the midpoint of each cut. These traces also enabled the cross-sectional shape and area of each cut to be determined. There was evidence of from 5 percent to 25 percent deflection of the grit relative to the work, depending on workpiece hardness and depth of cut. However, this was taken care of by using measured depths of cut t rather than the values set on the machine.

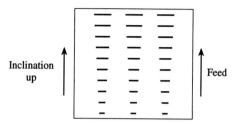

Fig. 3.40. A schematic plan view of a small specimen mounted on a high natural frequency piezoelectric dynamometer, showing plain fly milling cuts of increasing depth.

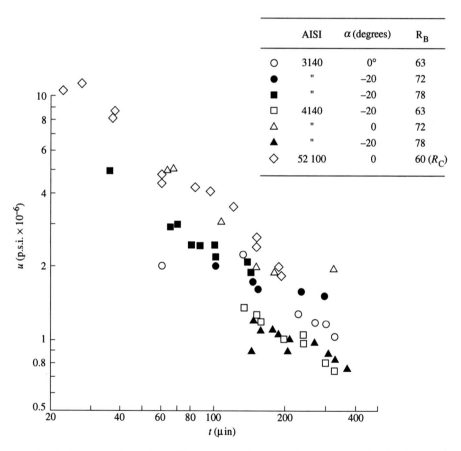

Fig. 3.41. The variation of specific energy (u) versus the maximum depth of cut (t) for three steels at different hardness levels.

Values of specific energy at the midpoint u versus t are shown in Fig. 3.41. Here test results for three steel compositions, several hardness values, and two rake angles are included. The specific energy is seen to increase substantially with decrease in t ($u \sim 1/t^{0.9}$ to a good approximation). Also, the hardness of the work had relatively little influence on the specific energy. This is a result that also pertains to complete grinding wheels, as well as to single abrasive particles (Backer *et al.* 1952). While it might be concluded that rake angle has little influence on the u versus t curve, such a conclusion is not justified. The effective rake angles in these tests at very small values of t are not the nominal values ground on the 3/16 in (4.76 mm) diameter spheres, but a very much more negative value, depending on the radius of curvature at the cutting edge (ρ in the inset of Fig. 3.39) and the value of t. It was found from the transverse traces that the cutting edges were subjected to frequent microchipping which would change the effective value of ρ. If ρ were of the order of 100 μin (2.5 μm), then ρ would be very important and the nominal rake angle (α) would be relatively unimportant.

In these tests there was a definite clearance in all cases. This rules out the suggestion that has been made in the past that the size effect in fine grinding is due primarily to rubbing.

References

Backer, W. R., Marshall, E. R., and Shaw, M. C., (1952). *Trans. ASME* **74**, 61.

Bielawski, E. J. (1966). *Proc. 9th midwinter Conf., Grinding Wheel Institute/Abrasive Gain Assoc.*, Buffalo, New York.

Brecker, J. N. (1967). Ph.D. dissertation, Carnegie-Mellon University.

Brecker, J. N. (1974). *J. Engng Ind.* **96**, 1253.

Brecker, J. N. and Shaw, M. C. (1974). *Ann. CIRP* **23**(1), 93.

Brecker, J. N., Komanduri, R., and Shaw, M. C. (1973). *Ann. CIRP* **22**(2), 219.

Brown, R. (1957). M.Sc. Thesis, MIT.

Coes, L. (1955). *Ind. Engng Chem.* **47**, 2493.

Coffin, L. F. (1961). *Trans. Am. Soc. Lub. Engrs* **1**, 108.

Crisp, J., Seidel, J. R., and Stokey, W. F. (1968). *Int. J. Prod. Res.* **7**, 159.

Dawihl, G. J. (1940). *Z. fur Metallkunde* **32**, 320.

Duwell, E. J. (1962). *J. appl. Phys.* **33**, 2691.

Duwell, E. J. and McDonald, W. J. (1961). *Wear* **4**, 372.

Duwell, E. J. and McDonald W. J. (1965). *ASM Metals Engng Q.* **5**, 56.

Glucklick, J. (1970). *NASA Report TS P71 10158* From Jet Propulsion Laboratory, CIT.

Goepfert, G. J. and Williams, J. L. (1959). *Mech. Engr.* **81**, 69.

Grisbrook, H., Moran, H., and Shephard, J. H. (1965). *Proc. Conf. Iron and Steel Institute*, London.

Ikawa, N. and Tanaka, T. (1971). *Ann. CIRP* **19**(1), 153.

Koenig, W., Steffens, K., and Ludewig, T. (1985). *ASME Grind Symp., PED 16*, p. 141.

Komanduri, R. (1976). *Ann. CIRP* **26**(1), 191.

Komanduri, R. and Shaw, M. C. (1975). *Nature* **255**, 211.

Komanduri, R. and Shaw, M. C. (1976a). *J. Engng Ind.* **98**, 1125.
Komanduri, R. and Shaw, M. C. (1976b). *Phil. Mag.* **34**, 195.
Lal, G. K. and Shaw, M. C. (1972). *Proc. Int. Grinding Conf.* Carnegie Press, Pittsburgh, Pennsylvania, p. 107.
Lal, G. K. and Shaw, M. C. (1973). *Int. J. Machine Tool Des. Res.* **13**, 131.
Lal, G. K., Matsuo, T., and Shaw, M. C. (1973). *Wear* **24**, 270.
Loladze, T. N. and Bokuchava, G. V. (1972). *Proc. Int. Grinding Conf.* Carnegie Press, Pittsburgh, 432.
Marshall, E. R. and Shaw, M. C. (1952). *Trans. ASME* **74**, 51.
Matsuo, T. (1981). *Ann. CIRP* **30**(1), 233.
Nakayama, K. and Shaw, M. C. (1967). *Proc. Inst. Mech. Engrs* **182**, 179.
Outwater, J. and Shaw, M. C. (1952). *Trans. ASME* **74**, 73.
Scrutton, R. F., Lal, G. K., Matsuo, T., and Shaw, M. C. (1973). *Wear*, **24**, 295.
Shaw, M. C. (1984). *Metal cutting principles*. Clarendon Press, Oxford.
Steijn, R. P. (1961). *J. appl. Phys.* **32**, 1951.
Takagi, J. and Shaw, M. C. (1983). *J. Engng Ind.* **105**, 143.
Takenaka, N. (1964). *Ann. CIRP* **13**(1), 183.
Tanaka, T. and Ikawa, N. (1964). *Tech. Rep., Osaka Univ.* **14**.
Tanaka, T., Ikawa, N., and Tsuwa, H. (1981) Ann CIRP **30**(1), 241.
Yoshikawa, H. and Sata, T. (1960). *Sci. Pap. Phys. Chem. Res.* (Japan) **54**, 389.

4

FORM AND FINISH GRINDING MECHANICS

Introduction

This chapter is concerned with the mechanics of chip removal, for fine grinding operations involving complete grinding wheels. The chip removal mechanism will be discussed first.

The generally accepted mechanism of chip formation in metal cutting (Merchant 1945) is one of concentrated shear, in which the metal removed is abruptly deformed as it crosses the shear plane A–B in Fig. 1.1. In this case, the rake angle (α) is near zero and there is little subsurface plastic flow, low residual stresses, and practically all of the energy consumed ends up in the chips.

In fine grinding most of the energy ends up in the workpiece (Sato 1961). Residual stresses are much more important and appreciable plastic deformation is evident below the finished surface. In addition, the mean rake angle is highly negative. This suggests that a mechanism which differs appreciably from that in cutting should pertain in grinding. An extrusion-like mechanism based on a plastic–elastic theory of Brinell hardness (Shaw and DeSalvo 1970) has been proposed (Shaw 1972).

An example of an extrusion-like cutting operation is to be found at the chisel point of a twist drill. While chips involving concentrated shear are formed at the two lips of the drill, the action at the chisel point is quite different, involving material removal by extrusion. Chips produced by these two cutting actions for a twist drill are illustrated in Fig. 16.35 on p. 467 of Shaw (1984).

Hardness Model of Chip Formation

In a Brinell test (Fig. 4.1(b)) the load (R) is supported by the workpiece over an area equal to πa^2. An abrasive grit will have a geometry at its tip which may be approximated by a sphere of radius (ρ) depending on grit size and type, dressing conditions, and wear (Fig. 4.1(b)). While the plastic zone beneath the indenter in the hardness test is completely confined, that with the inclined load will have a region that is unconfined. As the grit moves horizontally, metal will escape upward across the unconfined zone to form the grinding chip.

The hardness test involves essentially static deformation that is independent of friction between indenter and workpiece. The strain in the hardness test is very low (< 0.1, according to Tabor 1951) and the specific energy is consequently also relatively low. The fine grinding operation involves the deformation and removal of material at an extremely high strain and strain rate, unlike the hardness test, and the material being deformed is in the hot working regime, which means that the high strain rate is very important.

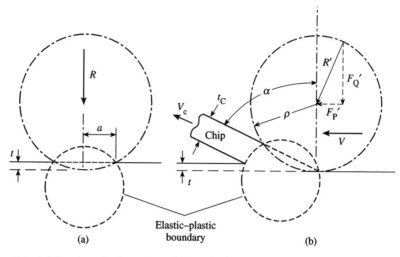

FIG. 4.1. (a) Hardness indentation with a spherical indenter. (b) Indentation at the tip of an abrasive grit with a radius of curvature at the grit tip $= \rho$ and resultant force R inclined at angle α to the vertical. The deformed chip thickness is t_c and the cutting ratio is t/t_c.

The undeformed chip thickness (t) in fine grinding is of the same order of magnitude as the effective grit-tip radius (ρ) and hence the effective rake angle will have a large negative value. This causes the material removal mechanism to shift from that of concentrated shear to an extrusion-like action. Figure 4.2 shows diagrammatically the change in metal cutting action as the rake angle of the tool goes from a positive value to a negative value. The following changes occur:

- the resultant force increases
- the shear angle decreases
- the cutting ratio (chip length ratio) decreases
- the coefficient of tool face friction ($\mu = \tan \beta$ in Fig. 4.2(b)) goes from about 1 to 0

The rake angle corresponding to that where the coefficient of tool face friction goes from a positive to a negative value ($\alpha = -45°$) is where the shift from a mechanism of concentrated shear to extrusion-like behavior should be expected. The specific energy in fine grinding unlike that for indentation hardness, is extremely high due to a combination of conditions to be discussed later.

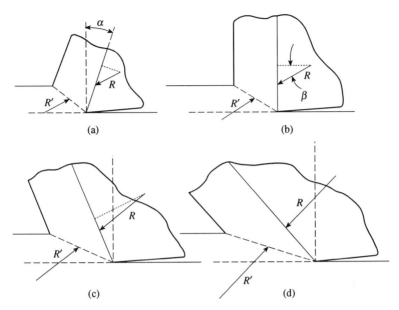

FIG. 4.2. A diagrammatic representation of concentrated shear type cutting with variable rake angle α. (a) $\alpha = +30°$, $\mu = \tan\beta = 1.0$; (b) $\alpha = 0°$, $\mu = 0.6$; (c) $\alpha = -30°$, $\mu = 0.15$; (d) $\alpha = -45°$, $\alpha = 0$.

Undeformed Chip Thickness

It is well known that the mean undeformed chip thickness in fine grinding plays a very much more important role relative to grinding forces, surface finish, surface temperatures, and wheel wear than does the wheel depth of cut d. The first paper published concerning the size of grinding chips was by Alden (1914). From the geometry of two intersecting circles, an equation for maximum chip thickness t was derived for external grinding. This equation is difficult to use, since a convenient method of finding the spacing of successive active grits is not given. However, Alden did direct attention to the importance of chip thickness in grinding. Guest (1915) discussed the importance of t and derived another equation involving the mean active grit spacing in external grinding. Several other workers (Chapman 1920; Krug 1925; Hutchinson 1938; Heinz 1943; Peklenik 1957) were concerned with maximum chip thickness (t) but all in terms of the mean active grit *spacing*. It was not until the work of Backer *et al.* (1952) that the more convenient measure of grit density C, equal to the number of active grits per unit area, was introduced, together with r, the ratio of mean scratch width to mean scratch depth, in the analysis for mean undeformed chip thickness.

The estimation of mean chip thickness in fine grinding ($\bar{t}$) has been discussed thoroughly by Reichenbach *et al.* (1956). In this analysis, the various possible sorts of chips were classified into five types (see Fig. 4.3) according to the

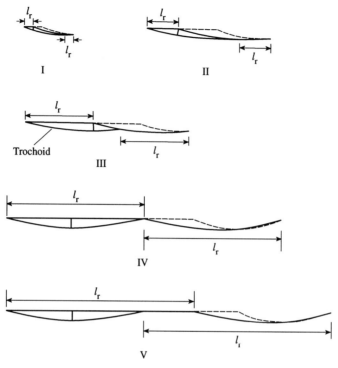

FIG. 4.3. Types of chips produced when the work speed (v) is increased progessively relative to the wheel speed (V). Quantity l_r is distance that work moves between successive grits cutting in the same groove (after Reichenbach *et al.* 1956).

magnitude of the work speed v relative to the wheel speed V. In the discussion that follows, horizontal surface grinding will be discussed, where V/v is large (typically 50–100). This is case I in Fig. 4.3. Similar results for all the other cases in Fig. 4.3 are derived in Reichenbach *et al.* (1956).

The equation for mean undeformed chip thickness $\bar{t}$ can be obtained by equating the mean volume of a single chip obtained in two different ways. All chips are assumed to have the same size and, in order to be sure that all metal to be removed is accounted for, the chips are assumed to be of constant width (Fig. 4.4(b)). Actually, grinding chips will be shaped as in Fig. 4.4(c), where the mean cross-section is the same as in Fig. 4.4(b)). However, it is not possible to be sure that all of the volume removed is accounted for unless Fig. 4.4(b) is adopted.

The magnitude of undeformed chip length (l) relative to t is always very much larger than that shown in Fig. 4.4(b), and hence the undeformed chip is actually a long slender triangle. To an excellent approximation, the volume B) of the chip shown in Fig. 4.4(b) will be

$$B = l\bar{t}b' = (l\bar{t}^2 r), \qquad (4.1)$$

FORM AND FINISH GRINDING MECHANICS

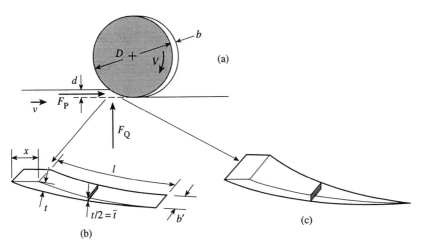

FIG. 4.4. Individual chips produced in surface grinding: (a) a schematic view of the process; (b) an idealized chip used to insure volume continuity; (c) the actual shape of the undeformed chip (after Lal and Shaw 1975).

where

$$r = b'/\bar{t} \tag{4.2}$$

and b' is the mean undeformed width of the chip.

If there are C active grits per square inch of wheel surface, then the volume per chip, B, can also be obtained by dividing the rate of metal removal (vbd) by the number of cuts made per unit time (VbC):

$$B = vbd/VbC = vd/VC, \tag{4.3}$$

where b is the width of cut in a surface plunge grinding operation (Fig. 4.4), d is the depth of cut, v is the table speed, and V is the wheel speed.

Equating the two values of B for volume continuity:

$$l\bar{t}^2 r = vd/VC \tag{4.4}$$

or

$$\bar{t} = [(v/VCr)(d/l)]^{0.5}. \tag{4.5}$$

All quantities in eqn (4.5) are readily determined except for C and r. Hence methods of estimating these values are discussed below.

Active Grit Density, C

Backer *et al.* (1952) and Reichenbach *et al.* (1956) determined C by rolling a dressed wheel over a soot-coated glass plate and counting the number of contacts indicated on an enlargement made, using the plate as a photographic

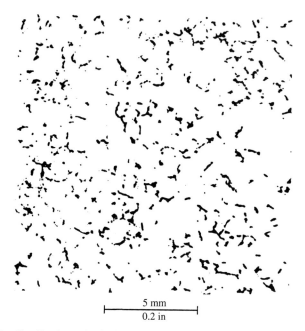

FIG. 4.5. The distribution of grits in the surface of a diamond-dressed 32A46-H8-VBE grinding wheel, obtained by rolling the wheel over a carbon black coated glass plate (after Backer et al. 1952).

negative (Fig. 4.5). Values of C determined from soot tracks for a special series of white Al_2O_3 wheels having vitrified bonds are given in Table 4.1. From this table, C is found to increase and then decrease as wheel hardness increases for a given grit size. Also, C increases with an increase in S (the screen number) for a given wheel hardness (see Fig. 4.6).

Soot-coated microscope slides were prepared by passing them over a gas burner. The gas feeding the burner was passed through benzene to obtain a flame of high soot content. The glass was coated until newsprint was just visible when viewed through the soot. It is estimated that the layer of soot thus produced was about 0.001 in (25 μm) thick.

This soot-coated glass plate method is rather crude and gives too high a value of C for several reasons:

1. Vibration was not included in this static determination.
2. Local wheel and work deflections were not included.
3. Imperfect dressing (deviations of the wheel face from a perfect circle) was not taken into account.
4. Some grits follow others too closely and cut air instead of metal in practice.

TABLE 4.1 *C values from soot tracks for white Al_2O_3 wheels, number per in^2*†

Grit Size S	Type of Dress‡	Wheel grading					
		F8V	H8V	J8V	L8V	N8V	P8V
24	Coarse	699	2 085	2 110	3 008	1 917	
	Fine	1 789	2 628	2 190	3 272	2 333	
36	Coarse	1 104	3 640	2 551	2 677	1 555	
	Fine	1 141	2 080	2 545	2 036	3 866	
46	Coarse	1 728	2 855	2 966	2 677	3 663	
	Fine	1 656	2 959	2 624	3 467	2 933	
60	Coarse	2 962	4 068	3 160	4 998	3 553	1 507
	Fine	3 960	3 891	4 089	6 208	3 241	2 504
80	Coarse		5 768	7 797	5 467	5 784	2 712
	Fine		7 637	7 629	9 165	3 428	7 733
120	Coarse		13 384	9 421	13 752	7 451	12 605
	Fine		13 019	14 022	13 117	11 329	12 676

† The values of C per mm^2 may be obtained b dividing C by 625.
‡ Coarse dress: the flat side of a Vickers diamond was swept across the wheel face with 0.001 (25 μm) in-feed at 12 i.p.m. (300 mm min^{-1}) cross-feed. Fine dress: the same flat side of the Vickers diamond was used as follows: two passes with 0.0005 in (12 μm) in-feed at 3 i.p.m. (76 mm min^{-1}) cross-feed; two passes with 0.0002 in (5 μm) in-feed at 3 i.p.m. (76 mm min^{-1}) cross-feed; one spark-out pass at 3 i.p.m. (76 mm min^{-1}) cross-feed.

In addition, it was not possible to determine the variation of C with the radial distance (y) from the outermost grit.

A somewhat improved technique was to roll a dressed wheel over a representative ground surface of the work material, coated with a thin (0.0001 in, or 2.5 μm) layer of soot which approximates t in thickness (Reichenbach et al. 1956). Advantages of this are as follows:

1. Unusually high points can penetrate the steel surface in a more characteristic manner than when glass is used.
2. The geometry of a ground surface more closely approximates the surface encountered by a wheel in practice than does the surface of a glass plate.

Another method of determining C is to count the number of wear flats produced on the grinding wheel by lighting the wheel surface at a glancing angle (Malkin and Cook 1971). The total number of grits with wear flats divided by the viewing area gives the number of active grits per unit area. An alternative method (Grisbrook 1962) is to count the number of chips produced. However, by this method, it is assumed that each contact point on the wheel removes one chip at each engagement with the workpiece.

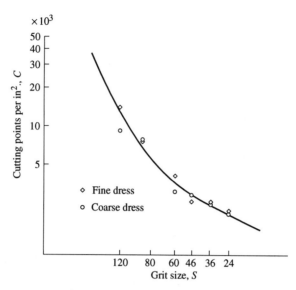

FIG. 4.6. The variation of C with grit size S for vitrified white Al_2O_3 wheels of J hardness, determined by rolling the dressed wheel slowly over a soot-coated glass plate.

A stylus method, in which a tracer instrument records the relative heights of abrasive grits in the surface of a grinding wheel, has been used by several workers. When a tracer having a radius of curvature at its tip is used, intepretation is complicated because the effective width of trace depends upon the depth of penetration (Shaw and Komanduri 1977). For this reason, Orioka (1957) employed a chisel-shaped tip having a straight edge to provide a fixed tracing width.

The importance of using C instead of the mean distance between active grits (l_r) in the wheel surface is that there is no way of determining l_r directly that does not involve a prior determination of C. However, the following equation may be used to determine l_r once C is known:

$$b' l_r C = 1, \qquad (4.6)$$

where l_r is the mean spacing between grits cutting in the same groove, and b' is the mean width of the scratch, equal to $\overline{rt}$.

It should be noted that, despite the fact that l_r is difficult to measure directly, it has been widely used in place of C in Europe and Japan (see, for example, Peklenik 1957).

A difficulty with many techniques is that they do not distinguish those grits at the same elevation that will actually cut metal from those that actually cut air because the metal they would cut has already been removed by a preceding grit. Values of C that take this factor into account are referred to as 'dynamic

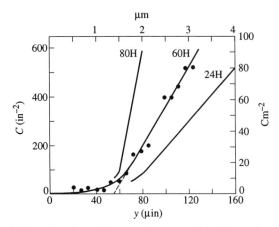

FIG. 4.7. The variation of active grits per unit area (*C*) in the surface of white Al_2O_3 wheels with the radial distance from the outermost grit (*y*). Practical values of *y* for fine ginding operations fall on the linear portions of the curves (after Nakayama and Shaw 1967).

values', to distinguish them from 'static values' where all grits of a given elevation are counted, whether or not they actually cut metal. While the static values are independent of grinding conditions, the dynamic values are not.

Three methods have been proposed for determining dynamic values of *C* directly.

1. Peklenik (1957) employed a small insulated platinum wire that penetrates through the surface being ground. When the wire is ground it smears on to the workpiece surface, thus forming a thermoelectric junction. As each grit traverses this junction it generates a thermoelectric emf which is recorded by an oscilloscope. By counting the voltage peaks and knowing the effective width, values of *C* may be determined for different grinding conditions.

2. Nakayama and Shaw (1967) ground a mirror-finished workpiece that was slightly inclined axially (650 μ in in^{-1} = 6.4 mm cm^{-1}). By using a relatively high work speed (2400 f.p.m. = 12 m s^{-1}) many individual scratches of variable depth were produced, which could be analyzed to obtain *C* versus *y* curves such as those shown in Fig. 4.7 (for examples, see Chapter 10).

3. Brecker and Shaw (1974) ground a very narrow workpiece mounted on a low-inertia piezoelecric dynamometer. From the number of peaks per unit area of work surface and the *V/v* values employed, values of *C* could be readily determined.

While all three of these methods give dynamic values of *C*, the first two are quite tedious and will not be discussed further. The third method employs a thin (0.004 in or 0.1 mm) razor blade cemented between plastic strips for stability (Fig. 4.8). In turn, this is mounted on a two-component, low-inertia

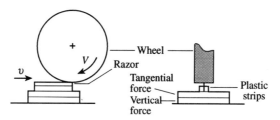

FIG. 4.8. A schematic showing a razor blade mounted on a low-inertia dynamometer (natural frequency $\simeq 100\,00$ Hz) for plunge surface grinding tests to determine the number of active grits per unit area in the wheel surface (after Brecker and Shaw 1974).

piezoelectric dynamometer (Crisp *et al.* 1968) and forces exerted by individual grits are recorded by a storage oscilloscope. The number of active cutting edges per unit wheel face area C may be obtained from the number of force peaks (N_o) as follows:

$$C = \frac{N_0}{lVb/v} = \frac{N_0 v}{lbV}, \tag{4.7}$$

where l and b are the wheel–work contact length and width of the razor blade respectively. In choosing the razor blade width (b), it is important that the width of individual grinding scratchs (b') be much less than b, in order to minimize edge effects. In practice, b should be at least $5b'$.

Figure 4.9 shows some typical results for a 60H Al_2O_3 wheel that was fine and coarse dressed as follows:

Dressing — single-point diamond:
  Coarse: two passes at 0.001 in (25 μm) at 12 f.p.m. (0.366 m min^{-1}) with no spark-out
  Fine: two passes at 0.0005 in (12 μm) at 3 i.p.m. (0.076 m min^{-1})
  two passes at 0.0002 in (5 μm) at 3 i.p.m. (0.076 m min^{-1})
  one spark-out pass

The abscissa in Fig. 4.9 is $(v/V)(d/D)^{0.5}$, which is closely related to $\bar{t}$ in eqn (4.5). Each point in Fig. 4.9 is the average of eight measurements. No difference was found between up and down grinding. Dressing diamond wear was found to have an important influence, particularly for softer wheels. The sharper the coarse dressing diamond, the lower were the values of C obtained. The chip size for cylindrical grinding will fall on the less steep (high $(v/V)(d/D)^{0.5}$) region of Fig. 4.9, while plunge surface grinding and internal grinding chip sizes generally fall on the steep region where C is not constant but decreases with decrease in chip size.

The two coarse dressing curves of Fig. 4.9 were obtained with freshly dressed wheels under identical conditions, except for the operator and the time at which the tests were run. The difference between the two curves is of unknown

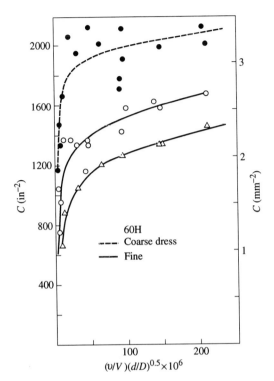

FIG. 4.9. The number of active grits per unit area (C) versus $v/V\,(d/D)^{0.5}$ for a 60H Al_2O_3 wheel operating under a wide range of conditions: $V = 4000$–6000 f.p.m. (20.3–30.5 m s^{-1}); $v = 0.25$–100 f.p.m. (0.075–30 m min^{-1}); $d = 0.0001$–0.002 in (5–25 μm) (after Brecker and Shaw 1974). △, Sharp diamond dressing tool; ○●, worn diamond dressing tool.

origin, but could be due to a slight difference in dressing or to a difference in relative humidity. The performance in dry surface grinding is found to be quite sensitive to dressing and to the relative humidity pertaining at the time of the test. On one occasion, the specific grinding energy was recorded daily for several days under carefully controlled supposedly identical conditions. However, a significant variation in the value obtained was observed and, upon checking with the weather bureau, this variation was found to correlate closely with the variation in relative humidity for the days on which the tests were made.

Figure 4.10 shows C versus $(v/V)(d/D)^{0.5}$ curves for four wheels of different hardness and grit size.

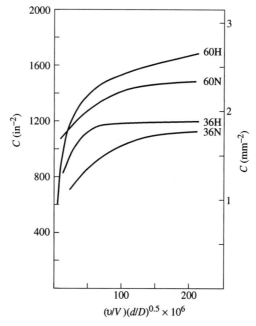

FIG. 4.10. Curves of C versus $V/v(d/D)^{0.5}$ for four coarse-dressed wheels, operating under a wide range of conditions as in Fig. 4.9 (after Brecker and Shaw 1974).

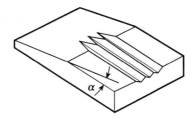

FIG. 4.11. A schematic of a taper section.

Ratio of Scratch Width to Scratch Depth

The ratio of the mean width b' to the mean depth $\bar{t}$ of a scratch in a grinding operation which appears in eqn (4.5) was originally estimated from taper sections of ground surfaces (Backer *et al.* 1952; Reichenbach *et al.* 1956). After plating a ground surface with nickel to protect it, a new flat surface was ground at a small angle α to the original surface, as shown in Fig. 4.11. The interface between the steel and nickel corresponds to the profile of a surface to be studied, with different magnifications in the horizontal and vertical

FORM AND FINISH GRINDING MECHANICS

FIG. 4.12. A taper section of a ground surface (surface ground) (after Backer *et al.* 1952). Wheel = 32A-46-H8-VBE; $V = 5200$ f.p.m. (26.5 m s^{-1}); $v = 12$ f.p.m. (0.061 m s^{-1}); $d = 0.001$ in $(25 \mu\text{m})$.

directions. For example, by making $\alpha = 2°17'$, a magnification of cosec $(2°17') = 25$ times is obtained. If the polished new surface is viewed at $100 \times$, then the horizontal magnification will be $100 \times$ but the vertical magnification will be $2500 \times$. Figure 4.12 shows a taper section of a typical ground surface for which the horizontal and vertical magnifications were $200 \times$ and $10\,000 \times$ respectively. The mean value of r for this surface was found to be 17.

In Chapter 3 it was found that, to a good approximation, the effective radius ρ at the tip of an abrasive particle is a function of the mean grit diameter g and, to a good approximation,

$$\rho \simeq g/4. \qquad (4.8)$$

From eqn (4.2),

$$r = b'/\bar{t} - [2(2\rho\bar{t})^{0.5}]/\bar{t} \simeq (2g/\bar{t})^{0.5}. \qquad (4.9)$$

The approximate equation relating grit diameter g and screen size S ($Sg \simeq 0.7$) was used with eqn (4.9) to obtain Fig. 4.13, which shows r versus $\bar{t}$ for different grit screen sizes (S).

From Fig. 4.13 it is evident that r increases as $\bar{t}$ decreases, but increases as grit diameter increases (S decreases). For SRG both the grit diameter and the mean undeformed chip thickness are large. The oppositite is true for FFG. Since changes in g and $\bar{t}$ have opposite effects on r, values of r for FFG and SRG tend to be about the same (somewhere between 5 and 20).

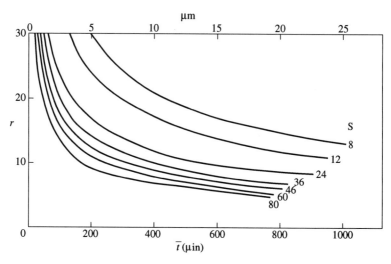

FIG. 4.13. The variation of r versus $\bar{t}$ for different grit sizes (S) from eqn (4.9).

Wheel–Work Contact Length

The wheel–work contact length l in horizontal spindle surface grinding (Fig. 4.4) is as follows, to a good appoximation, since $d \ll D$:

$$l = (Dd)^{0.5}, \tag{4.10}$$

where D is the wheel diameter and d is the wheel depth of cut.

This ignores wheel–work deflection, which will tend to increase D but decrease d. However, the value of d employed is generally a measured value of work material actually removed, rather than the value set on the machine. Hence contact length l will be somewhat larger than the value obtained from eqn (4.10) using the unloaded wheel diameter D. Wheel–work deflection is discusssed in Chapter 5.

For internal and external grinding, the conformity of the work relative to the wheel will be greater and lesser, respectively, than for surface grinding (Fig. 4.14). The difference in conformity may be taken into account (Reichenbach et al., 1956) by the use of the following effective wheel diameter, D_e in eqn (4.11) in place of the actual wheel diameter, D:

$$D_e = (D_w D_s)/(D_w \pm D_s), \tag{4.11}$$

where the plus sign is for external grinding and the minus sign is for internal grinding, and for surface grinding $D_e = D$.

The following equation covers all cases in which $t \ll d$:

$$l = (D_e d)^{0.5}(1 \pm v/V), \tag{4.12}$$

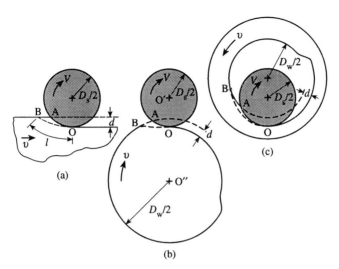

FIG. 4.14. Chips produced when $v \ll V$ for (a) surface grinding, (b) external grinding, and (c) internal grinding (after Reichenbach et al. 1956).

where the plus sign pertains to up grinding (v and V in opposite directions) and the minus sign pertains to down grinding (v and V in the same direction). Since V is usually 50–100 times as large as v, it is not necessary to distinguish between up and down grinding when considering l.

When eqn (4.10) is substituted into eqn (4.5), the following equation is obtained:

$$\bar{t} = [(v/VCr)(d/D_e)^{0.5}]^{0.5}. \qquad (4.13)$$

Two examples may be considered to indicate the large difference that normally exists between D_e and D.

1. For an external grinding situation, let D_w the work diameter, be 4 in, and let D_s, be wheel diameter, be 30 in. From eqn (4.11),

$$D_e = \frac{4(30)}{(30+4)} = 3.53 \text{ in.}$$

For this case, $D_e/D_s = 3.53/30 = 0.12$.

2. For a typical internal grinding operation, let $D_w = 4.0$ in and $D_s = 3.5$ in. From eqn (4.11),

$$D_e = 4(3.5)/0.5 = 28 \text{ in.}$$

For this case, $D_e/D_s = 28/3.5 = 8.3$.

These examples show that the effective diameter may be an order of magnitude smaller than the actual wheel diameter in external grinding, or nearly an order of magnitude larger in internal grinding.

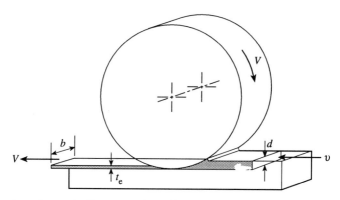

FIG. 4.15. A surface grinding operation, showing the equivalent chip thickness, t_e.

When grinding a given work material with a given wheel type under different kinematic conditions (different v, V, d, and D_e) and with wheels of different C and r, we should expect the same results (specific grinding energy, temperature, grit wear rate, and surface finish) provided that the mean chip thickness ($\bar{t}$) from eqn (4.13) is the same (see Backer *et al.* 1952; Reichenbach *et al.* 1956).

Equivalent Chip Thickness

A useful, but more approximate approach to mean undeformed chip thickness is illustrated in Fig. 4.15 for a horizontal spindle surface 'down' grinding operation. This shows material entering the wheel at the right at a rate vdb and an undivided sheet of chips of thickness t_e leaving at the left. Since there is no change in volume when a material is plastically deformed and chip formation is a plane strain operation ($r = 10$ or more in FFG),

$$vbd = Vt_e b \tag{4.14}$$

or

$$t_e = (vd)/V. \tag{4.15}$$

The quantity t_e is called the equivalent undeformed chip thickness.

Comparing eqns (4.15) and (4.5) it is seen that a major part of eqn (4.5) is present in eqn (4.15) and that t_e is approximately equal to vd/V. It has been found empirically (Snoeys 1971; Snoeys and Decneut 1971; Snoeys and Peters 1974) that t_e correlates quite well with grinding forces, specific energy, temperature, wear, and surface finish, just as $\bar{t}$ does, but offers a simpler, more approximate approach. This 'sheet of chips' concept was apparently first employed by Woxen (1933) and later by Colding (1959, 1971), before being adopted to correlate the grinding results in an international study (Snoeys and

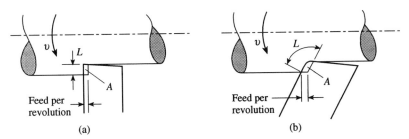

FIG. 4.16. (a) A turning tool with zero side cutting edge angle. (b) turning tool with a positive side cutting edge angle and greater L.

Decneut 1971) by CIRP (Collège International pour l'Etude Scienctifique des Techniques de Production Mécanique).

Woxen reasoned that the tool of Fig. 4.16(a) should have a shorter life than the tool of Fig. 4.16(b), since the work associated with removing the same chip area is distributed over a shorter length of cutting edge (L). He proposed a quantity termed the chip equivalent (q) to take into account differences in the length of cutting edge:

$$q = L/A, \qquad (4.16)$$

where L is the length of the cutting edge (see Fig. 4.16) and A is the chip cross-sectional area. He found that tool life increased with an increase in q. It is unfortunate that Woxen did not use the reciprocal of q as his parameter, since this corresponds to the effective undeformed chip thickness t_e, as shown below.

Considering the wheel to be a tool instead of a collection of individual grits, Colding wrote the corresponding chip equivalent for grinding as

$$q = L/A = (Vb)/(vbd) = V/(vd) = 1/t_e. \qquad (4.17)$$

Thus, the physical significance of t_e, which was arrived at empirically in the CIRP investigation, may be taken to be the reciprocal of the chip equivalent (q). The undeformed chip thickness of the layer removed, when grinding is considered as a continuum removal operation in which the material removed is considered to be a sheet of uniform thickness (the so-called chip sheet approach), corresponds to t_e.

At first, Fig. 4.15 appears to represent an overly simplified model of a fine grinding operation, since the swarf is not really in the form of a continuous sheet but actually consists of a large number of separate chips. However, this plane strain model turns out to be a useful one for correlating grinding performance and forms a good starting point for further analysis of FFG. This is because the behavior of individual abrasive grits is far less important than the collective action of the group of grits that act simultaneously over the area of wheel–work contact. The same situation exists in tribology, where the collective action of the contacting asperities in the surface of a slider are of far greater importance than the action of individual asperities. The peak temperature beneath an individual asperity will be much higher than the mean temperature over the apparent area of contact. However, the higher peak temperature is present for such a very short time (a few μs) that it is relatively ineffective in influencing the diffusion processes of practical importance (wear, subsurface changes, melting, etc.). The mean temperature over the apparent area of contact is present for several orders of magnitude longer than the localized peak temperature, and is the one that correlates with the performance of a friction slider or a grinding wheel. This is discussed further in Chapter 9.

References

Alden, G.I. (1914). *Trans. ASME* **36**, 451.
Backer, W. R., Marshall, E. R., and Shaw, M. C. (1952). *Trans. ASME*, **74**, 61.
Brecker, J. N. and Shaw, M. C. (1974). *Proc. Int. Conf. Prod. Engng*, Tokyo **1**, 740.
Chapman, W. H. (1920). *Trans. ASME* **42**, 595.
Colding, B. (1959). *A wear relationship for turning, milling and grinding*. KTH, Stockholm.
Colding, B. (1971). *Ann. CIRP* **20**(1), 63.
Crisp, J., Seidel, J. R., and Stokey, W. F. (1968). *Int. J. Prod. Res.* **7**, 159.
Grisbrook, H. (1962). *Adv. Mach. Tool Des. Res.*, 155.
Guest, J. J. (1915). *Proc. Instn. Mech. Engrs* 543.
Heinz, W. B. (1943). *Trans. ASME* **65**, 21.
Hutchinson, R. V. (1938) *SAE J.* **42**, 89.
Krug, C. (1925). *Maschinenbau* 875.
Lal, G. K. and Shaw, M. C. (1975) *Trans. ASME* **97**, 1119.
Malkin, S. and Cook, N. H. (1971). *J. Engng Ind.* **933**, 1120.
Merchant, M. E. (1945). *J. appl. phys* **16**, 267(a) and 318(b).
Nakayama, K. and Shaw, M. C. (1967). *Proc. Instn Mech. Engrs* **182**, 179.
Orioka, T. (1957). *Rep. Fac. Engng, Yamanishi Univ.*, Japan, No. 8.
Peklenik, J. (1957). Dissertation, Aachen, T. H.
Reichenbach, G. S., Mayer, J. E., Kalpakcioglu, S., and Shaw, M. C. (1956). *Trans. ASME* **78**, 847.
Sato, K. (1961). *Bull. Japan. Soc. Grind. Engrs.* **1**, 31.
Shaw, M. C. (1972). *Mech. Chem. Engng. Trans., Inst. Engrs, Australia* **MC8**, 73.
Shaw, M. C. (1984). *Metal cutting principles*. Clarendon Press, Oxford.
Shaw, M. C. and DeSalvo, G. J. (1970). *Trans. ASME* **92**, 480.
Shaw, M. C. and Komanduri, R. (1977). *Annals of CIRP*, 26/1, 139.
Snoeys, R. (1971). *Ann. CIRP* **20**(2), 183.
Snoeys, R. and Decneut, A. (1971). *Ann. CIRP*, **19**(1), 557.
Snoeys, R. and Peters, J. (1974). *Ann. CIRP* **23**(2), 227.
Snoeys, R., Peters, J., and Decneut, A. (1974). *Ann. CIRP* **23**(2), 227.
Tabor, D. (1951). *The hardness of metals*. Clarendon Press, Oxford.
Woxen, R. (1933). *Wheel wear in cylindical grinding IV*, Handlinger, Stockholm.

5

FORM AND FINISH GRINDING PERFORMANCE

Introduction

This chapter is concerned with a variety of topics involving the characteristics and performance of grinding wheels operating under form and finish conditions, where periodic wheel dressing is required to remove adhering work material and to provide new sharp cutting points.

Grinding Forces and Power

Grinding forces in plunge horizontal spindle surface grinding were measured as early as 1950 using the resistance strain gage dynamometer shown in Fig. 5.1. Here rings A with strain gages attached are used to record the radial component of grinding force (F_Q), while the ring at B records the tangential (power) component grinding force (F_P). The ball bearing pillow blocks at C provide a low friction support for the rings measuring F_Q. This relatively cumbersome dynamometer worked quite well and was used to make the following observations for fine FFG (Marshall and Shaw 1952; Backer et al. 1952):

1. Grinding forces are strongly influenced by the wheel dressing procedure.
2. The specific energy (u) is a very useful concept for estimating grinding forces and power in much the same way as shear and normal stresses are useful for estimating the strength of structures.
3. The specific grinding energy (u) is approximately independent of cutting speed (V) and grinding width (b), but is very much a function of the mean undeformed chip thickness ($\bar{t}$) and the strength of the material being ground.
4. The specific energy is essentially independent of the hardness of the work material.
5. Grinding forces are essentially the same for up and down grinding.
6. The depth of the plastically deformed layer beneath a FFG surface is approximately one half the wheel depth of cut ($\simeq d/2$).

The specific grinding energy is equal to the product of the shear stress (τ) and shear strain (ν) associated with chip formation. Materials of high hardness have high τ but low ν, while materials of the same chemistry of low hardness have low τ but high ν. Thus $u = \tau\nu$ tends to be the same for similar materials of different hardness (item 4 above).

During the 1950s a number of more convenient grinding dynamometers than

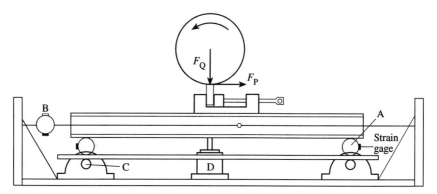

FIG. 5.1. An early strain gage dynamometer for measuring surface grinding forces (after Marshall and Shaw 1952).

that of Fig. 5.1 were designed using strain gages, such as that described on page 159 of Shaw (1984). For situations requiring dynamometers of higher natural frequency, such as for single-grit force measurements, piezoelectric transducers were employed (for example, see Shaw 1984, p. 160).

Higher wheel speeds can improve grinding productivity, wheel life, surface finish, and accuracy. However, in order to utilize higher wheel speeds, two conditions must be satisfied. Wheels capable of operating at higher speeds must be available and the machine tool on which the wheels are to be used must be capable of operating at the higher speeds. This is not merely a matter of providing the machine with a larger pulley. Sufficient power must be available to drive the grinding wheel at the higher speeds, and the machine tool must be capable of proportionately greater feed rates and have greater rigidity.

The key quantity in estimating grinding forces is the specific grinding energy. This is the work required to convert a cubic volume of metal into chips. This specific energy u (in-lb in^{-3} or J mm^{-3}) is a function of the work material and the mean chip size, and is approximately independent of other variables. Specific grinding energy is generally proportional to the toughness of the metal being ground, where toughness is measured by the energy required to rupture the metal. The tougher the metal is, the higher u will be.

The specific energy is also found to increase as the size of the chips decreases. For example, when mild steel (AISI 1020) is removed in a reciprocating surface grinding operation using a wheel containing white aluminum oxide of 40 grit size, the undeformed chips are approximately 50 μin (1.25 μm) thick and the specific energy is approximately 10^7 in-lb in^{-3} (69 J mm^{-3}). If, on the other hand, the same steel is cut by an abrasive cut-off wheel containing regular aluminum oxide of 24 grit size, the undeformed chip is approximately 0.001 in (25 μm) thick and the specific energy is about 2×10^6 in-lb in^{-3} (14 J mm^{-3}). This five to one ratio approximates the extreme range of variation in u with chip size in surface grinding.

FORM AND FINISH GRINDING PERFORMANCE

In general, two forces acting on a grinding wheel are of interest; a tangential or power component (F_P) and a radial component (F_Q), as shown in Fig. 5.1. The power required in a given grinding operation is $F_P V$, where V is the surface speed of the wheel (f.p.m. or m s^{-1}). Horsepower is obtained by dividing $F_P V$ in ft-lb min^{-1} by 33 000. In the SI system power is measured in watts. The power may also be expressed as the product of the specific energy and the volume rate of metal removal ($\dot{M}$). Thus,

$$F_P V = u\dot{M}. \tag{5.1}$$

For example, in a horizontal spindle surface grinder

$$\dot{M} = vbd \tag{5.2}$$

where v is the table speed, b is the width ground, and d is the wheel depth of cut. Therefore,

$$F_P = (uvbd)/V. \tag{5.3}$$

The following values are typical of a fine surface grinding operation of a construction steel: $u = 10^7$ in-lb in^{-3} (69 J mm^{-3}); $V = 6000$ f.p.m. (30 m s^{-1}), $v = 50$ f.p.m. (0.25 m s^{-1}); $b = 0.05$ in per stroke (1.27 mm per stroke) and $d = 0.002$ in (50 μm). From eqn (5.3),

$$F_P = [(10^7)(50)(0.05)(0.002)]/6000 = 8.33 \text{ lb } (37 \text{ N}).$$

The horsepower required at the wheel face is

$$F_P V/33\,000 = [(8.33)(6000)]/33\,000 = 1.51 \text{ h.p. } (1.13 \text{ kW}).$$

The size of the motor required would be approximately 2 h.p. if the efficiency of the drive from motor to wheel face is 75 percent.

The radial component of the grinding force (F_Q) is approximately twice the tangential component (F_P). In the foregoing example, F_Q is therefore about 16.7 lb (74 N).

The key quantity in the estimation of fine grinding forces is the specific energy, u. Table 5.1 gives values of u for FFG that are sufficiently accurate for engineering estimates.

Size Effect in FFG

One of the important questions to be answered is why the specific energy (energy per unit volume of material removed) is so high in fine grinding (about 30 times as high as for metal cutting). The first approximation model of Fig. 4.15 is useful for this purpose.

Figure 5.2 shows the wheel–work contact length l for a 'down' fine grinding operation, ignoring wheel curvature. Material of depth d enters the grinding zone with velocity v and leaves as a sheet of thickness t_e with wheel velocity V. Since the grinding width $b \gg t_e$, the material is deformed in plane strain and hence there will be no change in width b (as already inferred in going from

TABLE 5.1 *Approximate values of specific energy u for FFG*

Material	Brinell hardness (kg mm^{-2})	u p.s.i. × 10^{-6}	GPa
Soft Al	80	5.0	34.5
Hard Al	150	5.0	34.5
Cast iron	215	9.0	62.1
AISI 1020 steel	110	10.0	69.0
Soft A-6 tool steel	240	10.0	69.0
Hard A-6 tool steel	530	10.0	69.0
T-15 HSS	700	12.0	82.7
304 stainless steel	185	12.0	82.7
High-temperature alloy	340	12.0	82.7
Titanium alloy	295	10.0	69.0

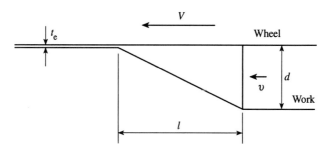

FIG. 5.2. The extrusion process involved in fine surface grinding.

eqn (4.14) to (4.15)). The strain involved in the deformation leading to the 'chip sheet' will be:

$$\epsilon = \ln(d/t_e) = \ln(V/v). \qquad (5.4)$$

This will have a very high value (i.e. for $V/v = 100$, $\epsilon = 4.6$). Furthermore, this strain will be introduced in the very short time, T, that it takes an active grit to traverse distance l:

$$T = l/V \simeq (D_e d)^{0.5}/V. \qquad (5.5)$$

The strain rate involved in FFG chip formation may be expressed conservatively as follows:

$$\dot{\epsilon} \simeq \epsilon/T. \qquad (5.6)$$

This will have a very high value, as the following example for plunge surface grinding illustrates. For:

FORM AND FINISH GRINDING PERFORMANCE

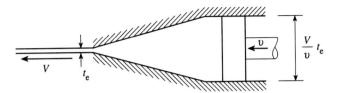

FIG. 5.3 The conventional extrusion counterpart of Fig. 5.2.

$V = 6000$ f.p.m. (30 m s^{-1}):
$v = 60$ f.p.m. (30 m s^{-1})
$D = 10$ in (254 mm)
$d = 0.0002$ in (5 μm)
$C = 1900$ in^{-2} (from Fig. 4.9 for a 60 H wheel) $= 2.95$ mm^{-2})
$t_e = (v/V)d = 2$ μin (0.05 μm)
$b' = 15 t_e = 30$ μin (0.75 μm)

From eqns (5.5) and (5.6) the strain rate for this case will be 124 000 s^{-1}.

The model shown in Fig. 5.2 is a very unusual extrusion operation, in which the extrusion energy is introduced locally at a very high strain rate, as each chip is formed at a high velocity V. Figure 5.3 shows the equivalent conventional extrusion operation corresponding to Fig. 5.2. Ignoring friction, redundant deformation, and strain inhomogeneity, the specific energy for this operation will be as follows:

$$u = \sigma_f \epsilon, \tag{5.7}$$

where σ_f is the flow stress of the material and ϵ is the natural (ln) strain.

The value of σ_f will depend upon the strain and the strain rate pertaining. However, to a first approximation, one of these variables may be considered negligible relative to the other, depending upon the nondimensional homologous temperature (θ_H):

$$\theta_H = \theta/\theta_{\text{melting point}}. \tag{5.8}$$

As a first approximation (Schey 1987),

$$\sigma_f \simeq K\epsilon^n \quad (\theta_H < 0.5 \text{ for cold working}), \tag{5.9}$$

$$\sigma_f \simeq C\dot{\epsilon}^m \quad (\theta_H > 0.5 \text{ for hot working}), \tag{5.10}$$

where K, n, C, and m are material constants, n is called the strain hardening index, and m is the strain rate sensitivity.

The deformation temperature in FFG will be in the hot working regime ($\theta_H > 0.5$) and hence eqn (5.10) will pertain. While values of C and m are given in Schey (1987) for specimens of normal size, there are no known values for specimens of extremely small size. Since hot working involves grain boundary deformation, it is to be expected that for very small deformation size the grain boundary layer will be very small, and both C and m will be greater than normal. Hence the following values would appear to be reasonable for FFG chip formation:

$$C = 30\,000 \text{ p.s.i.}, \quad m = 0.3.$$

Continuing the previous example, in which

$$\epsilon = 4.6, \quad \dot{\epsilon} = 124\,000 s^{-1},$$

the following value for specific energy is obtained from eqn (5.7):

$$u = (C\dot{\epsilon}^m)(\epsilon) = (30\,000)(124\,000)^{0.3}(4.6) = 4.7 \times 10^6 \text{ p.s.i. } (32.1 \text{ GPa}).$$

The expected experimental value for this operation would be about 10^7 p.s.i. (69 GPa), and the difference from the above estimated value might readily be explained by the fact that friction, strain inhomogeneity, and redundancy have been ignored in the above estimate.

The foregoing analysis is admittedly very approximate and is based upon a number of assumptions. The reason for including it is merely to demonstrate that the very large strains and strain rates, together with the very small specimen size, go a long way toward explaining the unusually high values of specific energy pertaining in FFG.

In metal cutting, a relatively modest increase in specific energy is observed with decrease in undeformed chip thickness t, and this 'size effect' is believed to be due to the decrease in the frequency of encountering defects in front of the tool with decrease in undeformed chip thickness t (Shaw 1950). When the relatively large increase in specific energy with decrease in undeformed chip thickness was first observed for surface grinding, this was believed to be due to the same cause (Backer *et al.* 1952). However, it now appears that, in fine FFG, the bulk of the relatively large increase in specific energy with decrease in $\bar{t}$ is due to the chip-forming process being a special high-strain extrusion process that involves a rapidly increasing strain rate with decrease in $\bar{t}$. In turn, this gives rise to very large values of specific energy in the hot working regime pertaining in the FFG chip-forming zone.

Mean Force Per Grit

The mean force per grit, F'_P, is a useful quantity for estimating the influence that operating variables have upon the effective hardness or grade of a grinding wheel. As F'_P increases, the effective grade of the wheel decreases. For surface grinding, the mean force per grit will be:

$$F'_P = F_P/lbC. \tag{5.11}$$

Substituting from eqns (4.5) and (5.3):

$$F'_P = u r \bar{t}^2. \tag{5.12}$$

For FFG,

$$u \approx 1/\bar{t}$$

and therefore, from eqn (5.12) and for constant r,

$$F'_P \sim (\bar{t}).$$

For SRG,

$$u \simeq 1/(\bar{t})^{0.3}$$

and therefore, from eqn (5.12) and for constant r,

$$F'_P \sim \bar{t}^{1.7}.$$

If a given grinding wheel is retaining grits too long (i.e. is too hard) it may be made to act more softly by increasing F_P, i.e. by increasing v/V or d/D, or decreasing C by use of a coarser dress. These changes are more effective in SRG than in FFG.

Measurement of Wheel Grade

The method most frequently used to measure grinding wheel hardness or grade is to determine the rate of penetration of a chisel as it repeatedly falls upon a wheel under standard conditions. This method is difficult to interpret in fundamental terms. In order to study different methods of grading grinding wheels relative to hardness, the series of wheels listed in Table 5.2 was specially prepared. These wheels, measuring $8 \times 0.75 \times 1.25$ in ($200 \times 19 \times 32$ mm) were used in all of the grading studies discussed below, as well as in the previous studies of C and r in Chapter 4. Typical impact penetration values for these wheels are given in Table 5.2.

A more or less static method introduced in Germany (Peklenik 1960) employs a V-shaped tool that plows out a groove in the face of a grinding wheel (see Fig. 5.4). The carbide or diamond tool is attached to a dynamometer which measures the force necessary to dislodge each grit at a speed of 0.5 f.p.m. (15 cm min^{-1}). The included angle of the tool is 52 °C. The output from the dynamometer is recorded on a chart, on which a peak is to be found for each grit encountered by the tool. The force peaks on the chart are of variable height, because the size and strength of bond bridges vary, as well as the size and shape of the grits. The mean force ($\bar{P}$) is computed along with the standard deviation of the individual force peaks. About 1000 peaks are usually considered in computing $\bar{P}$.

A second method (Colwell *et al.* 1962; Colwell 1963) is a dynamic one which employs a rotating steel cutter of 3 in (7.6 mm) diameter, as shown in Fig. 5.5. This cutter is fed across the face of the rotating grinding wheel at a rate f. The cutter is mounted at an angle of 45 ° to the wheel axis and the crushing force F (normal to the cutter axis) is measured and recorded. The cutter is mounted on a ball bearing shaft of low friction. The grinding wheel motion is transmitted to the cutter with very little slip, so that the surface speed of

TABLE 5.2 *The specifications of a carefully prepared set of vitrified white Al_2O_3 grinding wheels having a wide variety of grades and grit sizes, for use in fundamental performance studies*

Wheel marking	Designation	Bond w/o	v/o Bond	v/o Grit	v/o Pores	Impact penetration (in)†
V1156	24-F8-V	5	4	48	48	117
V1157	36-F8-V	5	4	48	48	104
V1158	46-F8-V	5	4	48	48	90
V1159	60-F8-V	5	4	49	47	70
V1160	24-H8-V	7	6	48	46	90
V1161	36-H8-V	7	6	48	46	85
V1162	46-H8-V	7	6	48	46	83
V1163	60-H8-V	7	6	48	46	65
V1164	80-H8-V	7	6	49	45	55
V1165	120-H8-V	7	6	50	44	42
V1166	24-J8-V	9	8	48	44	74
V1167	36-J8-V	9	8	48	44	71
V1168	46-J8-V	9	8	49	43	62
V1169	60-J8-V	9	8	49	43	51
V1170	80-J8-V	9	8	49	43	34
V1171	120-J8-V	9	9	50	41	24
V1172	24-L8-V	11	10	49	41	56
V1173	36-L8-V	11	10	49	41	52
V1174	46-L8-V	11	10	49	41	47
V1175	60-L8-V	11	11	49	40	39
V1176	80-L8-V	11	11	50	39	30
V1177	120-L8-V	11	11	51	38	18
V1178	24-N8-V	14	14	50	36	42
V1179	36-N8-V	14	14	50	36	39
V1180	46-N8-V	14	14	50	36	33
V1181	60-N8-V	14	14	50	36	31
V1182	80-N8-V	14	14	51	35	22
V1183	120-N8-V	14	15	51	34	16
V1184	60-P8-V	17	18	50	32	21
V1185	80-P8-V	17	18	50	32	19
V1186	120-P8-V	17	18	51	31	10

† 1 in drop of $\frac{1}{4}$ in tool.

the cutter and the grinding wheel are essentially the same. The depth of penetration of the cutter can be adjusted to any convenient value, but in most tests it corresponds to the mean grit diameter ($d = \bar{g}$). Peklenik *et al.* (1964) have compared these two methods using the wheels of Table 5.2.

Mean values of the bond force ($\bar{P}$), as determined by the quasi-static method (Fig. 5.4), are shown plotted against the bond volume in Fig. 5.6, where it is

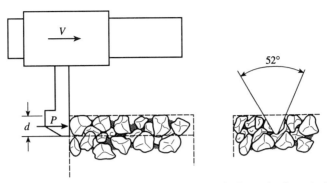

FIG. 5.4. The quasi-static method of measuring the grinding wheel grade in terms of force P (after Peklenik 1960).

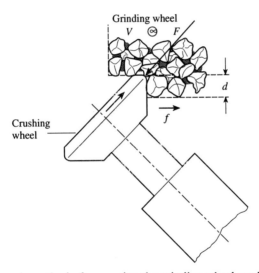

FIG. 5.5. The dynamic method of measuring the grinding wheel grade using a rotating crushing wheel mounted on a dynamometer to measure force F. With wheel velocity into paper. This device is called a dynagrader (after Colwell et al. 1962).

evident that $\bar{P}$ increases more rapidly with the bond volume for coarse grits (24) than for fine grits (80). While the mean force to break bonds ($\bar{P}$) is sufficient to specify the mean hardness of a grinding wheel, the method of Fig. 5.4 is capable of yielding additional information concerning the statistcal variation in bond strength.

Figure 5.7 gives mean values of the crushing force F as measured by the Dynagrader (Fig. 5.5) for the same wheels of different hardnesses and grit sizes as used in Fig. 5.6. The wheel speed was 1500 f.p.m. (7.62 m s^{-1}), the feed

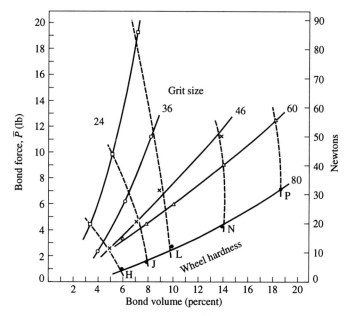

FIG. 5.6 The variation of the mean bond force ($\bar{P}$, Fig. 5.4) with bond volume percent (after Peklenik et al. 1964).

was 0.052 i.p.r. (1.3 mm rev.$^{-1}$), and the depth of cut was equal to the mean grit diameter in all tests. The abscissa is the bond volume, as in Fig. 5.6.

In order to gain some insight into the fracture mechanism pertaining in the dynamic test, screen analyses were made of the material crushed from the wheel. All the material removed from the 46 grit size, H hardness wheel was carefully collected on the last pass of the dynamic test of Fig. 5.5. When this material was classified by passing it through a series of screens, the results shown in Fig. 5.8 were obtained. A similar analysis was made on the material removed during a test of the 46 grit size, L hardness wheel. Particles 325 and finer were treated with HF for 4 h and all but 3 percent dissolved, indicating a glass content of approximately 97 percent. The material removed from the L hardness wheel contained fewer particles in the range 46–100. The H hardness wheel had many grits in the vicinity of size 46 and a smaller percentage of fines.

The results of these tests indicate that the material removed from a harder wheel in a dynamic test will consist of a number of grit clusters, considerable fractured grit, and a relatively large percentage of very fine particles, consisting primarily of glass bond. On the other hand, the grit of softer wheels is fractured to a smaller extent and very few grit clusters are obtained.

When the curves of Fig. 5.6 and 5.7 are compared, it is evident that the quasi-static method (Fig. 5.4) more widely distinguishes between wheels of

FORM AND FINISH GRINDING PERFORMANCE   117

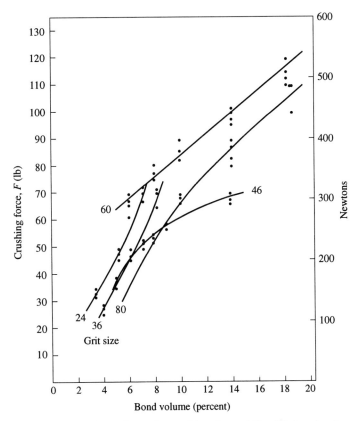

FIG. 5.7. The variation of the dynamic crushing force (F) with bond volume percent (after Peklenik *et al.* 1964).

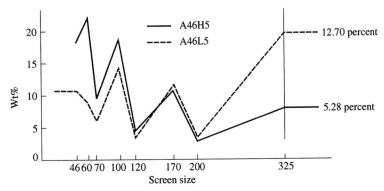

FIG. 5.8. A screen analysis of material removed in a dynamic test of wheels A46 H5 (solid curve) and A46 L5 (dotted curve) (after Peklenik *et al.* 1964). The material to the right of the vertical line (325 and finer) was shown to be 97 percent bond material by treatment with HF for 4 h.

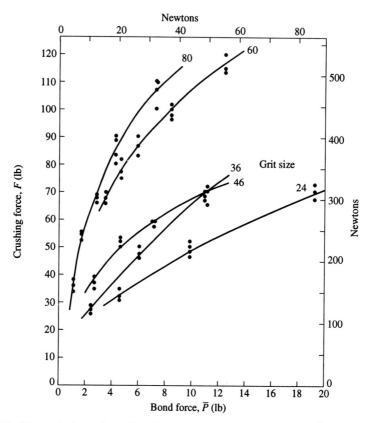

FIG. 5.9. The variation of crushing force F with mean bond force $\bar{P}$ (from Figs 5.6 and 5.7) (after Peklenik *et al.* 1964).

different grit sizes and hardnesses. Fig. 5.9, which shows F plotted against $\bar{P}$, enables the two sets of results to be more readily compared.

When attempting to explain the difference in the magnitude of the forces observed by the two methods, the following points should be kept in mind:

1. The method of Fig. 5.4 records the forces to remove one grit at a time, while the dynamic method (Fig. 5.5) measures the force required to remove several grits at one time.

2. The several grits removed at one time in the dynamic test may be subdivided into individual grits, or the grits may be removed in clumps, with some bond bridges left unbroken. The number of grits per clump will depend on wheel hardness (see Fig. 5.8).

3. The quasi-static method measures the force in the direction of crushing, while the force that is measured in the dynamic method is a variable com-

ponent of the resultant crushing force, which is dependent on wheel hardness and the geometry of the test.

4. The quasi-static method fractures very few grits, while the dynamic method ruptures not only bond bridges but also an appreciable number of grits.

5. The speed of the static test is sufficiently low that the dynamometer can record the force peaks and regions of zero force in true proportion. The frequency response of the Dynagrader, on the other hand, is not high enough to follow the variation in grit forces exactly and the values recorded will be *mean* values rather than actual instantaneous values. The instantaneous mean value will fluctuate somewhat. However, statistical analysis of Dynagrader force variations will not prove useful due to the complexity pertaining. The force variations will reflect bond force fluctuations only slightly due to insufficient frequency response and at the same time will be influenced by the variable number of bonds being broken at any one time.

It is not possible at the present time to evaluate the five effects listed above and thus compare results from the two methods on the same basis. In view of the large number of differences between the static and dynamic methods, it is surprising that the correlation between the two methods (Fig. 5.9) is actually as good as it would appear to be.

Despite the somewhat greater complexity of the dynamic method of testing grinding wheel hardness, it correlates quite well with the quasi-static method (Fig. 5.4). While the dynamic method may often be integrated with grinding-wheel finishing and hence is of greater potential for use as a production tool, it does not distinguish wheels of different hardnesses quite as clearly, and is not capable of yielding as much information. Since the static method records true instantaneous force values, statistical analysis of the force peaks can be used to describe grinding-wheel hardness characteristics more completely.

Wheel Grading Tests Based on Elastic Modulus

An interesting method for grading grinding wheels is a sonic test, in which the wheel is struck a blow and its natural frequency of vibration is measured. From this the value of Young's modulus (E) may be computed which, in turn, correlates with wheel hardness, since the increase in bond post size and strength causes an increase in E. Here the grinding wheel is considered as a uniform continuum, where the value of E that is measured is some sort of mean value of grit, bond, and pores. Average values of E were first found to correlate with the effective hardness of grinding wheels by Snoeys (1964). In performing a test, the wheel is supported vertically, as shown in the inset of Fig. 5.10, by a nondamping support that touches the bore at points on the nodal diameters shown. The wheel is then struck a standard blow by a pendulum consisting of a steel ball supported by a thread (Fig. 5.10). A microphone picks up the sound emitted by the wheel, which is then analyzed

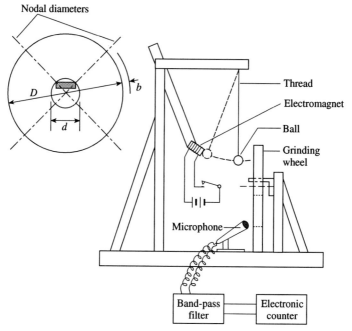

FIG. 5.10. The apparatus and instrumentation for determining the natural frequency of a grinding wheel.

and its frequency recorded. Peters *et al.* (1968) have discussed the sonic testing of wheels in order to characterize their grade.

The relationship between the frequency and the mean Young's modulus of the wheel is found, to a good approximation, to be as follows:

$$E/(1 - \nu^2) = [(\pi D^2 k)/0.875]^2 (\rho/b^2) f^2 = E', \qquad (5.13)$$

where f is the observed natural first mode frequency, D is the wheel diameter, ρ is the mean wheel density, b is the wheel width, E is the mean Young's modulus of the wheel, ν is the mean Poisson ratio of the wheel, and k is the the ratio of the frequency of a disc without a central hole to one with a central hole of diameter d (Fig. 5.11).

Since ν for grinding wheels ranges from 0.15 to 0.25, the quantity $E/(1 - \nu^2) = E'$ within less than 2 percent. The measured value was found to be insensitive to the intensity of the blow or to the exact position of impact.

Figure 5.12 shows the measured values of E' (average of 10 determinations for each point) for wheels of different grit size and hardness: $D = 8$ in (200 mm) and $d = 1.25$ in (32 mm). It is evident that this is a well ordered set of wheels and that E increases not only with grade but also with decrease in grit size.

It is not really surprising that the force required to pry grits from the surface of a wheel, the grade, and the density of a wheel all cause an increase in

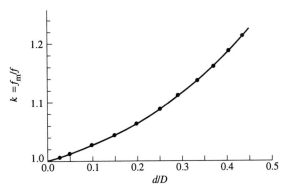

FIG. 5.11. The ratio of the natural frequency of a disc of diameter D without a central hole to that with a central hole of diameter d, versus d/D, (experimental).

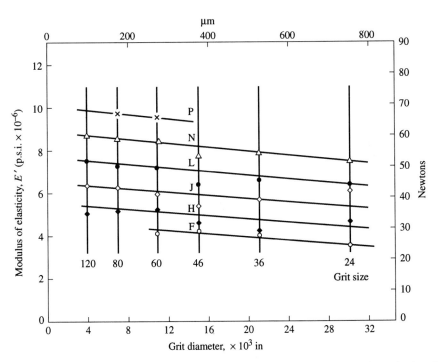

FIG. 5.12. A plot of the modulus of elasticity versus grit diameter for wheels of different letter grade.

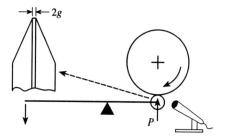

FIG. 5.13. The sonic crush grading apparatus.

Young's modulus, which is a measure of the elastic stiffness of the wheel structure.

This sonic method, which works well for vitrified wheels, gives poor results for resin-bonded wheels, apparently due to the unusually high damping for such wheels.

Another sonic test that has been used is the sonic crush grading test (Fig. 5.13). This test employs a steel crushing wheel that is slowly rolled over the surface of the grinding wheel under a constant radial load P. The grinding wheel is driven at about 1 rpm and rolls over the surface of the 1 in (2.54 cm) hardened steel crushing wheel with essentially no slip. The face of the steel wheel has a flat equal to twice the mean grit size in width (i.e. $2g$). A microphone placed near the point of contact picks up sounds emitted when a grit or bond is broken. The output from the microphone is filtered to remove extraneous noise, amplified, and recorded. The number of sound peaks corresponding to 1 in (2.54 cm) of travel is taken as a measure of wheel hardness (grade). Figure 5.14 shows typical results for wheels of different grades and grit sizes. Most curves are approximately linear with load and grinding wheels of different hardnesses are generally separated one from the other. However, the sonic crush method fails clearly to discriminate differences in grade when the grit size is relatively coarse (Fig. 5.14(e)). The sonic crush method holds promise as a routine test for constancy of wheel grade and a convenient load P to be used is 4 lb (22 N).

Another nondestructive method of measuring wheel grade in terms of Young's modulus is by measuring the elastic static deflection of a wheel when supported at three points near its periphery and loading it near the bore (Fig. 5.15). The deflection at the center of the disc relative to the periphery (δ) will be inversely proportional to the Young's modulus of the wheel, E, and the deflection will vary linearly with the applied load F. Figure 5.16 shows load–deflection data for 20 in (500 mm) diameter resinoid wheels of (a) different grades (P and T) but of the same grit size (24) and (b) for different grit sizes (46 and 24) but of the same grade (T). It is evident that wheel stiffness increases with increase in hardness (grade) and with decrease in grit size. This method is preferable to the sonic method for highly damped resin bonded

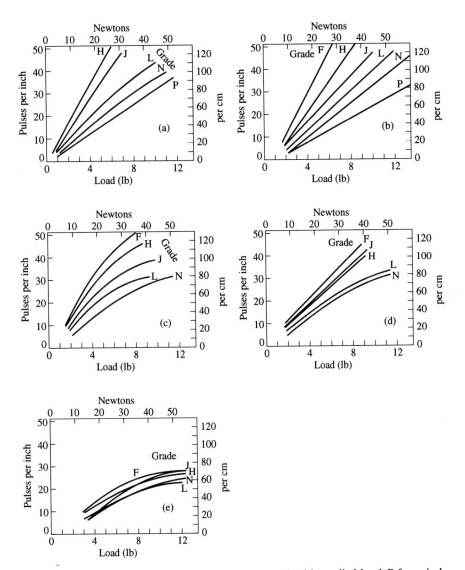

FIG. 5.14. The variation in pulses per inch (or per cm) with applied load P for grinding wheels of different grit sizes and grades. The wheels used were those listed in Table 5.2. (a) $S = 80$; (b) $S = 60$; (c) $S = 46$; (d) $S = 36$; (e) $S = 24$.

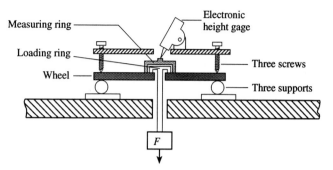

FIG. 5.15. The apparatus used to measure the bending deflection of resin-bonded grinding wheels as a measure of Young's modulus and hence of wheel grade.

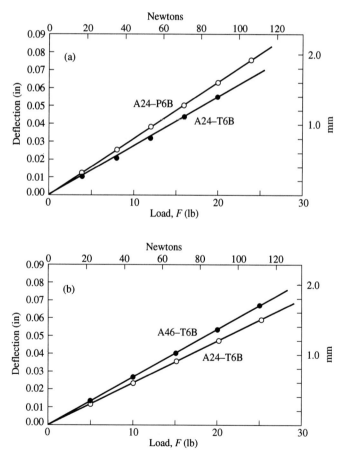

FIG. 5.16. Load–deflection data for 20 in (510 mm) diameter resin-bonded wheels (a) same grit size (24) and different hardness (P and T); (b) with same grade (T) and different grit sizes (24 and 46).

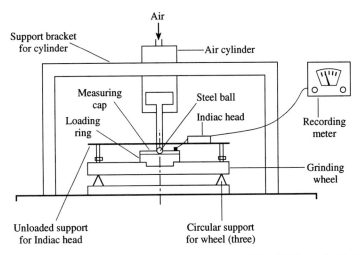

FIG. 5.17. An arrangement for applying the method of Fig. 5.15 to vitrified wheels, where dead weight loading is insufficient.

wheels. However, the time of loading of a resin-bonded wheel can be a concern due to room-temperature creep.

While the apparatus of Fig. 5.15 is satisfactory for thin resinoid wheels, a dead weight load is not convenient for thicker and stiffer vitrified wheels. In this case, the arrangement shown in Fig. 5.17, employing an air cylinder for loading, should be employed.

Another arrangement that has proven useful and which has potential for quality control testing is that shown in Fig. 5.18. The wheel is loaded diametrically by shoes extending circumferentially across about 20 percent of the wheel diameter and the deflection δ is measured under a standard load (about 500 lb = 2200 N). Typical results of deflection versus wheel hardness are shown in Fig. 5.18 for the wheels of Table 5.2 having a 36 grit size. Wheels of other grit sizes give similar results.

Wheel–Work Deflection

In deriving eqn (4.10) for wheel–work contact length, the elastic deflections of the wheel and work were neglected. Actually, under operating conditions there will be variations from the nominal values of wheel diameter (D) and wheel depth of cut (d) set on the machine due to deflection of the grinding machine, the wheel, the individual grits, and the work being ground. Some of this may be eliminated by taking the *measured* depth of layer removed for d. However, even in this case there will be some elastic recovery of the work and the measured value of d will be less than the value while cutting. However, the main difficulty lies with the value of D to be used. Under load, individual grits will recede from the surface and the shape of the arc of contact will

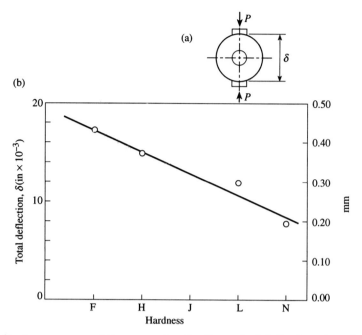

FIG. 5.18. An arrangement for measuring the diametrical deflection of a wheel, as a measure of Young's modulus and hence of wheel grade. (b) Typical results for wheels of 36 grit size but of different grade (hardness).

change, causing the effective diameter of the wheel to increase. These elastic deflections can have a significant influence on the rate of removal, on the effective value of C, on grinding forces and power, and on the wear rate but, most of all, on the contact length between wheel and work. For example, it has been found that in some cases the actual value of l under load can be as great as twice the value given by eqn (4.10), the factor increasing with decrease in wheel depth of cut d (Brown *et al.* 1971).

Hahn (1955, 1966) first considered the effects of static deflection assuming a relatively simple model of grinding grits mounted on individual springs. He suggested that differences in stock removal rate with conformity differences in internal and external grinding were consistent with the local wheel deflections to be expected. Peklenik (1957, 1961) mounted a small insulated platinum wire normal to a surface being ground in order to record transient grinding temperatures. The soft platinum was smeared across the insulation and formed a thermoelectric junction with the workpiece. As each successive grit passed through the thermal junction, a temperature spike was recorded and the spacing of these spikes was used to measure the number of cutting points. Peklenik found a difference in the effective number of cutting points with wheel grade, which he attributed to differences in wheel elasticity.

Schwartz (1959), using the same technique, showed the grit spacing to be less when grinding on the periphery of a wheel than when grinding on the side of the same wheel. This was presented as further indirect evidence of the importance of local wheel deflection.

Krug and Honcia (1964) estimated the amount of local wheel deflection to be expected in grinding and concluded that this would be of the order of 1 μm (40 μin) for a vitrified wheel and 2.5 μm (100 μin) for a resinoid wheel. They concluded that since these numbers are small relative to wheel depths of cut (10–100 μm) they are of negligible importance. However, when these deflections are compared with grit depths of cut (1–2 μm in fine grinding) they are found to be of the same order of magnitude.

Snoeys and Wang (1968) have carried out static loading tests to compare the 'grit mounted on springs' model with a model considering the grinding wheel as a simple continuous cylinder subjected to contact deflection at the wheel–work interface. They concluded that the 'grit mounted on springs' model gives the more realistic values for deflection. Brecker (1967) has considered the continuum deflection of the wheel and of the work and of one grit relative to its neighbor, and concluded that all three deflections are of importance.

Because elastic deflections of grits in the surface of a grinding wheel appear to affect grinding results significantly, it is important to establish the magnitudes of these deflections. The influence of elastic deflections in grinding have been extensively studied (for example, see Nakayama 1967; Verkerk 1975; Brown *et al.* 1977; Gu and Wager 1988; Salje and Paulmann 1988).

Rowe *et al.* (1993) point out that the actual contact length during grinding is greater than the geometrical length (eqn 4.10) not only due to wheel deflection but also due to the fact that the wheel–work contact surface is not smooth but rough. Actually, the ratio of real to apparent area of contact (A_R/A) in grinding is only about 1 percent and this plays an important role in wheel–work temperature analysis (Chapter 9). A_R/A also plays an important role in wheel–work contact mechanics. Instead of the elastic contact pressure distribution being Hertzian, as for smooth surfaces in contact, the effect of roughness is to provide several areas of concentrated contact, which results in a larger contact area than for smooth surfaces. This effect was first discussed for rough and smooth spheres by Greenwood and Tripp (1967). When applied to grinding it is evident the essential 'roughness' of the wheel–work contact should cause a significant increase in the contact length beyond the geometrical value.

Applying compressive loads to individual grits in the surface of a wheel is an effective method of establishing the deflection characteristics of individual grits. Initial tests were carried out to establish the magnitude of the normal deflection and the force–deflection relationship.

In these tests a simple lever mechanism (Fig. 5.19) was used to load individual grits in a 3/4 × 3/4 × 3 (19 × 19 × 75 mm) rectangular section of a vitrified grinding wheel. Loads were applied with a hardened steel cone having a flat tip. Since deflection of the grit tip was difficult to measure directly, deflections at the top of the loading tool were measured using a sensitive stylus

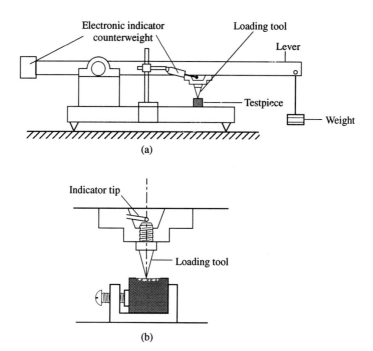

FIG. 5.19. A schematic of apparatus for single-grit loading tests; specimen size $\frac{3}{4} \times \frac{3}{4} \times 3$ in ($19 \times 19 \times 76$ mm). (b) The end view of a specimen with a single grit under load. (After Nakayama *et al.* 1971).

instrument. The measured displacement includes the contact deformation between the loaded grit and the loading tool as well as the deflection of the grit center.

The deformation of the loading tool could be reduced by using a very hard material such as tungsten carbide. However, such a hard material provides a very small contact area and thus a high stress concentration. This causes the grit tips to break during a test. Hardened steel was used since it gives about the same deformation and load distribution as occurs when grinding hardened steel.

Plastic deformation was accounted for by overloading the grit and then measuring displacements only during the unloading part of the cycle. Since bulk deformation of the tool will be small compared with the local contact deformation, it may be neglected. Therefore, the measured deflection included only the deflection of the center of the grit and the elastic deformation of the tool–grit contact. The wheels used were those described in Table 5.2. Two wheel hardnesses (H and N) were used to investigate the influence of wheel bond volume on grit deflection.

The displacement of the loading tool δ was found to be related to the applied load P by a power function of the form

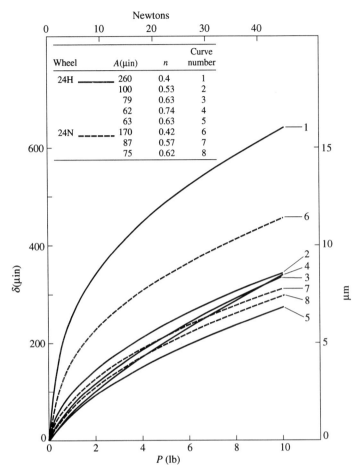

FIG. 5.20. Deflection (δ) versus load (P) curves for 24 grit size abrasive sections of H and N grade (after Nakayama et al. 1971).

$$\delta = AP^n, \qquad (5.14)$$

where A and n are constants for the system tested. Figure 5.20 shows a number of δ versus P curves for loaded grits. The deflection constants (eqn 5.14) are summarized in the inset in Fig. 5.20. From these results, the following observations may be made.

Scatter between the curves is too wide to enable discrimination of the two wheels. This is to be expected for multiphase materials such as grinding wheels. Considerable data and a statistical approach would be necessary in order to characterize wheels by this method. Since the test is slow and time consuming, it does not hold promise for this purpose.

Roughly speaking, the total local elastic deflection is 100 μin (2.5 μm) for

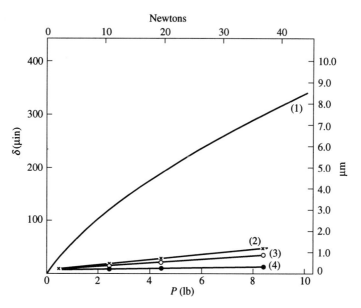

FIG. 5.21. δ versus P curves for grits near a loaded grit in a 24H abrasive section (after Nakayama et al. 1971).

a vertical load of 1.5 lb (6.67 N). This force is comparable to the force per grit in ordinary grinding.

The tangential force on an individual grit (F'_P) in grinding is given by the following equation:

$$F'_P = \tfrac{1}{2} u(r\bar{t})(\bar{t}), \qquad (5.15)$$

where u is the specific grinding energy, r is the ratio of width of cut to depth of cut, and $\bar{t}$ is the mean undeformed chip thickness.

When grinding under finishing conditions, u will be about 10×10^6 in-lb in^{-3} (69 J mm^{-3}). If r is 15 and $\bar{t}$ is 100 μin (2.5 μm), which are reasonable values in fine grinding then, from eqn (5.15), F'_P is found to be 0.25 lb (1.1 N). Since the mean radial load on a grinding wheel is very close to twice the tangential force, we may consider the load on the single grit in the foregoing example (F'_Q) to be 0.5 lb (2.2 N).

It is evident that, in fine grinding, the total elastic deflection associated with a grit will be of the same order of magnitude as the undeformed chip thickness. As a result, we should expect local grit deflections to play an important role in the mechanics of chip formation in fine grinding.

The exponents n were generally found to be between 0.4 and 0.7. The theory of elasticity (Hertz) for the deformation in the contact zone between a sphere and a flat plate gives an exponent of 0.67 (see Nakayama et al. 1971 for details), which is very close to the mean experimental value. Figure 5.21 shows the deflection for: (1) a loaded grit; (2) an adjacent grit 0.03 in (0.75 mm) from the loaded grit; (3) the fourth grit from the loaded grit; and (4) the specimen-holder. The last deflection should be subtracted from the first three. The deflection for the nonloaded grits was the deflection of the center of the grit alone and does not include the localized contact deformation. It should be noted that the deflection of the adjacent grit is only about 10 percent of that of the loaded grit. This means that most of the elastic deformation of a wheel is due to the grits that are actually loaded, and is not due to the deflection of adjacent grits. Thus, it would appear that there is ample opportunity for the number of cutting edges to increase due to local grit deformation.

Saini (1980) has determined single grit–workpiece deflections under grinding conditions with considerable accuracy by comparing the actual and theoretical groove shapes at different actual grit depths of cut. He found that grit–work deflection versus actual grit depth of cut gave an S-shaped curve with low slopes at low (2 μm) and high (20 μm) actual grit depths of cut. The grit work deflection δ for a single grit was a significant percentage of the actual grit depth of cut, as indicated by the following representative values:

Actual grit depth of cut		Wheel–work deflection		
μm	(μin)	μm	(μin)	%
4	(160)	0.7	(22)	18
8	(320)	2.1	(84)	26
12	(480)	4.7	(188)	39
20	(800)	5.4	(216)	27

In these tests it was not possible to separate the contact deformation from the grit center deflection. However, since the whole deformation system is similar to that in grinding, we can consider these results to establish the magnitude of the total local deformation present in grinding. A grinding wheel was next studied to allow determination of the relative magnitudes of the contact deformation and the grit center deflection.

An initial test series indicated that the total local deformation closely approximated the Hertz theory for elastic contact between a sphere and a flat surface or, more generally, between two spheres. Because the exponent of the power function representing total deformation (eqn 5.14) is so close to the Hertz exponent, the deflection of the grit center is either small or also Hertzian in nature.

Figure 5.22 shows a section of a grinding wheel. Kingery *et al.* (1963) have assumed that a grinding wheel can be approximated by a cubic structure because a wheel is only about 50 percent abrasive grit by volume. An analysis of the sintering process by Brecker (1967) using this model has indicated that the abrasive grits in a vitrified wheel are in direct contact. Thus a compressive load applied to an individual grit on the surface of a wheel will be absorbed largely by the Hertzian deformation of the contact zones between grits located

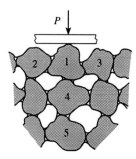

FIG. 5.22. A cross-section of a vitrified grinding wheel with load P applied to grit in the surface (after Nakayama *et al.* 1971).

directly beneath the loaded grit (grits 4 and 5 in Fig. 5.22). Since the elastic modulus for the aluminum oxide grit is much greater than the modulus for the glass bond, the grit will support almost all of the load. Some of the load will be transmitted to laterally adjacent grits (grits 2 and 3 in Fig. 5.22) by shear forces in the bond. However, the percentage of the load so transmitted to the side should be small. Since practically all of a compressive load is transmitted from grit to grit but across glass bond bridges in tension, the modulus of elasticity for a composite grinding wheel in compression will be greater than in tension (E for Al_2O_3 in compression $\simeq 60 \times 10^6$ p.s.i. (414 GPa), while E for glass in tension $\simeq 10^7$ p.s.i. (69 GPa)).

It has been shown by Nakayama *et al.* (1971) that the individual Al_2O_3 grits in a vitrified wheel deflect about half as much as the grit–steel contact zone. The static contact zone deflection of steel is about $34P^{2/3}$, while the deflection of the center of a grit is about $26P^{2/3}$. Both of these are insignificant when compared to the undeformed chip thickness, t.

In additional tests, individual grits in the surface of a grinding wheel were loaded using a beam platen to approach more closely the Hertzian assumptions of a semi-infinite platen. The apparatus is shown in Fig. 5.23.

The individual grits in the wheel surface were obtained by first dressing an 8 in (203 mm) wheel so that a band about four to five grit diameters wide and one grit diameter high remained in the center of the wheel face. The face of the band was coarse dressed and grits to be tested were selected using a stereo microscope (40×). The grits in the immediate neighborhood of the selected grit were ground down a few thousandths of an inch ($\sim 100 \mu$m) with a diamond hone. Care was taken not to disturb the bonds holding the grain selected for study. The remainder of the ridge was relieved 5–10 thousandths of an inch (125–250 μm) with a small diamond wheel. The deflections of the loaded beam were measured relative to the spindle on which the wheel was mounted.

Tests were conducted using the 24 and 46 grit size wheels described in Table 5.2. Wheel grades of F and N were used with annealed 52100 steel, 7075-T6

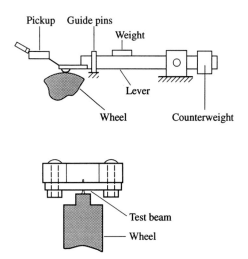

FIG. 5.23. A schematic of the apparatus for loading single grits in a grinding wheel (after Nakayama et al. 1971).

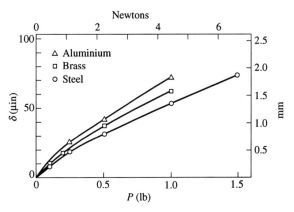

FIG. 5.24. Typical δ versus P curves for a loaded grit in a 24N grinding wheel (after Nakayama et al. 1971).

aluminium, and 70–30 brass platens. Typical test results are given in Fig. 5.24. The deflection varies as the load to the 2/3 power. This agrees with the results from the initial tests and with the assumed grinding wheel model. All results are summarized in Table 5.3, where the values of A are for the coefficient in eqn (5.14).

The average total deflection coefficient for the steel platen (50 μin, or 1.25 μm) is less than the coefficient observed in the initial tests (75 μin, or

TABLE 5.3 *A summary of the test results for the coefficients in eqn (5.14)*

Wheel	Measured coefficients			Separated coefficients for steel	
	A_S	A_A	A_B	A'_s	A''_s
24 F	52	82	69	43	9
	28	38	36	14	14
	48	72	54	34	14
24 N	52	72	62	30	22
46 N	70	100		43	27
	40	62	50	31	9
46 N	62	92		43	19
Average	50	74	54	34	16

A_S = total A for steel = $A'_s + A''_s$.
A_A = total A for Al.
A_B = total A for brass.
A'_s = grit deflection coefficient.
A''_s = grit center deflection coefficient.

1.9 μm). Part of the difference is probably due to the additional support (less deformation) in the contact zone when the beam platens were used.

The separated deflection coefficients show that twice as much deflection occurs in the grit–steel contact zone as in the center of the loaded grit. Although the grit center deflection is secondary to the contact deformation, it is significant. We cannot distinguish between deflections for wheels of greatly different grade; nor can we distinguish between deflections for the different grit sizes.

Wheel–work deflection studies have also been performed dynamically, as described below.

A problem in studying experimentally local elastic grinding deflections was the difficulty of separating local deflection from general deflections of the machine frame and the grinding wheel spindle. This was overcome by adopting a patch grinding technique (Brown *et al.* 1971). This was a procedure previously used by Tanaka *et al*, (1966) for an entirely different purpose. They used it to study the extent of grit rubbing prior to the start of cutting. In the present investigation, patch grinding was used essentially as a quick-start – quick-stop procedure.

The wheel was dressed to give a patch grinding region as shown in Fig. 5.25. In surface grinding at high table feed this produced a series of intermittent grinding cuts. The complete strip, dressed on the same wheel, was used simply to indicate the effect of spindle and machine frame deflections. Figure 5.26 is a plan-view sketch of the ground workpiece.

The patch and complete strip were dressed together to give the same depth of cut. Initially, the workpiece was ground flat to within a few microinches.

FORM AND FINISH GRINDING PERFORMANCE

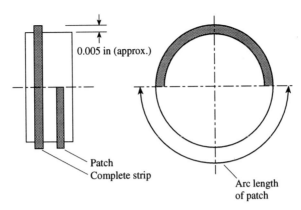

FIG. 5.25. The dressed shape of the wheel used for elastic deflection in patch grinding studies (after Brown *et al.* 1971).

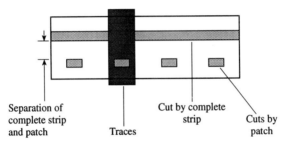

FIG. 5.26. The grinding pattern on the work surface under the conditions of Fig. 5.25. The dashed lines represent surface analyzer traces (after Brown *et al.* 1971).

After a test, depths of cut were measured from traverses by a surface analyzer as indicated in Fig. 5.26. These depths were plotted to give the longitudinal profile (Fig. 5.27).

Figure 5.27 is the profile for an A24-N8-V wheel cutting AISI 52100 steel of Rockwell hardness 50C, wheel diameter $7\frac{3}{8}$ in (187 mm), cutting speed 6080 f.p.m. (31 m s^{-1}), and work to wheel speed ratio 0.02. The widths of the patch and the complete strip were each 0.19 in (4.8 mm) and were 0.10 in (2.5 mm) apart; the patch arc was 185°. When the patch engages, it can be seen that the complete strip depth of cut decreases. This represents spindle and frame deflection, caused by the increased total force on the wheel. When the patch engages fully, the two sections of the wheel cut at approximately the same depth. Slight differences between the two suggest small inaccuracies in dressing.

The general profile shape of Fig. 5.27 was repeated in many tests, including tests at different speeds, depths of cut, and on 52100 steel of Rockwell hardness 88B. It was also obtained with an A24-H8-V wheel for several conditions.

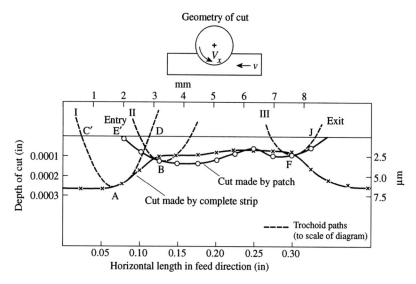

FIG. 5.27. Longitudinal profiles of cuts made by a patch and a complete wheel (after Brown *et al.* 1971).

The broken curves in Fig. 5.27 are trochoidal paths (to the scales of the figure) which the grits might be expected to follow, in the absence of deflection. The greater radii of curvature of the cut at patch entry and exit is direct evidence of the existence of local elastic deflection. The effect cannot be explained by machine frame or spindle deflections, because of the low natural frequencies of these systems and the short response time available. Considering the exit in Fig. 5.27 at the rotational speed of 3150 r.p.m., the last grit in the patch sweeps out of the cut from F in approximately 2.2×10^{-5} s. For spindle response, a natural frequency greater than $(0.25)(2.2 \times 10^{-5})$ or 11 400 Hz is required. Measured machine and spindle natural frequencies were only between 250 and 300 Hz. A similar argument applies to patch entry.

The gradual slope of the cut taken by the patch at entry indicates that a large arc length of patch is required before reaching a steady state depth of cut. Point A in Fig. 5.27, where the complete strip first starts to deflect upwards, must correspond to initial entry of the patch. Steady state cutting is not achieved until point B. The horizontal distance between A and B is approximately 0.06 in (1.5 mm). At the feed of 121 f.p.m. (0.615 m s^{-1}) the time to travel this distance is 2.4×10^{-3} s, which corresponds to a 47° rotation of the grinding wheel; i.e. a patch of approximately a 47° extent is required to achieve a steady state cutting depth. To confirm this somewhat unexpected result, tests were run with wheels having various patch lengths.

An example which demonstrates the influence of wheel–work elastic deflection on wheel–work contact length *l* and undeformed chip thickness *t* is presented in Brown *et al.* (1971) for a relatively coarse wheel (24 grit size). The plunge performance grinding conditions employed were as follows:

TABLE 5.4 *Typical values of l^*/l and t^*/t when grinding steel with an AA24-N8-V wheel in plunge horizontal spindle surface grinding*

Wheel dept of cut, d		F_Q		Total deflection		Wheel deflection		Work deflection		l^*/l	t^*/t
in × 10⁴	μm	lb	N	μin	μm	μin	μm	μin	μm		
1	2.5	24	106.8	265	6.63	185	4.63	80	2.00	2.85	0.58
2	5.0	36	160.1	360	9.00	260	6.50	100	2.50	2.32	0.65
3	7.5	48	213.5	440	11.00	325	8.13	115	2.88	2.06	0.70

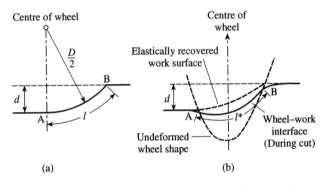

FIG. 5.28. The shape of the wheel–work interface in surface grinding: (a) without elastic deflection; (b) with wheel–work elastic deflection (after Brown *et al.* 1971).

Work: steel [$E = 30 \times 10^6$ p.s.i. (207 GPa), $\nu = 0.3$]
Wheel: A24-N8-V [$E = 6.66 \times 10^6$ p.s.i. (45.9 GPa), $\nu = 0.15$]
Wheel diameter: 7.50 in (191 mm)
Width ground, b: 0.19 in (4.83 mm)
Active grits per unit area, C: 400 in^{-2} (62 cm^{-2})

Using a dynamometer to measure forces, Hertzian analysis to obtain wheel and work deflections, and eqn (4.10), the values of chip–tool contact length l (without deflection) and l^* (with deflection), and values of undeformed chip thickness t (without deflection) and t^* (with deflection) given in Table 5.4 were obtained.

Most analyses of the mechanics of grinding assume that the curvature of the ground face equals the radius of the undistorted wheel and that, at any instant, the center of the wheel lies above the start of the ground face (point A in Fig. 5.28(a)). Consideration of local elastic deflection suggests that neither of these assumptions is correct.

Allowing for elastic deflection, the actual shape of the wheel–work interface will be as shown by the full line in Fig. 5.28(b) which, of course, is not to scale. If the operation is stopped suddenly and the workpiece and wheel allowed to recover elastically, they would be as shown by the broken curves in Fig. 5.28(b).

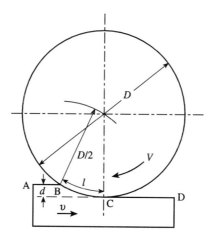

Fig. 5.29. Plunge surface grinding.

As an approximation, it may be assumed that the load distribution and wheel deflection are symmetrical about the center of AB and hence the wheel center lies on the centerline between A and B. Examination of the entry and exit cuts taken by the grinding patches indicates that this is a reasonable assumption.

An alternative to the patch approach is to employ a quick-stop device similar to that used in studying chip roots in metal cutting. However, grinding speeds are very much higher than those used in cutting and therefore a suitable quick-stop device for grinding must have a much higher disengagement speed.

Figure 5.29 illustrates the geometry of the contact zone and the critical parameters involved in surface grinding. In order to preserve arc BC, the wheel must move upward a distance d relative to the work before it has moved horizontally an appreciable percentage of distance l. For perfect disengagement, the center of the wheel must remain outside the arc of radius $D/2$ about point B. In any practical device, some cutting will occur at B during disengagement, which increases l by a small amount. However, the device shown in Fig. 5.30 results in an arc length that is too long by only a few tenths of a percent under typical plunge surface grinding conditions [($D = 10$ in (250 mm) and $v = 50$ f.p.m. (0.25 m s^{-1})].

The device shown in Fig. 5.30 was designed for high rigidity and speed of disengagement and utilizes a beam loaded in three-point bending that acts simultaneously as a spring and as the workpiece carrier. The beam is made from aluminum because of its high ratio of Young's modulus to specific weight and has a hollow square cross-section to provide a favorable ratio of bending stiffness to mass.

In use, the beam-mounted workpiece is bent upward against two rigid supports by a hydraulic jack exerting about 2000 lb (8.9 kN) and the knife edge is put in place. The bending displacement is about ten times the depth of cut. The jack is then removed and the device placed on the grinder. The workpiece

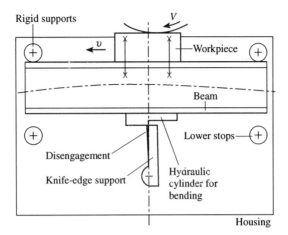

FIG. 5.30. A quick-stop device for surface grinding (after Sauer and Shaw 1974).

is first ground flat and then a lever is mounted on the shaft carrying the knife edge support. During the test pass, the support is quickly removed when the wheel is approximately over the center of the workpiece.

The vertical velocity of the beam is measured with a photoelectric transducer and recorded with an oscilloscope. From this trace the overcut length during disengagement is calculated, taking into account the elastic deformation of the grinding machine. From these measurements, it is concluded that the device acts fast enough to preserve the workpiece surface that exists during grinding.

During plunge surface grinding, the wheel and the workpiece are in close contact over an area corresponding to the width of the work and of length l. For simplicity, this contact surface is assumed to be part of a cylinder of radius R'. As shown in Fig. 5.31, this radius (R') is larger than the wheel radius ($R = D/2$) due to the elastic deflection of the grinding wheel caused by the grinding forces, but smaller than the radius of curvature (R'') found on the workpiece after a quick-stop test due to elastic recovery of the workpiece.

The workpiece radius (R'') is determined as follows from values of l_i and d_i measured on the quick-stop specimen by a tracer instrument:

$$R'' = l_i^2/2d_i. \tag{5.16}$$

A typical workpiece profile which closely approximates a circular arc is shown in Fig. 5.32. Every dot represents the average elevation of a trace across the workpiece surface perpendicular to the grinding direction. Tracer patterns were used in order to eliminate disturbances of the profile due to surface roughness.

In order to determine how R'' varies with grinding conditions, the Hertzian theory of contact stresses (Poritsky 1950) was applied to the wheel and the workpiece, neglecting individual grit deflections. For this purpose, the wheel

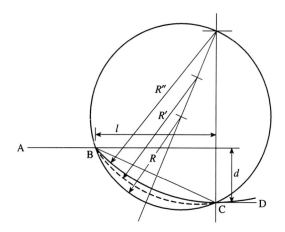

FIG. 5.31. Local elastic wheel–work deflection (after Sauer and Shaw 1974).

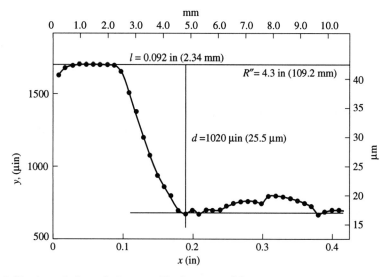

FIG. 5.32. A typical workpiece profile from a quick-stop test.

was considered to be homogeneous and the contact completely elastic. This resulted in the following equation:

$$R''/R - 1 = 4K_c F'_Q/d, \tag{5.17}$$

where F'_Q is the radial grinding force per unit grinding width and K_c is the bulk modulus for the wheel-work composite (for details, see Sauer and Shaw 1974).

TABLE 5.5 *The constants in eqn (5.18) for cold rolled AISI 1018 steel*

Wheel	K_0	m	n
AA60-H8-V	0.5×10^{-9}	1.0	2.2
AA36-H8-V	1.0×10^{-9}	1.1	2.0
AA36-N8-V	0.7×10^{-9}	0.9	2.0
Approximate average	0.7×10^{-9}	1.0	2.0

This theoretical prediction was compared with results from quick-stop tests using cold rolled AISI 1018 workpieces and the following three grinding wheels: A60-H8-V, A36-H8-V, and A36-N8-V. A constant wheel speed of $V = 4200$ f.p.m. (21 m s^{-1}) was used with the following ranges of table speed v and wheel depth of cut d:

> Table speed, v: 25–75 f.p.m. (0.13–0.38 m s^{-1})
> Depth of cut, d: 270–1300 µin (7–33 µm)

A strain gage dynamometer was used to measure grinding forces F_P and F_Q.

Measured values of $R''/R - 1$ were found to be much greater than values from eqn (5.17), and the difference between measured values and those from eqn (5.17) increased with increased values of F'_Q/d.

The discrepancies are believed to be related to the inhomogeneous structure of a grinding wheel and, in particular, to the fact that only a few points of contact exist between the wheel and the workpiece. Also, the workpiece material does not remain perfectly elastic but deforms plastically as grits penetrate into the surface and generate chips.

In order to improve the accuracy of the calculations without increasing the complexity of the equations to be used, an empirical equation is postulated based on eqn (5.17). The constant bulk modulus (K) of the wheel–workpiece combination is replaced by an effective bulk modulus K_O, and an equation of the following form was used:

$$R''/R - 1 = K_O F_Q'^m d^n. \qquad (5.18)$$

Table 5.5 gives the numerical values for the three constants K_Q, m, and n, calculated by the method of least squares, for each of the three wheels. The values of m and n determined for the three wheels of grade H and N and grain sizes 36 and 60 are almost identical, and therefore it can be concluded that, to a good approximation, the following equation represents their elastic deformation performance:

$$R''/R - 1 = K_O F'_Q / d^2. \qquad (5.19)$$

Figure 5.33 shows $R''/R - 1$ plotted against F'_Q/d^2, and it is to be seen that, to a good approximation, the data points for all wheels used follow a straight line, having the equation:

$$R''/R - 1 = 0.2 \times 10^{-8} F'_Q / d^2 \qquad (5.20)$$

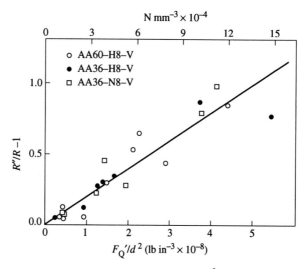

FIG. 5.33. The variation of $R''/(R-1)$ with F'_Q/d^2 for cold rolled AISI 1018 steel ground with three different wheels (after Sauer and Shaw 1974).

FIG. 5.34. A grit extruding soft Pt wire to establish a thermoelectric junction (after Peklenik 1957).

for F'_Q in p.s.i. and d in inches.

This empirical equation is useful in estimating the wheel–work contact length using eqn (5.18). Measured values of R'' are to be preferred to estimated values, since they include effects of lubricants, wheel wear, and other details that are difficult to model.

Zhou and van Luttervelt (1992) have introduced an interesting variation on the thermoelectric technique of Peklenik (1957) for estimating the actual wheel–work contact length in grinding. Instead of using the thermoelectric e.m.f. generated between a soft insulated platinum wire and the workpiece as an abrasive particle extrudes the platinum until it establishes thermoelectric contact with the work (Fig. 5.34), they employed e.m.f. from a battery. Figure 5.35(a) shows the arrangement used with a mica sheet between two blocks of workpiece material, while Fig. 5.35(b) shows the arrangement used with an insulated wire. Figure 5.35(a) gives the maximum contact length l_{max} for grits of all axial locations, while that in Fig. 5.35(b) gives the local contact length

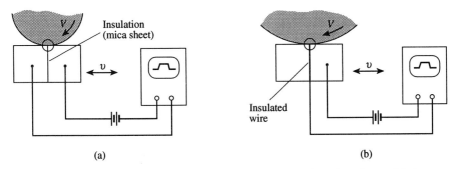

FIG. 5.35. (a) Blocks of work material separated by a thin mica sheet with battery terminals attached to each block for use in determining l_{max}. (b) Insulated wire and work material attached to battery terminals for use in determining l_a. (after Zhou and van Luttervelt 1992).

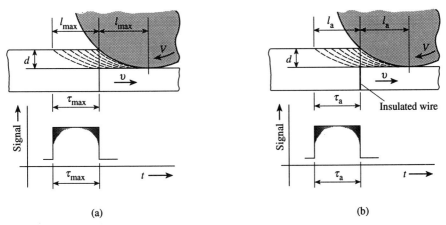

FIG. 5.36. The signals obtained from arrangements shown in (a) Fig. 5.35(a) and (b) Fig. 5.35(b) (after Zhou and Van Luttervelt 1992).

l_a for successive grits having a certain axial position in the wheel surface. Figures 5.36(a) and (b) show the types of signals obtained for the situations of Fig. 5.35(a) and (b) respectively. In all cases,

$$l = v\tau, \qquad (5.21)$$

where τ is the time for which current flows.

Figure 5.37 shows scratches left by individual grits of variable length. Figure 5.38 shows typical results for theoretical values of l assuming no deflection and for experimental values of l_{max} and l_a. In all cases the experimental values are greater than l and there is a significant difference between l_{max} and

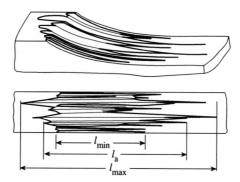

FIG. 5.37. Scratches left by individual grits in the wheel-work contact zone (after Zhou and Van Luttervelt 1992).

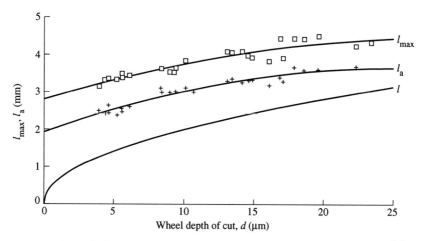

FIG. 5.38. The variation in wheel-work contact lengths l_{max}, l_a, and $l = (Dd)^{0.5}$ with wheel depth of cut d for typical grinding conditions: wheel, A 46M; wheel diameter, 398 mm (15.7 in); wheel speed, 22 m s^{-1} (4330 f.p.m.); work speed, 10 m min^{-1} (32.8 f.p.m.) (after Zhou and Van Luttervelt 1992).

l_a, which is consistent with the fact that not all active grits penetrate to the same depth.

Trueing, Dressing, and Conditioning

In FFG it is necessary to true and dress wheels periodically. It is also important that grinding wheels be well balanced, particularly those of large diameter that operate at a high surface speed. This is generally well attended to by the manufacturer, but should be checked under static or preferably dynamic conditions by the user. Wheel balance is achieved in wheel manufacture by

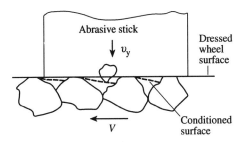

FIG. 5.39. The action of an abrasive conditioning stick to provide grit protrusion in the wheel surface of resin- and metal-bonded wheels for chip clearance (after Salje and Jacobs 1978).

removing material from the heavy side of the wheel or by adding powder of high density to the light side of the wheel. In some cases, wheel balance is carried out by the user after trueing, by means of special balancing means on the grinding machine.

Trueing consists of removing material from the grinding face of the wheel until it has a total indicator runout (TIR) of 100 μin (2.5 μm) or less in precision grinding. This is done by machining material from the wheel face on the grinding machine using a trueing/dressing tool.

Dressing consists of sharpening the active grits by removing wear flats and adhering metal.

Resin- and metal-bonded wheels have relatively few voids, and a freshly dressed surface will be too smooth and dense, with insufficient space between active grits to accommodate chips. This is remedied by eroding part of the bond away between active grits. The most widely used method of doing this, for superabrasive wheels, is to plunge an Al_2O_3 abrasive stick radially into the surface of a wheel operating at about 4000 f.p.m. (20 m s^{-1}). This causes the bond material to be worn away as shown in Fig. 5.39 to provide chip storage space. This treatment to provide storage space in the wheel surface is a special dressing procedure here referred to as conditioning. It is generally not required for wheels with a vitrified bond because of the relatively large number of pores in such wheels. The maximum depth of grit exposed after conditioning should not exceed about 30 percent of the grit diameter; otherwise, there is danger of losing whole grits due to insufficient bond support. Conditioning sticks use white Al_2O_3 with a grit size about one step below the grit size in the wheel ($S = 320$–500) and having a G or K grade. These sticks usually have a square cross-section 0.5–1 in (12–25 mm) on a side.

Metal-bonded superabrasive wheels are often conditioned to provide grit protrusion by electrochemical deplating of the bond. This is accomplished by insulating the wheel from the work and directing a copious supply of an aqueous electrolyte ($NaNO_3$ in water, for example) between the wheel and a copper block. A low-voltage d.c. source (~6 V) is introduced between the

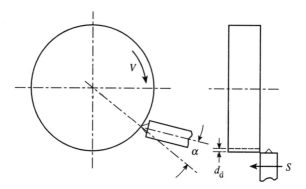

FIG. 5.40. The single-point diamond dressing procedure, showing the dressing depth of cut d_d and the traverse rate (pitch) S (i.p.r. or μm rev^{-1}). The diamond dressing tool is usually inclined as shown at an angle α of 10–20°. A fine dressing operation would correspond to a d_d value of 500 μin (12.5 μm) or less and S of 0.005 i.p.r. (125 μm rev^{-1}) or less; while a typical coarse dressing operation would have a d value of 1000 μin (25 μm) or greater and S = 0.020 i.p.r. (500 μm rev^{-1}) or greater.

wheel and the copper block with the wheel as the anode (+ ve) as the copper block is ground. This causes bond metal to be removed electrochemically to provide chip storage space between the grit tips and the bond of uniform depth, unlike the abrasive stick conditioning procedure that yields a sawtooth pattern of 'bondtails' (Fig. 5.39). The average protrusion for a freshly conditioned metal-bonded superabrasive wheel should be about 1200–1500 μin (30–40 μm).

Another method of conditioning resin- and metal-bonded wheels is to use a blast of air containing fine abrasive particles (50 μm Al_2O_3) that is directed radially against the active face of the wheel to provide grit protrusion for chip accommodation (Daimon and Ishikawa 1983). The distance between nozzle and wheel should be between 1.5 and 3 in (40 and 80 mm). This method has potential for continuous in-process conditioning.

Verkerk (1979) has identified two different approaches to dressing:

(1) where the dressing tool is moved axially across the wheel face but has zero velocity in the circumferential direction;
(2) where the dressing tool has a component of velocity in the circumferential direction on the wheel face and is fed radially into the wheel.

The first of these includes single diamond point dressing and multidiamond nib dressing. Multidiamond nibs provide less of a screw thread dressing pattern and allow high traverse rates. The second method includes rotary dressing tools.

In single-point diamond dressing, a ($\frac{1}{2}$ carat or smaller) diamond mounted in a holder is fed axially across the wheel face with a depth of cut d_d and a traverse rate of S per revolution (Fig. 5.40). The important variables for this operation are:

- dressing depth of cut, d_d
- traverse rate, S, (in rev^{-1} or $\mu\text{m rev}^{-1}$)
- number of dressing passes
- geometry of diamond tip
- number of spark-out passes

The traverse rate S is the pitch of the helical thread that is machined into the wheel face and is equal to the axial traverse velocity divided by the r.p.m. of the wheel. Pahlitzsch and Appun (1953) have demonstrated that the helical thread pattern on the surface of a coarsely dressed wheel is reproduced on the workpiece surface. This effect is reduced by use of a small value of S or by use of a multidiamond nib.

The geometry of the wheel face after dressing has two components (Verkerk 1979), a micro one on the tips of the active grits and a macro one. The micro component is due primarily to grit fracture and the macro one to 'screw thread' generation. The macro component increases with grit ductility, low bond strength (soft wheel) and large values of S and d_d (coarse dressing with high forces). The micro component increases with increased grit friability and wheel grade and decrease in S and d_d. The mean peak-to-valley roughness of the wheel face is primarily due to the macro component (fine dressing with low dressing forces). Coarse dressing (large S and d_d gives a distinct 'screw thread' pattern on the wheel face which, in turn, gives rise to low values of active grit density C, large values of $\bar{t}$, low values of specific energy u and low values of grinding forces, power, and temperature. On the other hand, a coarse dress gives relatively poor finish (large $\bar{t}$). A grinding wheel that receives a coarse dress will act harder than when the same wheel receives a fine dress.

As a grinding wheel wears, flats develop on the active grits, grinding forces and temperatures rise, the tendency for chatter to develop increases, and it becomes necessary to redress the wheel. When a wheel requires redressing may be monitored by measuring grinding forces or vibrational amplitude (Maris et al. 1975). A single-point diamond dressing tool will wear with use, and this tends to increase the radius of curvature at the tip of the diamond, which will cause the wheel to behave as though it had a finer dress (higher forces but better finish). A single-point diamond dressing tool should be rotated periodically to distribute wear and the diamond reset when the point becomes relatively blunt.

Radhakrishnan and Rahman (1981) have proposed the use of a turbulence amplifier or hot wire anemometer to measure changes in the intensity of turbulence of the air belt surrounding a rotating grinding wheel. Because of the inherent roughness of a grinding wheel, the air belt adjacent to the wheel surface will be turbulent at grinding wheel speeds. The intensity of this turbulence will change with wheel loading or dulling, and the point at which dressing is required will correspond to the point at which a certain intensity of turbulence is reached. The authors demonstrate that a relatively simple turbulence amplifier has potential for in-process monitoring of the surface condition of a grinding wheel when grinding with or without a fluid.

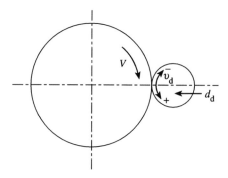

FIG. 5.41. A rotary dressing operation with the dressing wheel operating with (+) or against (−) the wheel speed V at a surface speed of v_d and an in-feed rate of d_d per revolution.

The number of passes of the diamond with feed depends upon the depth of loading, grit size, and the depth of the wear pattern to be removed A typical number is three. In precision grinding, where a very good finish is required, two or three dressing passes at low values of S and d_d are followed by two or three spark-out passes where the elastic deflection is exponentially relaxed instead of providing radial in-feed after each pass.

Roll dressing involves a hard steel or tungsten carbide roll, a star type hard steel dresser, a silicon carbide grinding wheel, or a diamond studded body of revolution (Fig. 5.41). The dressing tool spans the entire grinding wheel width and is fed radially inward at a rate d_d per revolution as the dressing tool rotates at a velocity $\pm v_d$ (the position sense being in the same direction as that of wheel velocity V). Schmitt (1968) has measured the peak-to-valley roughness of a dressed wheel (R_a) for different infeed rates d_d and different velocity rates v_d/V (Fig. 5.42). Normally wheels are crush dressed at $v_d/V = 1$ using a star dresser (an assembly of star-shaped hard steel elements) or a hard steel or tungsten carbide body of revolution. Silicon carbide (60/80 grit and L or M grade) and diamond-studded dressing wheels are normally operated in coarse dressing at about $v_d/V = +0.8$ using a friction brake to reduce the surface speed of the dresser from that of the grinding wheel. For fine dressing, a motor-driven dressing tool may be used at $V_d/V \simeq -0.8$ with a low rate of in-feed (d_d) followed by a few spark-out revolutions.

Wheel wear leading to wear flats on grit tips is of two types, adhesive wear and microchipping (Verkerk 1979). The former leads to wear flats, with an attendant increase in forces, temperatures, and the tendency to chatter. The latter leads to the development of new sharp cutting edges. While microchipping leads to a loss of accuracy and finish, it can postpone the need for dressing by a self-sharpening action. Tsuwa and Yasui (1972) have studied the progression of events in the life of abrasive particles in fine precision grinding using a number of microscopic techniques. They found that a freshly dressed grit has a network of microcracks beneath a thin layer of relatively weak

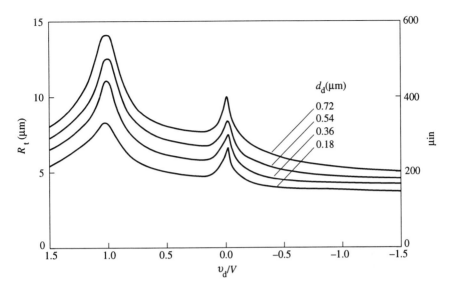

FIG. 5.42. The peak-to-valley roughness of the wheel surface R_d, produced by a rotary dressing tool operating with different in-feed rates d_d (μm rev^{-1}) and different speed rates v_d/V, where v_d is the speed of the dressing roller and V is the wheel speed. The dressing roller speed is positive when in the same direction as the wheel speed (after Schmitt 1968).

material (Fig. 5.43(a)). While this layer is present the grit surface is very smooth and grinding forces will be relatively high. This superficial layer is quickly removed, exposing a number of sharp cutting edges with a reduction in grinding forces (Fig. 5.43(b)). As grinding proceeds, flats develop on the sharp points (Fig. 5.43(c)), and when redressing is required essentially all of the points have merged into one flat surface (Fig. 5.43(d)). Figure 5.44 shows the variation of tangential grinding force F_P with the volume of metal removed for two different dressing depths per pass d_d. It is evident that the wear rate is more rapid and redressing is required sooner for the finer dressing condition ($d_d = 10\,\mu m$) than for the coarser one ($d_d = 50\,\mu m$).

Babu (1989) has used an unusual noncontact dressing technique—a high-powered pulsed laser. It was found that a carefully focused beam could be used to generate new cutting points on wear flats by a combination of vaporization and melting of the surface, the former producing minute craters and the latter a relatively weak resolidified layer. The resolidified material disappeared relatively rapidly when the wheel was put into service, leaving behind helical grooves that resembled those found when a grinding wheel is dressed by a single-point diamond. Laser-dressed wheels give essentially the same results (forces, finish, and wear rate) as those dressed by a single-point diamond. The advantages of laser dressing lie in its noncontact character, ease of control,

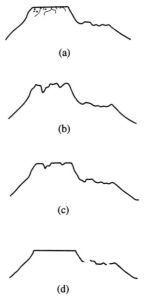

FIG. 5.43. A progressive wear sequence at the tip of an abrasive grit: (a) freshly dressed wheel with a soft superficial layer and subsurface microcracks; (b) the loss of the microcracked region, resulting in many sharp cutting points; (c) the development of flats on the cutting points; (d) the tip of the grit with a flat surface ready for redressing (after Tsuwa and Yasui 1972).

and the extreme shallowness of the disturbed layer ($\sim 10\,\mu$m). These characteristics make the method attractive for grinding with continuous dressing. Babu found that the dressing rate depended on the abrasive material (mainly reflectivity), the power density ($\sim 5 \times 10^{10}\,\text{W m}^{-2}$) of the beam, and the rate of feed of the beam across the surface ($\sim 0.8\,\text{mm rev}^{-1}$). Silicon carbide was removed about three times as fast as Al_2O_3 for the laser beam used (6 W, Nd/YAG, 1.06 μm wavelength, and 460 μs pulse width at 15 Hz) but laser-dressed Al_2O_3 wheels performed better than laser-dressed SiC wheels. It was found possible to remove pore-clogging chip material without weakening the grit or bond by properly adjusting the beam intensity and feed rate. Also, it was possible to introduce nearly axial grooves into the wheel surface by laser to provide a high-frequency interrupted grinding action (discussed in the next section). Little vitrified bond material was removed by laser dressing due to the low conductivity and high reflectivity of the mainly SiO_2 bond material. Therefore, some sort of conditioning treatment should follow the essentially grit sharpening laser treatment.

Only a few of the large number of articles on dressing have been referenced here. Traditionally, doctoral theses are an excellent source of a comprehensive list of references, and that is the case for the dissertations on dressing cited

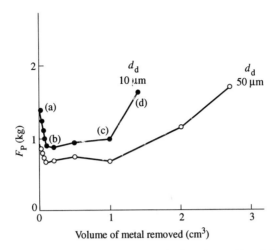

FIG. 5.44. The variation of circumferential grinding force F_P with the volume of material ground away for two different grinding conditions (after Tsuwa and Yasui 1972). The letters pertain to the wear sequence of Fig. 5.43. Dressing conditions: single-point diamond; $d_d = 10\ \mu m$ and $50\ \mu m$ (400 and 2000 μin); wheel, A46 8V; traverse rate, 170 μm rev^{-1} (0.0068 in rev^{-1}). Plunge surface grinding conditions: wheel speed V, 32 m s^{-1} (6300 f.p.m.); work speed v, 0.133 m s^{-1} (26 f.p.m.); wheel depth of cut d, 10 μm (400 μin); fluid, dry; work, 1 percent C steel, $R_c = 60$; width of cut, 0.8 mm (0.0032 in).

here. Also, additional descriptive details of dressing are to be found in Andrew et al. (1985) and King and Hahn (1986).

The Interrupted Grinding Principle

Several arrangements have been used to interrupt the grinding action at high frequency to allow heat to flow into the workpiece away from the surface, to allow coolant to be carried into the cutting zone, and better to accommodate chips. The aim of such techniques is to combine the good surface finish aspects of a fine dressing procedure with the low forces, specific energy, and surface temperature associated with a coarse dressing procedure.

One method of providing thermal relief by variable intensity grinding is by use of a 'Dalmatian' wheel (Fig. 5.45). This is a vitrified grinding wheel that contains abrasive grits with two different bond types (Lane 1968). In making such a wheel, clusters of grits are first sintered into pellets using a relatively weak but refractory bond system. This happened to give a black product in Lane's case. These pellets were then mixed with additional grit and a different bonding material, pressed, and fired. The second, much stronger, bond was white. The result is the spotted wheel shown in Fig. 5.45, which is referred to as a 'Dalmatian' wheel, since it resembles the famous spotted carriage dogs of Victorian times. In use, the black regions wear rapidly, producing depressions

FIG. 5.45. A 'Dalmatian' grinding wheel, having weakly bonded areas (black) in a matrix of more strongly bonded material (after Lane 1968).

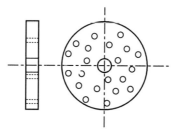

FIG. 5.46. A disc grinding wheel designed for face grinding, with transverse (axial) holes to provide space for chips and fluid and an intermittent grinding action.

for variable intensity grinding and more effective cutting fluid action. The result is a grinding wheel that gives unusually cool cutting and is useful where excessive temperature is a problem in FFG. By use of this technique it is possible to limit temperatures to reasonable values without reducing the rate of production.

There are some disc grinding wheels that grind on the face instead of the periphery (Fig. 5.46). These wheels are provided with holes extending through the thickness of the wheel to achieve benefits similar to those obtained with the 'Dalmatian' wheel.

Still another method of employing variable intensity grinding to provide thermal relief is to dress a pattern in the active surfaces of the wheel. An

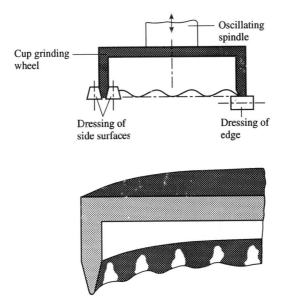

FIG. 5.47. Cup grinding wheels with sinusoidal edges for use in form grinding at a high removal rate (after Brecker 1970).

example of this is shown in Fig. 5.47. By the use of such a dressing procedure, grinding burn was eliminated even with an increase in removal rate.

Verkerk (1976) has shown that the advantages of both coarse and fine dressing may be combined in the same wheel. This was done by sawing 100 equally spaced 45° grooves in the surface of a grinding wheel to a depth of 3 mm (0.12 in). The tops of the grooves were then given a fine dressing operation. The resulting intermittent grinding operation gave better coolant action and time for heat to flow away from the surface when grinding was periodically interrupted. The result was good finish, reduced forces and temperatures, and better surface integrity, enabling an increase in production rate. It was subsequently shown (Nakayama *et al.* 1977) that 200–300 grooves could be crush dressed into the wheel face to a depth of 3 mm provided that the ratio of wheel speed to crush dressing roller speed was carefully controlled to be an integer.

Figure 5.48 shows how a wheel with surface grooves may achieve a high removal rate and good finish simultaneously. The leading edge of the flat surface will produce large chips which are accommodated by the grooves without interference, while the trailing edge of the flat produces small chips and good finish.

Warnecke and Gruen (1987) have devised a method of dressing axial grooves in the surface of a grinding wheel by a milling process (*Fraesabrichten*) using polycrystalline diamond cutters. Figure 5.49(a) shows the principle of the technique. Figure 5.49(b) shows the milled pattern at 1000× in the radial direction obtained by use of a CNC three co-ordinate measuring machine. The depth

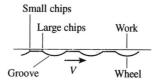

FIG. 5.48. The variation in chip size with a grinding wheel having grooves in the wheel face (after Shaw 1985).

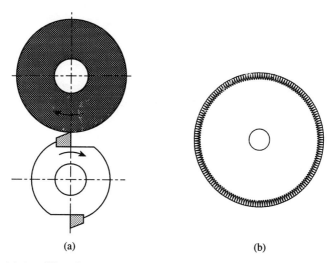

FIG. 5.49. (a) A milling dresser. (b) The milling pattern obtained from a CNC three-coordinate measuring machine for two cutters with a grinding wheel to cutter diameter ratio of 4:1. Radial magnification = 1000 ×. (after Warneke and Gruen 1987).

of each wave is about 1.8 μm (72 μin) and the pitch is 7.4 mm (0.29 in). One of the advantages of this technique is that it allows a smaller grit size, a greater bond content (harder wheel), and a lower structure number to be used without experiencing a chip accommodation problem. This results in a combination of better finish, longer wheel life, and lower forces and temperatures or, alternatively, an increase in material removal rate (greater productivity). Another possible advantage is that the inventory of wheels may be reduced by providing changes in wheel performance by dressing details rather than wheel characteristics (grain size, hardness, and structure number).

Grinding Swarf

Examination of grinding swarf under the electron microscope reveals important differences in behavior for common and superabrasive grits and for ductile and brittle work materials.

Spherical particles consisting of a large number of very small grinding chips (undeformed chip thickness 1 μm (40 μin) or less have been found in single-grit grinding tests using a 0.33 mm (0.013 in) diameter particle of Si_3N_4 as abrasive, operating at a wheel speed of 3000 f.p.m. (15 m s^{-1}). The material ground was hard ($R_c = 66.5$) AISI T-15 tool steel (Komanduri and Shaw 1975). It has been proposed that in these single-grit grinding experiments the individual platelets (chips) are so thin that they bend to accommodate the extra electrons in the newly generated surface rather than being absorbed in an oxidation reaction. The relatively unoxidized bent platelets then coalesce to form spheres consisting of a large number of platelets. When grinding chips of ordinary size are formed, they are initially at a relatively low temperature. They appear black on leaving the grinding wheel until they have moved through the air far enough for the exothermic oxidation that occurs in the absence of bending to be sufficient to raise their temperature to cause a dull red radiation. As they move farther from the wheel they continue to oxidize, and their temperature continues to rise until they either melt or explode like meteorites. Molten spheres have been found in grinding swarf (Tarasov 1953; Grof 1975; Malkin 1984), but these should not be confused with the spheres found in the experiments of Komanduri and Shaw, which consist of many extremely thin platelets that do not oxidize rapidly to form sparks but instead curl and coalesce to form composite spheres instead of molten ones. In the case of very thin chips, bending occurs to accommodate the extra electrons in the freshly formed surface instead of being absorbed by an oxidation reaction, as in the case of thicker nonbending platelets. The extra electrons establish strong bonds between thin platelets, which arrange themselves into spheres to minimize the area/volume ratio just as small spheres form for the same reason in the case of a molten metal.

Figure 5.50 shows scanning electron micrographs of grinding swarf produced when grinding very hard steel with common abrasive (Al_2O_3) and with cubic boron nitride (CBN) wheels under comparable surface grinding conditions. Two important differences may be seen:

1. The chips produced using Al_2O_3 are shorter and thicker than those produced by CBN.
2. The spherical particles are larger and more numerous in the case of Al_2O_3 grinding.

Figure 5.51 shows representative spherical particles at higher magnification. In both cases, the spheres are seen to consist of many small platelets. No evidence of melting is evident. The sphere in Fig. 5.51(a) (Al_2O_3) is about 520 μin (13 μm) in diameter, with individual platelets measuring 40–80 μin (1–2 μm). The spheres in Fig. 5.51(b) (CBN) are about 240 μin (6 μm) in diameter, with individual platelets measuring about 20–40 μin (0.5–1 μm).

By comparing the swarf when surface grinding hard steel using CBN and Al_2O_3, the following qualitative observations may be made:

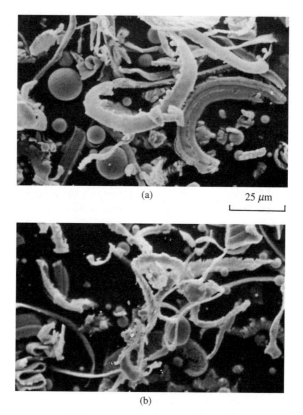

FIG. 5.50. SEM micrographs of up grinding chips obtained in horizontal surface grinding of case carburized AISI 8620 steel with a hardness of $R_c = 62$ (after Ramanath *et al.* 1987). Wheel diameter, 8 in 205 mm); wheel speed, 6400 f.p.m. (32 m s^{-1}); table speed, 28 f.p.m. (8.6 m min^{-1}; wheel depth of cut, 480 μin (12 μm; cross-feed, 0.040 in per pass (1 mm per pass). (a) Al_2O_3 wheel (A60 J8 VBE); (b) CBN wheel (CB150 TBA 1/8).

1. CBN grits are generally sharper than Al_2O_3 grits.
2. There is less rubbing and more cutting with CBN than with Al_2O_3 under comparable grinding conditions
3. The composite spheres observed in the grinding swarf for hard steel consist of wear particles in the form of platelets formed during the rubbing phase in up grinding.

Wetton (1969) has used optical and X-ray microanalysis to examine grinding debris and ground surfaces when a vitrified Al_2O_3 wheel was used in a plunge surface grinding mode. He found particles of bond material (SiO_2) in the ground surface but not in the loose grinding debris. It was speculated that fewer strongly adhering particles would remain on a ground steel surface if an effective grinding fluid were used.

FIG. 5.51. SEM micrographs of typical spheres shown in Fig. 5.50 at higher magnification: (a) Al_2O_3 wheel; (b) CBN wheel (after Ramanath *et al.* 1987).

A wide variety of workpiece materials were ground in a comparative study of Al_2O_3 and CBN wheels by Pai *et al.* (1989) under the following conditions:

Wheel speed, V): 6280 f.p.m. (32 m s^{-1})
Work speed, v: 10 f.p.m. (0.05 m s^{-1})
Aluminium oxide wheel (A): 38A 120M 8V,
$8 \times 0.5 \times 1.25$ in ($203 \times 6.3 \times 31.8$ mm)
Diamond wheel (D): SD 220J 100 B,
$8 \times 0.375 \times \times 1.25$ in ($203 \times 9.5 \times 31.8$ mm)

Forces were recorded using a two-component strain gage grinding dynamometer and representative chip samples were collected by placing a glass microscope slide coated with petroleum jelly in the spark stream.

Table 5.6 gives representative specific energy results, while Fig. 5.52 shows these values plotted.

The polymethylmethacrylate chips produced by the Al_2O_3 (A) wheel were

TABLE 5.6 *Specific energy values for a variety of materials, ground with A (Al_2O_3) and D (diamond) wheels (After Pai et al. 1989)*

Material	Wheel	d μin (μm)	b in (mm)	u p.s.i. × 10^{-6}	(GPa)
PMMA	A	1000 (25)	0.34 (8.5)	0.47	(3.24)
	D	1000 (25)	0.37 (9.4)	0.11	(0.76)
Steel (soft)	A	400 (10)	0.17 (4.3)	2.30	(15.9)
	D	400 (10)	0.17 (4.3)	2.80	(19.3)
WC	A	200 (5)	0.44 (11.2)	80.8	(557)
	D	200 (5)	0.36 (9.1)	43.6	(301)
Glass	A	400 (10)	0.40 (10.2)	1.8	(12.2)
	D	1000 (25)	0.36 (9.1)	0.9	(6.10)
Sialon	A	400 (10)	0.22 (5.5)	5.5	(37.9)
	D	400 (10)	0.17 (4.2)	1.9	(13.1)
LPB granite	A	400 (10)	0.41 (10.4)	8.1	(55.8)
	D	400 (10)	0.35 (8.9)	0.90	(6.20)

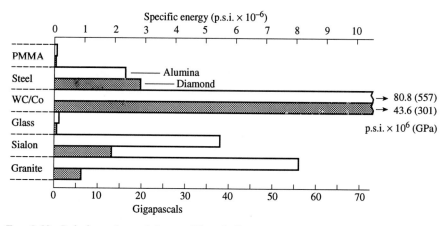

FIG. 5.52. Relative values of the specific grinding energy for a wide range of materials in plunge surface grinding (after Pai *et al.* (1989).

long folded filaments that fused together into large (150 μm) clusters. The diamond (D) wheel produced long straight unfused filaments. In both cases the extrusion chip-forming mechanism appeared to pertain, and the lower specific energy for the D case appears to be due to a lower extrusion pressure and more removal by fracture.

For low-carbon steel, many long (up to 100 μm) filaments were obtained with the A wheel. Shorter filaments (40 μm or less) and many more fracture particles were obtained with the D wheel. It is well known that diamond has a strong affinity for hot iron, which gives rise to short wheel life when soft steel is ground with diamond. This would tend to require a higher specific energy for the diamond than for the Al_2O_3. However, the fact that the harder diamond gives fewer shorter filaments and more fracture particles will tend to move the specific energy in the opposite direction. The net effect is practically the same specific energy for the A (2.3×10^6 p.s.i. = 15 859 MPa) and D (2.8×10^6 p.s.i. = 19 306 MPa) cases.

For tungsten carbide the debris consists of many fine (about 1 μm) particles and relatively few large (10–20 μm) particles for the A wheel. The opposite holds for the D wheel. There was no evidence of filament formation and hence no extrusion for either of the two wheel types when grinding tungsten carbide. The chip-forming mechanism appears to be primarily crushing. The finer crushed particles with the A wheel require about twice the specific energy as with the D wheel.

For glass, removal was mostly by crushing with the D wheel, while with the A wheel removal involves some extrusion (filaments) and less crushing. The specific energy for the A wheel was about twice as high as for the D wheel. However, the specific energy for both wheels was quite low because of the extreme brittleness of the glass.

Sialon is a Si_3N_4-based ceramic in which aluminium and oxygen are substituted for some of the silicon and nitrogen atoms respectively in Si_3N_4. This tool material is unusually tough relative to other ceramic tool materials. With the A wheel, the chips are a combination of extruded filaments and blocky fracture particles. For the D wheel, only relatively fine fracture particles were present. This suggests that the specific energy should be much higher for the A wheel than for the D wheel, which was found to be the case (u for the A wheel is more than twice u for the D wheel).

The filaments produced when Sialon is ground with Al_2O_3 are different from those produced by extrusion of polymethylmethacrylate (PMMA). Instead of being an homogeneous extruded filament, as in the case of PMMA, this filament consists of many small (about 3 μm) fracture particles bonded together. Bonding occurs between freshly generated fracture particles in much the same manner as that for metals, where spheres are produced. Apparently, the fine particles are first generated by crushing and then bonded together as they are extruded outward. The pressures and temperatures for the D wheel are not sufficient to give bonding as with the A wheel and hence $u_A > u_o$.

All granites are not alike with regard to their ease of diamond sawing. Birle and Ratterman (1986) have classified granitic rock into four broad categories, as discussed in Chapter 14, based on the wear performance of D saw blades when cutting these materials. In this classification scheme, type I granites, such as Lake Placid Blue, are the easiest to saw and type IV, such as Bright Red, are the most difficult. The type I granite gave unbonded particles that are small, while the mean particle size for the type IV granite was larger. This is undoubtedly due to type I granite having a greater flaw density and greater ease of accommodating fine particles in the space between bond and work surfaces over the arc of wheel–work contact. This suggests that the mean particle size of chip samples produced under a set of standard grinding

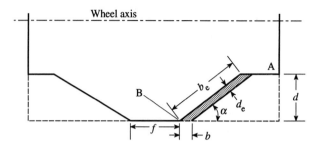

FIG. 5.53. Surface grinding with axial feed per pass (b) and wheel depth of cut d, showing tapers worn on the edges of the wheel, and with the area removed per pass shown cross-hatched.

conditions could serve as a means of rating the grindability of granites: the finer the mean particle size, the better the grindability.

Low-melting (PMMA) and soft ductile materials (mild steel) produce some chips in the form of long thin extruded filaments Brittle materials give blocky fracture particles. Al_2O_3 abrasive grits are not as sharp as diamond particles and hence produce more filaments and fewer fracture particles. The specific energy increases with the strength of the material and with the percentage of the chips that are filament-like. The presence of filaments indicates higher specific energy due to more large strain filament extrusion rather than relatively low strain crushing. An important difference between chips from ductile and brittle workpieces is that long ribbon-like chips tend to form with ductile materials, but small powder-like particles from with hard brittle material such as ceramics. At high removal rates there tends to be a chip storage problem relative to forces and energy required, particularly with metal- or resin-bonded superabrasive (diamond or CBN) wheels. An important consequence of this difference in chip types for ductile and brittle materials is that brittle materials tend to exhibit a smaller chip storage problem than ductile materials and hence allow a greater removal rate for the same available chip storage volume between bond and work.

Grinding with Axial Feed

Grinding with axial (cross-feed) is more complex than plunge grinding since the wheel depth of cut depends not only on the radial in-feed per revolution of the work, d (or downfeed per pass in the case of surface grinding), but also upon the slope of the surface worn on the wheel face. Figure 5.53 shows a wheel with an incline worn on its face. The axial feed per revolution is b and the wheel depth of cut is d. However, the effective wheel depth of cut (d_e) and the effective width of cut (b_e) are as follows:

$$d_e = b \sin \alpha, \qquad b_e = d \csc \alpha \qquad (5.22, 5.23)$$

and hence
$$b_e d_e = bd. \tag{5.24}$$

It is thus evident that in cross-feed grinding, the effective wheel depth of cut (d_e) is actually associated with the feed per pass (b), while the effective width of cut (b_e) is associated with the actual wheel depth of cut (d).

In practice, the situation is still more complex in that radii will form at A and B in Fig. 5.53.

This complex aspect of grinding with cross-feed was first identified by Purcell (1955), further studied by Hillier (1966), and still later was the subject of the dissertation of Banerjee (1969). Yet more papers on the subject are Banerjee and Hillier (1969a, b) and Banerjee (1975, 1979).

As long as a flat region remains on the wheel face (f in Fig. 5.53) good finish may be achieved without spark-out (grinding with gradually decreasing d). Characteristically, b_e/b will be about 4 and a practical wheel width (w) for a wheel fed in one direction is about eight times the axial feed per pass. For a wheel that cuts in both directions (as in horizontal spindle surface grinding), the width of wheel should be about 16 times the axial feed per pass for good finish. This accounts for the popularity of a feed of 0.05 in (1.25 mm) per pass when using a wheel of 0.75 inch (19 mm) width in ordinary horizontal spindle surface grinding with cross-feed.

When the width of flat 'f' becomes less than two or three times the cross-feed b, surface roughness will increase and the wheel must be dressed.

The specific grinding energy depends on the mean chip depth of cut ($\bar{t}$) which, in turn, depends on the effective wheel depth of cut (d_e) but is independent of the effective width of cut (b_e). To achieve low specific grinding energy, d_e should be large, which means that b should be large and the wheel selected so that α will be large. Low specific energy is of importance where the maximum removal rate possible is apt to be determined by limited horsepower.

Where finish is of major importance, the effective wheel depth of cut should be low. This calls for small values of b, and a wheel designed to give a low value of α (eqn 5.22).

Axial feed grinding may be characterized as a type leading to relatively fine finish, small chips, and relatively high specific grinding energy. It is therefore of greatest interest as a finishing operation. Where large rates of stock removal are desired, with less emphasis on finish, plunge grinding is of interest. In this latter type of grinding, chips tend to be larger and specific grinding energy relatively low.

In plunge grinding, the effective wheel depth of cut and the effective width of cut are the nominal values and there is no complication due to an inclined wear surface which gives rise to the additional variable α.

References

Andrew, C., Howes, T. D., and Pearce, T. R. A. (1985). *Creeb feed grinding*. Industrial Press, New York, Chapter 3.
Babu, N. R. (1989). Ph.D. thesis, IIT, Madras.
Backer, W. R., Marshall, E. R., and Shaw, M. C. (1952). *Trans. ASME* **74**, 61.
Banerjee, J. K. (1969). Ph.D. thesis, Waterloo University, Canada.
Banerjee, J. K. (1975). *16th MTDR, England*, p. 43.
Banerjee, J. K. (1979). *Trans. ASME* **101**.
Banerjee, J. K. and Hillier, M. J. (1969a). *Tool Mfg Engr (ASTME)*, Feb., 59.
Banerjee, J. K. and Hillier, M. J. (1969b). *Tool Mfg Engr (ASTME)*, June, 63.
Birle, J. D. and Ratterman, E. (1986). *Dimensional Stone* **2**, 12.
Brecker, J. N. (1967). Ph.D. dissertation, Carnegie-Mellon University.
Brecker, J. N. (1970). Unpublished report.
Brown, R. H., Saito, K., and Shaw, M. C. (1971). *Ann. CIRP* **19**, 105.
Brown, R. H., Wager, J. G., and Watson, J. D. (1977). *Ann. CIRP* **25**(1), 143.
Colwell, L. V. (1963). *J. Engng Ind.* **85**, 27.
Colwell, L. V., Lane, R. O., and Soderlung, K. N. (1962). *J. Engng Ind.* **84**, 113.
Daimon, M. and Ishikawa, T. (1983). *Proc NAMRC*, 305.
Greenwood, J. A. and Tripp, J. H. (1967). *J. appl. Mech.* **89**, 153.
Grof, H. E. (1975). *ZwF* **70**, 423.
Gu, O. Y. and Wager, J. G. (1988). *Ann. CIRP* **37**(1), 335.
Hahn, R. S. (1955). *Trans. ASME* **77**, 1325.
Hahn, R. S. (1966). *J. Engng Ind.* **88B**, 72.
Hillier, M. J. (1966). *Int. J. Mach. Tool Des. Res.* **6**, 109.
King, R. I. and Hahn, R. S. (1986). *Handbook of modern grinding technology*. Chapman and Hall, New York. Chapter 4.
Kingery, W. D., Sidhwa, A. P. and Waugh, A. (1963). *Ceram. Bull.* **42**, 297.
Komanduri, R. and Shaw, M. C. (1975). *Phil. Mag.* **32**, 711.
Krug, H. and Honcia, G. (1964). *Werkstattstech.* **54**, 53.
Lane, R. (1968). Private communication.
Malkin, S. (1984). *J. appl. Metalworking* **3**, 95.
Maris, M., Snoeys, R., and Peters, J. (1975). *Ann. CIRP* **24**(1), 225.
Marshall, E. R. and Shaw, M. C. (1952). *Trans. ASME* **74**, 51.
Nakayama, K. (1967). *Bull. Japan. Prec. Engng* **5**, 93.
Nakayama, K., Brecker, J. N., and Shaw, M. C. (1971). *J. Engng. Ind.* **93**, 609.
Nakayama, K., Takagi, J., and Abe, T. (1977). *Ann. CIRP* **26**(1), 133.
Pahlitzsch, G. and Appun, J. (1953). *Werk. u Maschinenbau* **43**, 296.
Pai, D. M., Ratterman, E., and Shaw, M. C. (1989). *Wear* **131**, 329.
Peklenik, J. (1957). D. Ing. dissertation, Aachen, T. H.
Peklenik, J. (1960). *Ind. Anzeiger* No. 28, 425.
Peklenik, J. (1961). *Ind. Anzeiger* No. 91, 23.
Peklenik, J., Lane, R., and Shaw, M. C. (1964). *Trans. ASME* **86**, 294.
Peters, J., Snoeys, R., and Decneut, A. (1968). *Adv. in machine tool design and research*, Pergamon Press, Oxford, p. 1113.
Poritsky, H. (1950). *J. appl. Mech.* **72**, 191.
Purcell, J. (1955). Unpublished note 132, Cranfield Aero College, U.K.
Radhakrishnan, V. and Rahman, J. F. (1981). *J. Engng Ind.* **103**, 99.

Ramanath, S., Ramaraj, T. C. and Shaw, M. C. (1987). *Ann. CIRP* **36**(1), 245.
Rowe, W. B., Morgan, M. W., Qi, H. S., Zheng, H. W. (1992). *Ann. CIRP* **42**(1), 409.
Saini, D. P. (1980) *Ann. CIRP* **29**(1), 189.
Salje, E. and Jacobs, U. (1978). *Ind. Anzeiger* No. 61, 24.
Salje, E. and Paulmann, R. (1988). *Ann. CIRP* **37**(2), 641.
Sauer, W. J. and Shaw, M. C. (1974). *Proc. Int. Conf. Prod. Engng.*, Tokyo, Part 1, p. 645.
Schey, J. (1987). *Introduction to manufacturing processes* (2nd edn). McGraw-Hill, New York.
Schmitt, R. (1968). D. Ing. dissertation, Braunschweig, T. H.
Schwartz, K. E. (1959). D. Ing. dissertation, Aachen, T. H.
Shaw, M. C. (1950). *J. appl. Phys.* **21**, 599.
Shaw, M. C. (1984). *Metal cutting principles*. Clarendon Press, Oxford.
Shaw, M. C. (1985). *Inst. of Engrs.* (India), **66**, 29.
Snoeys, R. (1964). Private communication.
Snoeys, R. and Wang, I. C. (1968). *Proc. 9th I MTDR Conf.*, p. 1133.
Tanaka, Y., Tsuwa, H., and Kawamura, S. (1966). *Bull, Japan Soc. Prec. Engrs.* **1**, 177.
Tarasov, L. P. (1953). *Machining Theory and Practice, ASM*, pp. 409–464.
Tsuwa, H. and Yasui, H. (1972). *Proc. Int. Grinding Conf.* Carnegie Press, Pittsburgh, Pennysylvania, p. 142.
Verkerk, J. (1975). *Ann. CIRP* **24**(1), 259.
Verkerk, J. (1976). *Ann. CIRP* **25**(1), 209.
Verkerk, J. (1979). *Ann. CIRP* **28**(2), 487.
Warnecke, G. and Gruen, F. J. (1987). *ZwF* **82**, 222.
Wetton, A. G. (1969). *Wear* **13**, 331.
Zhou, Z. X. and Van Luttervelt, C. A. (1992). *Ann. CIRP* **41**(1), 387.

6

ABRASIVE CUT-OFF

Introduction

Stock removal grinding involves processes in which the most important consideration is the rapid and low-cost removal of unwanted material. Form and finish are not primary considerations. The forces and energy required are high and wheel wear is rapid. In fact, wheel wear is so high that dressing and conditioning are not required. Grits are self-sharpening by microfracture or leave the wheel face when they become too dull. Chip accommodation is normally not a problem, since chips produced erode away the bond, making conditioning unnecessary.

In this chapter and the next two the following SRG processes will be discussed:

(1) abrasive cut-off;
(2) conditioning of slabs and billets;
(3) vertical spindle surface grinding.

The abrasive cut-off process (Fig. 6.1) is one of the more important abrasive machining (SRG) operations in which metal is removed at high rates (5 in^3 min^{-1} or 75 lb hr^{-1} for a 25 h.p. machine = 82 cm^3 min^{-1} or 334 N hr^{-1} for a 18.7 kW machine).

Chips must be large; otherwise, the specific energy will be excessive. Large chips call for large grits and abrasive cut-off wheels usually employ abrasives of from 20 to 46 grit size. Large chips also give large chip space but large grit forces.

The principal item of interest is cost per cut and the main items of cost are the wheel cost and 'labor' cost (the machine, the operator and overheads). Low wheel cost involves a large grinding ratio (G = volume of metal removed per volume of wheel consumed) and inexpensive wheels (low y = cents per unit usable wheel volume). Low labor cost involves low cutting time and a low 'labor' rate (x = cents per minute). The volume of wheel consumed in finding G is conveniently determined by measuring the change in wheel diameter after a few cuts (say five) and using this to compute the value of G.

Low cutting time requires a high removal rate. Since there is a critical space between grits to accommodate chips, a high removal rate requires a wheel speed (V) that is as high as practical.

The most widely used cut-off wheels have a resin bond, since this material may be operated at high speeds, is less brittle, creeps to relieve stress concentrations, and is better able to withstand the high tensile stress induced by high

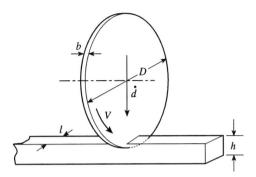

Fig. 6.1. An abrasive cut-off operation.

rotational speeds. Resin-bonded wheels are frequently reinforced with glass cloth and other fiber arrangements further to increase tensile strength and to prevent fragmentation should rupture occur at high speed. In practice, resin-bonded wheels are regularly operated at 12 000 f.p.m. (60 m s^{-1}) and in some cases as high as 16 000 f.p.m. (80 m s^{-1}).

The resin bond is very temperature sensitive. It loses its strength and undergoes degradation under equilibrium conditions at temperatures of 500–600 °F (260–315 °C). It is therefore important that the bonds holding the outermost grits always remain below this temperature range. Otherwise, the grinding ratio G will be excessively low (below 1).

Most abrasive cut-off is done dry to avoid the complications involved in the use of a coolant at high wheel speeds. Water penetrates the resin-grit interface and weakens the bond. While this tendency may be decreased by use of a water-repellent grit treatment (silane) which is not removed or degraded during curing of the resin bond at 300–400 °F (150–200 °C), most wheels designed to operate wet have a rubber bond. Rubber-bonded wheels are less expensive than resin-bonded wheels, but are normally operated at lower speeds (<10 000 f.p.m. = 50 m s^{-1}).

Important secondary considerations in the abrasive cut-off operation are:

- wheel wear
- cutting forces and power
- wheel dynamics and stability
- straightness of cut
- workpiece damage due to overheating

The simplest type of abrasive cut-off operation is one with radial feed (the so-called chop stroke operation; Fig. 6.1). The feed may be accomplished at:

- constant radial load, F_Q
- constant feed rate, $\dot{d}$

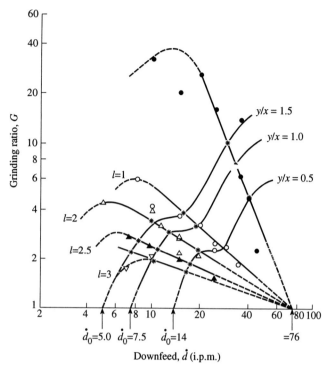

FIG. 6.3. Values of grinding ratio G versus downfeed $\dot{d}$ for square AISI 1020 steel workpieces of different width of cut $l = 1$–3 in (2.5–7.5 cm) using an A24-R6-B wheel having dimensions of 20 × 0.19 × 1 in (508 × 4.9 × 25 mm) at a speed of 12 500 f.p.m. (62.5 m s^{-1}) (after Shaw 1975b).

$\dot{d}$. Curves of constant G^* versus $\dot{d}^*$ are also shown for y/x 0.5, 1.0, and 1.5 in^3 min^{-1} ($y/x = 8.2$, 16.4, and 24.6 cm^3 min^{-1}).

The cost optimum downfeed $\dot{d}^*$ decreases with increase in l due to the limitation of chip space. As we shall see later, the volume per chip varies as the product $\dot{d}l$, and if there is a fixed optimum volume per chip, $\dot{d}$ must decrease as l increases. It is actually easier to pack long thin chips into the space available between grits than it is to pack short thick chips which are stiffer.

The value of G^* decreases with increase in l despite the fact that the optimum volume per chip remains roughly constant. This is because there is more time for heat to flow from the grit tip to the bond post with increase in l. The cutting time is in fact l/V, and as this increases the bond post temperature increases and the value of G drops.

Abrasive cut-off results (G versus $\dot{d}$) are shown in Fig. 6.4 for resin-bonded wheels of different grades and grit sizes using rectangular co-ordinates. The wheel size was 20 × 0.19 × 1 in (508 × 4.9 × 25 mm) and all tests were conducted on a 25 h.p. (18.6 kW) machine without coolant and at a wheel speed

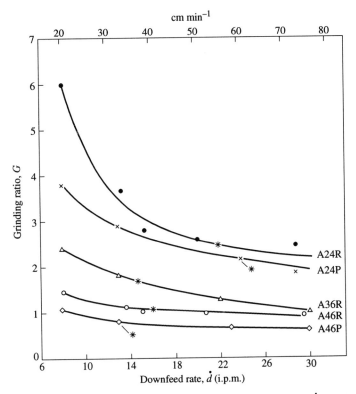

FIG. 6.4. Variation of the grinding ratio G versus the downfeed rate $\dot{d}$ for five wheels of different grit sizes and grades. Grinding conditions: work, hot rolled AISI 1020 steel, lxl in. (25 × 25 mm); wheel speed, 12 500 f.p.m. (62.5 m s^{-1}); grinding fluid, dry. The cost optimum combination of G and $\dot{d}$ is marked by an asterisk in all cases.

of 12 500 f.p.m. (62.5 m s^{-1}). The work material was 1 in (25 mm) square hot rolled AISI 1020 steel. The optimum feed increases with grit size and is about 22 i.p.m. (559 mm min^{-1}) for a grit size of 24 and about 15 i.p.m. (381 mm min^{-1}) for a grit size of 46. Figure 6.5 shows that the diameter of the wheel has a negligible influence on the rate of wheel wear. The G value is found to increase with cutting speed for a given removal rate, mainly because the space in which to accommodate chips increases with wheel speed.

Mechanics

The mean volume per chip (S) is important due to the limited volume between grits to accommodate chips. The mean volume per chip (S) will be equal to chip vol./time divided by cutting edges/time:

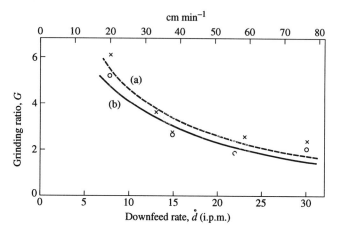

FIG. 6.5. The variation of G versus $\dot{d}$ for wheels of different diameters when cutting under the same conditions: wheel, A24-R6-B; work, hot rolled AISI 1020 steel, 1 × 1 in (25 × 25 mm); wheel speed, 12 500 f.p.m. (62.5 m s^{-1}).

$$S = (\dot{d}lb)/(VCb) = (\dot{d}l)/(VC), \qquad (6.6)$$

where b is the grinding width and C is the number of active cutting points per unit area.

S is seen to vary as the product $\dot{d}l$. Thus, for a constant S, $\dot{d}$ must decrease as l increases (unless V or C are also varied).

The undeformed chip thickness (t) may be found as follows (Shaw *et al.* 1967):

$$t = \dot{d}/(N\pi Db'C) = \dot{d}/(VrtC)$$

or

$$t = [\dot{d}/(VCr)]^{0.5}, \qquad (6.7)$$

where N is the wheel r.p.m. and $\pi DN = V$, b' is the width of an individual scratch $= rt$, and C is the number of active grits in the wheel face per unit area.

The quantities C and r may be readily measured, but this requires a different approach than was employed for FFG. If a piece of thin nylon fishing cord is drawn tightly over the surface of a used cut-off wheel and a parallel beam of light is projected across the combination, a magnified (75×) shadowgraph is produced, from which the number of effective cutting grits per unit length may be estimated (Fig. 6.6). The corresponding width on the wheel face is estimated by rolling the wheel across a piece of pressure-sensitive recording paper. These two measurements enable C, the mean number of cutting points per square inch, to be estimated (Shaw *et al.* 1967). When this was done for an A24-R6-B wheel, C was found to be 134 cutting points per square inch (0.21 per mm^2). This would be the number of grits acting in the wheel face,

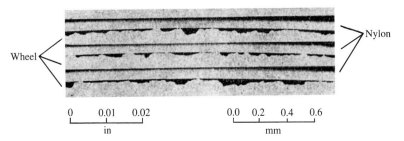

FIG. 6.6. The variation of the wheel face geometry of a used 20 in (508 mm) diameter A24-R6-B wheel operating dry at cost optimum downfeed rate $\dot{d}^* = 22$ i.p.m. (559 mm min^{-1}) and a wheel speed of 12 500 r.p.m. (62.5 m s^{-1}). Work material, hot rolled AISI 1020 steel (after Shaw et al. 1967).

provided that the wheel cuts over its entire periphery. In practice, it is found that a worn wheel face is wavy and contacts the work only about half of the time during one revolution. Thus, the effective value of C would be about 67 per square inch (0.11 per mm^2) in the above case. In practice, all wheels of the same grit size have about the same value of C, but C increases as grit size decreases.

The value of r may be found by tracing grits transversely and noting the width-to-depth ratio. For cut-off grits this value was found to be about 10.

We may thus estimate the mean undeformed chip thickness for an A24-R6-B wheel cutting with a 15 i.p.m. (381 mm min^{-1}) downfeed and wheel speed of 12 500 f.p.m. (62.5 m s^{-1}) as follows:

$$t = \left[\frac{15}{(12)(12\,500)(67)(10)}\right]^{0.5} = 390\,\mu\text{in}\ (9.75\,\mu\text{m}).$$

Cut-off chips may approach 0.005 in (125 μm) in undeformed thickness for cuts taken at a large downfeed with small l.

The energy per unit volume, which is essentially a constant like hardness for a given work material, is given by:

$$u = (VF_P)/(\dot{d}lb), \tag{6.8}$$

where F_P is the tangential force on the wheel.

When a table of values of u is available for different materials (Table 6.1), the h.p. (or kW) for a given cut-off operation may be readily estimated.

$$\text{h.p.} = (u\dot{d}lb)/[12(33\,000)] \quad \text{where } u \text{ is in p.s.i.} \tag{6.9a}$$

or

$$\text{kW} = u\dot{d}lb \quad \text{where } u \text{ is in MPa and } \dot{d}lb \text{ is in mm}^3\,\text{s}^{-1}. \tag{6.9b}$$

The tangential cutting force F_P may be estimated from eqn (6.8). The radial or feed force F_Q is usually about twice F_P and this may be used when

TABLE 6.1 *Values of u and G for several 1 × 1 in (25 × 25 mm) materials in abrasive cut-off:* $V = 12\,500$ *f.p.m.* (63 m s^{-1}), $d = 30$ *i.p.m.* (760 mm min^{-1})

Material	$u \times 10^{-6}$ p.s.i.	(mPa)	G	Wheel
Hard Al	0.7	(4 827)	105	A24-T-B
Soft Al	1.1	(7 585)	12	A24-T-B
Cast iron (class 40)	1.25	(8 619)	111	A24-T-B
A-6 tool steel (annealed)	1.25	(8 619)	14	A24-T-B
A-6 tool steel (hardened)	1.33	(9 170)	22	A24-T-B
AISI 1020	1.5	(10 343)	1.9	A24-T-B
304 SS	1.6	(11 032)	2.6	A24-T-B
T-15 tool steel	2.0	(13 790)	2.0	A24-T-B
Cast iron (class 40)	1.0	(6 895)	3.9	C46-P6-B
Rene 41	1.33	(9 170)	0.7	C46-P6-B
T-15	1.53	(10 549)	1.0	C46-P6-B
Titanium alloy[†]	1.8	(12 411)	0.25	A46-P6-B

[†] In this instance, wheel speed, $V = 7800$ f.p.m. (39 m s^{-1}).

an estimate of F_Q is required. The circumferential force component F_P is found, to a good approximation, to be independent of wheel speed V.

Equations (6.9) give the power at the wheel face and this must be divided by the efficiency of the power train from motor to wheel to obtain motor power.

The values of u and G given in Table 6.1 were obtained when cutting 1×1 bars with a downfeed of 30 i.p.m. (760 mm min^{-1}) and a speed of 12 500 f.p.m. (62.5 m s^{-1}).

As an example, the maximum h.p. at the wheel required to cut a bar of AISI 1020 steel 2 in (51 mm) in diameter at a downfeed of 20 i.p.m. (508 mm min^{-1}) with a 24 grit size wheel measuring $20 \times 3/16 \times 1$ in ($508 \times 4.8 \times 25$ mm) would be:

$$\text{h.p.} = \frac{(1.5 \times 10^6)(20)(2)(0.19)}{12(33\,000)} = 28.8 \text{ h.p. } (21.5 \text{ kW}).$$

Figure 6.7 shows that the specific energy u varies substantially with $\dot{d}$, but not very much with wheel specification.

The energy per unit volume varies inversely with downfeed rate approximately as $u \sim 1/\dot{d}^n$. For a wheel diameter of 20 in (508 mm) and $l = 1$–3 in (25–75 mm), $n = 0.5$ for $\dot{d} = 10$ in min^{-1} (254 mm min^{-1}), and $n = 0.25$ for $\dot{d} = 30$ in min^{-1} (762 mm min^{-1}).

Another quantity of importance is the mean force per grit (F'_P):

$$F'_P = (u\dot{d}lb)/(VlbC) = (\dot{d}u)/(VC). \tag{6.10}$$

If this quantity increases the wheel will behave more softly, and vice versa. This equation may be used as a guide for changes to produce a change in effective wheel grade. Thus, for a wheel to exhibit the same relative hardness when V is increased, $\dot{d}$ should be increased in the same proportion.

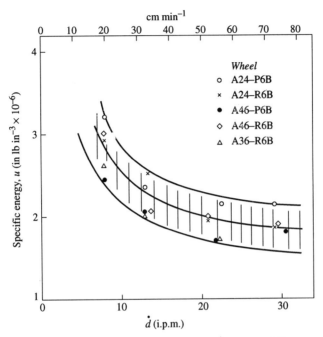

FIG. 6.7. The variation of the specific energy u versus $\dot{d}$ for five different wheels grinding hot rolled AISI 1020 steel without a cutting fluid. Wheel speed = 12 500 f.p.m. (62.5 m s^{-1}).

Wheels with grits of large size require a smaller bond volume for the same relative grade and hence provide more chip space. In general, large grit (24) cut-off wheels perform better than finer grit (36) wheels.

When a round bar is cut, the cutting length l varies continuously and, since $\dot{d}$ changes with l, $\dot{d}$ should likewise be varied. In fact, the volume per chip S (eqn 6.6) will have an approximately constant value S^* under cost optimum conditions. It is evident from eqn (6.9) that h.p. (kW) will be constant if $\dot{d}l$ is constant. Thus, the proper value of $\dot{d}$ will automatically be provided as l changes if $\dot{d}$ is controlled by maintaining the power constant.

An alternate way of keeping the product $\dot{d}l$ constant is to use a constant feed force F_Q instead of a constant downfeed rate $\dot{d}$ or a constant power level. From eqn (6.8) and the fact that $F_Q \cong 2F_P$, we obtain

$$F_Q = [2u(\dot{d}lb)]/V. \tag{6.11}$$

Thus, if V, b, and u are constant and F_Q is held constant $\dot{d}l$ will be constant and hence the volume per chip S will be constant. The value of F_Q may be adjusted to give the optimum volume per chip S^* and hence the minimum cost per cut (¢*).

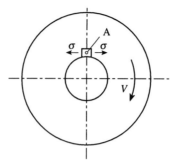

FIG. 6.8. A grinding wheel showing the point of maximum tensile stress (σ).

Wheel Speed

Abrasive cut-off productivity is found to increase appreciably, whether measured by reduced cost per cut or reduced time per cut, with increased wheel speed. This is because the limited chip space available between active grits in the wheel surface is more effectively used at high wheel speeds.

Until about 1960, conventional grinding wheel speeds were 6500 f.p.m. (32 m s^{-1}) for vitrified wheels and 9500 f.p.m. (40 m s^{-1}) for resin-bonded wheels. These speeds were limited by safety considerations. The resin-bonded wheels were capable of higher speeds, despite the lower strength of the bonding material, due to their greater ductility and ability to relieve points of local stress concentration by plastic flow.

The major problems in high-speed grinding are associated with:

- wheel strength
- safety
- machine tool design

Surface temperatures and surface integrity are usually not a problem for SRG.

Conventional grinding wheels consist of a disc with a central hole (see Fig. 6.8). The maximum stress that a wheel is subjected to is due to centrifugal acceleration. This stress is tensile in character, which is unfortunate since grinding wheels are brittle, and all brittle materials are much weaker in tension than in compression. The critical stress in a grinding wheel is the tangential tensile stress σ, which is a maximum at the bore (point A in Fig. 6.8). This stress increases as the square of the wheel speed V.

Grinding wheels fail by radial tensile cracks running from the bore to the periphery, due to centrifugally induced tensile stresses. These tensile stresses may be avoided by dividing the wheel into sector-shaped segments that are kept from moving out radially by providing them with tapered faces (Fig. 6.9). As the segments tend to move radially outward, heavy compressive retaining stresses develop at F. The key feature of this design is that the wheel is now

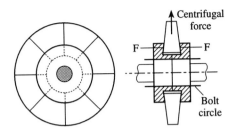

FIG. 6.9. A segmental wheel design (after Shaw 1971).

held together by compressive instead of by tensile stresses. Since grinding wheels are about six times as strong in compression as in tension, this arrangement enables very much higher safe wheel speeds. Wheels measuring 24 × 3 × 12 in (61 × 7.6 × 30.5 cm) of the design of Fig. 6.9 have been speed tested at 37 000 f.p.m. (185 m s^{-1}) and operated in conditioning stainless steel billets in a steel mill at 25 000 f.p.m. (125 m s^{-1}). The U.S. Safety Code requires that grinding wheels be speed tested at 1.5 times operating speed.

The design shown in Fig. 6.10 was used with straight-sided 1/8 in (3.2 mm) thick resin-bonded segments to provide a 20 in (508 mm) diameter cut-off wheel. Fluid introduced at A at atmospheric pressure gave a flow rate of 5000 g.p.h. (18 950 /h^{-1}) at an operating speed of 22 000 f.p.m. (110 m s^{-1}). The parallel-faced wheel segments and steel wedges W used in the design of Fig. 6.10 enables the holding chuck to be made in one piece, and the changing of segments is greatly facilitated.

Normally, grinding wheels that operate at high speed must have a small grit size, a large bond content, and a low structure number for added strength. Use of the segmental wheel concept enables softer wheels of large grit size and higher porosity to be used, all of which provides greater chip storage space and thus results in a greater cost optimum removal rate. The higher speeds thus made possible have a negligible influence on grinding forces and specific energy but do require machines of greater power in proportion to the increase in wheel speed V. Machines that operate at abnormally high speeds are bound to cost more, not only because of the greater power requirement but also due to greater attention to safety and stability in machine design, greater maintenance costs, and the greater costs associated with a more sophisticated operator. This is bound to increase the magnitude of x (the cost of the machine, operator, and overheads per unit time) in the economic analysis, and care must be taken to insure that the net effect of going to a high-speed machine results in a reduction of total cost.

General Observations

A number of useful observations may be made relative to abrasive cut-off operations:

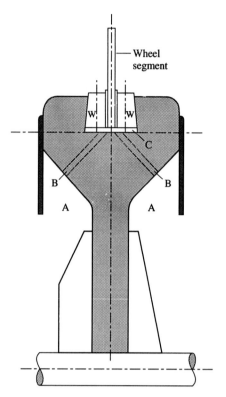

FIG. 6.10. A sectional view of a segmental grinding wheel with steel clamping wedges W and fluid distribution system A-B-C (after Shaw 1971).

1. Those materials forming long stringy chips are more difficult to cut than those giving short broken chips. As a result, soft A-6 tool steel requires more power to cut than hardened A-6 tool steel.
2. Those materials having a high strength level require a high energy per unit volume in the cut-off operation. High-temperature alloys such as Rene 41 are difficult to cut for this reason.
3. Materials containing hard abrasive particles, such as T-15 tool steel, give a relatively low value of grinding ratio.
4. Relatively weak, brittle materials such as cast iron are easy to cut and have a high value of the cost optimum downfeed rate $\dot{d}^*$ and a correspondingly high value of the grinding ratio G^* (30–40). The cost optimum downfeed rate for such materials is so high that unusually high values of horsepower (kW) are required.
5. Materials which have a strong tendency to bond to cut-off wheel abrasives and bonding materials (such as titanium and aluminum alloys) give unusually low values of the grinding ratio under the best grinding conditions.

6. Cut-off wheels tend to cut more on one part of the periphery than another and, as a consequence, feed marks are evident on the cut surfaces. Good grinding conditions are usually associated with clearly evident, uniformly spaced, feed marks and poor grinding conditions give feed marks of variable spacing.
7. In cutting mild steel, the hardest wheels of the largest grit size yield the lowest cost per cut.
8. From eqns (6.5) and (6.3), it is evident that under cost optimum conditions the ratio of the wheel cost to the total cost per cut ($¢_W/¢$) will be

$$¢_W/¢ = n/(1 + n), \qquad (6.12)$$

where n is the exponent in the wheel wear equation (6.4).

9. For short cuts [$l = 0.5$ in (50 mm)], the value of n will normally be about 0.5 and hence ($¢_W/¢$) $\simeq 0.6$ under cost optimum conditions. Since n increases with l a greater percentage of the total cost will be due to the wheel cost as the size of the section cut increases.
10. Of the grit types tested for the abrasive cut-off of mild steel (regular, white, 40 percent ZrO_2, and sintered Al_2O_3) regular aluminum oxide gave the best results relative to the grinding ratio.
11. Screen analysis reveals that cut-off wheels with coarse grits tend to experience more microchipping and hence are sharper and give lower grinding forces.
12. Thin wheels are to be preferred over thick cut-off wheels as long as breakage does not occur. The power consumed and heat generated will be less with thin wheels and a better grinding ratio will be obtained.
13. Grinding temperatures discussed in Chapter 9 are usually not important with respect to surface integrity in abrasive cut-off, but can lead to bond softening and low G if the downfeed or the cutting speed is too low or the wheel-work contact length is too large.

Machine and Operating Considerations

For each wheel diameter D, there is a preferred wheel thickness b. This is the thickness that will limit the wheel deflection to a reasonable value under radial load F_Q. In practice, wheels are made thicker in proportion to their diameter, and current practice appears to be as follows:

$$D/b = 150. \qquad (6.13)$$

The horsepower required for a cut-off machine appears to vary as the square of the wheel diameter, and a useful guide for the horsepower required using wheels of conventional width (eqn 6.13) is

$$\text{h.p.} = D^2/25 \ (D \text{ in inches}) \qquad (6.14a)$$

or

$$\text{kW} = 46D^2 \quad (D \text{ in m}). \tag{6.14b}$$

For example, a machine having a 25 in (0.64 m) diameter wheel should have a minimum of 25 h.p. (18.6 kW).

Small cut-off machines ($D < 20$ in) may use a flywheel to advantage (Nakayama and Shaw 1972). The cutting time is so short in such cases that energy stored in the flywheel when the machine is not cutting is useful in preventing the wheel from slowing down when taking a cut. By use of a flywheel weighing between 100 and 200 lb (450–900 N), it is possible to use a smaller motor to do the same work.

The method of clamping the workpiece is important, and this should be done so that the expansion of the work at the end of a cut does not cause deflection of the work that will cause it to grip the sides of the wheel.

Abrasive cut-off machines are designed to operate at a constant downfeed rate or at a constant downfeed force (F_Q). When constant force is used, the feed rate will automatically adjust to a value close to the cost optimum value as the length of cut changes. Thus, when cutting a circular bar the cost optimum downfeed will be approximated as the wheel penetrates the bar. The downfeed rate $\dot{d}^*$ will be high at the beginning of the cut, low at the point of maximum diameter, and high again at the end of the cut.

Optimization

An adaptive control unit has been developed (Gall 1968; Gall et al. 1969) that automatically measures the change in wheel diameter after a certain number of cuts in production (say, five) and then computes the wheel and labor cost per cut using analog components and stores the sum. The downfeed is then increased by a small incremental amount. After the next series of five cuts the mean cost per cut is again computed and compared with the previous one. If the mean cost per cut is less for the second series than for the first, the downfeed is again increased by a small incremental amount. Otherwise, the downfeed is reduced by the equivalent amount. In this way, cost optimum operating conditions will be automatically achieved after a relatively small number of cuts.

Values of specific wheel cost y and labor cost x may be dialed into the unit to take care of different operating conditions, and the corresponding value of cost optimum grinding ratio G^*, downfeed $\dot{d}^*$, and optimum cost $¢^*$ may be read directly from gages.

Another simpler method of approaching cost optimum conditions incrementally is the Manual Adaptive Control (MAC) technique (Shaw and Komanduri 1977; Stelson et al. 1977; Shaw and Avery 1979). The operator is provided with a programmable calculator containing a program that has been generated by an engineer. Equation (6.2) is used for abrasive cut-off. For a given job, the following quantities would be constant: x, h, y, A, and b. These would be in the magnetic tape provided to the operator with the work order, together with the engineer's best estimate of the initial downfeed to be used. The operator would begin with this value of $\dot{d}$ and enter this together with the starting wheel diameter into the calculator. After five cuts the operator would determine the diameter of the wheel and enter this. When key A is pressed, item I in eqn (6.2) would appear; a second key (B) give item II in eqn (6.2) and a third key (C) would give the mean cost per cut and automatically put this value into storage. The downfeed would then be increased by a small amount (say 5 percent) and the procedure repeated. By pressing a fourth key (D) the new total cost would be subtracted from the previous value in storage and if the answer were positive $\dot{d}$ would again be increased; otherwise, $\dot{d}$ would be decreased. After a few entries, the value of ¢ when key D is pressed should cycle between positive and negative, and this would indicate that the cost optimum downfeed $\dot{d}^*$ had been reached. Of course, keys A, B, and C would not need be pressed each time unless it was desired to know the total cost per part and how this was distributed between x (machine, operator, and overheads cost) and y (wheel cost). The application of MAC to centerless grinding is given in Chapter 13.

It is evident that both of these approaches would be very useful not only in assuring that a machine is being applied under optimum conditions but also in rating the performance of wheels of different specifications or from different manufacturers.

Oscillation and Rotation

As the size of the work increases l will increase and performance of the cut-off operation will fall off. Two ways of reducing the effective length of cut l_e from the contact length l are the following kinematic arrangements:

- oscillating the work relative to the wheel
- rotating the work as well as the wheel

When the work size exceeds about 3 in (75 mm) diameter for a 20 in (508 mm) diameter wheel, it is important to decrease the length of cut. This may be done by oscillating the work relative to the wheel. Vertical motion of the work relative to the wheel in the downfeed direction is not effective.

Horizontal oscillation of the work (Fig. 6.11) is an effective way of decreasing the mean cutting length or the length of a chip before it is discharged from the wheel. This may be done at high frequency and low amplitude, or at low frequency and high amplitude. The latter is more practical. The offset position of the work relative to the centerline of the wheel in its lowest position is then a variable of importance and it appears advantageous to have this at about 45°. This causes the wheel to operate intermittently, thus also invoking the principle of intermittent stock removal.

Under optimum oscillation conditions, it is possible to reduce the mean cutting length to about 10 per cent of the nominal length, which considerably

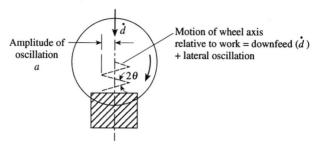

Fig. 6.11. A schematic diagram of a cut-off operation with horizontal oscillation.

extends the capacity of a given cut-off machine (Murata and Shaw 1976). Decreasing the effective length of cut is not only important from the chip accommodation point of view, but it decreases the time of heat flow up the grit before cooling begins.

The cutting force per grit is significantly higher with oscillation, and hence wheels that are used with oscillation should be of higher grade than those used without oscillation.

The main advantage of transverse oscillation appears to lie in extending the range of a given machine. While a conventional 20 in (508 mm) wheel diameter machine without oscillation is capable of cutting stock only up to $l = 3$ in (76 mm), this range may be considerably extended by use of transverse oscillation. However, such a machine should be provided with a much larger drive motor 50 instead of 25 h.p., or 38 instead of 17 kW) and be equipped with an oscillating mechanism with an amplitude of $D/4$. The oscillating mechanism should also have a variable frequency in the range of 20–240 c.p.m.

For work diameters above about 9 in (230 mm) it appears best to rotate the work relative to the wheel instead of oscillating the wheel back and forth. This introduces a new variable to the cut-off process—work r.p.m. (N_W). In analyzing the planetary cut-off operation it is found that the important variable is not radial infeed rate $\dot{d}^*$ or work speed but, instead, undeformed chip thickness t. It is found that the same optimum performance may be achieved from several combinations of operating variables as long as t is the same.

The length of cut for the planetary cut-off operation is

$$l = [(D_W \dot{d})/N_W]^{0.5} \qquad (6.15)$$

where D_W is the work diameter, $\dot{d}$ is the radial in-feed rate, and N_W is the r.p.m. of work relative to wheel. The mean volume per chip will be

$$S = (\pi D_W \dot{d})/(VC), \qquad (6.16)$$

where V is the wheel speed and C is the number of active cutting points per unit area.

The horsepower required at the wheel face will be:

TABLE 6.2 *Kerf loss for different cut-off methods*

Cut-off method	Material lost	
	in	mm
Burning	3/8	9.5
Shearing	1/4	6.4
Abrasive cut-off	3/8	9.5
Hack sawing	1/8	3.2
Band sawing	1/16	1.6
Cold sawing	1/4	6.4
Friction sawing	1/4–3/8	6.4–9.5
Precision crack-off	0	0

$$\text{h.p.} = (u\pi D_\text{w} b\dot{d})/[12(33\,000)] \qquad (6.17\text{a})$$

or

$$\text{kW} = u\pi D_\text{w} b\dot{d} \qquad (u \text{ in MPa}; \pi D_\text{w} b\dot{d} \text{ in mm}^3). \qquad (6.17\text{b})$$

For optimum performance S should be maintained at a constant value S^*. This means that a given wheel running at given speed V the product $D_\text{w}\dot{d}$ should be maintained constant. Since this quantity also appears in the equation for horsepower, S^* will be held constant as D_w changes, if the horsepower (kW) is held constant. Thus, a planetary cut-off machine should be equipped with a device that maintains the horsepower constant at a preset value throughout the cut.

Competition for Abrasive Cut-off

In addition to the abrasive cut-off operation, the following methods of cutting stock to length are used:

- burning
- shearing
- hack sawing
- band sawing
- cold sawing (milling)
- hot sawing
- friction sawing
- precision crack-off

The first two methods give relatively rough cuts and are not competitive with the remaining methods, including abrasive cut-off, from the point of view of quality of cut and of metallurgical damage.

The lost work material associated with different cutting methods will usually correspond to values given in Table 6.2.

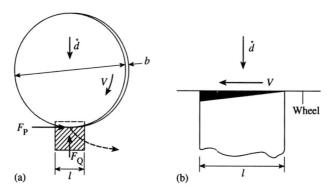

FIG. 6.12. (a) A friction sawing operation. (b) An enlargement of the interface between the wheel and the work, showing the black molten zone.

The material lost per cut (kerf loss) does not represent an important part of the total cost per part unless small bars are cut into very small pieces or the work material is of especially high value, such as titanium or phosphor bronze.

All of the methods in Table 6.2 may be familiar except for the last two, and so a brief discussion of these methods follows.

Friction Sawing

This process is a very old one, having been described in the literature early in the nineteen century (Lardner 1833). At that time it was reported that a soft iron disc could cut through a file when rotating at a very high speed, and would similarly cut steel of any hardness, but not cast iron. In modern times this method has been employed in steel mills to remove the ends of ingots and, in this case, very large wheels up to 5 ft (1.5 m) in diameter are rotated at speeds as high as 25 000 f.p.m. (125 m s^{-1}). In the case of these large saws the rate of penetration is relatively high, as is the power employed (up to 200 h.p. ≡ 150 kW). Another modern application of friction sawing is in cutting off individual parts from the central stem in a precision casting (lost wax) operation. In this case a high-speed soft steel band operating at speeds up to 15 000 f.p.m. (80 m s^{-1}) is used.

It has been suggested (Shaw 1988) that friction sawing involves melting of the workpiece (Fig. 6.12). This is consistent with the fact that the wheel may be softer than the work, which is in contrast to grinding, where it is essential that the abrasive be harder than the material ground (at least 3×).

Figure 6.12(b) shows a molten heat source (solid black) moving downward at a velocity $\dot{d}$ and a horizontal velocity V at the upper surface, where V is about $10^4\, \dot{d}$. The total power at the wheel face will be

$$P = (F_P V + F_Q \dot{d}) \cong F_P V. \qquad (6.18)$$

The amount of heat flowing from the molten heat source into the wheel and work will depend on:

- the area traversing the heat source per unit time $[\sim V(A_R/A)]$
- the time of contact with heat source $\sim 1/V$)
- the value of thermal diffusivity $k/\rho C = \alpha$ for the wheel and work, as discussed in Chapter 9

where A_R/A is the ratio of the real to the apparent area of contact of the upper and lower surfaces of Fig. 6.12 (=1 in this case), k is the coefficient of thermal conductivity, and ρC is the volume specific heat.

The first and second items cancel and in the third α will be about the same for wheel and work. Therefore, the rate of heat flow will be the same to wheel and work, although much more superficial in the case of the wheel. The specific energy will be

$$u = (RP)/(\dot{d}lb), \qquad (6.19)$$

where R is the fraction of heat going to the work (=0.5).

The grinding ratio for friction sawing will be

$$G = \frac{\text{volume melted from work}}{\text{volume melted from wheel}}. \qquad (6.20)$$

It is reasonable to assume that the ratio of the volume melted from the work to the volume melted from the wheel will approximately equal the ratio of the penetration of the thermal front into the work to the penetration of the thermal front into the wheel.

From the theory of heat transfer (Chapter 9),

$$(\theta_s - \theta_x)/(\theta_s - \theta_o) = \text{erf}[y/[2(\alpha t)^{0.5}], \qquad (6.21)$$

where θ_s is the constant temperature of the surface (melting temperature), θ_x is the temperature at distance y from the surface after elapsed time t, θ_o is the ambient temperature, α is the thermal diffusivity ($=k/\rho C$), and erf is the error function. From this, the depth of melting, y_m, should be proportional to $(\alpha T_C)^{0.5}$, where T_C is the contact time, which varies as $1/V$. Thus, since α is approximately the same for wheel and work (both steel),

$$G = (lby_{mw})/(ly_{ms}) = (V/\dot{d})^{0.5}. \qquad (6.22)$$

For a wheel speed of 25 000 f.p.m. (127 m s^{-1}) and a downfeed rate $\dot{d}$ of 30 i.p.m. (0.0127 m s^{-1}), the G value would be

$$G = (V/\dot{d})^{0.5} = 100,$$

which is a high value relative to what might be expected for abrasive cut-off.

Large friction saw discs usually have rectangular grooves in the wheel surface, about 1 in (25 mm) deep, and with a slot width and spacing also of about 1 in (25 mm). The purpose of the slots is to employ the interrupted grinding

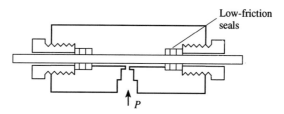

FIG. 6.13. The high-pressure apparatus in which the 'pinch-off' effect was discovered (after Bridgman 1912).

principle (see Chapter 5). However, a major disadvantage of friction sawing is noise produced by the high-speed disc, which is intensified by the slots, which make the wheel act as though it were a siren.

Precision Crack-off

There are many situations in materials processing in which brittle or quasi-brittle material must be cut in two or divided into wafers. Examples include the following:

- cutting of glass sheet, rods, or tubing
- cutting of structural ceramics
- cutting of rock (marble, granite, etc.)
- slicing of tungsten carbide or ceramic bars of any cross-sectional shape into cutting tool inserts
- slicing of silicon and gallium arsenide into wafers for integrated circuits
- production of residual stress free flat surfaces of magnetic ceramics (ferrites) for use in recording heads

Requirements for cut surfaces range from the absence of microchipping in glass rod or tubing to residual stress free surfaces in the case of magnetic and electronic materials (Chandrasekar *et al.* 1985*a,b*).

Almost all slicing and cutting of ceramic materials today is done by diamond sawing. In all cases, diamond sawing involves loss of the material (kerf loss) as well as plastic deformation of the cut surface. The first of these is an economic loss, while the second alters the functional characteristics of the surface. An alternative method of cutting brittle materials, called precision crack-off, offers a way of overcoming many of the disadvantages of conventional cutting and slicing techniques as well as improving productivity.

The basic procedure involved in the precision crack-off method was discovered accidentally by Bridgman (1912) when he was investigating the flow and fracture of solids under very high pressures. His experiments involved a solid cylinder exposed to pressure over the external curved surface only (Fig. 6.13), the ends being free to expand with negligible restraint. At a certain

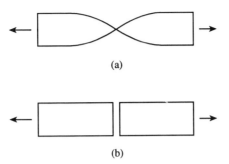

FIG. 6.14. (a) Pinch-off of ductile metal. (b) Pinch-off of brittle material. (After Bridgman 1912).

value of fluid pressure, the cylinder fractured into two parts somewhere in the region between the seals, as if it had been subjected to a uniaxial tensile test. For a ductile material such as a mild steel, the specimen necked down completely just as though it were lead, whereas for a brittle material, such as glass, fracture occurred without plastic flow (Fig. 6.14). Bridgman called this type of failure 'pinch-off'. When the rod is a perfectly brittle material such as glass or ceramic, where fracture stress in tension is less than yield stress, fracture occurs on a perfectly plane surface perpendicular to the axis. Cherepanov (1979) has discussed the mechanism of pinch-off and his views differ from those of Bridgman. Bridgman (1912) attempted to explain pinch-off using a maximum tensile strain approach. Cherepanov (1979), using subsequently developed fracture mechanics, explained that pressure entering an open surface crack produces a stress intensity factor equal to that produced by an axial stress equal to the pressure. To Bridgman and Cherepanov, pinch-off was a nuisance, since it was frequently the cause of failure of high-pressure equipment.

Sato and Naoyuki (1976) employed pinch-off to cleave circular rods and tubes made of brittle materials such as glass, marble, and graphite. Their contribution was the use of a surface defect in the form of a Knoop indentation, scratch, or small radial hole to determine the exact location of the fracture as well as to reduce the pressure required for fracture. Fracture always occurred instantaneously at the defect and the surfaces were often mirror-like.

Sato, like Bridgman, attributed failure to the sample being expanded along the axis due to biaxial compression and employed a maximum tensile strain criterion. However, in view of the fact that brittle materials fail according to a maximum tensile stress criterion (Griffith 1924), this explanation is not correct.

Figure 6.15 shows a schematic of the pressure set-up used to crack off glass rods and tubes. The annular region between the O-rings is pressurized by means of a hand pump. A similar arrangement which uses a neoprene cord in a rectangular groove instead of an O-ring has been used to crack off specimens of other cross-sectional shapes (squares, rectangles, triangular rods, etc.) The pressurized fluid used is cutting oil.

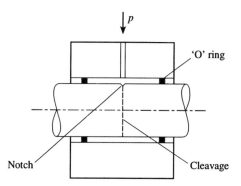

FIG. 6.15. The schematic apparatus used to crack-off brittle rods and tubes (after Chandrasekar and Shaw 1988).

When notched specimens are exposed to fluid pressure, fluid penetrates the notch and exerts a pressure normal to the notch walls. This pressure has an axial component which, in turn, will induce an intensified tensile stress σ_i at the tip of the notch. The value of σ_i depends on the geometry of the notch and the fluid pressure. When σ_i reaches a critical value, a sharp crack initiates and runs spontaneously across the specimen, in accordance with the maximum intensified tensile stress criterion proposed by Griffith (1924). This criterion is extensively discussed in Takagi and Shaw (1983). When the stress concentration is high (deep or sharp notch), the pressure p required to induce a given value of σ_i is lower and hence fracture occurs at a lower pressure.

When a thick glass specimen is scribed and cracked off by bending in the usual way, the fracture plane changes direction as it crosses the neutral axis, producing a lip. This does not occur in the precision crack-off method, since the crack-initiating tensile stress is not accompanied by a compressive stress below the centerline. A possible explanation for this change of crack growth direction on entering a compressive field of stress is that the intensified tensile stresses on the surface of void are orthogonal to each other for nominal unintensified tensile and compressive stresses (Takagi and Shaw 1983). Once a sharp crack is initiated, it will run in a straight line clear across the specimen, provided that there is sufficient stored energy in the system, and the greater the stored energy the smaller will be the deviation from a straight line and the better will be the finish. Thus, it is important to employ a dull notch so that more energy is stored before sharp crack initiation. An example of the effect of pressure required for crack-off on surface finish is given in Fig. 6.16.

There appear to be but two ways of increasing the crack-off pressure so that intragranular fracture is promoted:

(1) use of a blunt defect at the point of desired crack-off;
(2) partial axial constraint on the ends of the specimen.

If either or both of these techniques are employed, the surface finish of

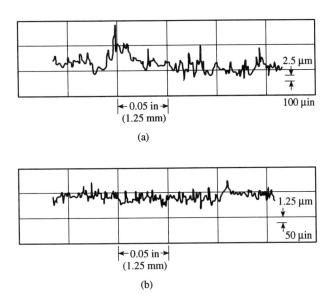

FIG. 6.16. Traces of cracked-off Ni–Zn ferrite surfaces using notches requiring different pressures (p): (a) $p = 11\,500$ p.s.i. (79 MPa); (b) $p = 15\,000$ p.s.i. (103 MPa) (after Chandrasekar and Shaw 1988).

cracked-off polycrystalline materials will be improved. The reason why the surface finish is so poor for polycrystalline graphite is that this material is so weak that a very low pressure is required to initiate a sharp crack. Hence, the stored energy at crack initiation is so low that fracture planes follow weaker grain boundaries instead of a straight line. Apparently, the only hope of successful crack-off for such inherently weak brittle materials is partial axial restraint.

Perhaps the greatest advantage that precision crack-off offers is the finishing of electronic and magnetic materials. Since cracked-off surfaces are produced by brittle fracture, there is minimal plastic deformation of the cleaved surface, which is not the case in diamond sawing or grinding. Cracked-off surfaces should have electrical and magnetic properties that are unaltered by any form of residual stress. This should be beneficial in the production of magnetic recording heads, photovoltaics, and in other electronic material applications (Chandrasekar et al. 1985a). Since all materials behave in a more brittle fashion with increased strain rate, a possibility for the extension of the precision crack-off process to ductile materials is to apply pressure in the notch with explosive speed. This would eliminate the need for the confining chamber required with a static pressure system and could thus also prove to be an important simplification for brittle material application.

An interesting method of cracking-off hardened steel bars has been described by Oertenblad and Appell (1976). Two opposing curved tools load a hard steel workpiece (>850 Vickers) to produce a tensile stress sufficient to induce a crack at the center of the bar which then propagates to the surfaces (Fig. 6.17).

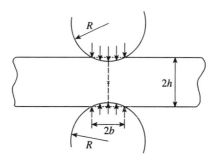

FIG. 6.17. Crack-off of hardened steel (after Oertenblad and Appell 1976).

The authors state that friction is unimportant, as in a Brinell hardness test, and that

b/h should be between 0.3 and 1.0,
R/h should be between 0.2 and 0.6.

References

Bridgman, P. W. (1912). *Phil. Mag.* **6**, 63.
Chandrasekar, S. and Shaw, M. C. (1988). *J. Engng Ind.* **110**, 187.
Chandrasekar, S., Shaw, M. C., and Bushan, B. (1985a). *ASME PED 17*, 45.
Chandrasekar, S., Shaw, M. C., and Bushan, B. (1985b). *ASME PED 17*, 69.
Cherepanov, G. P. (1979). *Mechanics of brittle fracture.* McGraw-Hill, New York, p. 878.
Farmer, D. and Shaw, M. C. (1967). *J. Engng Ind.* **89**, 514.
Gall, D. A. (1968). *Ann. CIRP* **71**(1), 395.
Gall, D. A., Sauer, W. J., and Atkins, N. (1969). *Proc. ASTME, MR 69-228.*
Griffith, A. A. (1924). *Proc. 1st Int. Congr. Appl. Mech., Delft,* p. 55.
Lardner, D. L. (1833). *The cabinet cyclopaedia, Vol. II.* Longman, Rees, Ormer, Brown, Longman, London, p. 156.
Murata, R. and Shaw, M. C. (1976). *Trans. ASME* **98**, 410.
Nakayama, K. and Shaw, M. C. (1972) *Abrasive Engng* May/June, 32.
Oertenblad, B. and Appell, B. (1976). *Ann. CIRP* **25**(1), 159.
Sato, Y. and Naoyuki, S. (1976). *J. Japan Mech. Soc.* **42**, 2250.
Shaw, M. C. (1971). *Mech. Engng* **93**, 19.
Shaw, M. C. (1975a). *Ann. CIRP* **24**(2), 539.
Shaw, M. C. (1975b). *J. Engng Ind.* **96**, 138.
Shaw, M. C. (1988). *Ann CIRP* **37**(1), 331.
Shaw, M. C. and Avery, J. P. (1979). *Proc. 5th World Congr. Machines and Mechanisms* **1**, 777. Concordia University, Montreal.
Shaw, M. C. and Komanduri, R. (1977). *Am. Machinist* **121**, 81.
Shaw, M. C., Farmer, D. A., and Nakayama, K. (1967). *Trans. ASME* **89B**, 495.
Stelson, T. S., Komanduri, R., and Shaw, M. C. (1977). *Proc. 4th Int. Prod. Eng. Conf., Tokyo.* Taylor and Francis, London.
Takagi, J. and Shaw, M. C. (1983). *J. Engng Ind.* **105**, 143.

7

CONDITIONING OF SLABS AND BILLETS

Introduction

Defects such as ingot cracks, folds, scabs, and cinder patches in steel ingots are brought to the surface in primary hot rolling. These defects, and those such as laps introduced in primary rolling, must be removed before the semi-finished steel is worked further. This removal of surface defects from slabs and billets is called conditioning or snagging.

The degree of conditioning depends on the type of steel and its ultimate use. Carbon steel for large structural, merchant quality bars and reinforcing bars are not conditioned except to remove obvious ingot cracks. However, deep-hardening alloy steels and other high-quality products require complete and careful conditioning.

Prior to the First World War conditioning was practically unknown. Large scabs were first removed by cold chisel and hammer in the early 1920s. Pneumatic hammers were used around 1930 and at about the same time machine chippers (similar to milling cutters) and scarfing were also introduced. In scarfing, an oxygen fuel torch is used to melt the surface of the steel. The oxygen has sufficient velocity to sweep the molten layer from the surface.

Scarfing replaced chipping, and today we have automatic billet scarfing machines capable of conditioning a 4 in (100 mm) × 4 in (100 mm) × 40 ft (12 m) billet weighing 1 ton (910 kg) on all four sides simultaneously in about 1 min. However, not all of the molten metal is removed, and patches of poorly bonded material are frequently found. Also, cold scarfing of alloy and high-carbon steels produces a hot surface which is quenched rapidly by the cold billet, giving rise to a skin of brittle martensite which may crack in further processing.

Snagging (abrasive conditioning) was introduced many years ago to overcome these difficulties with alloys and stainless steels. The need for high quality in stainless steels, in addition to the fact they do not burn well, led to the use of swing frame grinders. This grinding method is slow and fatiguing for the operator, who must swing a grinding wheel mounted on a pivoting arm over the work surface. Pressure on the wheel is largely due to the weight of the operator.

In the mid-1950s, semi-automatic machines were introduced which have enabled grinding to make serious inroads on scarfing. The operator rides in a cab and moves with the wheel as it traverses the length of the work. These machines use 24 in (615 mm) diameter × 3 in (77 mm) resinoid wheels of 8 to 16 grit size operating at speeds up to 16 000 f.p.m. (81 m s^{-1}), head pressures up to 2000 lb (910 kg), and traverse table speeds of 150 f.p.m. (0.76 m s^{-1})

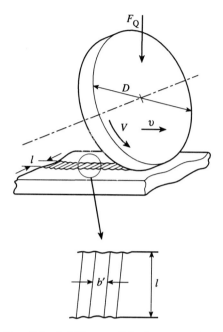

Fig. 7.1. A billet conditioning operation.

and upward, and employ 150 hp (112 kW) and more. After a path has been cut over the length of the work, the wheel is moved a short distance toward or away from the cab and a second overlapping path is cut. The wheel can be tilted so that the area of contact is reduced and particularly deep defects can be removed. Tilting the wheel can also be used to resharpen the wheel (equivalent to dressing it).

In many grinding operations, the entire billet or slab surface will first be ground away to a given depth and then local defects will be removed by spotting (grinding away deep defects) after inspection. The uniform removal operation which accounts for most of the material removed will be considered here.

Mechanics

The objective of conditioning is to remove the faulty material from the surface at the lowest cost per pound weight (or kg) possible. Finish is of secondary importance as long as there are no slivers of metal or ridges which would result in surface defects in subsequent rolling operations.

Most machines operate with a preset radial wheel load W (Fig. 7.1). The wheel usually vibrates vertically at wheel rotational frequency with an amplitude sufficient for the wheel to come out of the cut. High-speed motion pictures show that the wheel grinds over only one-third to one-half of its circumference. This may be further verified by painting the wheel face and observing the gradual removal of the paint. The grinding pattern left on the work also provides evidence of the wheel vibration (Fig. 7.1). Feed marks having a pitch b' equal to the feed per revolution of the wheel are clearly evident on the work. These marks indicate that a single point on the wheel circumference takes the deepest cut.

This rate of precession is a useful quantity in conditioning studies, since it is related to the rate of wheel wear. It is also found that as the wheel speed is increased the amplitude of wheel vibration decreases, so that the wheel is in contact with the work over a greater percentage of the wheel periphery.

The vertical motion of the wheel surface due to out of roundness and vibration, which may be as much as 0.015 in (0.384 mm), determines the quality of the surface more than the grit size of the wheel. The ridges left by the vertical motion of the wheel surface are much higher than the local surface finish amplitude and can be rolled into laps in subsequent rolling if this amplitude is too great.

The important parameters for this process, in addition to the wheel specifications, are the wheel speed (V), head load (F_Q), wheel feed (v) along the work, and cross-feed (l'). The depth of cut is increased by increasing head load and/or decreasing wheel feed or cross-feed, and also by tilting the wheel to reduce the wheel–work contact area.

Wear

The grinding ratio (G = volume of metal removed/volume of wheel wear) is a convenient measure of wheel life. At a given wheel speed this grinding ratio is found to vary with the volumetric metal removal rate ($\dot{M}$, in^3 min^{-1} (cm^3 min^{-1}) regardless of the combination of W, v, and l' used to obtain $\dot{M}$ (Farmer and Shaw 1967). As $\dot{M}$ is increased at 12 500 f.p.m. (64 m s^{-1}), wheel speed G will first rise, pass through a maximum and then fall (Fig. 7.2).

At low metal removal rates the wheel face becomes dull and more heat flows into the wheel, producing high bond temperatures and degradation of the bond, and thus wear increases (G decreases) as metal removal decreases. At high removal rates there is insufficient space in the wheel surface to accommodate the large chips and grits will be pried from the wheel face. Wheel wear then increases (C decreases) as the metal removal rate increases.

On both branches of the G versus $\dot{M}$ curve, it is observed that

$$\dot{M}G^n = C, \qquad (7.1)$$

where n and C are constants for that particular branch of the curve. Since costs depend on wheel life (G) and productivity ($\dot{M}$), the minimum cost point will generally lie on the right-hand branch of the G versus $\dot{M}$ curve, where a compromise exists between maximum wheel life and high productivity (large $\dot{M}$).

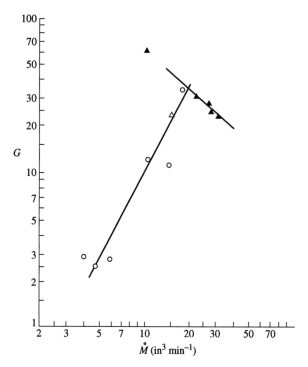

FIG. 7.2. The variation of G with $\dot{M}$ for the conditioning of stainless steel with a 75 h.p. (59.2 kW) fixed load machine. Wheel diameter, D, 24 in (610 mm); wheel speed, 12 500 f.p.m. (64 m^{-1}) (after Brecker et al. 1972).

Economics

A cost analysis by Farmer and Shaw (1967), which considered the costs of labor, machine, and overheads and the cost of the wheel, showed that the optimum metal removal rate, $\dot{M}^*$, for an SRG operation will be:

$$\dot{M}^* = C_2^{1/(n_2+1)} (y/xn_2)^{n_2/(n_2+1)} \qquad (7.2)$$

where y is the wheel cost, ¢ in^{-3} (¢ cm^{-3}), x is the labor, machine, and overheads cost (¢ min^{-1}), and n_2 and C_2 are constants for the right-hand branch of the G versus $\dot{M}$ curve.

The constants for eqn (7.1) for the left-hand branch of the curve (small value of $\dot{M}$ in Fig. 7.2) are

$$C_1 = 2.87 \text{ in}^3 \text{ min}^{-1} \ (47.1 \text{ cm}^3 \text{ min}^{-1}), \qquad n_1 = -0.55$$

and for the right-hand branch (large value of $\dot{M}$) are

$$C_2 = 1040 \text{ in}^3 \text{ min}^{-1} \ (16\,750 \text{ cm}^3 \text{ min}^{-1}), \qquad n_2 = 1.11.$$

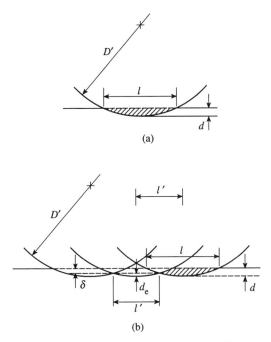

FIG. 7.3. A section of metal removed in conditioning: (a) first pass; (b) subsequent passes (after Brecker et al. 1972).

For a wheel cost of 8.5 ¢ in^{-3} (0.52 ¢ cm^{-3}) and a 'labor' cost of 16.7 ¢ min^{-1} the optimal metal removal rate was found to be 40.5 in^3 min^{-1} (665 cm^3 min^{-1}), which was beyond the limits of the machine. This indicates that this machine should be used at the maximum metal removal rate possible.

As seen in Fig. 7.1, grinding with the plane of the wheel perpendicular to the direction of the wheel feed (v) leaves a scalloped surface on the material being ground. Thus, excess material (in the scallop) must be removed in order to insure removal of a uniform layer containing defects of depth δ. As the value of the material being ground increases, it becomes important to minimize the material removed beyond the uniform layer required for a quality product. Thus, an economic analysis of this process should consider the value of the excess material removed.

During the first pass over the work the length of cut, l, will be

$$l = 2(D'd)^{0.5}, \qquad (7.3)$$

where D' is the effective wheel diameter, considering vibration, and d is the maximum depth of cut (Fig. 7.3(a)). In subsequent passes the cross-feed, l', will be less than l, and the total material removed is shown cross-hatched in Fig 7.3(b).

The depth of the layer that must be removed to eliminate defects is δ. However, to do this it is necessary to grind a maximum depth d, where

$$d = \delta + d_e. \tag{7.4}$$

The quantity d_e is the extra depth that must be removed. Therefore, the metal removed can be divided into two parts: the useful rate $\dot{M}_\delta$ and the extra rate $\dot{M}_e$. From Fig. 7.3(b), it is evident that

$$\dot{M} = \dot{M}_\delta + \dot{M}_e = \delta l' v + (\tfrac{2}{3} l' d_e v), \tag{7.5}$$

where

$$l' = 2(D'd_e)^{0.5}. \tag{7.6}$$

The total removal rate $\dot{M}$ can also be expressed in terms of the specific grinding energy u (in lb in^{-3} or J mm^{-3}):

$$\dot{M} = (F_P V)/u, \tag{7.7}$$

where F_P is the tangential grinding force. The force component F_P is about one-third of the normal force component (F_Q) for a wide range of grinding operations, so that

$$F_P/F_Q = \mu = 0.33. \tag{7.8}$$

Therefore, from eqns (7.5)–(7.8),

$$\dot{M} = \mu F_Q V/u = (l'v)(\delta + l'^2/6D') \tag{7.9}$$

which may be solved for F_Q

$$F_Q = [(ul'v)/(\mu V)](\delta + l'^2/6D') = (u/\mu V)(\dot{M}_\delta + \dot{M}_e). \tag{7.10}$$

When the values of l' and v have been chosen for cost optimum performance, eqn (7.10) gives the value of F_Q to be used for the removal of the defective layer of material δ.

The removal necessary to eliminate defects can vary from 1 to 2 percent of the weight of the billet for carbon steel, and up to 6 percent for high-quality stainless steel. The extra material removal rate $\dot{M}_e$ may be as high as 20 percent of the useful removal rate $\dot{M}_\delta$ (Brecker et al. 1972). This percentage is particularly high for a light degree of conditioning. For billet snagging, the following degrees of conditioning pertain:

- light — 0.5 percent of total billet weight ($\delta \simeq 0.03$ in or 0.8 mm)
- medium — 1.5 percent of total billet weight ($\delta \simeq 0.10$ in or 2.5 mm)
- heavy — 3.0 percent of total billet weight ($\delta \simeq 0.20$ in or 5.1 mm)

There is less total material waste in slab and billet conditioning if the wheel diameter is small and the degree of spindle vibration is low (small D' in eqn (7.6).

When the extra material removed is included in the cost optimization analysis, the total conditioning cost per unit length, L, becomes:

$$\text{¢}/L = x/v + (\dot{M}y)/(Gv) + (\dot{M}_e z)/v, \tag{7.11}$$

where x is the value of the labor, machine, and overheads (¢ min^{-1}) y is the

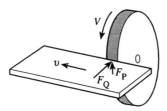

FIG. 7.4. A schematic of the physical simulation of a conditioning operation for laboratory studies of different abrasive types (after Matsuo et al. 1980).

value of the useful grinding wheel volume (¢ in^{-3} or ¢ cm^{-3}), and z is the value of the work material (¢ in^{-3} or ¢ cm^{-3}).

The quantity ¢/L will obviously be least when v has the highest value possible (v_m) consistent with the horsepower available and the value of δ and l pertaining.

A detailed analysis of conditioning cost optimization based on actual production data, including the cost of the extra material that must be removed with conventional conditioning kinematics using a 75 h.p. (56 kW) machine, leads to the following results (Brecker et al. 1972):

1. The cost optimum cross-feed (l^*) decreases with an increase in the maximum table speed v_m, particularly for inexpensive work material (carbon steel) and for light levels of conditioning (low δ).
2. For carbon steel, the cost optimum cross-feed (l^*) may be as high as 1.0 in (25 mm), but as low as 0.05 in (1.25 mm) for stainless steel.
3. The value of l^* increases by a factor of about 1.5 when the effective wheel diameter is doubled.

Role of Abrasive Type

Matsuo et al. (1980) have studied the conditioning performance of several types of abrasive grits using the physical simulation of the process shown in Fig. 7.4. Resin-bonded wheels of 205 mm (8.07 in) OD and 25 mm (1 in) width, of T grade, 4 structure number, and 20 grit size, were tested at a wheel speed (v) of 53 m s^{-1} (10 500 f.p.m.) and a traverse speed (v) of 40 mm s^{-1} (7.9 f.p.m.) under constant conditioning load (F_Q) of 13–35 kg (29–77 lb). The machine was an altered cylindrical grinder. Workpieces were 300 mm (11.8 in) long and 6 mm (0.24 in) thick. The three work materials tested were:

AISI 1055 steel, $H_v = 220$
304 stainless steel, $H_v = 215$
430 stainless steel, $H_v = 195$

Typical results are shown in Fig. 7.5 for 304 stainless steel. Figure 7.5(a) shows the removal rate ($\dot{M}_w$) versus the load F_Q, where it is seen that all grits give about the same results. This corresponds to a threshold value at

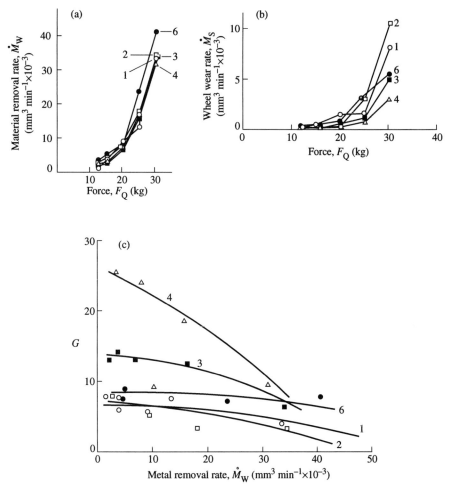

FIG. 7.5. Experimental results for several abrasive types in simulated conditioning of 304 stainless steel under operating conditions given in the text: (a) removal rate $\dot{M}_W$ versus load F_Q; (b) wheel wear rate $\dot{M}_S$ versus load F_Q; (c) grinding ratio G versus removal rate $\dot{M}_W$ (after Matsuo et al. 1980). The abrasives used were as follows: (1) fused regular Al_2O_3; (2) (1) after roasting; (3) 25 w/o ZrO_2-Al_2O_3 (4) 40 w/o ZrO_2-Al_2O_3 (5) mixed grit (1 + 4); (6) sintered Al_2O_3.

which removal begins at a load (F_Q) of about 10 kg (22 lb) and an exponential increase in removal rate with increase in load. Figure 7.5(b) shows the wheel wear rate ($\dot{M}_s$) versus the load F_Q, where an appreciable difference in wear rate was found for different grit types. The 40 w/o ZrO_2/Al_2O_3 grit gave the best wear performance, while the roasted and unroasted regular fused Al_2O_3 grit gave the poorest results. Figure 7.5(c) shows the grinding ratio (G)

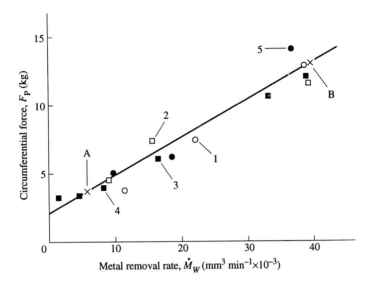

FIG. 7.6. Varation of the circumferential grinding force (F_P) with the removal rate ($\dot{M}_W$) for AISI 1055 steel. For cutting conditions see the text and for grit types see the caption to Fig. 7.5 (after Matsuo et al. 1980).

versus the metal removal rate for several grits grinding 304 stainless steel. The 40 w/o ZrO_2/Al_2O_3 gave a relatively high G value (~ 25) at a low removal rate, but this falls rapidly to values in the range of the other abrasive types ($G = 3$-5) with an increase in the removal rate.

Figure 7.6 shows the variation of the circumferential force (F_P) with the metal removal rate when snagging AISI 1055 steel. To a good approximation, all grits are seen to lie on the same linear curve.

The specific energy may be obtained as follows:

$$u = F_P V / \dot{M}_W. \tag{7.7}$$

The specific energy for points A and B on Fig. 7.6 may be found from this equation, with the following results:

$$u_A = 18.82 \text{ GPa} (2.73 \times 10^6 \text{ p.s.i.}),$$
$$u_B = 9.79 \text{ GPa} (1.42 \times 10^6 \text{ p.s.i.}).$$

This corresponds to about a 2:1 decrease in u for an 8:1 increase in removal rate, which is a manifestation of the size effect normally observed in cutting and grinding.

In general, it is found that the specific energy increases with decrease in undeformed chip thickness (t) as follows:

$$u \sim 1/t^n, \tag{7.12}$$

where n is an exponent that depends upon the range of chip size as follows:

- For cutting, $n \simeq 0.2$
- for SRG, $n \simeq 0.3$
- for FFG, n approaches 1.0

The value of n for the data of Fig. 7.6 may be estimated as follows. To a good approximation for constant wheel speed, grit size, wheel type, and wheel diameter, referring to eqn (4.13)

$$t \sim (\dot{M})^{0.5}.$$

Therefore, for the present example,

$$t_B/t_A \simeq (8)^{0.5}$$

and from eqn (7.12)

$$u_B/u_A \simeq 1/(8^{0.5n}) = 9.79/18.82 = 0.52$$

From this, n is found to be 0.314, which is a value to be expected for an SRG operation such as the one under consideration.

Matsuo *et al.* (1980) have also presented values of surface roughness, but these are only of academic interest, since in practice the predominant geometrical surface defect is the scallop shown in Fig. 7.3, having a depth d_e. The surface roughness measured with a tracer instrument applied to the specimen of Fig. 7.4 will be a small fraction of d_e.

In summary, the steel conditioning study of Matsuo *et al.* (1980) indicates little difference in removal rate due to grit type of the same size but an appreciable difference in wear rate. The 40 w/o ZrO_2/Al_2O_3 grit gives the best result. The specific grinding energy and size effect exponent derived from Fig. 7.6 are in excellent agreement with other SRG investigations, and may be used to estimate forces and power, and how these vary with removal rate for this and other SRG operations.

High-speed – Constant-power Conditioning

Conditioning wheel speeds have increased to 16 000 f.p.m. (81 m^{-1}) and higher-power, constant-horsepower machines have been introduced. Wheel feeds have increased with the advent of stationary wheel heads and cabs and moving work tables. More efficient grinding wheels also have been developed. Together, these changes have had a strong effect on the metal removal and wear characteristics of the conditioning process.

G versus $\dot{M}$ curves (Fig. 7.7) become flat instead of having a distinct minimum wear point (maximum G) with varying metal removal rates for high-speed constant power conditioning. This simplifies the cost analysis considerably, leading to the following value for cost optimum cross-feed (Brecker *et al.* 1972):

$$l'^* = \sqrt[3]{\frac{3D'(x/y)}{v_m(\dot{z}/y + 1/G)}}. \tag{7.13}$$

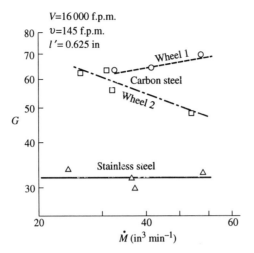

FIG. 7.7. The variation of G with $\dot{M}$ for constant-power conditioning of carbon steel and stainless steel at $V = 16\,000$ f.p.m. (81 m s^{-1}) (after Brecker *et al.* 1972).

The grinding ratio G is relatively high for these test conditions and has little influence on l'^* for high-cost material such as stainless steel. A striking difference between this result and that for lower-speed constant wheel-load conditioning is that the optimum cross-feed is unaffected by the depth of the layer removed. The useful depth removed, δ, is important only with respect to the removal rate $\dot{M}$ and the required power. The power level available limits the possible removal rate $\dot{M}$.

For defective depths δ which require more power than is available at the cost optimum cross-feed l'^*, it is better to reduce the cross-feed rate than to take multiple passes over the work. It was found that conditioning costs are less for machines with a large wheel diameter, a high table speed, high wheel speeds, and constant-power operation (Brecker *et al.* 1972).

The Future

The segmental grinding wheel design (Fig. 6.10) with thicker segments offers promise for the future development of more cost-effective high-power, large wheel diameter, conditioning machines. The advantages of the segmental design for wheels of large diameter include the following:

- greater safety at high operating speeds
- less scrap in wheel manufacture
- ease of replacing grinding elements (less down time)
- easier shipment of abrasive elements than for large monolithic wheels, with less shipping damage
- greater potential for automation of wheel manufacture

- better possibility of employing a fluid should this prove advantageous (the use of a fluid should be studied relative to the possibility of providing less oxidized swarth with a greater value as a by-product)
- possible use of alternate hard and soft segments to provide controlled intermittent cutting — this should give a cooler cutting action, which may become important at the very high speeds required for high removal rates
- the possibility of using vitrified bonded segments that are retained in place primarily by compressive stresses rather than tensile stresses, as with conventional monolithic resin-bonded wheels

The segmental wheel design provides the possibility of removal rates that are an order of magnitude greater than at present. To do this effectively will require a considerable increase in horsepower and machine stiffness. Higher head loads and higher table speeds will also be required to assure use of this concept under cost optimum operating conditions. Constant power grinding appears to be most effective.

Another possibility is condition grinding, with the wheel oriented perpendicular to the conventional direction. At first thought, this idea may be objected to because of the greater difficulty of detecting seams at final inspection. However, the availability of very effective inspection tools (magnetic particle and eddy current NDT techniques) appear to make this objection obsolete. The greatly improved surface finish possible and the elimination of extra material removal with the changed wheel orientation are two of its most attractive possibilities.

It appears to be feasible to conjecture that a future conditioning machine might have the following specifications:

>Wheel size: $48 \times 5 \times 24$ in ($1200 \times 125 \times 600$ mm), with 12–24 segments
>Power: 500 hp (370 kW)
>Table speed: 200 f.p.m. ($1.2 \, \text{m s}^{-1}$)
>Head pressure: 3000 lb (13 300 N)
>Wheel orientation: perpendicular to traverse direction
>Removal rate: $3000 \, \text{lb hr}^{-1}$ ($1360 \, \text{kg h}^{-1}$) for cold work; $4000 \, \text{lb hr}^{-1}$ ($1818 \, \text{kg h}^{-1}$) for hot carbon steel

References

Brecker, J. N., Sauer, W. J., and Shaw, M. C. (1972). *Proc. Int. Grinding Conf.*, Carnegie Press, Pittsburgh, Pennsylvania, p. 562.

Farmer, D. and Shaw, M. C. (1967). *J. Engng Ind.* **89**, 514.

Matsuo, T., Ueda, N., Sonoda, S., and Oshima, E. (1980). *Proc. 4th Int. Conf. Prod. Engng*, Tokyo, p. 667.

8

VERTICAL SPINDLE SURFACE GRINDING

Introduction

Vertical spindle surface grinding (VSSG) is an abrasive machining operation for heavy stock removal grinding (SRG). It should be mentioned, however, that the process is also used for finishing parts to required tolerances. This can be achieved with the same set-up used for SRG by reducing the downfeed of the wheel and sparking-out. However, it is the SRG aspects of the operation that are discussed here.

A common characteristic of most SRG processes is that the grinding wheel is self-dressing and hence wheel wear is relatively high under properly adjusted operating conditions. For the abrasive cut-off operation (Chapter 6), the controlling parameter was found to be the mean chip size (Shaw *et al.* 1967) for the following two reasons. First, the newly produced chips must be accommodated in the surface pores of the wheel until they can be released at the end of the cut. If large chips are produced and not enough chip clearance is available in the wheel face, the chips will break out grits and thus produce the required clearance. Second, when cutting specimens with a small arc of cut at a high removal rate, the cross-sectional areas of the chips become very large and the grits are no longer able to withstand the high grinding forces. Similar relationships between chip size and wheel wear hold for vertical spindle surface grinding.

The goal of stock removal grinding is usually to remove the unnecessary material as fast or as cheaply as possible. Thus, minimum cost per unit volume of material removed is the desired goal.

Chip Geometry

The mean chip size depends on operating and wheel parameters as well as the workpiece configuration. The following analysis is for machines having a rotating work table and Fig. 8.1 shows schematically the kinematic arrangement.

The mean volume of the average chip produced in VSSG is given by

$$V_C = A_C l, \tag{8.1}$$

where A_C is the cross-sectional area of the chip and l is the length of the chip.

The chip volume can also be found by considering the metal removal rate, $\dot{M}$, and the number of chips produced per unit time, $\dot{C}$:

$$V_C = \dot{M}/\dot{C}. \tag{8.2}$$

The metal removal rate is given by

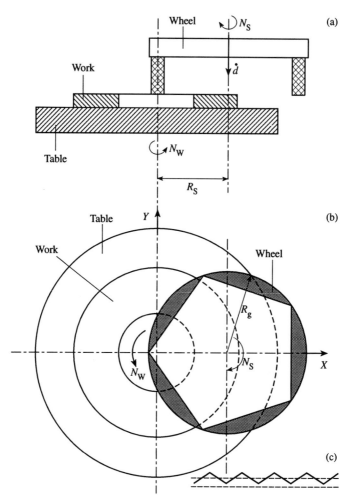

FIG. 8.1. A schematic of a vertical spindle surface grinding operation: (a) side view; (b) plan view; (c) criss-cross ground surface pattern.

$$\dot{M} = A_W \dot{d}', \tag{8.3}$$

where $\dot{d}'$ is the downfeed rate corrected for wheel wear, and A_W is the area of the work.

The number of chips produced per unit time may be found from

$$\dot{C} = N_S A_S C C', \tag{8.4}$$

where N_S is the wheel speed (r.p.m.), A_S is the wheel area, C is the number of active grits per unit area of wheel face, and C' is the number of chips per grit per revolution of the wheel.

The average cross-sectional area of a chip is found to be

$$A_C = \frac{\dot{M}}{\dot{C}} = \frac{A_w \dot{d}'}{N_S A_S CC' l} \tag{8.5}$$

The mean volume per chip is obtained by combining eqn (8.1) and (8.5):

$$V_C = (A_w \dot{d}')/(N_S A_S CC'). \tag{8.6}$$

In eqns (8.5) and (8.6), the chip length l and the number of chips per grit per wheel revolution (C') are two parameters influencing the volume per chip, which can be changed by varying the arrangement of workpieces on the table. A number of published papers provide experimental evidence that, by rearranging workpieces on the table, different grinding performance can be obtained. Tarasov (1962) used different workpiece arrangements consisting of six bars, each measuring $18 \times 3 \times 3$ in ($457 \times 76 \times 76$ mm), made from a class 40 gray cast iron. The grinding wheel consisted of 12 1.5 in (38 mm) wide segments (32A 24H 12V) held in a 26 in (660 mm) diameter chuck. The value of A_S was 83.4 in^2) (538 cm^2). Water with a rust inhibitor was the coolant and the machine employed was of 100 h.p. (74.6 kW). Tarasov reported grinding ratios which varied by as much as a factor of two depending on the configuration used (Fig. 8.2). In these experiments, the individual chip volume (V_C) correlated well with wheel wear parameter G (Sauer and Shaw 1973).

Very often in heavy stock removal vertical spindle grinding, the grinding wheel axis is tilted from the vertical so that grinding takes place only as the work enters under the wheel. In this case a unidirectional scratch pattern is produced and the undeformed chip length equals the scratch length. However, if the wheel surface is parallel to the table surface, a distinct criss-cross pattern may be obtained. Such a pattern is produced when the wheel is grinding at both front and rear, and when the average scratch depth is greater than the downfeed per half wheel revolution. In this case chips are formed at an angle of approximately 90° and two sets of scratches should be evident (Fig. 8.1(c)). The lengths of individual chips are then quite short and can move to the side, making it possible to accommodate a larger volume of chips in a given space in the surface of the wheel than when long continuous chips are involved. When the undeformed chip length is large, as it usually is in VSSG, and only one set of scratches is evident in the finished surface, this indicates that the wheel and work surfaces are not parallel, that the surface roughness is too low, or that the downfeed is too great. VSSG under such conditions will usually lead to high temperatures and wear rates (low G, values).

Figure 8.3 is a plan view of a VSSG wheel grinding without tilt and cutting a ring of work material located on the rotating work table. Here the work is ground twice, first at A and then at B. The two sets of scratches produced will intersect at approximately 90°. Three arcs are shown, where $\alpha_1 < \alpha_2 < \alpha_3$:

α_1 = cutting arc for the work,
α_2 = noncutting arc for the wheel,
α_3 = noncutting arc for the work.

When VSSG is performed with a criss-cross pattern, the total undeformed

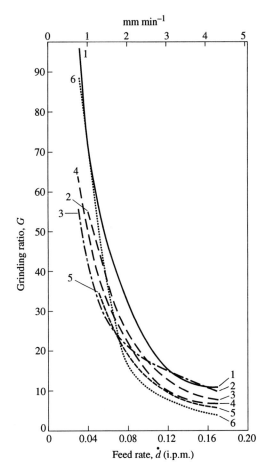

FIG. 8.2. The influence of the workpiece arrangement on the grinding ratio G versus the downfeed rate $\dot{d}$ (after Tarasov 1962). 1, Concentric squares; 2, open square; 3, parallel bars; 4, compact square; 5, star; 6, ring.

chip length is not important relative to the grinding ratio but only the chip cross-sectional area A_C (Sauer and Shaw 1973). Thus, since the Tarasov experiments described above were conducted with a criss-cross surface pattern, the grinding ratio G should correlate well not only with the volume per chip but also with the chip cross-sectional area (A_C), and this is found to be the case.

A convenient measure of when a criss-cross pattern is obtained is the roughness ratio:

$$\lambda = \frac{\text{mean peak-to-valley roughness of single cut surface}}{\text{downfeed between cuts}}.$$

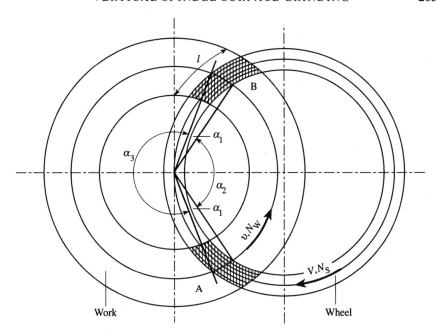

FIG. 8.3. A schematic view of a VSSG wheel and workpiece arrangement with two cutting zones.

When $\lambda < 1$ a continuous chip is to be expected, but when $\lambda > 1$ a criss-cross pattern is obtained. When G is plotted against A_C, a single curve is obtained independent of chip length when a criss-cross pattern pertains. However, G is a function of A_C and l when there is no criss-cross pattern. Wheels having fewer coarse grits give a rougher surface and promote a criss-cross pattern. Higher wheel speeds tend to give too smooth a finish for a criss-cross pattern to be produced. These are important reasons why VSSG is normally performed with coarse grit wheels and at a relatively low wheel speed (4000 f.p.m. (20 m s^{-1}) or less).

Cost Optimization

The cost of removing a layer of depth t will be

$$¢/t' = x/\dot{d} + A_W y/G + A_W x/G A_S \dot{d} \qquad (8.7)$$
$$\text{I} \text{II} \text{II}$$

where x is the cost of the machine, operator and overheads per minute, $\dot{d}$ is the downfeed rate (i.p.m. or cm min^{-1}), A_W is the area of work, A_S is the wheel area, y is the mean cost per unit usable volume of wheel, and G is the nondimensional grinding ratio.

The three items on the right-hand side of eqn (8.7) represent:

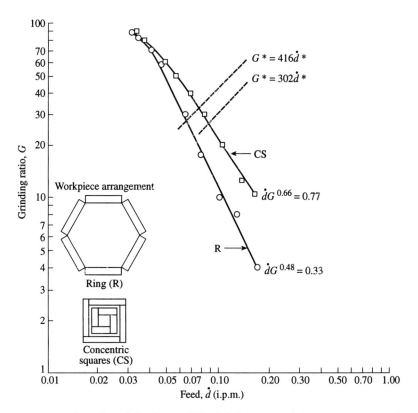

FIG. 8.4. A log–log plot of the data of Fig. 8.2 for curves 1 (concentric squares) and 6 (rings). Work area, $A_W = 324$ in^2 (2090 cm^2); wheel area, $A_S = 83.4$ in^2 (528 cm^2) (after Shaw and Farmer 1965).

    I = machine labor and operator cost,
    II = abrasive wheel cost,
    III = machine labor and operator cost to downfeed the additional amount required by wheel wear.

The third term III, originally introduced by Tarasov (1962), is usually small compared with the other terms and therefore is ignored in the following analysis.

Figure 8.4 shows curves for the ring and concentric squares workpiece arrangements of Fig. 8.2 replotted on log–log co-ordinates. Except for low values of downfeed, the data are seen to be well approximated by a straight line corresponding to:

$$\dot{d}G^n = C, \tag{8.8}$$

where n and C are constants for a given wheel–workpiece–machine combination.

The minimum value of ¢/t' may be found by substituting G from eqn (8.8) into eqn (8.7), differentiating the result with respect to $\dot{d}$ and setting the result equal to zero (Shaw and Farmer 1965). When this is done, the following result is obtained, where the (*) designates cost optimum values:

$$G^* = (A_W y / nx) \dot{d}^*. \tag{8.9}$$

For the R and CS workpiece arrangements of Fig 8.4, the following relations are found, and these are shown plotted in Fig. 8.4 as dashed lines. The values of x and y employed were 16.7 ¢ min^{-1} and 10.3 ¢ in^{-3} (0.63 ¢ cm^{-3}) respectively. For English (Imperial) units:

$$\text{for CS,} \quad G^* = 416 \dot{d}^*; \quad \text{for R,} \quad G^* = 302 \dot{d}^*.$$

The values of n required in evaluating eqn (8.9) in the two cases are obtained from Fig. 8.4 (the inverse slope of the G versus $\dot{d}$ curves; for CS, $n = 0.66$, and for R, $n = 0.48$). The values of G^* and $\dot{d}^*$ correspond to the intersection of the G versus $\dot{d}$ and G^* versus $\dot{d}^*$ curves. For the CS configuration, the values of G^* and $\dot{d}^*$ (print A in Fig. 8.4) are

$$\dot{d}^* = 0.089 \, \text{i.p.m.} \, (0.226 \, \text{cm min}^{-1}), \quad G^* = 27.$$

The optimum (minimum) cost per depth removed may be found by substituting values of $\dot{d}^*$ and G^* into eqn (8.7), ignoring the small term III:

$$(¢/t'')^* = x/\dot{d}^* + (A_W y)/G^*. \tag{8.10}$$

When the values of $\dot{d}^*$ and G^* for the CS configuration are substituted into eqn (8.10), the following is obtained:

$$(¢/t')^* = 188 + 144 = 312 \, ¢ \, \text{in}^{-1} \, (123 \, ¢ \, \text{cm}^{-1}).$$

The value assumed for x equal to 16.7 ¢ min^{-1}, is on the low side for a 100 h.p. (74.6 kW) machine. Assuming a value for x that is twice as large (33.4 ¢ min^{-1}) the following values are obtained for the CS case:

$$\dot{d}^* = 0.117 \, \text{i.p.m.} \, (0.23 \, \text{cm min}^{-1}), \quad G^* = 17.7, \quad (¢/t')^* = 475 \, ¢ \, \text{in}^{-1} \\ (187 \, ¢ \, \text{cm}^{-1}).$$

In Shaw and Farmer (1965) a simplified approach to the foregoing calculations employing special charts is presented that should enable an operator on the shop floor quickly to determine cost optimum values for any VSSG operation.

The foregoing procedure may be used to answer a number of important questions. For example, the amount one is justified in paying for a better performing grinding wheel may be determined. To a good approximation, the justified increase in wheel cost is found to be proportional to the resulting improvement in the G^* value (Shaw and Farmer 1965). Thus, if the value of G^* that results from the use of an improved grinding wheel doubles, approximately twice the value of y is justified.

The two most important parameters from the economic point of view are the grinding ratio G and the metal removal rate $\dot{M}$. For low cost, both should be as large as possible. To obtain a large G, the chip cross-sectional area A_C should be small. For a constant metal removal rate $\dot{M}$, the chip cross-sectional area can be reduced by increasing the wheel area, the wheel speed, and/or the number of scratches per revolution per grit, C'. Increases in C' can be achieved by placing more workpieces on the table and reducing downfeed $\dot{d}$ to keep the removal rate constant.

Either removal rate $\dot{M}$ or downfeed rate $\dot{d}$ may be used as the independent variable, since they are proportional to each other (Farmer and Shaw 1967). When $\dot{M}$ is used, eqn (8.8) becomes

$$\dot{M}G^n = C', \qquad (8.11)$$

where n has the same value as in eqn (8.8), but C' is different from C.

Optimum Work Area

Varying the position of the pieces on the table can increase the total scratch length. There are two areas on the table where no scratch pattern is produced: these are the center and the very outside rim of the table. It is especially important not to place workpieces directly over the center of the table, since not only will there then be no scratch pattern but the average scratch length will be long. The number of pieces which can be placed on the table is limited only by the power available, the rigidity of the machine, and the size of the table.

When the work area consists of a number of small parts, a cost optimum work area exists. This will be illustrated by a series of test results performed under the following conditions:

Work: AISI 1020 steel pieces measuring 7×3.94 in (17.8×10 cm)
Wheel: five 8A 30G 12V segments with total area $A_S = 93$ in^2 (600 cm^2)
Power: 40 h.p. (29.8 kW)
Work chuck: 22 in (559 mm) diameter, magnetic
Fluid: water-based
Total depth removed: 0.1 in (2.5 mm)
Workpiece configurations: Fig. 8.5 for 2-16 pieces

Figure 8.6 shows G versus $\dot{M}'$ curves for the four workpiece arrangements tested. Figure 8.7 shows values of C' and n (eqn 8.11) versus the number of parts. The extrapolated curves shown dotted in Fig. 8.7 were used to obtain the extrapolated curves of Fig. 8.6.

The cost optimum points are seen to lie on a smooth curve (Fig. 8.6), with values of G^* ranging from 4.8 to 6.3. Values of $\dot{d}^*$ and ¢* per part are given in Table 8.1.

In making these calculations, x was set at 16.7 ¢ min^{-1} and 10.3 ¢ in^{-3} (0.63 ¢ cm^{-3}). From Table 8.1, the cost optimum number of parts is seen to be about the same, from 16 to 28 parts. Also from Table 8.1, the cost optimum downfeed rate $\dot{d}^*$ is seen to decrease appreciably, from 2 to 16 parts, and then remain about constant.

VERTICAL SPINDLE SURFACE GRINDING 209

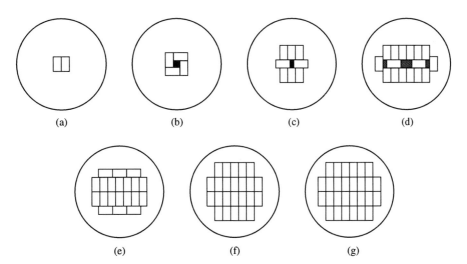

FIG. 8.5. The workpiece loading configuration used in the area study. Cases e, f and g were not tested but are included to show how they would have been loaded had they been included (after Farmer and Shaw 967).

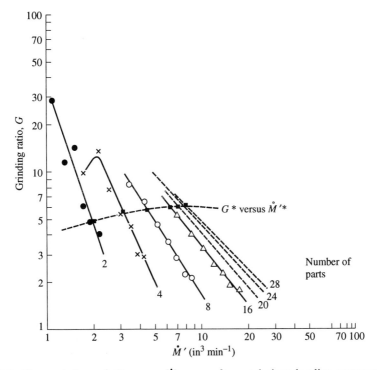

FIG. 8.6. The variation of G versus $\dot{M}$ curves for workpiece loading arrangements shown in Fig. 8.5. The dotted curves were obtained by extrapolation (after Farmer and Shaw 1967).

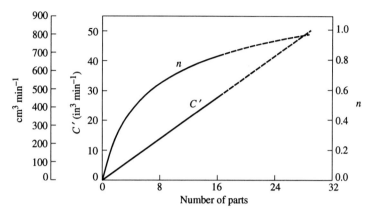

FIG. 8.7. The variation of n and C' in eqn (8.11) with the number of parts ground (after Farmer and Shaw 1967).

TABLE 8.1 *Values of $\dot{d}$ and ¢* for different work loads in numbers of parts in the example of Figs 8.4 and 8.5 (after Farmer and Shaw 1967)*

No. of parts	$\dot{d}^*$		¢* per part
	i.p.m.	cm min^{-1}	
2	0.036	0.091	318
4	0.028	0.071	231
8	0.021	0.053	193
16	0.015	0.038	178
20	0.013	0.033	175
24	0.012	0.030	176
28	0.001	0.028	178

This analysis has not taken into account a difference in work handling cost per part with the number of parts ground, which will favor large chuck loads and make it even more advantageous to grind a large number of parts at one time.

This example indicates that the size of the chuck is about right for the size of the wheel when grinding AISI 1020 steel. The ratio of the optimum work area ($16 \times 7 \times 3.94 = 441$ in^2, or 2847 cm^2) to the wheel area (93 in^2 = 600 cm^2) is nearly five. The optimum removal rate (M_W^*) was found to be about 6.62 in^3 min^{-1} (2708 cm^2 min^{-1}). Since the energy per unit volume for AISI 1020 steel in the abrasive machining regime is experimentally found to be about 2×10^6 in lb in^{-3} (13.8 J mm^{-3}), the horsepower required at the wheel surface when grinding at the cost optimum rate ($\dot{M}_W^*$)) should be about

$$(6.62)(2 \times 10^6)/(12 \times 33\,000) = 33.4 \text{ h.p. } (25 \text{ kW}).$$

This is seen to be generally consistent with the 40 h.p. (29.8 kW) motor with which the machine was provided.

The paradoxical observation may be made that VSSG machines designed for use on gray iron should be provided with from two or three times the horsepower of comparable machines designed for use on steel. The explanation for this paradox lies in the fact that although the grinding energy per unit volume for the cast iron will only be about half of that required for steel, the cost optimum downfeed rate d^* will be much higher than that for steel. In addition, machines designed for grinding gray cast iron should have a downfeed capacity about five times that of machines for use in grinding mild steel (Farmer and Shaw 1967).

Wheel Wear

The main causes of wheel wear in VSSG are:

- chip crowding
- high grinding force per grit

The first of these is less likely to be predominant when grinding with a criss-cross pattern. The grinding force per grit (F'_P) is directly related to the chip cross-sectional area (A_C) and the specific energy (u):

$$F'_P = A_C u. \qquad (8.12)$$

Grits of different size wear in different ways. Small grit wheels have relatively low bond strength but relatively high fracture strength. Therefore, they tend to wear primarily by loss of whole grits. Large grit wheels have relatively high bond strength but relatively low fracture strength and tend to wear primarily by microchipping. Both actions tend to result in a self-sharpening action. Paradoxically, wear debris from large grit grinding is smaller than with fine grit grinding. SRG is best carried out with relatively large grits, not only because there is then more chip storage space between active grits but also because there is then less whole grit loss under the relatively high values of F'_P pertaining.

When VSSG is performed under high stock removal conditions with a criss-cross pattern, the chip area (A_C) is the most important wear rate parameter. Wheel wear occurs because the force necessary to form a chip is greater than the strength of either bond or grit as a wear flat develops. If the wheel is tilted or the table speed is small, the scratch pattern will be unidirectional and the chip then equals the complete scratch length. Wheel wear is then primarily due to insufficient chip storage space between active grits. It is advantageous to perform VSSG operations without wheel tilt and at a high table speed in order to obtain a criss-cross pattern. This is particularly important when grinding large compact areas. Also, the grinding force per grit will than be smaller since material is removed at two locations. This allows the use of soft wheels with large grits, resulting in lower specific energy without lowering the grinding ratio at high removal rates.

The hardness of the wheel segments should be such that the bond forces are slightly higher than the forces necessary to fracture the grits. Segments which are too hard give rise to an increase in specific energy due to rubbing of excess bond material against the work and because the excess bond material also fills the pores required to accommodate chips. If the segments are too soft, bond fracture occurs before grit fracture and wheel wear will increase dramatically. Abrasives for VSSG should be friable and fracture into small particles to give sharp edges.

In the literature it has been suggested that under certain conditions loose grits may become trapped between the wheel and the work, and then form more chips in the trapped position. Pollock (1966) calls these loose particles 'secondary grits' and shows that they may account for a sizeable percentage of the produced chips. By contrast, Farmer (1968) did not observe any secondary grits and the wear areas observed on the segments were different in the two investigations. The major difference between the two investigations appears to be in grit size. Pollock used 36 grit wheels, which will tend to wear by bond failure and produce large wear particles, which can then become trapped under the wheel. Farmer used 24 grit wheels, which will wear primarily by grit microfracture. Farmer's particles will therefore be smaller than the grits and should not become trapped as easily in the interface between wheel and workpiece.

There is relatively little variation in roughness of a ground SRG surface with change in grit size. This is because the number of active grits is about the same, because more whole grits are removed from a fine grit wheel than from a coarse grit wheel under SRG conditions (Sauer and Shaw 1973). However, this is not an important difference, since in all SRG operations surface finish is usually not important.

References

Farmer, D. A. and Shaw, M. C. (1967). *ASTM Conf. Paper MR67-595.*
Farmer, D. A. (1968). Private communication.
Pollock, C. (1966). *J. Engng Ind.* **88.**
Sauer, W. J. and Shaw, M. C. (1973). *Proc. 3rd. NAMRC*, p. 131.
Shaw, M. C. and Farmer, D. A. (1965). *J. Engng Ind.* **87**, 349.
Shaw, M. C., Farmer, D. A., and Nakayama, K. (1967). *Trans. ASME* **89B**, 495.
Tarasov, L. P. (1962). *Am. Machinist*, Apr. 16, 25.

9

GRINDING TEMPERATURES

Introduction

As previously discussed, it is convenient to distinguish two regimes of grinding:

(1) Form and finish grinding (FFG), in which the undeformed chip thickness t is small, the specific energy and the grinding ratio G are relatively high, and periodic dressing is required.

(2) Stock removal grinding (SRG), in which the undeformed chip thickness t is large, the specific energy and the grinding ratio G are relatively low, and periodic dressing is unnecessary.

Similar subdivisions have been useful in other areas of engineering, such as between:

- laminar and turbulent behavior in fluid flow
- elastic ($\varepsilon < 10^{-3}$), plastic ($\varepsilon < 1$), and large-strain ($\varepsilon > 1$) deformation in materials processing
- cold working ($\sigma_f \sim \varepsilon^n$ for $\theta_H < 0.5$) and hot working ($\sigma_f \sim \varepsilon^m$ for $\theta_H > 0.5$) in deformation processing
- conduction, convection, and radiation transport in heat transfer
- forced and self-excited vibrations in dynamics
- hydrodynamic and boundary lubrication in tribology
- ductile and brittle failure in fracture mechanics
- solid and liquid phase sintering in powder metallurgy

In each of these subdivisions there are dominant models or parameters that should be considered, while other less important ones may be ignored in the interest of simplification.

High surface temperatures are involved in FFG operations, which sometimes lead to surface damage (overtempering, structural transformations, oxidation and other degradation, unwanted residual stresses, thermal cracks, and excessive rates of abrasive wear). Since it is relatively difficult accurately to measure grinding temperatures and to obtain a broad appreciation of the roles that material and operating variables play from experimental studies, an analytical approach will be taken first. This will be followed by a discussion of experimental techniques.

While surface temperatures in SRG are relatively unimportant, these will also be discussed mainly to indicate why surface temperatures are not as important for this type of grinding as in FFG.

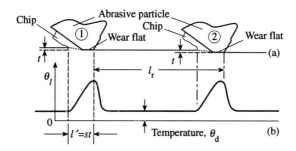

FIG. 9.1. (a) Successive active grits cutting the same groove. (b) Corresponding localized temperature profiles.

Form and Finish Grinding Temperatures

Introduction

It is important to distinguish two types of grinding temperatures:

(1) a localized temperature (θ_l) due to the action of a single, isolated abrasive particle;
(2) a temperature due to the collective action of all of the particles operating on the wheel–work contact area (θ_d).

The distributed temperature θ_d is the one of importance for friction sliders and in fine grinding where the temperature involved is that corresponding to a uniformly distributed energy source, even though in fact energy is being dissipated at a discrete number of points constituting the real area of contact, as opposed to the apparent area of contact. Temperature θ_l corresponds to that obtaining at a single isolated asperity.

Figure 9.1 shows one isolated abrasive particle followed by the next particle that will cut in the same groove. The distance between these successive grits (l_r) is (Kumar and Shaw 1979):

$$Cb'l_r = 1, \qquad (9.1)$$

where b' is the width plowed out by a single abrasive particle and C is the number of active grits in the wheel surface per unit area.

Figure 9.1 shows the grits operating with an undeformed chip thickness (depth of cut) equal to t. It is assumed that the total length over which energy is being dissipated (including the chip-forming region and the rubbing region associated with the formation of wear flats) will be proportional to t (i.e. this distance $l' = st$, where s depends on the effective rake angle of the abrasive particle and its state of wear). The area over which energy is dissipated will therefore be $b'st$.

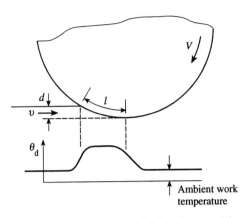

FIG. 9.2. The temperature profile for several abrasive particles dispersed over the wheel–work contact area.

Since the bulk of the workpiece will be at a relatively low temperature, heat transfer into the work will be extremely rapid in the case of a metal. Successive grits cutting the same groove will be sufficiently far apart so that the surface temperature will fall to temperature θ_d between cuts. Temperature θ_l will be a maximum at the trailing edge of the contact zone, and the time at which a given point in the surface will be at the mean temperature $(\bar{\theta}_l)$ will be

$$T_1 = st/V. \quad (9.2)$$

The thermal flux into the workpiece (energy (time)$^{-1}$ (area)$^{-1}$) will be

$$q_1 = (RuVb't)/(b'st) = RuV/s, \quad (9.3)$$

where u is the specific grinding energy, V is the wheel speed, b' is the mean width of scratch, t is the undeformed chip thickness, and R is the fraction of the total energy going to the workpiece.

Figure 9.2 shows the situation for energy dispersed uniformly over the wheel–work contact. The area of contact will be

$$A_d = lb \simeq (Dd)^{0.5}b, \quad (9.4)$$

where l is the length of wheel–work contact and b is the total width of cut.

The time at which a given point on the workpiece at a level d below the original surface will be at the mean temperature $(\bar{\theta}_d)$ will be

$$T_d = l/v \simeq (Dd)^{0.5}/v, \quad (9.5)$$

where v is the work speed.

The thermal flux for this case will be

$$q_d = (Ruvbd)/(lb) = (Ruvd)/l. \quad (9.6)$$

When assessing the consequence of a high temperature, it is important to know how long it must act to be significant. If time at temperature is too short, such as when there is an induction period in a chemical reaction, or the resulting action involves atomic diffusion then what equilibrium experience suggests should happen will not occur. It is well known that organo-metallic reactants have to be at temperature for some time before a reaction starts (induction period) but, once the reaction begins, it accelerates to completion (autocatalytic behavior). Examples of this type of action for potentially useful cutting fluid additives are given in Shaw (1984a, Chapter 13). In considering reaction kinetics it is important to distinguish between the rate of diffusion of vibrational amplitude associated with a temperature rise and the rate of atomic diffusion, which is several orders of magnitude slower. Thus, it is possible for fleeting temperatures well beyond the equilibrium melting temperature to occur without the structural change involved in melting taking place, since this involves substantial atomic diffusion.

Equations (9.2) and (9.5) may be used to estimate the time at temperature for the localized temperature θ_l and for the distributed temperature θ_d. When this is done the time associated with θ_l is about a microsecond while the time associated with θ_d is a couple of orders of magnitude larger. Since the time associated with θ_l is so very short and atomic diffusion is involved in the consequences of the temperature of interest here, the mean value of θ_d is considered to be of major importance. In the following discussion, the symbol θ will be used, but $\bar{\theta}$ is the temperature being considered.

Calculation of FFG temperatures is a very complex problem, which is made particularly difficult by a number of assumptions that must be made which can be postulated only very approximately. Estimation of the real area of contact between abrasive grits and work material in the grinding zone is but one of these. It has been determined experimentally (Hahn and Lindsay 1967; and Malkin 1968) that this real area of contact is about 2 percent of the apparent area of contact when wear flats on active abrasive grits become sufficiently large to require dressing. The heat transfer between wheel and work is directly dependent on this ratio. On the average, we might assume that this ratio is about 1 percent, but it is obvious that, based on this assumption alone, we should expect estimates of *absolute* FFG grinding temperatures to be very approximate.

A more practical approach is to seek a solution for the *relative* temperature rise pertaining in FFG: that is, to obtain a relation that is proportional to the temperature rise which may be used to determine which variables to change and by how much in order to lower a temperature that is obviously excessive. Most past approaches to grinding temperatures have involved making the necessary assumptions and then proceeding to carry out more and more sophisticated heat transfer calculations that are usually far more elaborate than are justified by the assumptions. A number of such papers are cited in the references and marked by asterisks, since they will not be discussed here. In what follows the objective is to employ the simplest analytical approach possible, so that physical reality is not obscured by analytical complexity.

TABLE 9.1 *Values of the error function* (erf)

Ψ	erf Ψ
0	0
0.2	0.223
0.4	0.428
0.6	0.604
0.8	0.742
1.0	0.843
1.5	0.966
2.0	0.995
∞	1.000

Conductive Heat Transfer

In conductive heat transfer, two very important applications are:

(1) the nonsteady state problem of determining the temperature a distance y below the surface a certain time T after the surface is suddenly brought to a temperature that is maintained constant;

(2) the steady state problem of determining the temperature of the interface between a slider and an extensive surface moving at constant velocity with a constant energy flux q.

A solution to the first of these is to be found in every elementary text on heat transfer, and will not be covered here except to give the well known result. If a surface is suddenly brought to a temperature θ_s at time $T = 0$, and held constant at this value, then the temperature rise a distance y below the surface will have a value θ_y at time T, and

$$\frac{\theta_y}{\theta_s} = 1 - \mathrm{erf}\left(y \Big/ 2\sqrt{\frac{k}{\rho C}T}\right) \qquad (9.7)$$

where k is the coefficient of thermal conductivity, ρC is the volume specific heat, and erf is the error function (see Table 9.1 for a few values).

The above solution assumes that all of the thermal energy flows in the y direction and that there is no flow of heat in the x or z directions. Both k and ρC are involved in this problem, but they appear only as the ratio $k/\rho C$, because of its importance, is called the coefficient of thermal diffusivity and is usually represented by $\alpha = k/\rho C$.

A simplified approach to the second (steady state) problem has been thoroughly discussed by Jaeger (1942) and this is abstracted in the following section.

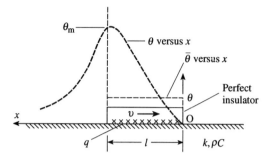

FIG. 9.3. A two-dimensional slider (perfect insulator) moving over a semi-infinite body with velocity V (after Jaeger 1942).

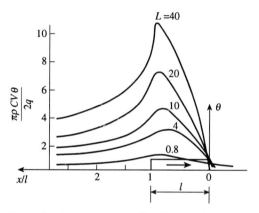

FIG. 9.4. The nondimensional temperature distribution for the slider of Fig. 9.3 for different Peclet numbers $L = (Vl\rho C)/k$ (after Jaeger 1942).

Jaeger

Figure 9.3 shows a two-dimensional perfect insulator of length l moving over a semi-infinite body with velocity V, a ratio of real to apparent area of contact $A_R/A = 1$, and with q units of thermal energy dissipated at the surface per unit area per unit time. The thermal properties of the extensive conducting member are k and ρc. The temperature rise goes from zero near the leading edge of the slider of length l to a maximum near the trailing edge. Figure 9.4 shows the variation of a nondimensional temperature quantity with distance x/l from the leading edge for different values of the nondimensional Peclet number L:

$$L = (Vl\rho C)/k = (Vl)/\alpha. \qquad (9.8)$$

Figure 9.4 is for a two-dimensional slider of infinite width, referred to as a band heat source by Jaeger. For values of $L > 20$, the temperature varies linearly to a good approximation along the slider and it may be shown that, to a good approximation, the mean surface temperature rise θ is as follows:

$$\theta = 0.754\, ql/\sqrt{Vl(k\rho C)}. \tag{9.9}$$

Since most applications of Jaeger's work to tribology, cutting, and grinding involve values of Peclet number greater than 20, this is a useful simplification.

It is important to note that the thermal quantity of importance is again a composite one of k and ρC, but this time it is the geometric mean of these two quantities (i.e. $(k\rho C)^{0.5}$) that is involved. This important quantity has no generally accepted name or symbol but has been called the coefficient of *heat diffusion* (Chvorinov 1940; Chalmers 1964), or the coefficient of heat activity (Luikov 1968) to distinguish it from α. As a matter of convenience, we shall refer to it as the coefficient of heat diffusion and use the symbol β, where

$$\beta = \sqrt{k\rho C}. \tag{9.10}$$

The difference between the two applications discussed above is of course one of boundary conditions. In the first application the boundary condition pertaining is one of constant surface temperature θ_s and the solution involves α as the sole composite thermal property. In the second application, the boundary condition is one of constant thermal flux q, and the solution for surface temperature involves β as the sole composite thermal property. The boundary condition leading to α is called the first kind, while that leading to β is called the second kind (Luikov 1968).

Before proceeding to apply eqn (9.9) to FFG it is considered useful to discuss two alternative approaches to this equation, in order to see more clearly why β is the thermal item of importance and also to present a greatly simplified solution.

Linear Jaeger Model

From Fig. 9.4, it is evident that if $L > 20$, then the temperature along the interface may be considered to vary linearly, as shown by the dotted curve in Fig. 9.5(a). Figure 9.5(b) shows the variation of temperature beneath the trailing edge with depth below the surface y. The solid curve is based on the nonsteady state solution involving the error function, while the dotted curve is the linear equivalent such that the areas associated with the solid and dotted curves are equal (i.e. for equal total thermal flux for the two cases). The temperature rise beneath the surface will vary with y in accordance with eqn (9.7), where T is the time of heat flow at the trailing edge and

$$T = l/V. \tag{9.11}$$

For the linear model shown in Fig. 9.5(c), the total heat flux (Q) may be expressed as heat conducted or heat absorbed, which are equal:

$$Q = qlb = k(bl)(\theta_m/\hat{y}) = (\rho C)\theta(b\hat{y}V), \tag{9.12}$$

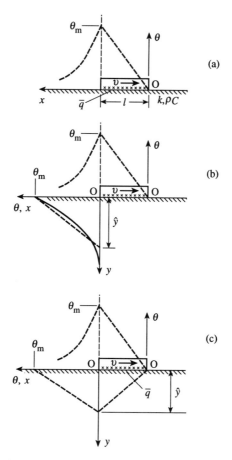

FIG. 9.5. A linearized model for the slider of Fig. 9.3 for $L > 20$ (after Shaw 1990a). (a) Variation in the surface temperature along the slider. (b) Variation of the temperature at the trailing edge beneath the surface: solid line, error function solution; dashed line, linear equivalent such that total heat flow penetration is the same for the solid and dashed line cases. (c) A composite, showing the linear temperature distribution and the depth of heat penetration along the slider.

where θ_m is the maximum surface temperature rise, $= 2\theta$, and $\theta_m/\hat{y}$ will be constant along l for the linear model. Solving for $\hat{y}$,

$$\hat{y} = \sqrt{2(l/V)(k\rho C)}. \tag{9.13}$$

Substituting this into eqn (9.12) gives

$$\theta = 0.707 \frac{ql}{\sqrt{Vl(k\rho C)}}. \tag{9.14}$$

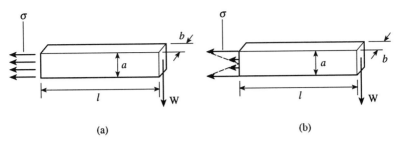

FIG. 9.6. (a) The free body diagram incorrectly assumed by Galileo. ($\sigma = 2wl/ba^2$)
(b) The correct free body diagram. ($\sigma = \sigma\,wl/ba^2$)

Comparing eqns (9.9) and (9.14), it is seen that they are the same except for the coefficients, which differ by only 6 percent.

The Galileo Principle

In his second book *Two New Science* (Galileo 1638), written as a dialog between a professor and two students, Galilee makes an error that suggests a useful concept. In discussing the strength of beams, Galileo assumed the uniform stress distribution shown in Fig. 9.6(a) rather than the correct one shown in Fig. 9.6(b). The solutions for the two cases are given in the figure caption. Here it is evident that the absolute values of the maximum stress (σ) are quite different, although the relative values ($\sigma \sim (Wl/[ba^2])$) are the same. Galileo solved a large number of problems based on this incorrect distribution of stress, but in each case obtained the correct solution. This is because in all cases he sought the strength of one configuration relative to another instead of absolute values. This suggests that in engineering, where we are often more concerned with relative values than with absolute ones, approximate models may be used with less loss in accuracy provided that absolute solutions are not insisted upon. This concept of seeking relative solutions rather than absolute ones has been referred to as the 'Galileo Principle' (Shaw 1992). The approximate solution of the Jaeger problem discussed in the previous section (Fig. 9.5) is an example of how a very approximate model may lead to a useful relative solution. Another example of how a correct result may be obtained with very little analytical effort, provided that a relative solution is considered acceptable, is considered in the next section.

Dimensional Analysis

Dimensional analysis is a powerful form of engineering analysis, based on the principle of dimensional homogeneity, which recognizes that each term in any physically correct equation must have the same dimensions. Details of this type of analysis may be found in Shaw (1984a).

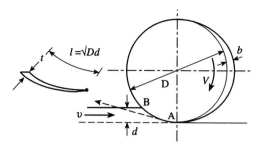

FIG. 9.7. The horizontal spindle surface grinding operation (up grinding), showing the undeformed chip in inset to the left.

For the Jaeger problem (Fig. 9.3), the following quantities are obviously of importance to the main dependent variable (the mean surface temperature, θ):

q = thermal flux per unit area per unit time $[FL^{-1}T^{-1}]$
V = sliding speed $[LT^{-1}]$
β = heat diffusivity of extensive member = $(k\rho c)^{0.5}$ $[FL^{-1}T^{-0.5}\theta^{-1}]$
l = wheel- work contact length $[L]$

Here F, L, T and θ (force, length, time, and temperature) are the set of fundamental dimensions employed.

Before dimensional analysis:

$$\theta = \psi(q, V, \beta, l), \qquad (9.15)$$

where ψ is some unspecified function.

Since, in this case, it takes a combination of all five variables to form a nondimensional group, this group must equal a nondimensional constant:

$$(\theta\beta V^{0.5})/q(l)^{0.5} = \text{nondimensional constant, } [F^0L^0T^0\theta^0]$$

or

$$\theta \approx (q/\beta)\sqrt{l/V}. \qquad (9.16)$$

This is consistent with eqns (9.9) and (9.14), but was derived with very much less effort. However, this is a relative solution that depends on the knowledge that for this problem the only thermal property of importance is β (heat conductivity).

FFG Temperatures

For horizontal spindle surface grinding (Fig. 9.7):

$$q \approx uvdb/lb. \qquad (9.17)$$

Substituting into eqn (9.16), assuming that all energy goes to the work,

$$\theta \approx uvd/\beta\sqrt{vl}. \qquad (9.18)$$

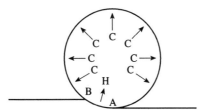

FIG. 9.8. Heating (H) versus cooling (C) of abrasive grits.

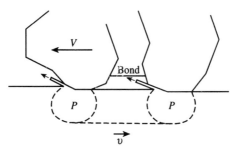

FIG. 9.9. Active abrasive grits with plastic deformation beneath their tips and chips being extruded from the workpiece surface (after Shaw 1990*b*).

The model of Jaeger (Fig. 9.3) that leads to eqn (9.18) must be altered to take care of several differences between Figs 9.3 and 9.7.

In fine grinding (Fig. 9.7) the wheel depth of cut (d) is very much smaller than the wheel diameter (D), and the length of wheel–work contact (l) is large relative to the undeformed chip thickness, but very small relative to the peripheral length of the wheel surface (πD). Since πD will normally be two or three orders of magnitude greater than l, it is found (Ramanath and Shaw, 1988) that all of the energy that enters the grits during grinding will be removed by the grinding fluid or by air during the nongrinding portion of a revolution (Fig. 9.8). This is important, since it enables the wheel as well as the work to be considered a semi-infinite body.

Most of the energy involved in fine grinding is not dissipated at the wheel–work interface but within the workpiece, as shown in Fig. 9.9. Also, the heat source is moving with a velocity v in this case. The fine grinding operation may be modeled as in Fig. 9.10(a), where the work is moving past the stationary heat source at a velocity v while the wheel is moving past the heat source with a much higher velocity V.

In a fine grinding operation only a small fraction of the wheel surface makes contact with the work, and hence a distinction must be made between the real (A_R) and apparent (A) areas of wheel–work contact in heat transfer calculations. For dry grinding, the effective rate of wheel–work contact area will be (A_R/A)Vb and hence, for dry grinding, Fig. 9.10(b) is a more appropriate model than Fig. 9.10(a).

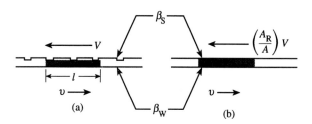

FIG. 9.10. A heat transfer model for fine horizontal spindle surface grinding: (a) with an interrupted wheel surface moving with velocity V; (b) with a corresponding surface moving with equivalent velocity $(A_R/A)V$ (after Shaw 1990b).

Assuming the temperature of the chips leaving the wheel to be the same as the wheel-work surface temperature (θ), the fraction of the total energy convected away by the chips will be approximately as follows:

$$R' = \theta(\rho C)_w/u. \tag{9.19}$$

This may be considered to be included in the fraction of the total energy going to the work (R). In any case, R' will be only about 0.05 for FFG and may be ignored.

An approximate value of R may be found by use of the technique of Blok (1937) in tribological temperature studies. This involves first assuming the wheel and then the work to be a perfect insulator and equating the results. The transfer of heat to the wheel (grits) depends upon the rate at which the grit contact area A_R moves along l, which corresponds to $V(A_R/A)$.

When the mean temperature rise θ is calculated from eqn (9.18), first as a point in the equivalent wheel surface of Fig. 9.10(b) and then as a point in the work surface, and the two are equated:

$$0.754(1-R)uvd \bigg/ \left(\beta_s \sqrt{V\frac{A_r}{A}l}\right) = 0.754 Ruvd/(\beta_w\sqrt{vl}) \tag{9.20}$$

and, solving for R,

$$R = \left(1 + \frac{\beta_s}{\beta_w}\sqrt{\frac{V}{v}\frac{A_R}{A}}\right)^{-1}. \tag{9.21}$$

Since A_R/A will approach zero for a freshly dressed wheel, but will be about 0.02 for a wheel about to be dressed (Hahn and Lindsay 1967; Malkin 1968), a reasonable mean value of A_R/A is about 0.01. On the other hand, V/v will be about 100 for most grinding operations. Therefore the value $(v/V)(A/A_R)$ in eqn (9.21) will be close to unity, and hence eqn (9.21) will be approximately as follows for dry grinding:

$$R = \left(1 + \frac{\beta_s}{\beta_w}\right)^{-1} \tag{9.22}$$

GRINDING TEMPERATURES 225

and the mean wheel-work interface temperature will be given by the following proportionality:

$$\theta \approx Ruvd/(\beta_w \sqrt{vl}). \tag{9.23}$$

While V and wheel topography as influenced by dressing, grit size, bond content, and concentration do not appear explicitly in eqn (9.23) they are indirectly included in terms of their influence on specific energy u due to their effect on undeformed chip thickness t and hence on u. The specific energy varies inversely with t^n, where n is close to 1 for FFG.

From Chapter 4 for FFG (see eqn 4.5),

$$\bar{t} = \sqrt{vd/VCrl}$$

and therefore if

$$u \approx u_0/\bar{t} \approx u_0 \sqrt{VCrl/vd}, \tag{9.24}$$

where u_0 is a constant for a given grit, work, and environment (fluid) with units $[FL^{-1}]$. Substituting eqn (9.24) into eqn (9.23):

$$\theta \approx (u_0 R \sqrt{VCrd})/\beta_w. \tag{9.25}$$

From eqn (9.22) it is found that

$$\frac{R}{\beta_w} \approx \left[\left(1 + \frac{\beta_s}{\beta_w}\right)\beta_w\right]^{-1} \approx \frac{1}{\beta_w + \beta_w}. \tag{9.26}$$

Thus, in FFG for dry grinding,

$$\theta \approx \frac{u_0 \sqrt{VCrd}}{\beta_s + \beta_w}. \tag{9.27}$$

From this it appears that the surface temperature in FFG increases with:

- an increase in $V^{0.5}$ and an increase in $d^{0.5}$
- a fine grit size or fine dress (large $C^{0.5}$)
- decreases in the values of β_s and β_w

Equation (9.27) is in good agreement with experiment.

It also appears that, to a first approximation, the mean surface temperature (θ) is independent of the work speed v and the wheel diameter D_e. However, it is not suggested that wheel *life* is independent of either v or d. It is found experimentally that wheel life does depend on v and d, and the reason for this is discussed at the end of Chapter 11.

In the foregoing analysis, it is assumed that the active grits extend beyond the bond a sufficient amount so that there is relatively little contact between bond and work and ample space between active grits to accommodate the chips without wheel loading. When the removal rate exceeds the chip storage capacity in the wheel face, a substantial rise in power and temperature results, leading to generally unsatisfactory surface integrity.

It turns out that the energy partition coefficient R is not only a very convenient quantity to use in converting the Jaeger relationship for a perfect insulator to a form applicable to FFG, but it also provides a check on the validity of the analysis. R may be readily determined experimentally by a simple calorimetric experiment. Such experiments have been made by several workers (Sato 1961; Malkin 1968; Ramanath 1986) and the results have been found to be in excellent agreement with values from eqn (9.22) for dry FFG.

A question often asked is why diamond (D) and cubic boron nitride (CBN) abrasives give much lower grinding temperatures than common abrasives (SiC and Al_2O_3). It is sometimes stated that this is because D and CBN have such a high coefficient of thermal conductivity (k) relative to common abrasives. It is also sometimes attributed to the fact that values of $\alpha = k/\rho C$ for D and CBN are so high. Both of these statements are incorrect. The important thermal property of the abrasive is actually the composite thermal property $\beta_s = [k\rho C]_s^{0.5}$ and not k_s or α_s.

Grinding with Coolant

In proper wet grinding, the space between active grits will be filled with liquid. Since A_R/A for the abrasive particles will be about 0.01, the value of A_R/A for the coolant will be $1 - 0.01 \simeq 1$ when the space between active grits is completely filled with liquid coolant. Even though water has a low value of $\beta = \beta_C$, it will increase the energy going to the wheel and hence decrease R and hence θ. Following steps similar to those employed for dry grinding and employing the linearized analysis (Fig. 9.5) and an energy approach, the following value of R is obtained for FFG with a coolant (subscript c):

$$R = \left(1 + \frac{\beta_s}{\beta_w}\sqrt{\frac{V}{v}\frac{A_R}{A}} + \frac{\beta_c}{\beta_w}\sqrt{\frac{V}{v}}\right)^{-1}. \qquad (9.28)$$

It is found that $\beta_s/\beta_w \simeq 0.10$ for a water/steel combination: so, if $V/v = 100$, then the third term in the denominator of eqn (9.24) will be about unity. Therefore, the approximate value of R for complete wet grinding of steel that is equivalent to eqn (9.22) for dry grinding again with $(V/v)(A_R/A \simeq 1)$ will be:

$$R \simeq \left(2 + \frac{\beta_s}{\beta_w}\right)^{-1} \qquad (9.29a)$$

and eqn (9.26) for wet grinding becomes

$$\frac{R}{\beta_w} \simeq \frac{1}{2\beta_w + \beta_s}. \qquad (9.29b)$$

Representative Applications

Approximate thermal properties are given in Table 9.2 for a variety of materials, and values of β ratios are given in Table 9.3 for several work/abrasive combinations, together with the corresponding values of R_{dry} and R_{water} obtained from eqn (9.22) and (9.29a). Here it is evident that substantially less heat goes to the work with CBN than with Al_2O_3 or SiC and that a complete layer of water over the wheel–work contact also decreases the fraction of total energy going to the wheel (R). Values of R_{dry} and R_{wet} for Al_2O_3 were found experimentally to be 75–85 percent and 25–35 percent, respectively, by Lee *et al.* (1972).

From Table 9.3 based on eqn (9.22) and (9.29a), values of R_{dry} and $R_{coolant}$ for Al_2O_3/steel are found to be 0.56 and 0.36 respectively, which are not in good agreement with the range of values of Lee *et al.* (1972) quoted above. If values of $V/v = 50$ and $A_R/A = 0.005$ (equally plausible values compared to the ones used to obtain eqns (9.22) and (9.29a)) had been used, the values of R_{dry} and $R_{coolant}$ obtained using eqn (9.21) and (9.29a) would be 0.71 and 0.47 respectively. These are in much better agreement with the experimental range of values of Lee *et al.* (1972). This illustrates the importance of the values substituted into an analytical solution when predicting performance. It appears to be unnecessary to derive an expression to appreciably greater accuracy than that to which the data substituted is known. The values of R_{dry} and $R_{coolant}$ given in Table 9.3 are not very accurate in an *absolute* sense but are useful in a *relative* sense. For example, the value of R_{dry} for Al_2O_3/St is about one-half of that for CBN/St, while $R_{coolant}$ for Al_2O_3/St is about two-thirds of that for R_{dry}. Also, $R_{coolant}/R_{dry}$ for CBN/St is about 0.8. By virtue of the Galileo Principle, these relative results should be substantially more accurate than the absolute values given in Table 9.3.

Example

The mean grinding temperature and power at the wheel are estimated below for a dry fine plunge surface grinding operation when using a 10 in (254 mm) diameter vitrified Al_2O_3 wheel. The grinding conditions are as follows:

Wheel speed, V: 30 m s^{-1} (5904 f.p.m.)
Work speed, v: 0.5 m s^{-1} (98.4 f.p.m.)
Wheel depth of cut, d: 10 μm (0.0004 in)
Width ground, b: 12.7 mm (0.500 in)
Work: AISI 1020 steel
Specific energy, u: $68\,950 \times 10^6$ J m^{-3} (10^7 p.s.i.)
$R = 0.8$
From Table 9.2, $\beta = 15.7 \times 10^{-3}$ J m^{-2} S$^{-0.5}$ C^{-1}
From eqn (4.10), $l = (Dd)^{0.5} = [(0.254)(10 \times 10^{-6})]^{0.5} = 0.00161$ m
From eqns (9.9) and (9.17),

TABLE 9.2 *Approximate thermal coefficients*

Material	Density (g cm^{-3})	Melting point (°C)	Thermal conductivity, k (W m^{-1} K^{-1})	Volume specific heat, ρc (J m^{-3} K^{-1})	$\beta = \sqrt{k\rho c}$ (J m^{-2} s$^{-0.5}$ K^{-1})	$\alpha = k/\rho c$ (m^2 s^{-1})
Cu	9.0	1082	390	3.5×10^6	36.8×10^3	111.4×10^{-6}
Al	2.7	660	220	2.4×10^6	23.0×10^3	91.7×10^{-6}
1020 st.	7.9	1520	70	3.5×10^6	15.7×10^3	20.0×10^{-6}
430 s.s.	7.8	1510	26	3.6×10^6	9.7×10^3	7.2×10^{-6}
303 s.s.	7.8	1420	16	3.9×10^6	7.9×10^3	4.1×10^{-6}
Ni	8.9	1453	92	3.9×10^6	18.9×10^3	23.6×10^{-6}
Ti	4.5	1668	15	2.3×10^6	5.9×10^3	6.5×10^{-6}
W	14.7	3410	170	2.7×10^6	21.4×10^3	63.0×10^{-6}
D	3.5	>3500	2000	1.8×10^6	60.0×10^3	11.11×10^{-6}
CBN	3.5	3300	1300	1.8×10^6	48.4×10^3	722×10^{-6}
SiC	3.2	>2300	100	2.3×10^6	15.2×10^3	43.5×10^{-6}
Al$_2$O$_3$	4.0	2050	50	3.1×10^6	12.5×10^3	16.1×10^{-6}
Si$_3$N$_4$	3.2	>1900	33	2.3×10^6	8.7×10^3	14.3×10^{-6}
PSZ†	6.1	—	2	3.8×10^6	2.8×10^3	0.5×10^{-6}
Average ceramic	3.9	—	7	3.0×10^6	4.6×10^3	2.3×10^{-6}
Average glass	2.4	1000	1.0	1.8×10^6	1.3×10^3	0.6×10^{-6}
Water	1.0	32	0.6	4.2×10^6	1.6×10^3	0.14×10^{-6}

† Partially stabilized zirconia.
St. = steel; s.s. = stainless steel

TABLE 9.3 *Typical Values of R (FFG)*

Material combination	β_s/β_w	R_{dry}	$R_{coolant}$
Al$_2$O$_3$/st.	0.80	0.56	0.36
SiC/st.	0.98	0.51	0.34
CBN/st.	3.08	0.25	0.20
Al$_2$O$_3$/303 s.s.	1.58	0.39	0.39
Al$_2$O$_3$/Ti	2.12	0.32	0.24
SiC/Ti	2.61	0.28	0.22
CBN/Ti	8.2	0.11	0.10

st. = steel; s.s. = stainless steel.

$$\theta = 0.754 \frac{Ruvd}{\beta\sqrt{vl}}$$

$$= \frac{(0.754)(0.8)(6895 \times 10^6)(0.5)(10 \times 10^6)}{(15.7 \times 10^{-3})\sqrt{(0.5)(0.0016)}} = 467C = (873F)$$

The power at the wheel is $(68\,950 \times 10^6)[(0.5(12.7 \times 10^{-3})(10 \times 10^{-6})] \times 10^{-3} = 4.378$ kW (5.87 h.p.).

These values for mean wheel–work temperature and power consumed by the wheel are consistent with experimental experience. However, estimation of such absolute grinding values should not be expected to be very accurate, because of the number of assumptions employed. The *relative* influence of the several variables involved will be much more useful and reliable.

Surface Melting in Grinding

From time to time it has been suggested that, in the fine grinding of metals, melting of a layer of metal at the surface is involved. However, when ground surfaces are examined at high power in the scanning electron microscope, there is no sign of melting. The surface shows many sharp edges characteristic of a crystalline material along the grooves plowed out by the abrasive grits. In dry surface grinding, high surface temperatures are reached that may approach or even exceed the melting point of the metal, and yet there is no sign of melting. The answer to this paradox lies in the fact that melting involves a substantial change in structure and therefore is a time-temperature reaction.

It is well known that crystalline materials exhibit a substantial change in density at constant temperature (melting point) in going from the liquid to the solid state (Fig. 9.11). On the other hand, a glass-like material has no melting point but merely becomes gradually more rigid with decrease in temperature without undergoing a change in structure. The absolute temperature (K) of a body is a measure of the vibrational amplitude of its atoms. The homologous temperature scale (θ_H) of Fig. 9.11 measures temperature in nondimensional terms:

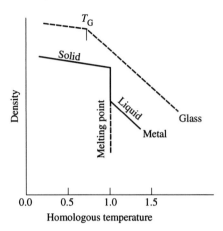

FIG. 9.11. The variation of density with temperature for a metal (solid line) and a glass (dotted line) under equilibrium conditions. T_G is the glass transition temperature.

$$\theta_H = \frac{\text{temperature (K)}}{\text{melting temperature (K)}}.$$

A liquid contains many vacancies relative to the number found in a crystalline solid and, as a consequence, has a significantly lower density than that of the corresponding solid. A latent heat of fusion is associated with the melting of a metal. Vacancies move through a solid by diffusion and, since this involves essentially the same breaking and rearrangement of bonds as self-diffusion, it seems reasonable to use the self-diffusion coefficient of a material as a measure of the rate of diffusion of vacancies through the same material. When this is done (Shaw 1984b) it is estimated to take about 0.25 s to melt a layer 1 μm (40 μin) thick. This is very much greater than the time at temperature in grinding. However, the time required for *temperature* to penetrate a distance of 1 μm will be about eight orders of magnitude shorter than that for vacancy migration.

Malkin (1984) has postulated that the total specific grinding energy u consists of three components:

$$u = u_{\text{chip}} + u_{\text{plow}} + u_{\text{rub}}, \quad (9.30)$$

where u_{chip} is the specific energy to generate chips, u_{plow} is the specific energy for flow to the sides of the main groove, and u_{rub} is the specific energy due to rubbing manly by wear flats.

It is further suggested that u_{chip} is limited to the specific energy content of the material at the melting point. Evidence for this thesis is presented in Fig. 9.12, where experimental minimum specific grinding energies for a variety of materials, for sharp grits, excellent lubrication (oil), and presumably large

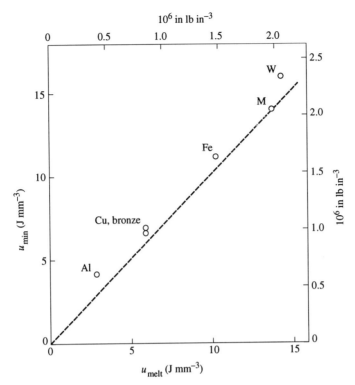

FIG. 9.12. The minimum specific grinding energy versus energy content of liquid metal at the melting point for several materials (after Malkin 1984).

undeformed chip thickness t is plotted against the melting energy (the equilibrium enthalpy difference between room temperature and the melting point in the *liquid* state). The data points lie only slightly above the 45° dashed line, showing that the experimental values are only slightly above the corresponding u_{melt} values.

Malkin stresses that it is not inferred that melting is actually involved in grinding chip formation, but merely that the minimum specific grinding energy correlates with the difference in enthalpy in the *liquid* state between room temperature and the melting temperature. What is not clear is why the enthalpy difference for the *liquid* state is involved if there is no change of phase (melting) in grinding.

An alternative explanation is that the specific extrusion energy at very high strain and strain rate increases as the material becomes more refractory. The additional observations that minimum specific grinding energy is independent of room temperature, hardness, and alloy content may be explained by the observation that grinding chip formation occurs in the hot working regime (homologous temperature > 0.5) where there is little or no strain hardening

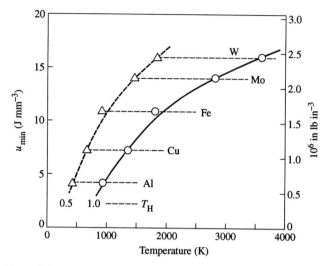

FIG. 9.13. The minimum specific grinding energy from Fig. 9.12 versus the absolute temperature for several materials. Solid line, homologous temperature, $\theta_H = 1.0$; dashed line, $\theta_H = 0.5$.

or resistance to deformation due to second-phase particles (deformation being mainly due to grain boundary rearrangement rather than dislocation transport, as in cold working).

Reichenbach and Coes (1972) have pointed out that a major difference between wear in cutting and grinding is that in cutting, the wear rate is influenced in a major way by cutting speed, whereas in grinding the wear rate is essentially independent of grinding speed. They suggest a 'constant-temperature hypothesis' to explain this. It is assumed that, in grinding, the temperature must rise to the point at which there is thermal softening, and that this is related to the melting temperature of the material (actually, homologous temperature somewhat > 0.5). It would appear that once the hot working regime is entered a further increase in grinding speed is relatively unimportant with respect to wear. While an increase in grinding speed will result in an increase in temperature and hence an increase in wear rate, at the same time an increase in speed causes a compensating decrease in flow stress and hence a decrease in wear rate due to thermal softening.

It is now suggested that the specific energy for grinding correlates with the melting temperature of the workpiece not because of a limiting energy content of the *liquid* material at the melting point but because chip formation in grinding is a hot working phenomenon, and that this requires a specific energy high enough to provide an homologous temperature greater than 0.5. Figure 9.13 shows Malkin's minimum specific grinding energy values (u_{min}) plotted against the absolute melting temperature for several materials for $\theta_H = 1$ as well as for values of $\theta_H = 0.5$. The experimental data fall upon reasonably smooth curves, as in the case of Fig. 9.12. Figure 9.13 merely supports the suggestion of Reichenbach and Coes that a precondition for chip formation is thermal softening (i.e. that chip formation in grinding is a hot working phenomenon).

Sparks and Chips

Little can be inferred with regard to surface temperatures involved in grinding by observing either the sparks or chips after they have cooled. The chips leaving the workpiece do not glow immediately. Glowing chips or sparks are first observed a short distance beyond the wheel-work interface. This indicates that the sparks are due to an exothermic reaction with oxygen in the air. To show this, a large observation balloon was placed around the wheel and workpiece to provide an impervious envelope (Outwater and Shaw 1952). When the air was displaced from the bag with nitrogen and steel was ground in the inert atmosphere no sparks were observed. This demonstrates the importance of oxygen in the formation of sparks.

Not all metals will produce sparks even in air. To show this, a number of pure metals were ground in air using a silicon carbide wheel, and the approximate mean temperature at the chip-wheel interface was measured and the presence of sparks noted in offhand grinding. These results are presented in Table 9.4. Here it is evident that of the 19 pure metals tested, only seven produced sparks. When these tests were repeated in a nitrogen atmosphere, no sparks at all were visible.

When energy is supplied to a body at a rate much faster than vacancies can diffuse into the material to limit the vibrational amplitude by melting, the vibrational amplitude may climb to the point at which the material becomes unstable and explodes. This is what happens when an electrical fuse wire is overloaded and 'explodes' to interrupt current flow. It is also what happens to a chip when it explodes down stream from the point of grinding to give a pattern of sparks characteristic of the alloy ground relative to pattern and color. This provides a convenient means for sorting materials relative to alloy content. Chips leaving a grinding wheel are at relatively low temperature and essentially invisible for a centimeter or more. As these chips pass through the air they oxidize; the exothermal reaction energy quickly brings them to the point of incandescence, and shortly thereafter they explode. On a grander scale, meteorites behave in a similar fashion. These are a few of the many cases in which melting is bypassed during rapid heating because there is no time for the structural change associated with melting to occur.

Modeling and Simulation

The models that are used in cutting and grinding analysis are abstract representations that are usually simplified to emphasize major behavior while suppressing

TABLE 9.4 *Sparking characteristics of pure metals*

	Melting point		Grit-chip temperature		
Metal	°F	°C	°F	°C	Sparks
Aluminum	1215	657	—		No
Antimony	1167	631	—		No
Bismuth	519	271	—		No
Cadmium	604	318	—		No
Calcium	1567	853	—		Yes
Cobalt	2715	1491	2300	1260	Yes
Copper	1981	1083	1150	621	No
Chromium	2822	1550	2900	1593	Yes
Gold	1945	1063	—	—	No
Iron	2950	1632	2300	1260	Yes
Lead	621	327	—	—	No
Magnesium	1204	651	—	—	No
Molybdenum	4748	2620	3460	1904	Yes
Nickel	2646	1952	2600	1427	No
Silver	1761	961	1150	621	No
Tin	449	232	250	121	No
Titanium	3272	1800	—	—	Yes
Tungsten	6098	3370	6500	3593	Yes
Zinc	787	419	750	399	No

minor behavior. Such models are of two types—physical and empirical. Physical models deal with behavior in fundamental physical terms so that a given solution may be applied to a variety of situations. Empirical models are specific to a given operation and range of operating variables, and are derived for experimental results by regression analysis. Physical models are designed to handle new situations and hence are extrapolative, while empirical models are designed to handle regularly encountered situations and hence are interpolative. The approximate models presented throughout this monograph are of the physical type.

Simulation involves the use of a model to predict performance. Just as there are two types of models, so there are two types of simulation—analytical and physical. Analytical simulation involves the use of a model to predict the performance of a process, often with the aid of a computer. The results obtained with this type of simulation are only as good as the model adopted. Simulation of this type is often not successful due to adoption of an oversimplified or over-extended model, or of one that is simply incorrect. Physical simulation involves doing experiments on a system that is more easily managed because of size or force and power requirements, but which gives results that may be interpreted in terms of the prototype. The study of flow characteristics in forming processes by use of Plasticene models is one example of physical simulation. The study of scale models with interpretation by application of

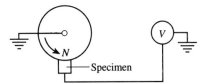

FIG. 9.14. A grit–chip thermocouple circuit (after Outwater and Shaw 1952).

dimensional similitude is another example (wind tunnels, towing tanks, dust bins, etc.).

Analytical simulation is best adapted to empirical models, but is of limited value relative to physical models. Since physical modeling is based on gaining physical insight into the behavior of a process, it is not surprising that different authors come up with very different solutions. Just how complex a problem this is can be illustrated by the variety of physical models that have been proposed for the many important aspects of grinding (temperature, surface finish, wear, stability, and force and power requirements, as discussed by Toenshoff *et al.* (1992)). After reviewing Toenshoff *et al.* (1992), it is evident that physical modeling is at best very approximate. It further reinforces the idea that, in general, only relative solutions are justified and that the pursuit of absolute solutions is frequently unjustified.

Thermoelectric Measurements

The measurement of surface temperatures in grinding is a very difficult problem. Since the temperature decreases extremely rapidly from the surface into a metal, it is next to impossible to obtain accurate absolute values at the surface, although valuable relative values of temperature obtained under different operating conditions are quite valuable, as presently discussed.

A technique for measuring temperatures in cutting that has been quite useful is the tool–work thermocouple method, in which the tool–work interface constitutes the hot junction of a thermoelectric circuit. The application of this technique to estimate the mean surface temperature in grinding was one of the earliest methods used for measuring grinding temperatures (Outwater and Shaw 1952). To do this, a vitrified SiC wheel having a relatively low contact resistance was employed.

While the thermoelectric power of a silicon carbide – steel combination is very much higher than that for ordinary metal combinations, the high impedance of the grinding wheel compared with that of ordinary metals makes it difficult to obtain accurate values. The advantage that exists in the high thermoelectric power is more than offset by the difficulty introduced by the high impedance. By the use of a vacuum-tube voltmeter, in the manner indicated in Fig. 9.14, it was possible, however, to obtain some approximate data.

The wheel–work couple was calibrated in a furnace against a Chromel–

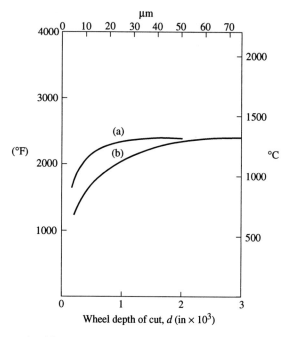

FIG. 9.15. The grit–chip temperature at two work speeds v: (a) = 1 f.p.m. (0.3 m min^{-1}); (b) = 4 f.p.m. (1.22 m min^{-1}) (after Outwater and Shaw 1952). Wheel speed, 5000 f.p.m. (25.4 m s^{-1}); D = 8 in (203 mm); grit = SiC.

Alumel couple to a temperature of 1600 °F (857 °C). The straight calibration line was extrapolated linearly beyond 1600 °F (857 °C), inasmuch as the results of Busch et al. (1947) showed the thermal e.m.f. of silicon carbide against metals to be linear over a wide range of temperatures. The slope of the curve corresponded to 0.01 volts per 100 °F (56 °C). Representative grinding results are given in Fig. 9.15, where each data point represents the mean of ten independent readings. These mean temperatures are seen to reach values in excess of 2000 °F. However, they were obtained for relatively low work velocities v, which tend to yield unusually high values of θ. There was no sign of melting, even for temperatures well above the equilibrium melting point.

The first application of thermocouples buried beneath the surface for measuring subsurface grinding temperatures was by Littman and Wulff (1952). They studied the influence of grinding temperatures on structural changes beneath ground surfaces. Takazawa (1966) also extensively studied the influence of grinding temperatures on subsurface structural changes of hardened carbon steel, using the apparatus shown in Fig. 9.16. A 0.6 mm (0.024 in) constantan wire was spring-loaded against the bottom of a 1 mm (0.040 in) diameter hole drilled into the workpiece. Beryllium oxide was used

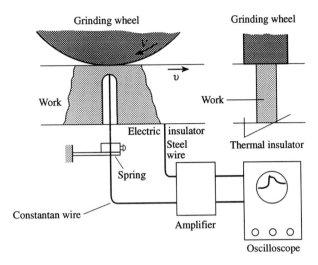

FIG. 9.16. A apparatus for measuring the temperature beneath the ground surface (after Takazawa 1966).

to insulate this wire, since it has the same thermal conductivity as steel. The temperature was recorded for each grinding stroke, and was observed to rise as the thermal junction approached the ground surface (Fig. 9.17). Figure 9.18 shows isothermal curves for the grinding conditions of Fig. 9.17.

The structural change beneath the ground surface was evaluated by making microhardness traverses on taper sections. Figure 9.19 shows the variation of hardness beneath the ground surface for the plunge grinding conditions of Fig. 9.17. The upper three curves show a decrease in hardness with increasing wheel depth of cut as the rate of removal (vdb) increases. This is due to over-tempering of the hardened steel. The fourth curve is for severe grinding conditions, leading to burn and rehardening of material at the surface. In this case, the surface layer is heated to above the transformation temperature and is converted to autensite, and then to martensite, as heat flows rapidly into the workpiece when the wheel passes from the work.

Although buried thermocouples are useful in studying subsurface phenomena such as over-tempering, rehardening and burn, it is very difficult to estimate the temperatures right at the grit–work interface. Attempts to extrapolate subsurface temperatures to the surface are generally not successful because of the very steep temperature gradient at the surface and the difficulty of establishing the exact location of the thermocouple junction beneath the surface. An alternative is to have the hot junction right in the surface.

The method of doing this is the technique of Peklenik (1957), discussed in Chapter 5. Peklenik mounted a soft insulated platinum wire in a steel workpiece, with the wire extending through the surface. When the grinding wheel traversed the hole and wire, the soft wire was smeared over the insulation and

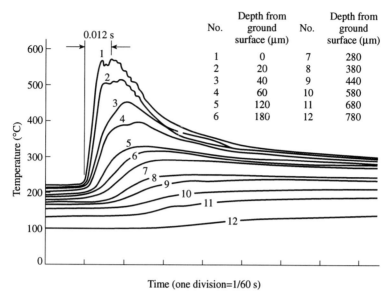

FIG. 9.17. Measured temperatures beneath the ground surface in plunge surface grinding (after Takazawa 1966). Conditions: wheel, WA60 MV; work, hardened carbon steel; wheel speed, 30 m s^{-1} (5900 f.p.m.); work speed, 6 m min^{-1} (19.7 f.p.m.); depth of cut, 20 μm (0.0008 in).

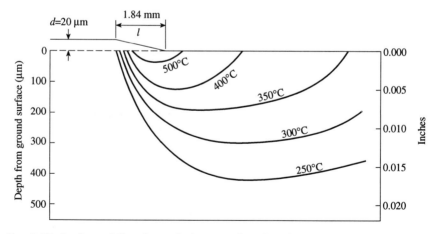

FIG. 9.18. Isothermal lines beneath the ground surface for the conditions of Fig. 9.17 (after Takazawa 1966).

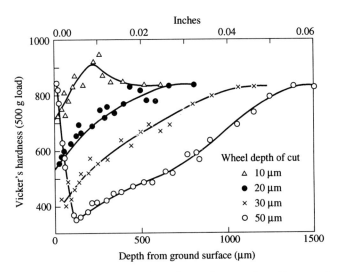

FIG. 9.19. The variation in microhardness beneath ground surfaces under grinding conditions which are the same as in Fig. 9.17 except for the wheel depth of cut, d (after Takazawa 1966).

made contact with the workpiece, thus establishing a platinum/steel hot junction. Being a noble metal, platinum does not oxidize and makes good contact with the steel workpiece. Since Peklenik's wire was very small in diameter, the hot junction responded to the passage of individual grits, and a series of emf peaks were recorded from which an estimate of the spacing of active grits in the wheel surface could be determined. In fact, the estimate of grit spacing was the initial purpose of Peklenik's experiments.

Since Peklenik's original experiments, the concept of establishing a hot thermoelectric junction by having one metal smear over a thin insulating layer on to a second metal has been rather widely used, but mostly for estimating mean surface temperatures. For this purpose, it is better to have the soft (smearing) element in the form of a sheet rather than a wire, so that the mean of many individual encounters is automatically recorded. Figure 9.20 shows a split workpiece arrangement used by Alvi (1988) to measure mean surface grinding temperatures by a variation of the Peklenik technique better suited to mean surface temperature measurements.

Radiation Measurements

The development of photoconducting cells of short response time has made it possible to measure temperature flashes of very short duration by the radiation method. The first application of this technique to grinding was by Mayer and

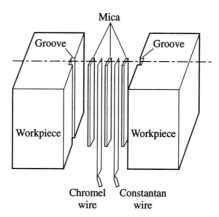

FIG. 9.20. A split workpiece with thin sheets of iron and constantan ($\frac{1}{2}$ × 2 mm, or 0.020 × 0.080 in) mounted in grooves in a split workpiece (after Alvi 1988).

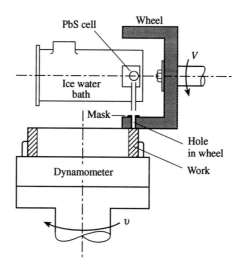

FIG. 9.21. An apparatus for measuring surface temperatures in grinding by radiation (after Mayer and Shaw 1957).

Shaw (1957). Figure 9.21 shows the method employed using an A46-J5-V cup wheel with a hole masked to reveal a square area 0.063 in (1.59 mm) on a side once per revolution. The circuit used to record the output from the cell is shown in Fig. 9.22.

While this initial study was relatively crude, it did reveal some interesting results. Figure 9.23 shows the variation in the mean surface temperature θ with wheel speed when AISI 52100 steel was ground without fluid under conditions to give a constant undeformed chip thickness t of 30 μin (0.75 μm). This is seen

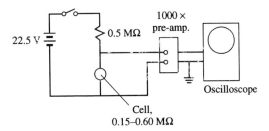

FIG. 9.22. A lead sulfide cell measuring circuit (after Mayer and Shaw 1957).

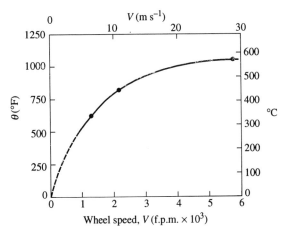

FIG. 9.23. The variation of surface temperature in the dry grinding of AISI 52100 steel with wheel speed for constant undeformed chip thickness $t = 30\,\mu\text{in}$ (0.75 μm) (after Mayer and Shaw 1957).

to be in good qualitative agreement with theoretical results. Figure 9.24 shows specific energy and mean surface temperature curves versus undeformed chip thickness t for several grinding fluids. The shapes of the two sets of curves in parts (a) and (b) are seen to be similar except at very small values of t ($< 25\,\mu\text{in} = 0.6\,\mu\text{m}$). This is as it should be, since the mean surface temperature has analytically been found to be directly proportional to u.

Ueda et al. (1985) have used an optical fiber of 50 μm (0.002 in) diameter to transmit infrared radiation to an InAs cell, as shown in Fig. 9.25. The fiber was located 100 μm (0.004 in) from the wheel face and at a setting angle θ of 45° or greater. When AISI 1055 steel was ground with an A36-K wheel, single grit temperatures from 500 °C to 1400 °C (932–2552 °F) were recorded at $\theta = 45°$, which corresponds to 4.2×10^{-3} s after grinding. The dry grinding conditions were as follows: $V = 29\,\text{m s}^{-1}$ (5200 f.p.m.), $v = 0.167\,\text{m s}^{-1}$

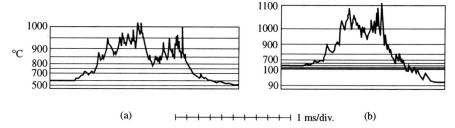

FIG. 9.26. Infrared radiation from the bottom of a 0.4 mm (0.016 in) hole located 20 μm (800 μin) below the surface being ground: (a) dry; (b) with a water-based coolant (after Ueda *et al.* 1986).

compared the results of the three methods are in relatively good agreement. It should be noted that the 1985 article estimated the *mean* wheel–work temperature to be 1150 °C, whereas all three methods used in the 1986 article indicate a mean temperature of about 800 °C.

When a 1:50 soluble oil–water mixture was used, the temperature profile at the bottom of the 0.4 mm hole, 20 μm below the surface being ground, had the same general appearance as for dry grinding (Fig. 9.26). However, the maximum and mean temperatures were actually higher with the fluid than without. This unusual result could be due to exclusion of air by the fluid, causing a greater increase in temperature due to adhesion than the cooling effect of the fluid.

Indium antimonide (InSb) cells are well suited for estimating grinding and cutting temperatures because of their low time constant (~ 7 μs) and because they operate without contact; however, calibration is a problem. Calibration is normally performed by comparing the radiation from a heated specimen with that from a black body at the same temperature to determine the emissivity of the specimen (Ueda *et al.* 1985). Emissivity compensation may be achieved by making radiation measurements at different wavelengths, using filters before amplification. The method of doing this is given by Smith *et al.* (1957), Weichert and Schonert (1978), and Doeblin (1983). Corrections must also be made for the radiation acceptance angle of the sensor.

Chandrasekar *et al.* (1990) have measured grinding temperatures for a single diamond particle and a resin-bonded diamond wheel when grinding a Ni–Zn ferrite (ceramic) and heat-treated AISI 1045 steel. A multiple element infrared sensor consisting of four InSb cells and a high-speed thermal monitor (time constant = 7.4 μs) was used. The cells were maintained at 77 K (in liquid N_2 in a Dewar flask) to provide longer-werelength sensitivity and decreased thermal noise. Fiber optics were used to transmit the radiation from the source to the detector. The spot size was 30 μm (1200 μin) and the output from each detector was amplified and recorded in a multichannel digital storage oscilloscope. The detectors were calibrated statically and also dynamically by exposing the detector to the source through a hole near the periphery of a spinning disc. The response time (time to reach 63 percent of final value) was found to be

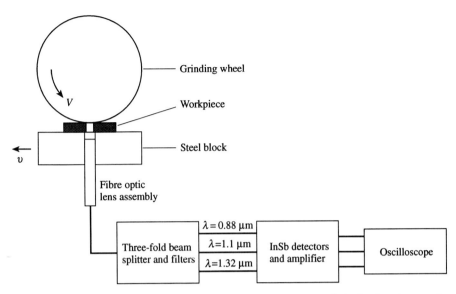

FIG. 9.27 An arrangement used to measure the temperatures of single abrasive particles by observing them through a small hole in the workpiece immediately after grinding. Emissivity compensation was used in calibration and fiber optics were used to conduct infrared radiation to InSb cells (after Chandrasekar et al. 1990).

about 7 μs for all four cells. After 2.5 time constants, the static and dynamic calibrations were the same. Multiwavelength emissivity compensation was employed.

Figure 9.27 shows the experimental arrangement used. The surface of the single point diamond or of the grinding wheel was observed through a 2 mm (0.080 in) diameter hole in the specimen. Figure 9.28 shows the measured average temperature of a single diamond tip when grinding Ni–Zn ferrite or AISI 1045 steel at different speeds. The depth of cut (d) was 10 μm (0.0004 in) and the table speed was 0.023 m s^{-1} (0.092 in s^{-1}) in all tests. The diamond was a 90° cone with a tip radius of 15 μm (600 μin). There was considerabe cut-to-cut variation and the vertical bars show the range. The mean grit-tip temperature increased with wheel speed (V) for both materials. Figure 9.29 shows the decrease in grit-tip temperature for positions A, B, C, and D when grinding the Ni–Zn ferrite under the conditions of Fig. 9.28 but with a wheel speed of 37 m s^{-1} (7280 f.p.m.). The temperature is seen to reach essentially the ambient value before it again engages the workpiece.

Hebbar et al. (1992) have presented temperature results for full resin-bonded diamond wheels grinding several ceramics. The technique used for measuring temperatures is shown in Fig. 9.27. Since different active grits on the wheel surface remove different volumes of material, and also since the sensor covers an area equivalent to a 30 μm (1200 μin) diameter, the grinding temperature for a full wheel is best characterized by a histogram (Fig. 9.30). The mean wheel

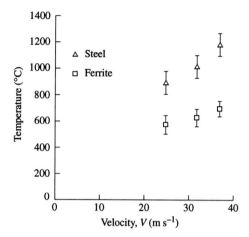

FIG. 9.28. Single point diamond temperatures when grinding steel and Ni–Zn ferrite. Depth of cut, 0.01 mm (0.0004 in); table speed, 0.023 ms^{-1} (0.92 in s^{-1}) (after Chandrasekar et al. 1990).

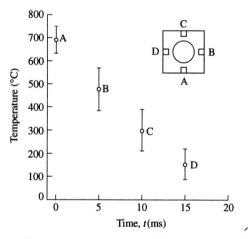

FIG. 9.29. The decay of temperature for single diamond when grinding Ni–Zn ferrite under the conditions of Fig. 9.28; except for the wheel speed, which was 37 m s^{-1} (7280 f.p.m.). The faces of the fiber optics at A ($\theta = 0°$), B ($\theta = 90°$), C ($\theta = 180°$), and D ($\theta = 270°$) were 3.5 mm (0.138 in) from the diamond tip in all cases (after Chandrasekar et al. 1990).

temperature may be obtained from the weighted average of the temperature distribution from Fig. 9.30.

Typical results are given in Table 9.5 for three different ceramics ground with representative resin-bonded diamond wheels. Values of the fraction of energy going to the work (R) for the three tests of Table 9.5 are given in Table 9.6.

The values of R for all three ceramics ground with a diamond wheel are

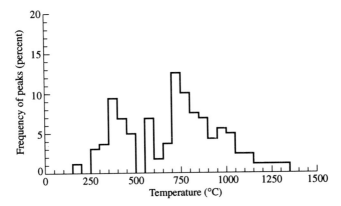

FIG. 9.30. A histogram of the frequency of temperature peaks versus the temperature of grinding of hot pressed Si_3N_4 with a 220 grit resin-bonded diamond wheel (after Hebbar et al. 1992). Wheel speed, 32 m s^{-1} (6300 f.p.m.); table speed, 0.023 m s^{-1} (4.53 f.p.m.); wheel depth of cut, $12.5 \, \mu\text{m}$ (0.0005 in). The values shown are for a sampling time of $2 \, \mu\text{s}^{-1}$ and for one full wheel revolution. The mean wheel temperature in this case was $692 \, °C$ ($1278 \, °F$).

TABLE 9.5 *Mean grinding temperature for several ceramics*

Work	D grit size	V		d		θ	
		m s^{-1}	f.p.m.	μm	in	°C	°F
Si$_3$N$_4$	220	32	6300	12.5	0.0005	692	1278
PSZ†	220	32	6300	12.5	0.0005	592	1098
Ni–Zn ferrite	320	32	6300	20	0.0008	620	1148

† Partially stabilized zirconia.

seen to be relatively low, and hence it may be concluded that, to a first approximation, when a ceramic is ground with a diamond wheel, R may be taken to be zero. Differences in the mean grinding temperatures for different ceramics are due primarily to differences in specific grinding energy and are not due to differences in their thermal properties.

Infrared Photographic Film

Infrared photographic film has been used to photograph radiation from grinding operations performed in a darkroom (Kops and Shaw 1982) Figure 9.31(a) shows the side view of a surface grinding operation. In addition to the main spark stream, some sparks trapped in the wheel face boundary layer are carried halfway around the wheel. Figure 9.31(b) is a close-up view of the grinding zone in up grinding, showing the increase in radiation as the undeformed chip thickness increases. For fine grinding (low removal rate), the surface

TABLE 9.6 Thermal properties of materials of Table 9.5 (after Hebbar et al. 1992)

Material	K (W m^{-1} K^{-1})	ρC (J m^{-3} K^{-1})	$\beta_w = (k\rho C)^{0.5}$ (J m^{-2} s$^{-0.5}$ K^{-1})	$R = (1 + \beta_s/\beta_w)^{-1}$
Si$_3$N$_4$	33	2.32×10^6	8.75×10^3	0.17
PSZ†	2.2	3.84×10^6	2.91×10^3	0.06
Ni–Zn ferrite	8.7	3.76×10^6	5.72×10^3	0.12
D	1000	1.84×10^6	42.9×10^3	—

† Partially stabilized zirconia.

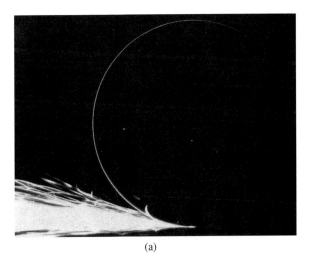

(a)

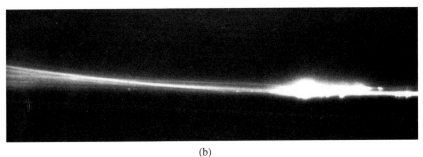

(b)

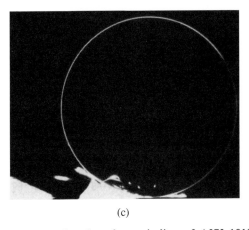

(c)

FIG. 9.31. Infrared photographs of surface grinding of AISI 1018 steel (after Kops and Shaw 1983). The exposure time was equivalent to three revolutions of the wheel ($\sim \frac{1}{15}$ s). White Al_2O_3 wheel of 60 grit size and 8 mm (200 mm) diameter, operating at 6000 f.p.m. (30 m s^{-1}). (a) An axial view for relatively low downfeed; (b) a close-up of the grinding zone for (a); (c) an axial view for a relatively high downfeed, showing piled up chips at A that were carried around in the boundary layer.

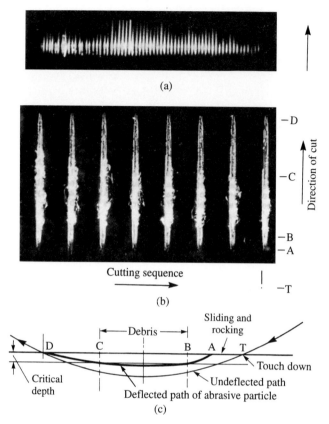

FIG. 9.32. (a) An infrared photograph of a glass plate ground by a single SiC grit at 6000 f.p.m. (30 m s^{-1}) as seen from below, using a mirror to reflect the radiation emitted to the camera during grinding in a dark room. (b) A photomicrograph of portion of the glass surface in (a), taken from above. (c) An interpretation of (a) and (b). (After Kops and Shaw 1983).

of the wheel drops to ambient temperature before reentering the work (Fig. 9.31(a)). However, at high removal rates hot chips travel clear around the wheel circumference, and when they are scraped off and pile up at the point where the wheel reenters the work, they constitute an unwanted added heat source to the work (A in Fig. 9.31(c)). The solution to this problem is to provide a scraper-nozzle near the point at which the wheel leaves the work, to peel off the boundary layer of air and inject a high rate of fluid flow.

A glass microscope slide was ground using a single SiC grit mounted in the periphery of an 8 in (200 mm) metal disc. The workpiece was fed only in the axial direction of the wheel. Figure 9.32(a) shows an infrared photograph for successive cuts, using a mirror below the glass plate to reflect the radiation emitted to the camera lens. Figure 9.32(b) is a visible-light photomicrograph of the glass surface after grinding, and Figure 9.32(c) is an interpretation of

Figs 9.32(a) and (b). The grit is moving upward in Figs 9.32(a) and (b). The grit first touches the glass surface at T, but cutting does not begin until A is reached and continues to D. Cutting is at full depth from B to C. This corresponds to the region of most intense radiation in Fig. 9.32(a) and the region of maximum side flow of debris in Fig. 9.32(b). The lines of radiation in Fig. 9.32(a) are longer than those of Fig. 9.32(b) extending to E, beyond the point at which the grit leaves contact with the work as the grit cools. Distance AB is less than CD. This is consistent with the observation that a grit not only deflects but rotates in its bond at the beginning of a cut (Saini *et al.* 1982).

Stock Removal Temperatures

Introduction

While surface temperatures are usually very important in FFG relatve to surface integrity and wheel wear, this is generally not the case in SRG, mainly because of the following:

- the specific energy is lower because of the large undeformed chip thickness (t) in SRG
- a larger percentage of the total energy is convected away in the chips
- wheel speeds are unusually high, which results in a reduction in the time at high temperature

In SRG the most important consideration concerning wheel wear is not temperature but chip storage space during wheel–work contact. If there is insufficient space between adjacent active grits to accommmodate the large volume of chips involved in SRG, then over-crowding will result in abnormally high temperatures and low values of the grinding ratio (G). The principal ways of increasing chip storage are:

- high wheel speed
- large grit size
- limiting the length of individual chips

Wheel speed is usually limited by safety considerations (wheel fracture under high centrifugally induced tensile stresses). Wheels of conventional design that must operate at unusually high speeds to provide sufficient chip storage space in SRG have a resin bond. The resin is less brittle and flows at points of stress concentration to relieve tensile stresses. Resin-bonded wheels are normally used in abrasive cut-off and conditioning SRG operations. In the vertical spindle surface grinding SRG operation, segments have a vitrified bond and the chip storage problem is solved by limiting the individual chip length. This is done by using a downfeed that provides a criss-cross pattern in the ground surface, which results in discontinuous chip formation, as discussed in Chapter 8. In the conditioning operation, which is the ultimate in stock removal rate, chip

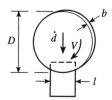

Fig. 9.33. An abrasive cut-off operation.

storage is provided by a combination of high wheel speed, use of a resin bond, and limited individual chip length. The latter is provided by the fact that such wheels move in and out of the cut periodically at the rotational frequency, as discussed in Chapter 7.

With resin-bonded wheels it is important that little of the heat flowing into the wheel during wheel–work contact reaches the first layer of bond posts before reversing direction and leaving the wheel face during the noncontacting portion of a revolution. Resin-bonded material disintegrates at a temperature of about 500 °F (260 °C), the exact value depending on the time at temperature before cooling. The use of a resin bond in SRG depends on the use of a relatively large grit size and an abrasive material that has a low thermal diffusivity (α).

Even though temperature is not a major concern relative to surface integrity in SRG, it is useful to discuss the surface temperature in the abrasive cut-off operation. Because of the complexity of the problem, this will be done employing a semi-quantitative approach. We will begin by considering a few experimental results, and we will follow this by an interpretation of the results.

Experimental Results

Figure 9.33 shows a typical abrasive cut-off operation, in which a thin resin-bonded wheel is fed radially downward into the work.

Figure 9.34 shows an experiment in which an abrasive cut-off operation was performed with thermocouples mounted at A and B to measure the temperature as the wheel passed through these points. The measured temperatures at A and B were 2150 °F (1177 °C) and 2250 °F (1232 °C) respectively. The fact that the temperature at B is higher than at A suggests that the velocity of the thermal front into the work exceeds the rate of downfeed in this case. On the other hand, the bulk of the downward flowing heat is captured by the downfeed and convected away by the chips. Otherwise, the difference in the two temperatures would have been much greater.

Temperature-sensitive paints that melt at different temperatures have been used to estimate the temperature at the wheel–work interface in abrasive cut-off operations on 1 in (25.4 mm) square hot rolled AISI 1020 steel bars, using 20 × 0.19 × 1.0 in (508 × 4.76 × 2.54 cm) wheels. Figure 9.35 shows 1/8 in

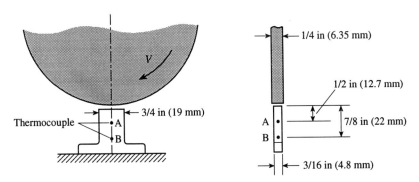

FIG. 9.34. An abrasive cut-off test with thermocouples mounted at A and B. Wheel, A24 R6 B; wheel speed V, 12 250 f.p.m. (62 m s^{-1}); width ground b, 0.188 in (4.76 mm); wheel diameter D, 20 in (508 mm); downfeed rate, 91 i.p.m. (229 mm min^{-1}); grinding force F_P, 17 lb (75.6 N); horsepower, 6.31 (4.71 kW); specific energy u, 1.97×10^6 p.s.i. (13.58 GPa).

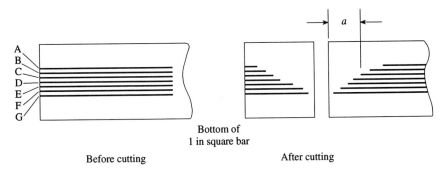

FIG. 9.35. Stripes of temperature-sensitive paint having different melting points, ruled on the bottom of a 1 in (25.4 mm) square bar before and after cutting.

(3.18 mm) wide stripes before and after cutting. All tests were at $V = 12\,500$ f.p.m. (63.5 m s^{-1}) using a 25 hp (18.65 kW) machine. Figure 9.36 shows results for five wheels versus melting temperature for seven paints plotted against distance a (Fig. 9.35) for two of the four values of downfeed rate investigated. The wheel-work contact temperature should correspond approximately to the value of θ at $a = 0$ (θ_0). This is difficult to estimate from Fig. 9.36. However, when the log θ is plotted against $a^{0.5}$, as shown in Fig. 9.37, straight lines are obtained, making it possible to extrapolate the data points to $a = 0$. When this was done, the values given in Table 9.7 were obtained for values of θ_0. Also given in Table 9.7 are values of cost optimum downfeed rate (G^*), cost optimum grinding ratios (d^*), and relative costs per cut under cost optimum conditions obtained by techniques outlined in Chapter 6.

Values of θ_0 as high as 2200 °F (1204 °C) were obtained under unfavorable conditions. However, the values corresponding to cost optimum feed rates

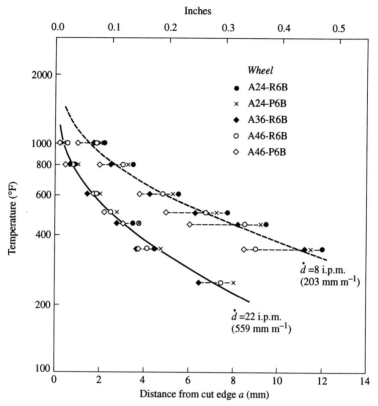

FIG. 9.36. Temperature versus distance a for five wheels cutting at $V = 12\,500$ f.p.m. (62.25 m s^{-1}) and two downfeed rates.

($\dot{d}^*$) were found to be remarkably constant for different values of grit size and wheel grade (1600–1800 °F = 870–980 °C).

Calorimetric tests were performed for several abrasive tests to determine the fraction of total energy going to the workpiece on either side of the cut. A plastic bowl with a cover to prevent chips from entering the bowl was fitted around the workpiece (1 × 1 in × 25 × 25 mm AISI 1020 steel). Immediately after a dry cut-off operation, 2.4 l of water was placed in the insulated bowl and the equilibrium temperature rise was carefully measured. The apparatus was calibrated relative to the cooling rate at different temperatures and a correction made (~ 1 percent per minute). When a dry cut-off test was performed under the following conditions, the fraction of heat flowing axially into the work from both sides of the wheel (R'') was found to be 0.31:

> Wheel: A24 R6 B
> Size: 20 × 0.190 × 1 in (508 × 4.8 × 25 mm)
> Wheel speed, V: 12 500 f.p.m. (63.5 m s^{-1})
> Downfeed $\dot{d}$: 20 i.p.m. (508 mm min^{-1})

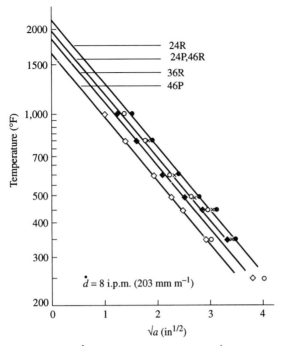

FIG. 9.37. A plot of data for $\dot{d} = 8$ i.p.m. (203 mm min^{-1}) to enable extrapolation to $a = 0$.

When this test was replicated several times, the value of R'' was found to vary only slightly about 0.31.

When other similar wheels from a different maker (A243 R6 B) were tested at the same speed but at different values of downfeed, the values of R'' indicated in Fig. 9.38 were obtained. Here it is evident that less of the total energy ends up in the work on either side of the cut as $\dot{d}$ is increased. When similar tests at the same wheel speed were performed on somewhat softer wheels containing a grinding aid (A24 S G) the values of R'' were only about 60 percent of those for the A243 R6 B wheels. Also, when the wheel speed was increased from 12 500 f.p.m. (63.5 m s^{-1}) to 16 000 f.p.m. (81.3 m s^{-1}), the values of R'' for the wheels containing the grinding aid were virtually unchanged.

From the foregoing experimental results, the following values were found to be representative for a cost optimum cut-off operation:

Work: 1 × 1 in (25 × 25 mm) AISI 1020 steel
Wheel: A24 R6 B
Wheel size: 30 × 3/16 × 1 in (508 × 4.8 × 25 mm)
Wheel speed, V: 12 500 f.p.m. (63.5 m s^{-1})
Downfeed, $\dot{d}$: 20 i.p.m. (508 mm min^{-1})

TABLE 9.7 *Temperature*, θ_0, °F (°C)

Wheel	$\dot{d}$, i.p.m. (cm min^{-1})					$\dot{d}^*$		θ_0^*		G^*	Relative cost per cut
	8 (20.3)	13 (33.0)	22 (55.9)	30 (76.2)	i.p.m.	(m s^{-1})		°F	(°C)		
24R	2200	2050	1600	1400	22	9.3×10^{-3}		1600	871	2.65	1.00
24P	2000	2000	1600	1600	24	10×10^{-2}		1600	871	2.65	1.09
36R	1900	1850	1400	1400	15	6.4×10^{-2}		1800	982	1.65	1.55
46R	2000	1850	1400	1400	16	6.8×10^{-2}		1800	982	1.1	1.99
46P	1700	1650	1400	1400	13	5.5×10^{-2}		1650	899	0.80	2.51

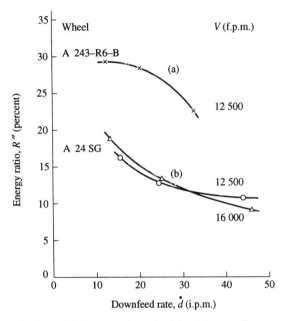

FIG. 9.38. The variation of R'' versus downfeed for (a) an ordinary cut-off wheel and (b) a wheel with an impregnated grinding aid.

Power: 22 h.p. (16.4 kW)
Specific energy, u: 2×10^6 p.s.i. (13.79 GPa)
Fraction of total energy to work on both sides of cut: 0.31
Mean surface temperature, θ: 1600 °F (871 °C)

Analysis

If the fraction of total energy to the work and chips is R, the fraction of the total energy going to the wheel, $1 - R$, may be approximated as follows. From eqn. (9.9) for the Jaeger slider:

$$\theta = 0.754\, ql/\beta\sqrt{Vl},$$

where, for this case,

$$q = (1 - R)ulb\dot{d}/lb = (1 - R)u\dot{d}. \tag{9.31}$$

Hence,

$$\theta_s = \frac{0.754(1 - R)u\dot{d}}{\beta_s}\sqrt{\frac{l}{V}\frac{A}{A_R}}. \tag{9.32}$$

For the example just considered, $\theta = 871\,°C$ (1600 °F), $u = 13.79$ GPa (2×10^6 p.s.i.), $\dot{d} = 8.47 \times 10^{-3}\,\text{m s}^{-1}$ (20 i.p.m.), $l = 2.54$ cm (1 in), $V =$

63.5 m s^{-1} (12 500 f.p.m.), and $\beta_s = 12.5 \times 10^3$ J m^{-2} S$^{-0.5}$ °C^{-1} for Al$_2$O$_3$, from Table 9.2.

The ratio of real to apparent contact area (A_R/A) should be expected to be smaller than the value assumed in FFG (10^{-2}) because of the relatively large grit size ($S = 24$) and the fact that wheel wear is rapid, giving rise to continuous self-dressing. The ratio A_R/A is therefore taken to be 5×10^{-3} in this case. When the above values are substituted into eqn (9.32), the value of $1 - R$ is found to be 0.44, and hence the fraction of total energy associated with the work and chips, $R = 0.56$.

For the example being considered, the fraction of the total energy convected away by the chips (R_C) may be estimated as follows, assuming that the mean temperature of the chips equals the mean wheel–work interface temperature:

$$R_C = \theta \rho C / u. \tag{9.33}$$

For $\theta = 871$ °C, $\rho C = 3.5 \times 10^6$ J m^{-3} °C^{-1} (Table 9.2 for AISI 1020 steel), and $u = 13.79$ GPa:

$$R_C = \frac{(871)(3.5 \times 10^6)}{13.79 \times 10^9} = 0.22.$$

Thus, of the fraction of total energy associated with the work and chips ($R = 0.56$), approximately 0.22 will be convected away by chips, 0.31 will flow into the work on either side of the cut, and only the remaining $0.56 - 0.22 - 0.31 = 0.03$ is unaccounted for.

Since a resin bond is involved, it is important – in order to avoid bond softening or disintegration – that the distance that the thermal front travels into an abrasive particle during wheel-work contact be a fraction of the grit diameter. In the foregoing example, the time during which a single grit will be in contact over the contact arc will be 4×10^{-4} s. The distance (y) that the thermal front will travel into an abrasive particle in 4×10^{-4} s when the surface temperature is constant may be estimated from eqn (9.7) and Table 9.1. When (θ_y/θ_s) is 0.005 (essentially zero),

$$y/2\sqrt{\alpha T} = 2$$

or the distance to the thermal front will be

$$y_f = 4\sqrt{\alpha T}. \tag{9.34}$$

For $\alpha = 16.1 \times 10^{-6}$ m^2 s^{-1} (0.025 in^{-1} s^{-1}) from Table 9.2 for Al$_2$O$_3$ and $T = 4 \times 10^{-4}$ s, the distance to the thermal front $y_f = 0.013$ in (0.32 mm).

From eqn (2.1), the diameter of a 24 grit is approximately 0.029 in (0.74 mm). Hence, during wheel–work contact the thermal front should extend only about 0.013/0.029, or about halfway up the grit. If all of this heat is removed during the noncontact portion of a revolution, there should be no problem with grit softening or disintegration.

A rule of thumb concerning the abrasive cut-off process for a resin-bonded wheel is that the length of cut (l) should not exceed 5 percent of the circumference of the wheel. For a 20 in (508 mm) diameter wheel, this corresponds to a maximum cutting length l of 3.14 in (79.8 mm) and a cooling : heating ratio of 19:1. For a wheel speed of 12 500 f.p.m. (93.5 m s^{-1}) the corresponding distance that the thermal front travels into an Al$_2$O$_3$ grit before cooling would be 0.022 in (0.56 mm), or about 80 percent of the grit diameter for a 24 grit size. Thus, the limiting value $l/\pi D = 0.05$ appears to be a reasonable limiting ratio.

A thorough discussion of the mechanics, economics, and performance characteristics of grinding wheels in abrasive cut-off processes is given in Chapter 6.

References

Alvi, S. S. (1988). PhD thesis, University of Roorkee, India.
Blok, H. (1937). *Proc. General Discussion of Lubrication and Lubricants, Inst. Mech. Engrs* **2**, 225.
Chalmers, B. (1964). *Principles of solidification*. John Wiley, New York.
Chandrasekar, S., Farris, T. N., and Bhushan, B. (1990). *J. Tribology* **112**, 535.
Chvorinov, N. (1940). *Die Glesserei* **10**, 17.
*Des Ruisseaux, N. R. and Zerkle, R. D. (1970). *Trans. ASME* **92**, 4.
Doeblin, E. O. (1983). *Measurement systems*. McGraw-Hill, New York.
Galileo (1638). *Two New Sciences*. McMillan Co. (1914); Dover reprint.
Hahn, R. and Lindsay, R. (1967). *Ann. CIRP* **15**(1), 197.
Hebbar, R. R., Chandrasekar, S., and Farris, T. N. (1992). *J. Am. Ceram. Soc.* **75**, 2742.
Jaeger, J. C. (1942). *Proc. R. Soc. New South Wales* **76**, 203.
*Kopalinski, E. M. (1984). *Wear* **94**, 295.
Kops, L. and Shaw, M. C. (1982). *Ann. CIRP* **31**(1), 211.
Kops, L. and Shaw, M. C. (1983). *Proc. IIth NAMRC*, 390. SME, Dearborn, Michigan.
Kumar, K. V. and Shaw, M. C. (1979). *Ann. CIRP* **28**(1), 205.
Lee, O. G., Zerkle, R. D., and Des Ruisseaux, N. R. (1972). *Trans. ASME* **94**, 1206.
*Lavine, A. S. (1988). *J. Engng Ind.* **110**, 1.
*Lavine, A. S. (1990). *NSF Grantees Conf.*, SME, Dearborn, Michigan.
*Lavine, A. S. and Jen, T. C. (1989). *ASME HTD 123*, 267.
*Lavine, A. S. and Malkin, S. (1986). *Proc. 3rd Int. Grinding Conf.*, MR 88-618, SME Dearborn, Michigan.
*Lavine, A. S., Malkin, S., and Jen, T. C. (1989). *Ann. CIRP* **36**(1), 557.
Littman, W. E. and Wulff, J. (1952). *Trans. ASM* **47**, 692.
Luikov, A. V. (1968). *Analytical heat diffusion theory*. Academic Press, New York.
*Maksond, T. M. A. and Howes, T. P. (1989). *ASME PED 39*, 297.
*Malkin, S. (1968). PhD dissertation, MIT.
*Malkin, S. (1978). *Ann. CIRP* **27**(1), 233.
*Malkin, S. (1984). *J. appl. Metalworking* **3**, 95.
*Malkin, S. and Anderson, R. B. (1974). *J. Engng Ind.* **96**, 1177.
Mayer, J. and Shaw, M. C. (1957). *Lub. Engr* **13**, 21.
Outwater, J. O. and Shaw, M. C. (1952). *Trans. ASME* **74**, 73.

Peklenik, J. (1957). D. Ing. dissertation, Aachen, T. H.
*Powell, J. W. and Howes, T. D. (1978). *Proc. 19th MTDRC*, Manchester, p. 629.
Ramanath, S. (1986). PhD thesis, Arizona State University.
Ramanath, S. and Shaw, M. C. (1988). *J. Engng Ind.* **110**, 15.
Reichenbach, G. S. and Coes, L. (1972). *New developments in grinding.* Carnegie Press, Pittsburgh, 538.
Saini, D. P., Wager, J. G., and Brown, R. H. (1982). *Ann. CIRP* **31**(1), 215.
Sato, K. (1961). *Bull Japan Soc. Grinding Engrs* **1**, 31.
*Sauer, W. J. (1971). PhD dissertation, Carnegie-Mellon University.
*Shafto, G. R., Howes, T. D., and Andrew, C. (1975). *Proc. 16th MTDRC*, Manchester, p. 31.
Shaw, M. C. (1984a). *Metal cutting principles.* Clarendon Press, Oxford.
Shaw, M. C. (1984b). *Ann. CIRP* **33**(1), 221.
Shaw, M. C. (1990a). *ASME HTD 146*, 17.
Shaw, M. C. (1990b). *Ann. CIRP* **39**(1), 345.
Shaw, M. C. (1992). *Ann. CIRP* **41**(1), 303.
Smith, R. A., Jones, P. E., and Chasmar, R. P. (1957). *The detection and measurement of infrared radiation.* Oxford University Press, Oxford.
*Snoeys, R., Maris, M., and Peters, J. (1978). *Ann. CIRP* **27**(1), 571.
Takazawa, K. (1966). *Bull. Japan Soc. Prec. Engrs* **2**, 14.
Toenshoff, H. K., Peters, J., Inasaki, I., and Paul, T. (1992). *Ann. CIRP* **42**(2), 1.
Ueda, T., Hosokawa, A., and Yamoto, A. (1985). *J. Engng Ind.* **107**, 127.
Ueda, T., Hosokawa, A., and Yamoto, A. (1986). *J. Engng Ind.* **108**, 247.
Weichert, R. and Schonert, K. (1978). *J. Mech. Phys. Solids*, **26**, 151.

10

SURFACE INTEGRITY

Introduction

Surface integrity is a term coined in the 1960s to include all aspects of the quality of surfaces, including surface finish, metallurgical damage, and residual stresses (Field and Kahles 1964). These three aspects of ground surfaces will be discussed in order in this chapter.

The importance of surface integrity depends upon its impact on product performance. Performance characteristics that are usually sensitive to surface integrity include the following:

- fracture strength
- fatigue strength
- corrosion rate
- stress corrosion cracking
- lubrication, friction, and wear (tribological behavior)
- magnetic properties (memory discs)
- dimensional stability

Surface Finish

Introduction

One of the items of particular importance in fine grinding is the geometry of the surface produced. This is usually measured in terms of peak-to-valley roughness (R_t) on the European Continent and in Japan or in terms of the centerline average or arithmetic average roughness R_a, in the U.K. and the U.S.A. A tracer instrument having a radius of curvature of 0.0005 in (12.5 μm) at the tip of the diamond stylus is normally used to make such measurements.

To distinguish between waviness (long wave length) and roughness (short wavelength), the output is filtered to remove variation of long wavelength when roughness is desired, and vice versa. The wavelength that separates these two regimes is called the 'cut-off' which, in the U.S.A., is usuall 0.03 in (0.75 mm). Beginning in the 1930s, the output from a tracer instrument in the U.S.A. was given in terms of the square root of the mean of the vertical stylus motion (r.m.s. roughness), because it was easily measured. However, the arithmetic average roughness (AA) is more easily interpreted and has now replaced r.m.s. roughness. The AA roughness is the mean deviation of the peaks from the centerline of a trace. In the U.K. this is called the centerline averge value (CLA). For a sine wave, r.m.s. roughness = 0.90 AA roughness, which is a relatively small difference.

The performance of ground surfaces is not a linear function of the AA value but more closely approximates a geometric progression. Hence, a standard sequence of roughness values for ground surfaces would be 4, 8, 16, 32, and 64 μin (0.10, 0.20, 0.40, 0.80, and 1.60 μm) AA.

In determining the peak-to-valley roughness it is customary to ignore the five highest peaks and the five lowest valleys to obtain a more characteristic value. For a sine wave $R_t = \pi R_a$; but for actual ground surfaces the ratio R_t/R_a is closer to 5, while for honed surfaces the ratio is about 10.

In some applications the orientation of the grinding scratches relative to the loading direction (the lay of the surface) is important, and this will be different for different grinding processes. For example, the fracture strength of a brittle material will be lower when a tensile loading stress is across the lay than when along the lay. An important item not covered by R_a or R_a is the sharpness of the valleys and of the peaks. The former is more important to fracture strength, while the latter is more important to load-bearing capacity.

There are several methods of producing finely ground surfaces and each has its own particular characteristics. Cylindrical and internal grinding have one set of characteristics, while vertical spindle surface grinding has another, and horizontal spindle reciprocating grinding yet another. These differences are due primarily to relative differences in work speed and length of cut. In addition, we may consider plunge grinding, in which there is no axial feed, and cross-feed grinding in which an axial feed is present. These also represent significantly different conditions. The type of grinding to be considered here is reciprocating horizontal surface grinding without cross-feed. The width of the workpiece is less than the width of the wheel. The results and discussion presented should not be applied to other types of grinding without due consideration of the basic differences involved in the different modes of grinding.

The geometry produced in horizontal spindle surface grinding is influenced by a great many variables including the following:

- Wheel characteristics—wheel diameter (D_s), grit type, grit spacing, grit diameter (g) and tip radius (ρ), wheel grade and elastic modulus (E_s), structure number, bond type, dressing method, and degree of wheel balance.
- Work characteristics—work diameter (D_w), workpiece chemistry, workpiece hardness and structure, and the elastic modulus (E_w).
- Machine characteristics—spindle and table stiffness, damping, dynamic characteristics, and isolation.
- Operating conditions—wheel speed (V), work speed (v), wheel depth of cut (d), the relative directions of vectors V and v (i.e. up versus down grinding (as shown in Fig. 10.1), and grinding fluid.

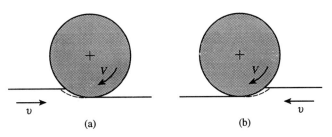

FIG. 10.1. (a) up versus down (b) horizontal spindle surface grinding.

Factorial Experimental Design

When the influence of a large number of potentially important variables upon a single factor such as surface finish is to be studied, the effect of changing one variable at a time is often used. However, this is not the most efficient procedure for studying large systems. The method known as the factorial design of experiments is an alternative procedure that is frequently useful in such instances. By this method, all of the variables to be studied are changed at once in a systematic manner. In a 2^n factorial design, each variable factor is assigned two values—one high and one low. The range spanned by these values should be as large as possible, consistent with an assumed linear increase or decrease of the main dependent variable (surface finish in this case) over the range. Tests are then performed using all combinations of the high and low values of all factors. The total number of tests will be 2^n where n is the number of factors investigated and 2 is the number of levels studied for each factor—hence the term 2^n factorial design.

In order to keep track of the individual tests a special naming system is employed. Each factor is assigned a letter, such as a, b, or c in the case of a three-factor experiment. When the high value of the factor is used, a letter appears in the symbol for this experiment, but if the low value is used, unity appears. The letters are simply multiplied together to obtain the symbol designating the experiment. For example, if the high values of variables a and b are used with the low value of c in a three-factor experiment, this experiment would be referred to as ab. It should be noted that multiplication by unity for c is inferred here. On the other hand, if all variables had their respective low values, the symbol for this experiment would be 1. Experiments should be performed in a random order.

After all of the experiments have been performed, the average may be computed as follows:

$$\text{Av.} = \frac{abc + ab + ac + bc + a + b + c + 1}{8}. \tag{10.1}$$

Here we have assumed a three-factor experiment, for which there will be eight tests.

The direct effect of variable a may be found as follows:

$$DE_a = \pm \tfrac{1}{8}[(abc - bc) + (ab - b) + (ac - c) + (a - 1)] \quad (10.2a)$$

and, similarly, the direct effect of variable b will be

$$DE_b = \pm \tfrac{1}{8}[(abc - ac) + (ab - a) + (bc - c) + (b - 1)]. \quad (10.2b)$$

The relative effects of variables a and b may be obtained by comparing the magnitudes of these two results. The $\pm$ signs in these and subsequent equations refer to the variation due to the variable in question above and below the mean.

It should be noted that the effect of variable a will be more representative as determined in this way than in the method in which one variable is changed at a time, since variables b and c cover a range of values instead of corresponding to a fixed combination.

The factorial design of experiments also enables the estimation of interactions between variables. By 'interaction' is meant the effect that variable a has upon the effect that variable b has on the main dependent variable (say, roughness).

When the number of factors to be investigated is large, it frequently takes too much effort to do a complete factorial experiment. Techniques for fractional factorial experiments have been developed, and computer programs are available to obtain the direct effects of each variable, as well as second and higher order interactions if these are of interest. Further details are available in many references (see, for example, Montgomery 1991).

Surface Finish Experiments

A comprehensive study of the arithmetic average roughness produced in surface grinding was carried out using combinations of the high and low values of the variables listed in Table 10.1. Details of the study are presented in Farmer *et al.* (1967).

In making a test the wheel was first balanced (to $<5\,\mu$in = $1/8\,\mu$m spindle displacement) and then dressed according to the procedure called for in Table 10.1. The wheel was fed downward at the end of each stroke and grinding was continued for ten passes. All tests were without spark-out and without cross-feed (plunge grinding), and the specimen measured 1/2 in wide by 3.125 in long (12.5 × 79.4 mm). In all instances the CLA (AA) roughness was measured perpendicular to the grinding direction. The work material was AISI 52100 steel.

Instead of doing a complete 2^n factorial set of experiments (256 tests) a one-quarter replicate (64 tests) was performed.

The surface roughness values ranged from a low value of $10\,\mu$in ($0.25\,\mu$m) to a maximum value of $105\,\mu$in ($2.6\,\mu$m). When initially processing these data it was observed that two of the values of run 1 (*efgh* and *acegh*) were far out of line relative to other values. In these two particular instances, chips had

TABLE 10.1 *Variables investigated. The wheels used were AA24 F8 V, AA24 L8 V, AA60 F8 V, and AA60 L8; all $8 \times 0.75 \times 1.25$ in $(203 \times 19 \times 31\,mm)$ (after Farmer et al. 1967)*

Symbol	Variable	Low value	High value
a	Dressing technique	Fine: two passes with 0.0005 in in-feed at 3 i.p.m.; two passes with 0.0002 in in-feed at 3 i.p.m.; one pass spark-out	Coarse: two passes with 0.001 in in-feed at 12 i.p.m.; no spark-out
b	Size	24	60
c	Grinding type	Down	Up
d	Wheel hardness	F	L
e	Work hardness	$R_B = 93$	$R_c = 63$
f	Wheel speed, (m s^{-1})	4500 (22.9)	6000 (30.5)
g	Work speed, f.p.m.	13	83
h	Down-feed, in per pass	0.0002	0.0004

TABLE 10.2 *Computer analysis—run 1. The average for all tests = 29.7 μin (0.74 μm) (after Farmer et al. 1967)*

Direct effect	D	For fine finish	Relative value
Work speed, g	9.65	Slow	1.00
Grain size, b	−4.91	Fine	0.55
Down-feed, h	3.26	Low	0.33
Dressing, a	3.22	Fine	0.33
Type grind c	−2.91	Up-grind	0.3
Wheel speed, f	−2.84	Fast	0.3
Wheel hardness, d	−2.66	High	0.27
Work hardness, e	−1.41*	High	0.15

First-order interactions	I	For fine finish
dg = Wheel hardness, work speed	−2.62	d high, g low
ag = Dressing, work speed	2.06	a fine, g low
bg = Grain size, work speed	−2.02	b fine, g low
cg = Type grind, work speed	−1.72	c up-grind, g low
ch = Type grind, down-feed	−1.72	c up-grind, h low
fg = Wheel speed, work speed	−1.63	f high, g low
hg = Down-feed, work speed	1.12*	h low, g low

All values listed are significant at 1 percent level (1.59) except those marked *.
* = significant at 5 percent level (1.19).

been observed to weld to the ground surface, thus causing these values of roughness to be unusually high. It was therefore decided to use interpolated values in place of the two anomalous values in the final computer analysis.

The results of the computer analysis for run 1 are presented in Table 10.2, where the direct effects (D) and significant first order interactions (I) are given in decreasing order of importance.

Significance was established by considering all interactions higher than the first to represent error and applying the Students' *t*-test. Only those first-order interactions that were at the 0.95 or 0.99 significance level are listed in Table 10.2. The value referred to as 'average' is the average roughness for all 64 tests performed.

Auxiliary tests revealed that the R_a roughness heights for replicated tests varied by as much as ±15 percent when chip welding was present, but only by ±5 percent when there was no chip welding.

It should be noted that in a 2^n factorial experiment the effect of a variable or interaction is assumed to be linear throughout the range employed. Tests in which only one variable was varied at a time indicated that in two instances a maximum roughness occurred at values intermediate between the high and

TABLE 10.3 *Computer analysis—run 2. The average value for all tests = 26 μin (0.65 μm) (after Farmer et al. 1967)*

Direct effect	D	For fine finish	Relative value
Work speed, g	4.95	Slow	1
Type grind, c	−2.96	Up-grind	0.6
Grain size, b	−2.66	Fine	0.53
Work hardness, e	−2.44	High	0.5
Down-feed, h	2.28	Low	0.46
Dressing, a	2.00	Fine	0.40
Wheel speed, f	−1.25	High	0.25
Wheel hardness, d	−1.19	High	0.25

First-order interactions	I	For fine finish
ag = Dressing, work speed	1.99	a fine, g low
cg = Type grind, work speed	−1.31	c up-grind, g low
fg = Wheel speed, work speed	−1.30	f high, g low
dg = Wheel hardness, work speed	−1.22	d high, g low
df = Wheel hardness, wheel speed	1.12	d high, f high
ce = Type grind, work hardness	0.96*	c up-grind, e high

All values listed are significant at 1 percent level (1.59) except those marked *.
* = significant at 5 percent level (1.19).

low values used in the tests of run 1. These variables were grit size and wheel hardness.

In order to investigate the possible influence of nonlinear behavior on the 2^n factorial analysis, it was decided to repeat run 1 with different high values for grit size (46 instead of 60) and wheel hardness (J instead of L). In addition, the high value for work speed was also changed to 48 f.p.m. instead of 83 f.p.m. (14.6–25.3 m min^{-1}), since the work speed had been found to play such an important role in run 1. The high and low values of the other five variables were the same as in run 1, and the test procedure was the same except that the ten strokes were taken in rapid succession without stopping to measure the amount removed after each stroke. Only the total amount removed after ten strokes was measured in run 2. The results of the computer analysis for run 2 are given in Table 10.3. The relative values of Table 10.3 are in close agreement with those of Table 10.2 with three exceptions: type of grind, work hardness, and downfeed.

The main results of this surface finish study may be summarized as follows:

1. Work speed is by far the most important variable and should be low for good finish.

2. The following variables are but one-half to one-third as important as work speed in providing good finish:

Variable	For good finish
Dressing	Fine
Wheel grade	High
Grit size	Fine grit
Type of grind	Up
Wheel speed	High
Downfeed	Low
Work hardness	Hard

The relative importance of these variables will vary depending on the specific combination of variables employed. As a first approximation, all of these quantities may be considered to be of equal importance.

3. Work hardness is least important, but should be hard.

4. It cannot be said that any of the variables tested are completely unimportant to surface finish, but merely that their relative importance falls into the categories suggested above.

Measurements of R_t used on the European Continent instead of R_a used in the U.S.A. and the U.K. were also made for all the tests of run 1. Values of R_t against R_a are shown plotted in Fig. 10.2, where it is evident that practically all points lie between the dotted lines corresponding to R_t/R_a–6.

Therefore, to a good approximation, it may be concluded that for surface grinding of the type studied here:

$$h_t \simeq 5h_a. \tag{10.3}$$

Analytical Approach

There have been several attempts to estimate the roughness of a ground surface analytically. Sato (1950) expressed the CLA roughness of a ground surface in terms of the mean spacing of abrasive grits, the mean spacing of scratches on the ground surface in a transverse direction, and the grit-tip radius. This analysis incorrectly assumed different values of roughness in the longitudinal and transverse directions, and predicted values of R_a that were too low by more than an order of magnitude.

Yang and Shaw (1955) derived an expression for R_t in horizontal spindle surface grinding without cross-feed based upon the following assumptions:

- the number of active cutting points in the wheel surface per unit area (C) is the static value obtained by rolling a dressed wheel over a soot-covered glass plate
- there is metal available to be cut by every active grit
- all active grits are at the same mean elevation
- all active grits are at the same uniform spacing $\simeq C^{-0.5}$
- there is no built-up material on the tool tip and no chatter

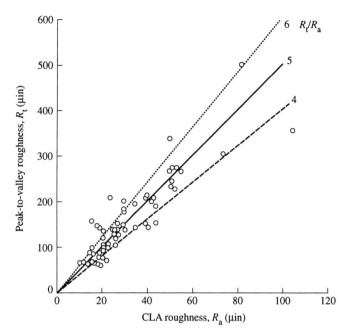

FIG. 10.2. The relation between peak-to-valley roughness (R_t) and arithmetic average roughness (R_a) for all 64 tests of run 2. (after Farmer et al. 1967).

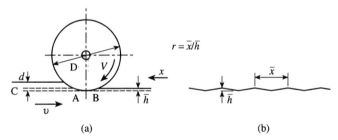

FIG. 10.3. A diagram for use in determining the first approximation value for R_t: (a) side elevation; (b) end view of idealized surface finish looking in the x direction (after Yang and Shaw 1955).

In Fig. 10.3b the mean height of the scratches left behind is $\bar{h}$, and the mean distance AB is then

$$\overline{AB} = 2\sqrt{D\bar{h}}. \tag{10.4}$$

The time for the work to feed a distance AB will be $2\sqrt{D\bar{h}}/v$ and the number of grits, N_{AB}, which will sweep through a given line on the surface of an axial width $C^{-0.5}$, as it moves through distance AB, will be:

$$N_{\overline{AB}} = (2\sqrt{D\bar{h}}/v)\, V\sqrt{C} \tag{10.5}$$

where $C^{-0.5}$ is the mean distance between cutting points and C is the number of active cutting points per unit area of wheel surface.

The distance $\bar{x}$ shown in Fig. 10.3b will be:

$$\bar{x} = 1/(N_{AB}^{-1}\sqrt{C}) = r\bar{h} \tag{10.6}$$

where $r = \bar{x}/\bar{h}$, and hence from eqns 4.5, 4.10 and 10.6:

$$\bar{h} = (v/[2VrC\sqrt{C}])^{2/3}. \tag{10.7}$$

From eqns (4.5) and (10.7), we obtain

$$\bar{h} = \bar{t}^{4/3}/(2^{2/3}d^{1/3}). \tag{10.8}$$

The actual peak-to-valley distance (R_t) will differ from $\bar{h}$ owing to the fact that the abrasive grits are not spaced uniformly and are all at the same elevation in a grinding wheel, and also because of the possible formation of built-up material and chatter. However, we should expect R_t to be proportional to $\bar{h}$ and hence, to a first approximation,

$$R_t \approx t^{4/3}/d^{1/3}. \tag{10.8a}$$

This tells us that the surface roughness produced in a grinding operation should be expected to vary a little more rapidly than linearly with mean grit depth of cut $\bar{t}$, but only slightly with wheel depth of cut.

Orioka (1957, 1961) presented two analyses for surface roughness in plunge surface grinding without spark-out. In these analyses the variation of the height of individual grits was taken into account and the grits were assumed to be randomly distributed over the wheel face. Based on tracer measurements using a knife-edge stylus, Orioka assumed that the grit population density C varied parabolically with the distance from the outermost grit in the wheel face. Ignoring vibration, built-up edge formation and side flow during chip formation, he obtained the following result for surface grinding without cross-feed or spark-out (Orioka 1961):

$$R_a = 0.33(v/[VrC\sqrt{D}])^{2/9}H_0^{2/3}, \tag{10.9}$$

where H_0 is the maximum grit depth of cut for any grit.

From this analysis it is quite clear that the radial distribution of grit population density plays a very important role, and that relatively few grits in the outer periphery of the wheel are responsible for the final finish. The quantity H_0 (the grit depth of cut taken by the outermost grit in the wheel face) is seen to be dominant relative to surface roughness. However, Orioka presents no simple way of determining H_0 for a given wheel and finally assumes that it lies in the range of 1–4 μm (40–160 μin).

Two methods based upon a dynamic determination of the active grit density (C) versus the radial depth from the outermost grit (h), which was briefly mentioned in Chapter 4, will now be presented. Both methods involve estimation of the peak-to-valley roughness (R_t). Both methods also assume a uniform distribution of active cutting points (C) at a given depth of cut (h). The first

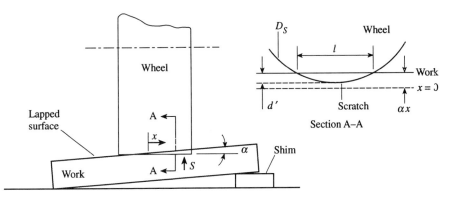

FIG. 10.4. The method of determining the C value of a grinding wheel under dynamic conditions.

method is based on area continuity, i.e. that the total area removed at a given level equals the sum of the surface areas of the individual undeformed chips. The second method predicts the transverse profile of the surface for a given C versus h distribution and from this derives R_t. Before discussing these two methods the dynamic determination of the C versus h curve on which each is based will be discussed in greater detail than in Chapter 4.

Dynamic Active Grit Density, C

Figure 10.4 illustrates how C values may be obtained. A carefully lapped plate is mounted on a surface grinder using a 0.001 in, (25 μm) shim, to provide a tapered surface having a slope of about 1500:1. The surface is then ground once, using a high table feed (125 f.p.m., or 38 m min^{-1}) and a low wheel speed (2400 f.p.m., or 12.2 m s^{-1}). Surface scratches such as those shown in Fig. 10.4 are thus produced and photographed at 10×. Figure 10.5 shows only a small portion of an actual photograph, which will cover about 1 in (25 mm) in the grinding direction.

The profile of the ground surface of Fig. 10.5 is shown in Fig. 10.6 as obtained by a tracer instrument. This is used to determine the precise inclination of the ground surface which, in this instance, is 650 μin in^{-1} (6.4 μm cm^{-1}). It is then possible to compute the depth of cut (d') corresponding to the length of each scratch (l) observed in Fig. 10.5 as follows (see the insert to Fig. 10.4):

$$d' = l^2/4D, \qquad (10.10)$$

where D is the wheel diameter.

The surface of Fig. 10.5 is divided into intervals of 0.1 in (0.25 mm) width, and the depth, d', of each scratch is plotted at the value of x corresponding to the center of the appropriate interval (Fig. 10.7). When more than one

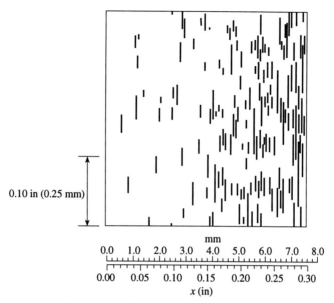

FIG. 10.5. A tracing of the plan view of the surface ground in Fig. 10.4, showing individual scratches of different lengths (l). Wheel, white Al_2O_3, 60H (after Nakayama and Shaw 1967).

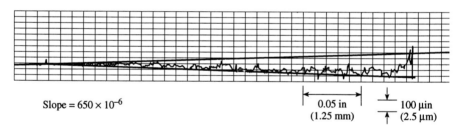

FIG. 10.6. A trace in the x direction of the surface shown in Fig. 10.5, to obtain on accurate slope of work (α). The upper solid straight line is the original surface of the work (after Nakayama and Shaw 1968).

scratch has the same length (depth d') it is represented by a dot beside the first one. A line is next drawn having a slope equal to that obtained from Fig. 10.6 (650 μin in^{-1}, or 6.4 μm cm^{-1}) and just touching the outermost grits (AB in Fig. 10.7). To a good approximation, the deepest grits in each x interval are seen to lie at about the same level along AB.

The abscissa is next shifted vertically in Fig. 10.7 so that it goes through the lowest points. The distance from this new abscissa to the workpiece surface line (AB) represents the sampling depth (h) for each interval in x.

The number of cutting points per unit area (C) may then be plotted against

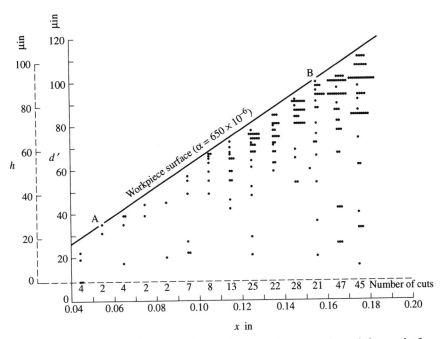

FIG. 10.7. A plot used to determine the number of grits cutting in each interval of x, and for establishing the zero of the d' scale, where d' is the radial distance inward from the outermost grit on the wheel face (after Nakayama and Shaw 1968).

the radial distance from the outermost grits (h), as shown in Fig. 10.8. To illustrate the procedure for obtaining these curves consider point P for the 60H wheel of Fig. 10.8. This point corresponds to the first interval in Fig. 10.7 ($x = 0.04$–0.05 in). The value of h for this interval is seen to be 20 μin (0.5 μm) and four cutting points were observed for this interval. The area on the work surface for this sample corresponds to the interval width (0.01 in, or 250 μm) times the sample length in the grinding direction (0.835 in, or 21.1 mm) in this case. The sample area on the work surface is then 0.01×0.835 in² (5.3 mm²), while the corresponding sample area on the wheel surface will be $V/v = 2400/125$ times this area. The number of cutting points for the first interval will therefore be

$$C = \frac{4}{(0.01)(0.835)(2400/125)} = 25 \text{ in}^{-2} \ (3.87 \text{ cm}^{-2}).$$

Other points in Fig. 10.8 were obtained in a similar manner.

The curves of Fig. 10.8 approximate the parabolic distribution suggested by Orioka (1957), but are more conveniently represented by the straight lines shown, since practical values of h fall in the straight-line regions. It is interesting to note that all three curves have approximately the same intercept

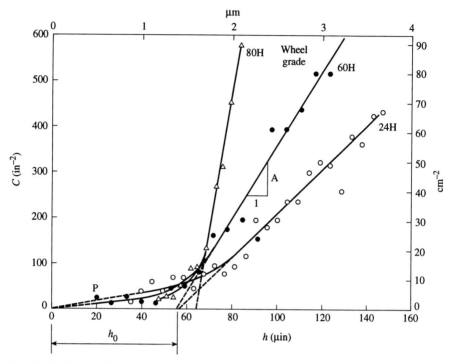

FIG. 10.8. The variation of the effective number of cutting points (C) with the distance from the outermost grits (h) for three white aluminuim oxide wheels (80H, 60H, and 24H) (after Nakayama and Shaw 1968).

but quite different slopes (A). The varation of C with h is well approximated by the following equation in the practical region:

$$C = A(h - h_0). \tag{10.11}$$

Values of A and h_0 for the three white Al_2O_3 wheels are given in Table 10.4.

The results shown in Fig. 10.8 and Table 10.4 are for wheels of typical surface condition. The wheels were carefully balanced and then dressed using the flat face of a pyramidal diamond, passed tangentially across the wheel surface. The dressed wheels were then used for a short time to insure that loose grits were removed.

Area Continuity Approach

If one observes a ground surface from above, it is found to consist of many individual scratches that are very long and narrow. They appear to be needle-shaped. Figure 10.9 represents the shape of an individual scratch, with width shown exaggerated where it is assumed that points A, B, C, and D are all at

SURFACE INTEGRITY

TABLE 10.4 *Values of A and h_0 in eqn (10.11) for three vitrified white Al_2O_3 wheels of H grade and different grit sizes (after Nakayama and Shaw 1968)*

Wheel	Imperial units		Metric units	
	A (in^{-3})	h_0 (μin)	A (mm^{-3})	h_0 (μm)
80H	30×10^6	65	1831	1.63
60H	7.5×10^6	55	457	1.38
24H	5.0×10^6	55	305	1.38

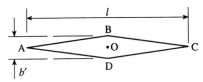

FIG. 10.9. An idealized model of a grinding scratch as viewed from above.

the same elevation. Furthermore, O is assumed to be the deepest point of the scratch and to be located a distance h below plane ABCD. It is then evident that, to a good approximation,

$$l = 2\sqrt{Dh}, \qquad b' = 2\sqrt{2\rho h}, \qquad (10.12, 10.13)$$

where D is the wheel diameter and ρ is the radius of the grit tip in a transverse plane.

Dividing eqn (10.12) by eqn (10.13),

$$\frac{l}{b'} = \sqrt{\frac{D}{2\rho}}. \qquad (10.14)$$

From this equation it is evident that l will be much larger than b'. For example, if $D = 8$ in (203 mm) and $2\rho = 0.015$ in (0.38 mm) for a 46 grit size, then l/b' will be 23, which is truly a needle-like scratch. This explains why the sides of the scratches on ground surfaces appear parallel.

If the local number of effective cutting points per unit area is C, then one grit will be associated with an area $1/C$ on the wheel face, but with an area of v/VC on the work surface. The projected area of the single scratch shown in Fig. 10.9 will be $lb'/2$. In accordance with area continuity, the two expressions must be equal:

$$\frac{lb'}{2} = \frac{v}{VC}. \qquad (10.15)$$

Equations (10.14) and (10.15) may now be solved simultaneously for l and b'.

$$l = \sqrt{\frac{2v\sqrt{D}}{VC\sqrt{2\rho}}}, \qquad b' = \sqrt{\frac{2v\sqrt{2\rho}}{VC\sqrt{D}}}. \qquad (10.16, 10.17)$$

Upon substituting eqn (10.16) into eqn (10.12) and solving for h, we obtain:

$$h = \frac{v}{VC\sqrt{2\rho D}}. \qquad (10.18)$$

This will be the local peak-to-valley roughness at point O of section BOD (see Fig. 10.9). The value of C to be used in this equation is that corresponding to h. Substituting eqn (10.11) into eqn (10.18), we obtain:

$$h = \frac{h_0}{2}\left(1 + \sqrt{\frac{2v}{AVh_0^2\sqrt{2\rho D}}}\right). \qquad (10.19)$$

It would appear that h_0 is most important in eqn (10.19) and that the other variables (v, V, A, ρ, and D) play a secondary role. However, h_0 tends to be rather constant, as may be seen in Fig. 10.8, and therefore in practice the secondary variables will have the greatest influence on the value of h pertaining. The fact that h_0 is so constant with grit size explains why grit size does not have a greater influence on h than might naturally be expected. The chief influence of grit size will be relative to A and ρ. However, it may be seen from eqn (10.19) that the influence of these two variables will tend to compensate, for as grit size is increased A will decrease (Fig. 10.8) but ρ will increase.

Example

When a vitrified white aluminum oxide grinding wheel of 60 grit size and H grade was operated in plunge surface grinding under the following conditions,

$V = 6000$ f.p.m. (30.5 m s^{-1}),
$V = 48$ f.p.m. (0.24 m s^{-1}),
$d = 0.0002$ in per stroke (5 μm per stroke),
$D = 8$ in (203 mm),

the resulting values of finish were obtained:

$R_t = 100$ μin (2.5 μm)   and $R_a = 22$ μin (0.55 μm).

A surface finish trace for this surface is shown in Fig. 10.10.

Assuming the grit-tip radius to equal half the mean grit diameter,

$$\rho = 0.011/2 = 0.0055 \text{ in } (137.5 \, \mu m).$$

From Table 10.5 (60 H wheel)

$$A = 7.5 \times 10^6 \text{ in}^{-3}, \qquad h_0 = 55 \, \mu\text{in},$$

and hence, from eqn (10.19),

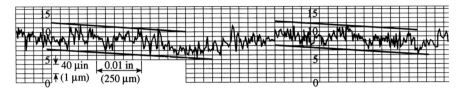

FIG. 10.10. The trace of the surface profile for the example.

$$R_t = \left(\frac{55 \times 10^{-6}}{2}\right)\left(1 + \frac{2 \times 48}{(6000)(7.5 \times 10^6)(55 \times 10^{-6})^2\sqrt{0.0108}}\right)$$

$$= 78\,\mu\text{in }(1.95\,\mu\text{m}).$$

This value is to be compared with the measured value of $100\,\mu\text{in}$ ($2.5\,\mu\text{m}$). The agreement is remarkably good when it is realized that the analytical value does not include built-up edge, side flow during cutting, and the maximum value of C pertaining, all of which would tend to raise the analytical value.

From Fig. 10.8, the value of C corresponding to $h = 78\,\mu\text{in}$ ($1.95\,\mu\text{m}$) is 185 grits per in^2 (0.29 grits per mm^2), and this point is found to lie on the straight portion of the curve, in accordance with our assumption in using eqn (10.11).

While a constant value of C has been used, in reality C will vary stochastically, and at some particular point C will be smaller than the mean value given by Fig. 10.8. Since h (R_t) varies inversely with C (eqn 10.18) the local value of C required to bring the calculated value of R_t in agreement with the measured value would be

$$(78/100)(185) = 144\,\text{in}^{-2}\ (22\,\text{cm}^{-2}).$$

This is a reasonable value and it appears that the stochastic character of C may be taken into account approximately by using about 80 percent of the value of C from Fig. 10.8 in eqn (10.18) instead of the mean value.

Equations (10.16) and (10.17) may be used to estimate the mean length (l) and width (b') of the scratches in the surface:

$$l = \sqrt{\frac{(2)(48)\sqrt{8}}{(6000)(185)\sqrt{0.011}}} = 0.048\,\text{in }(1.22\,\text{mm}),$$

$$b' = \sqrt{\frac{(2)(48)\sqrt{0.011}}{(6000)(185)\sqrt{8}}} = 0.0018\,\text{in }(0.046\,\text{mm}),$$

and

$$l/b' = 27.$$

From the surface trace in the transverse direction, 15 grooves were found

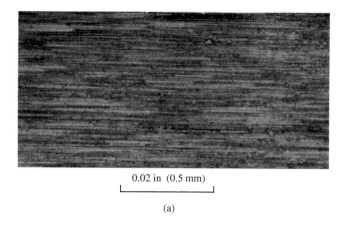

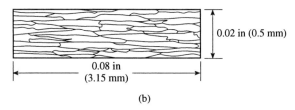

FIG. 10.11. A plan view of the ground surface of Fig. 10.10: (a) a photomicrograph; (b) an overlay, showing the outlines of individual scratches in (a).

in a distance of 0.01 in (0.25 mm). Since this will correspond to the mean groove spacing, we may write:

$$b'/2 = 0.01/15 \quad \text{or} \quad b' = 0.00133 \text{ in } (33.25 \, \mu\text{m}),$$

which is found to be in fairly good agreement with the calculated value of 0.0018 (46 μm).

Figure 10.11(a) shows a portion of the ground surface of the example, while Fig. 10.11(b) shows scratches that may be identified in this surface on an overlay.

A rough check on the value of C pertaining for this surface may be obtained by counting the number of scratches and dividing by the corresponding area on the photograph. This will give the scratch density on the work which, in turn, may be converted to the grit density on the wheel surface by multiplying by v/V. Thus, for Fig. 10.11(b),

$$C = [35/(0.08)(0.02)](48/6000) = 175 \text{ in}^{-2} \ (0.27 \text{ mm}^{-2}).$$

This value is in good agreement with the value from Fig. 10.8 of 185 in^{-2} (28.7 cm^{-2}).

The spacing (l_r) of active cutting edges on this wheel that cut in the same groove of width b' may be obtained as follows:

SURFACE INTEGRITY

TABLE 10.5 *Values of grits per unit area (C) and mean transverse spacings (b') for different levels h from the 60H wheel curve of Fig. 10.8 and eqn (10.17) (after Nakayama and Shaw 1968)*

h (μin)	C (in^{-2})	b' (in)
10	5	11.2 × 10^{-3}
30	15	6.43
50	35	4.21
70	120	2.27
90	280	1.49
110	440	1.19
130	600	1.01

$$C(l_r)(b') = 1 \quad (10.20)$$

or, for this example,

$$l_r = 1/Cb' = 1/[(185)(0.0018)] = 3 \text{ in } (76.2 \text{ mm}).$$

Similar examples were worked for the 24H and 80H wheels and comparable results were obtained.

Transverse Profile Approach

A second method of estimating the peak-to-valley roughness R_t is presented in this section. An idea of the transverse profile of a ground surface, such as that obtained with a tracer instrument, may be obtained by considering grits at a given level to be equally spaced. The transverse spacing of grits (b') is given (see eqn 10.17) by

$$b' = \frac{2v}{VC}\sqrt{\frac{2\rho}{D}}.$$

For the case considered in the previous example

$$b' = \frac{2(48)}{6000C}\sqrt{\frac{0.011}{8}} = \frac{2.44 \times 10^{-3}}{C},$$

where C is grits per in^2 and b' is in inches.

The values in Table 10.5 may be readily obtained by use of eqn (10.17) for the 60H curve of Fig. 10.8. Table 10.5 indicates that at an elevation h of 10 μin (0.25 μm) there will be 5 grits per in^2 (0.775 cm^2) and their mean transverse spacing (b') will be 0.0112 in (280 μm). In Fig. 10.12, a series of grit tips is drawn at level $h = 10$ μin (0.25 μm) at a uniform spacing of 11.2 × 10^{-3} in (280 μm). These grit tips (A, B, etc.) are actually spheres of radius 0.0055 in (0.140 mm), but since there is a 20:1 difference in the vertical and horizontal scales, they appear as parabolas.

Next, a series of equally spaced grit tips are drawn at $h = 50$ μin (1.25 μm) with $b' = 0.00421$ in (0.107 mm) (C, D, E, F, etc.). Similarly, grit tips are

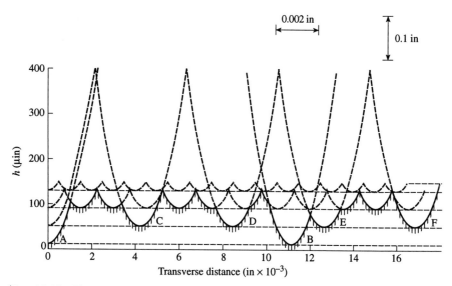

FIG. 10.12. The construction of a transverse profile of a ground surface based on the radial grit population density of Fig. 10.8 for a 60N wheel and assuming all grits at a given level (h) to be equally spaced. Grit tip = sphere of radius $\rho = 0.006$ in (150 μm) (after Nakayama and Shaw 1968).

drawn at levels of $h = 90\,\mu$in and $130\,\mu$in ($2.25\,\mu$m and $3.25\,\mu$m). It is now possible to sketch in the resulting transverse profile on the surface to be expected (shown cross-hatched in Fig. 10.12). It should be noted that the grit tips at a level higher than $130\,\mu$in ($3.25\,\mu$m) do not contribute to the generation of this surface. Thus the maximum peak-to-valley value (R_t) is seen to be about $120\,\mu$in ($3.0\,\mu$m).

When this value ($120\,\mu$in $= 3.0\,\mu$m) is compared with the measured value of peak-to-valley roughness of $100\,\mu$in ($2.5\,\mu$m). It is seen to be too high. This undoubtedly results from the assumed uniform spacing of grits at each level. The actual distribution will be random and should lead to a smaller peak-to-valley value. The random spacing will not change the depth of the deepest valley but may decrease the height of the highest peak.

From this transverse profile approach to surface roughness it is evident that the extreme elevations of a very few outermost grits play the major role in determining the surface finish of a ground surface. If these few grits could be eliminated the surface finish could be greatly improved. Even the finest dressing technique leads to considerable variation in grit elevation.

This graphical approach to surface roughness may be repeated analytically, and this is to be found in Nakayama and Shaw (1968).

Several attempts have been made to treat the generation of surface geometry in terms of randomly distributed grits. Basuray *et al.* (1980, 1981) present one of the better of these efforts, where references to other similar attempts are

SURFACE INTEGRITY

to be found. Here it is recognized that not every potentially active grit will produce a chip, as in the treatment associated with Fig. 10.12. However, instead of all active grits at a given level being uniformly spaced as in Fig. 10.12, they are assumed to be distributed stochastically according to a probability density function that is experimentally determined. Agreement between theory and experiment is relatively good since the key item (the probability density function) is experimentally derived.

Spark-out

The spark-out procedure, in which grinding is continued without in-feed, not only improves accuracy of size and shape but also surface finish.

When a surface is ground a second time without additional in-feed, this is equivalent to doubling the number of effective cutting points per square inch of wheel surface. If the C versus h curve is as shown by the solid line in Fig. 10.13, then the curve corresponding to the first spark-out pass will be the dotted curve marked 1 (which has all the ordinates double those for the solid curve) and that for the second spark-out pass will be the curve marked 2, etc.

Upon substituting eqn (10.16) into eqn (10.22) and solving for h,

$$h = \frac{v}{2VC\sqrt{2\rho D}} \tag{10.21a}$$

For constant values of v, V, ρ, and D:

$$hC = \text{constant}. \tag{10.21b}$$

This will hold for the equilibrium in-feed case (the solid curve marked O in Fig. 10.13) as well as for all spark-out passes. A curve corresponding to eqn (10.21b) is shown as a centerline in Fig. 10.13.

Point A, at which the curve of eqn (10.21b) intersects the O curve, gives the value of h pertaining with uniform in-feed. Point B, at which the curve of eqn (10.21) intersects curve 1, gives the value of h pertaining after the first spark-out pass. The finish is seen gradually to improve with successive spark-out passes.

The actual rate of finish improvement with spark-out will be somewhat less than that indicated, since in some instances there will be no metal to be cut at a given point when a grit again traverses the work surface. This will cause the actual curve of surface finish improvement to lie somewhat above the line AE.

The curves of Fig. 10.13 correspond to actual data for the example previously worked and hence points B, C, D, etc. show the relative surface improvement to be expected for successive spark-out passes.

Hahn (1961) was the first to discuss the role of spark-out and rounding up in internal grinding, with emphasis on the role of wheel–work elastic recovery. Verkerk and Pekelharing (1978) have discussed the role of spark-out

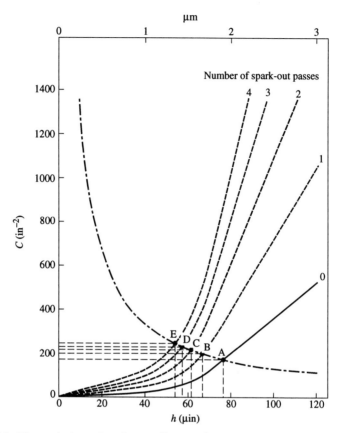

FIG. 10.13. The variation of peak-to-valley roughness with the number of spark-out passes (after Nakayama and Shaw 1968). Wheel: 8 in (203 mm) Al_2O_3, 60H; wheel speed, 6000 f.p.m. (30.5 m s^{-1}); work speed, 48 f.p.m. (0.24 m s^{-1}); wheel depth of cut, 200 μin (5.08 μm); cross-feed, none; fluid, none.

in cylindrical grinding. Shen (1983) has also discussed the significance of spark-out in cylindrical grinding with regard to forces, finish, and residual stresses.

Tracing Direction

The fact that different values of CLA roughness (R_a) are obtained when a given surface is traced longitudinally and transversely using a cut-off (sampling length) of 0.03 in (0.76 mm) is apt to be incorrectly interpreted as meaning that there is in fact a different peak-to-valley roughness in the two directions. For example, Sato (1950) actually computed two components and added them together. In reality, the roughness must be the same in the two directions and the apparent difference is caused by the difference in wavelength in the two directions relative to the cut-off length. The wavelength of individual scratches

TABLE 10.6 *Values of longitudinal and transverse AA roughness for different cut-off lengths (after Nakayama and Shaw 1968)*

Cut-off length (in)	Centerline average roughness (μin)	
	Longitudinal	Transverse
0.100	27	31
0.030	19	26
0.010	8	20

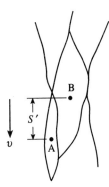

FIG. 10.14. The spacing of two adjacent grinding scratches on a workpiece in the grinding direction (S').

is very much shorter in the transverse direction than in the longitudinal direction. A cut-off of 0.03 in (0.76 mm) is satisfactory for the transverse direction but is entirely too short a sample length for the longitudinal direction.

Table 10.6 gives CLA (R_a) values for the same surface using different cut-off values. It is seen that when the cut-off length is 0.1 in (2.54 mm) there is little difference between the longitudinal and transverse values. However, when the cut-off length is very short 0.01 in (0.25 mm), the difference between the two values is large. It therefore appears safe to assume that the CLA roughness will be independent of tracing direction if a long cut-off length is used and there is no waviness.

Chatter

If a grinding wheel moves in a radial direction relative to the workpiece, in chatter vibration, this may or may not contribute to surface roughness. Roughness will be induced only if adjacent scratches are produced at times which are long relative to the period of the vibration. Figure 10.14 shows two adjacent scratches on a ground surface which are spaced a distance S' apart (points A and B are the deepest points of each groove). The time interval between cuts A and B will be

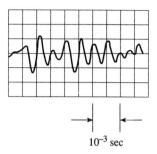

Fig. 10.15. A oscillograph trace of the vibration of a 6 × 18 in (150 × 450 mm) horizontal spindle surface grinding machine.

$$T_{AB} = S'/v. \qquad (10.22)$$

If this time is long compared with the vibrational period, the amplitude of the vibration will influence the surface roughness.

In the previous example of surface grinding with a 60H wheel the mean scratch length was found to be 0.048 in (1.22 mm). The time interval between the formation of two adjacent grits under such conditions, where the table speed is 48 f.p.m. (14.6 m min^{-1}), will be

$$(0.048/48)(60/12) = 5 \times 10^{-3} \text{ s}.$$

The wheel rotational period for this example is equal to $(8\pi/6000)(60/12) = 0.021$ s. Thus, the vibration must have a frequency several times the rotational frequency if the vibration is to contribute to roughness.

The vibrational frequency of the spindle of the surface-grinding machine used in the foregoing example was determined using a vibration pick-up and oscilloscope. The wheel, which was carefully balanced, was rotated at 2900 r.p.m. but was not grinding during the test. Figure 10.15 shows the resulting trace, which indicates a complex variation in amplitude but a constant frequency of about 2200 Hz. When grinding the frequency will be even higher (Pahlitzsch and Cuntze 1966). Thus the vibrational period will be $1/2200 = 0.45 \times 10^{-3}$ s or less. It should be noted that the oscillograph trace does not include possible grit and local work deflection since the wheel was not grinding and, even if it were, the pick-up is not capable of including such components.

Since this time is an order of magnitude smaller than the time to produce adjacent scratches on the surface, the vibration of the wheel should be expected to influence surface roughness and not just cause waviness.

In order to estimate the vibrational amplitude of importance to roughness, it is necessary to measure the relative displacement between wheel and work while grinding. This cannot be done using a pick-up attached to the wheel or work, only since both may move. A convenient way of making the required measurement is to use a single abrasive point as a measuring device in the

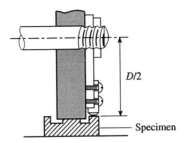

FIG. 10.16. A method of measuring the relative motion between the grinding wheel and the work in surface grinding.

manner shown in Fig 10.16, where a single aluminum oxide abrasive grit mounted in a metal holder with ceramic cement is shown adjustably mounted on a metal disc attached to the same spindle as the abrasive wheel. The specimen to be ground has three flat areas. The center region is the test specimen to be ground by the wheel, while the right-hand outer region is lapped and serves as the recording 'chart' for measuring the vibrational motion. The third left-hand surface, which is at the same higher elevation as the right-hand one, is present as a matter of convenience in lapping the right-hand surface.

Each time the wheel makes one revolution the auxiliary grit cuts a single groove in the 'chart' surface. The length of the scratch produced is a measure of the depth of the scratch and hence of the relative radial motion between wheel and work. The scratch length (l) is related to scratch depth (d') as follows:

$$1 = 2\sqrt{Dd'}, \tag{10.23}$$

where $D/2$ is the radial distance to the auxiliary grit.

This is a particularly attractive measuring technique, since measurement of l provides a 'gain' over the direct measurement of d'. This may be seen by differentiating eqn (10.23) with respect to d':

$$\partial l/\partial d' = \sqrt{D/d'}, \tag{10.24}$$

which is the gain in sensitivity when l is measured in place of d'. In an actual case D will be about 8 in (203 mm) and d' will be about 5×10^{-4} in (12.5 μm), in which case the 'gain' (dl/dd') is 127.

When a white Al_2O_3 24 H wheel was used to grind A-6 tool steel at $V = 6000$ f.p.m. (30.5 m s^{-1}), $d = 400$μin (10 μm), $v = 48$ f.p.m. (0.24 m s^{-1}), the mean amplitude of vibration (d') was found to be 35 ± 10 μin (0.88 $\pm$ 0.25 μm). As shown in Nakayama and Shaw (1968), the surface roughness should be increased by $d'/2$ (~ 18 μin = 0.45 μm) when wheel–work vibration is present.

Methods of reducing chatter include the following:

- the use of a softer wheel
- frequent dressing
- reduction in the work speed v and grinding width b
- a change of wheel speed V
- an increase in wheel diameter D
- the use of a grinding fluid with good lubricity
- an increase is the damping of the system

Snoeys and Brown (1969) have discussed sources of chatter and chatter reduction.

Metallurgical Damage

Introduction

The importance of metallurgical grinding damage to tool steels was first discussed by Tarasov and coworkers (see Tarasov and Lundberg 1946; Tarasov 1950, 1951; Tarasov *et al*. 1957). Hahn (1956) also emphasized the importance of this topic. Field, Kahles and their colleagues published many papers, beginning in the 1960s, concerning the importance of metallurgical damage to ground structural members, and some of this work is considered below. Malkin has stressed the importance of 'burn' to ground surfaces (Malkin 1984, 1989). In addition, there has been considerable interest in metallurgical damage to ground surfaces in Europe, and most of this work has been summarized by Brinksmeier *et al*. (1982).

Metallurgical damage may be detected by change of microstructure, change in hardness at the surface, and the presence of microcracks (very brittle materials) or by changes in fatigue strength, fracture strength, stress corrosion cracking, or the rate of wear. All of these are related to excess surface temperatures in grinding.

In the case of hardened steel, metallurgical damage usually involves an alteration of surface structure that is microscopically visible in an etched section perpendicular to the ground surface. This requires sectioning, mounting with edge protection, polishing, and etching. Edge protection can be provided by use of a copper strike followed by electrodeposition of nickel. A simpler method is to use a polymer mounting material that contains finely divided Al_2O_3 particles (Gatto and Di Lullo 1971) and to cast the polymer in vacuum to eliminate air bubbles. Either of these methods prevents the surface of interest from being eroded during polishing. Littman (1967) has described an effective etching technique for detecting grinding burn. This involves etching in 5 percent HNO_3 in ethanol until the surface is black (5-10 s) followed by washing in warm water and displacing water from the surface with methanol. Then the specimen is etched in 5-10 percent HCl until the black smut is removed (5-10 s). The surface is then neutralized (in 2 percent Na_2CO_3), rinsed with methanol, and dried in warm air. Dark areas correspond to overtempered martensite (OTM)) and light areas are untempered martensite (UTM), while gray areas indicate no thermal injury.

Grinding burn may also be detected by a microhardness traverse. In this case OTM will give microhardness values lower than the base material, while UTM will give values that are higher than the base material. This method must be used with care, since plastic flow at the surface will give a small increase in surface hardness that may be confusing, and a thin OTM layer may be missed if the microhardness indenter penetrates this layer without detecting it.

With very brittle materials (ceramics and glass) surface damage is often in the form of microcracks oriented perpendicular to the lay. Surface damage of this sort is best detected by a liquid penetrant inspection using a developer and ultraviolet light.

Sometimes grinding burn refers to the presence of interference colors visible to the naked eye. However, such films represent oxides of different thickness that depend on time at temperature as well as temperature and the availability of oxygen which may be excluded by a coolant. In most cases, colored films are only of esthetic significance.

A change in metallurgical structure in a ground surface is frequently referred to as a heat affected zone (HAZ).

Metallurgical damage is associated with high surface temperatures and involves one or more of the following surface defects:

- untempered martensite (UTM)
- overtempered martensite (OTM)
- oxidation and decarburization
- superficial microcracks

Collectively, these problems are sometimes lumped together and referred to as grinding burn.

Martensitic Transformations

Since both UTM and OTM obtained when grinding steel are associated with martensitic transformation, a brief review of this important metallurgical topic is presented next. This is based on the extensive monographs of Nishiyama (1978) and Krauss (1990).

Figure 10.17 is the phase equilibrium diagram for iron and carbon, in which weight percent carbon (w/o C) is plotted against temperature. Pure iron will have the body-centered cubic (b.c.c.) structure shown in Fig. 10.18(a) until a temperature of 910 °C is reached. At this temperature the structure changes to face-centered cubic (f.c.c.), as shown in Fig. 10.18(b). The α and γ phases are solid solutions of C in Fe. The solubility of C in α is essentially zero (< 0.03 w/o), while the solubility of C in γ is much greater (up to 1.7 w/o). The small amount of C in solution in α will be randomly distributed interstitially

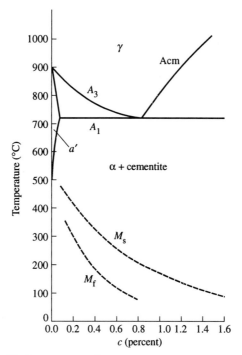

FIG. 10.17. An equilibrium phase diagram for Fe–C. The dashed lines indicate temperatures at which martensitic transformation starts (M_s) and finishes (M_f) when rapidly quenching γ from A_3 with C in solution.

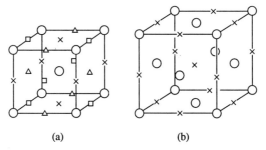

FIG. 10.18. Unit cells for (a) ferrite (α) and (b) austenite (γ). Large circles represent Fe atoms, and $\square$, $\triangle$, and x indicate sites for interstitial C atoms in α and γ solid solutions.

at the points marked $\square$, $\triangle$, and X in Fig. 10.18(a) and the structure of α will remain cubic. The C in solid solution in the γ phase will be at locations such as x in Fig. 10.18(b). This requires an increase in the atomic spacing of C atoms in the vertical direction to accommodate the C. The γ phase with all the carbon in solid solution will have a body-centered tetragonal (b.c.t.) structure, when the temperature falls rapidly below the A_1 line.

For a steel containing 0.45 w/o C the α-γ transformation will begin at temperature A_1 (Fig. 10.17) and will be complete on crossing the A_3 line as the temperature is increased under equilibrium conditions. In heat treating steel, the first step is to heat the material 50–1000 °C above the A_3 line and wait for about 1 h until the part is heated uniformly and all of the carbon has an opportunity to be distributed uniformly throughout the γ iron already formed. The waiting period is necessary since atomic diffusion of C in Fe is involved, which is a relatively slow process. If the material is then rapidly quenched to a relatively low temperature (below the M_f temperature in Fig. 10.17) the b.c.t. structure will be retained with all of the carbon in solid solution. This is a very hard and brittle metastable material (untempered martensite $= \alpha'$) which is subsequently tempered by a time–temperature treatment in which the carbon is precipitated as cementite (Fe_3C) as the b.c.t. structure goes to body-centered cubic α iron (ferrite).

In order to obtain untempered martensite (α') after quenching, the steel must contain at least 0.25 w/o C. If C < 0.25 w/o the transformation on quenching is directly from austenite (γ) to relatively soft ferrite (α) and no hard brittle α' phase is involved. Thus, low-carbon steels cannot be hardened by heat treatment.

Surface Grinding

When a metal is ground and gives a surface temperature above the A_3 line (Fig. 10.17), a layer of material may be transformed to UTM (α'). This is highly undesirable, since α' is metastable and extremely brittle. The action that occurs is the same as in heat treatment of steel with two important differences:

1. There is a very short time at temperature to allow C atoms to diffuse uniformly throughout the γ phase to form an f.c.t. $= \alpha'$ structure.
2. There is no tempering treatment to convert the very brittle α' to α plus Fe_3C.

An important question to be answered is how α' can be obtained in grinding where the time at temperature is a small fraction of a second while it takes a very much longer time in heat treatment. The very long time required to form α' in heat treatment is not due to the time required for the α-γ transformation. It is well established that the time required for the γ to α reaction is extremely short. This is because diffusion is not involved, but there is a cooperative coordinated shift of atoms that occurs in approximately 0.1 μs (Nishiyama 1978). Although not discussed in the literature, probably because it is not involved in heat treatment (Shewmon 1969), the time for the α-γ transformation should also be of the order of 0.1 μs, since it involves the same cooperative coordinated shift of atoms. The long time required to form α' in heat treatment is due to the relatively long time for carbides to dissolve and diffuse to γ unit cells and be absorbed.

In grinding the matrix material (γ) is not stationary as in heat treatment but

TABLE 10.7 *Typical gentle and abusive grinding conditions for horizontal spindle surface grinding of hardened AISI 4340 steel with cross-feed (after Field and Kahles 1971)*

Conditions	Gentle grind	Abusive grind
Wheel grade	H	M
Wheel speed (V), f.p.m. (m s^{-1})	2000 (10.1)	6000 (30.5)
Work speed (v), f.p.m. (m s^{-1})	40 (0.2)	40 (0.2)
Cross-feed per pass (b'), in (mm)	0.050 (1.25)	0.050 (1.25)
Wheel depth of cut (d), in (μm)	0.0002 (5)	0.002 (50)
Fluid	Oil	Dry

is being plastically deformed at a very high rate during heating. The transport of C atoms in grinding is due not only to diffusion but is augmented by displacement of γ relative to C in shear.

Analogy

A useful analogy to the above problem is that of bringing a cup of coffee to a uniform state of sweetness. If crystals of sugar are placed in the coffee without stirring, it will take a long time to reach an homogeneous state of sweetness. However, with vigorous stirring, the desired state may be reached in a few seconds. The displacement due to stirring augments the role of diffusion. The high rate of subsurface plastic flow that accompanies chip formation in grinding provides an action akin to stirring in the coffee analogy. This analogy may be carried one step further, which will be useful in later discussion. If the crystals of sugar are relatively large, it will take a long time to achieve homogeneous sweetness even with vigorous stirring.

Example 1

In order to illustrate some of the points that have been made, a typical example of horizontal spindle surface grinding with cross-feed will be considered. Field and Kahles (1971) refer to gentle and abusive grinding and Table 10.7 gives representative grinding conditions for each of these when grinding AISI 4340 steel ($R_c = 51$). Figure 10.19 shows photomicrographs of transverse sections of material ground under the two conditions of Table 10.7. While the gentle condition shows no evidence of a phase change, the abusive grind shows a substantial white layer. This is UTM (α'), which is very brittle and which lowers the reverse bending endurance limit by about 32 percent (Field and Kahles 1971).

For a wheel diameter of 10 in (254 mm), the wheel–work contact length (l) for the abusive case will be approximately $10(0.002)^{0.5} = 0.141$ in (3.58 mm) and the time of wheel–work contact will be:

(a)

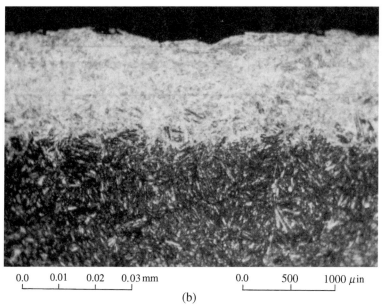

0.0   0.01   0.02   0.03 mm   0.0   500   1000 μin

(b)

FIG. 10.19. Representative polished and etched cross-sections for quenched and tempered AISI 4340 ($R_c = 51$) steel: (a) gentle grinding conditions (Table 10.7); (b) Abusive grinding conditions (Table 10.7) (after Field and Kahles 1985).

$$l/v_w = 0.00359/0.20 = 18 \text{ ms}.$$

The mean surface temperature required for a UTM layer of approximately 10^{-3} in (25 μm) depth, as shown in Fig. 10.19(b), may be estimated from the well known heat transfer equation found in most elementary heat transfer or materials science texts (VanVlack 1970, for example):

$$\theta_y/\theta_s = 1 - \text{erf}\left\{y \bigg/ \left[2\left(\frac{k}{\rho C}T\right)^{0.5}\right]\right\}, \tag{10.25}$$

where θ_y is the mean temperature at a distance y below the surface at time T after the surface reaches a mean temperature θ_s and remains constant at this value; erf is the Gaussian error function; and $\frac{k}{\rho C}$ is the thermal diffusivity.

From an equilibrium phase diagram for AISI 4340 steel, the temperature for the α–γ transition is found to be 800 °C. The thermal diffusivity for this material is approximately $0.127 \text{ cm}^2 \text{ s}^{-1}$ (VanVlack 1970). Substituting into eqn (10.25) with $T = 18$ ms (time of wheel–work contact) the value of the mean surface temperature required for a UTM depth of 25 μm (0.001 in) and an assumed temperature (θ_y) of 800 °C may be found:

$$\frac{800}{\theta_s} = 1 - \text{erf} \frac{25.4 \times 10^{-4}}{2\sqrt{(0.127)(18 \times 10^{-3})}} = 1 - 0.024^3)]^{0.5}\}$$

or

$$\theta_s = 820 \text{ °C}.$$

Thus, to obtain a layer of UTM approximately 0.001 in (25 μm) thick will require a mean surface temperature that is only about 20 °C above the temperature at which the α–γ transition occurs with rapid heating. For this to end up as a layer of UTM there must have been time for the decomposition of the cementite present in the original structure and then for distribution and absorption of the C atoms into the austenite to convert it to the f.c.t. (UTM) structure. An estimate of the distance over which atomic C will diffuse in austenite during the short time of wheel–work contact (18 ms) is of interest in considering whether there is time to obtain UTM.

The rate of diffusion of atomic C in γ iron depends upon the temperature and the concentration gradient. Atomic diffusion theory gives (for example, VanVlack 1970):

$$\frac{C_s - C_y}{C_s - C_0} = \text{erf} \frac{y}{2\sqrt{DT}} \tag{10.26}$$

where C_s is the w/o C at the surface, C_y is the w/o C a distance y below the surface, C_0 is the w/o C of the base material, D is the diffusion coefficient ($10^{-6.5} \text{ cm}^2 \text{ s}^{-1}$) for C diffusing in f.c.c. iron at 1000 °C (VanVlack 1970), and T is the elapsed time (18 ms).

For $C_y = C_0 = 0$, and using an error function table,

$$y/[2(DT)^{0.5}] = 2. \tag{10.27}$$

Thus, the distance over which C will diffuse in 18 ms will be only

$$y = 4\sqrt{DT} = 4\sqrt{(10^{-6.5})(18 \times 10^{-3})} \text{ cm} = 3 \,\mu\text{m}.$$

From the quenched and tempered structure of the base material of Fig. 10.19, it appears that a diffusion distance of $3\,\mu$m during the time at temperature should be sufficient when augmented by 'stirring' due to the subsurface plastic flow that accompanies chip formation in fine grinding (Chapter 4). Figure 10.19(b) confirms that the layer of γ was in fact converted to UTM.

The above diffusion calculations are based on a total time of diffusion at the mean surface temperature which is thought to be a good approximation. In reality, the temperature varies along the wheel-work arc of contact, and the thermal and atomic diffusion coefficients will vary from point to point rather than being constant as assumed. The temperature for the α-γ transformation has been taken as the equilibrium value (800 °C). With rapid heating it is possible that the α-γ transition will not occur until a temperature higher than the 800 ° equilibrium value is reached, just as the α-γ transition occurs at M_s instead of A_1 with rapid cooling. If this is the case, the temperature at the bottom of the HAZ in Fig. 10.19(b) would be greater than 800 °C, but the temperature at the surface would be less than 3 percent greater than the temperature at the bottom of the HAZ.

The presence of a layer of UTM on a ground surface is very detrimental to product performance. Figure 10.20 shows reverse bending (zero mean stress) fatigue results for the specimens of Fig. 10.19.

Annealed Steel

If a material being ground is soft and consists of patches of pearlite in a ferrite matrix, there will normally not be time for the carbon in the pearlite to be dispersed and absorbed completely in the layer of austenite that forms with abusive grinding. Even with the substantial shear pertaining in fine grinding to augment diffusion, there will not be sufficient time at temperature for the carbon in the pearlite patches to be reduced to atomic particles and to be absorbed homogeneously in the layer of austenite that forms in abusive grinding. In the coffee analogy this is equivalent to having sugar crystals that are so large that homogeneous sweetening cannot be achieved in a short time even with vigorous stirring.

Example 2

Eda *et al.* (1993) described an interesting study of HAZ formation by analytical simulation. In this study the dispersion of atomic C in γ iron was assumed to occur by atomic diffusion alone. Figure 10.21 is a photomicrograph presented

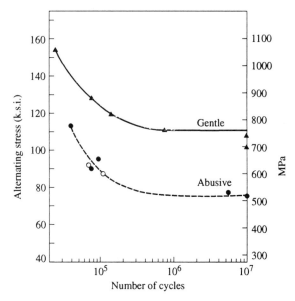

FIG. 10.20. S/N curves of ground AISI 4340 ($R_c = 51$) steel for gentle and abusive grinding. The grinding conditions were the same as in Fig. 10.19 (after Field and Kahles 1985).

FIG. 10.21. The heat-affected zone for annealed AISI 1045 steel after surface grinding with a relatively low work speed, v (after Eda et al. 1993).

to verify the results of the simulation. The grinding conditions are given in Table 10.8 together with those of the previously discussed Example 1. The AISI 1045 steel employed by Eda et al. was in the annealed condition before grinding and had the pregrinding structure evident below the HAZ in Fig. 10.21.

Figure 10.21 is relatively complicated. The total depth of the HAZ is about

TABLE 10.8 *A comparison of grinding conditions for Examples 1 and 2*

Conditions	Example 1	Example 2
Wheel	M grade	WA46KMV
Work	AISI 4340 ($R_c = 51$)	Annealed AISI 1045
Wheel diameter, D_s	10 (254)	350 mm (13.8 in)
Wheel speed, v_s	6000 f.p.m. (30.5 m s^{-1})	5510 f.p.m. (28.0 m s^{-1})
Work speed, v_w	40 f.p.m. (0.20 m s^{-1})	0.08 m s^{-1} (16.4 f.p.m.)
Wheel depth of cut, d	0.002 in (50 μm)	100 μm (0.004 in)
Width ground, b	0.050 in (1.25 mm)	2 mm (0.08 in)
Fluid	Dry	Coolant
Unit removal rate, vd	0.960 in^2 min^{-1}	0.787 in^2 min^{-1}
Arc length, $l = (D_s d)^{0.5}$	0.1414 in	0.235 in
Contact time, $T = l/v_w$	18 ms	72 ms
θ_s	820 °C	851 °C
Diffusion distance	3 μm	6 μ
Depth of HAZ	25 μm (Fig. 10.19b)	130 μm (Fig. 10.21)

130 μm (0.005 in) and contains a variety of structures that are not present in the simulation results. When the HAZ of Fig. 10.21 is compared with the HAZ for the abusive grinding case of Example 1, the depth is seen to be five times as great. Based on this alone, Example 2 might be classified as superabusive grinding.

While the rate of removal per width ground (vd) for Example 2 is actually somewhat less than that for Example 1, the wheel-work contact length l for Example 2 is unusually large, while the work speed v_w is unusually low. These conditions give rise to a relatively long wheel-work contact time T and a small mean undeformed chip thickness (due primarily to low v_w). The low undeformed chip thickness gives rise to a high specific energy and consequently a high value of surface temperature.

Using the same analysis as for Example 1, the mean surface temperature required to give an α-γ temperature of 800 °C at the bottom of the 130 μm (0.005 in) HAZ was found to be 851 °C. Thus, as in the case of Example 1, the temperature of the HAZ is essentially constant from top to bottom.

The diffusion distance of atomic C in γ during wheel-work contact was estimated to be about 6 μm for example 2. The base structure of the material of example 2 consists of patches of fine pearlite and ferrite, each averaging about 25 μm in diameter.

The ferrite in this structure, which transforms essentially instantaneously on reaching the dynamic transformation temperature, is of two types — that in the pearlite patches and that in the ferrite patches.

From electron micrographs of fine pearlite, the extent of austenite between cementite platelets is found to be about 1 μm. Thus, it appears likely that all of the γ that forms within the pearlite patches will end up as UTM after rapid quenching. However, the patches of ferrite between pearlite patches will be about 15 μm across and it is unlikely that a mean diffusion distance of 6 μm for the time at temperature involved for Example 2 (72 ms) is sufficient to cause all of the cementite to be distributed within the austenite. Thus, the white layer in the structure following rapid quenching should contain some cementite. This is seen to be the case for the band at the bottom of the HAZ in Fig. 10.21. The structure in this band appears to consist of a small amount of cementite dispersed in a matrix of UTM.

The other complexities of the HAZ in Fig. 10.21 appear to be due to a variable cooling rate from the bottom to the top of the HAZ, which results from two quenching sources. At the bottom of the HAZ self-quenching is predominant, while at the top of the HAZ quenching by coolant is predominant. Apparently, the resulting cooling rate is a maximum at the bottom of the HAZ and is a minimum just above the white band, but at the surface it is intermediate between that for the light band and the minimum. For points above the light band, the cooling rate is sufficiently low to cause transition of γ to bainite or fine pearlite, the fineness and hardness of the structure increasing as the surface is approached.

While it is well known that an HAZ as shown in Fig. 10.19(b) will cause a significant decrease in performance, the consequences of a complex one such as that in Fig. 10.21 have not been established. However, this may be only of academic interest since, on the basis of wheel life and excessive temperatures and forces, it is expected that grinding under such superabusive conditions would prove to be impractical.

Heat-affected zones in ground steel may be classified as follows with increasing reduction in surface integrity:

- with annealed material or when the C content is < 0.25 w/o
- overtempered martensite
- untempered martensite of quenched and tempered steel where a layer of UTM is produced

In the first case if the A_3 temperature (or some higher critical dynamic transformation temperature) is exceeded, the α phase will be converted to γ instantaneously, but the coarseness of the carbides may allow very little of the C to be absorbed into the γ iron to produce a b.c.t. structure (UTM). The result is the same as if the C content were so low (< 0.25 w/o) that heat treatment is impossible. Of course, no OTM will be involved in the first case.

OTM, which consists of an increase in the size and spacing of precipitated carbides, will lead to softening of the surface. This gives rise to an increase in friction, wear, and the tendency for metal build-up. The extent of surface softening due to OTM will depend upon the surface temperature and the time at temperature ($T = l/v_w$). When formed below an UTM layer, an OTM layer will make the layer of UTM at the surface even more brittle than if the softer OTM were not there. This is because the OTM layer provides a less rigid support for the UTM layer above.

UTM on a ground surface is extremely brittle and drastically reduces the service performance of loaded parts. Such a layer should always be avoided by adjusting grinding conditions or should be removed by a finishing operation.

The surface transformations that occur in grinding are more complex than those in heat treatment, since they involve not only the dynamic effects of variable rates of cooling but also those associated with relatively short times at elevated temperature before quenching. In the case of steel the significance of the short times at temperature in grinding is not associated with time required for the α-γ transformation but rather with the time required for decomposition of carbides and homogenization of C in γ before quenching.

Issues still to be resolved include the following:

- the degree of superheat required above the A_3 temperature for the α-γ transformation under nonequilibrium heating conditions
- the time that it takes to decompose cementite into iron and C before diffusion begins
- the role that plastic deformation in grinding chip formation plays in augmenting ordinary atomic diffusion in the transport of carbon during homogenization of C in γ
- the relative roles of self-quenching and of quenching due to the application of coolants at the ground surface on structures that develop in a HAZ
- the role that surface energy associated with a freshly generated surface has upon the rate of atomic diffusion of C in γ iron

Detrimental thermal effects are also involved with some materials that do not undergo a phase transformation or overtempering at the surface in grinding (titanium alloys, nickel-based alloys, maraging steels, etc.). In these cases high temperatures may induce welding of chips to the ground surface or on grits in the wheel face, cause softening due to recrystallization, lead to the development of microcracks due to excessive surface strains, or cause chemical interactions.

A very important aspect of high grinding temperatures relative to brittle fracture, fatigue, stress corrosion, and other surface integrity concerns is the development of residual stresses in ground surfaces.

Residual Stresses

Introduction

The residual stresses in any static body are elastic and give rise to a balanced set of forces in accordance with static equilibrium. Figure 10.22 shows a free body diagram of a rod. For static equilibrium, the integrated compressive stress in the surface must just equal the integrated tensile stress in the core. Residual stresses are usually associated with localized plastic flow, differential thermal expansion, localized phase transformation or the localized development of microcracks.

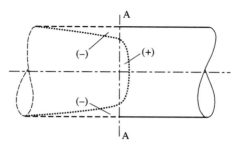

FIG. 10.22. A free body diagram for a rod having compressive residual stress in the surface and statically balanced tensile residual stress in the core.

Shot peening is a process in which hard spherical particles impact a surface, leaving behind a small Brinell impression. Collectively, many impacts produce a thin plastic zone that elongates the surface relative to the bulk of the material below. Since the plastic and elastic zones are connected and must end up having the same length, a compressive residual stress is induced in the surface and a corresponding tensile stress in the core material that remains elastic. Since most stress-induced failures (fracture, fatigue, stress corrosion cracking, and wear) are associated with *tensile* stress, the presence of a compressive residual stress in the critical region is beneficial and a tensile residual stress is detrimental. Shot peening is a treatment that is frequently used to produce residual compressive surface stress in á part, to improve its life or strength (Alman 1943; Moyan 1984). The magnitude and depth of the surface residual stress produced by shot peening depends upon the hardness of the part as well as the size and density of the particles and their velocity and angle of impact. Some ductility is required for shot peening to be effective. Shot peening is effective on quenched and tempered steel ($R_c \sim 60$), but not on glass or ceramics. Figure 10.23 shows representative residual stress patterns produced when shot peening a steel surface.

In the case of glass, shot peening leads to surface roughening. Instead of producing a thin layer of plastically elongated material, small particles are chipped out by the impacting hard spheres. Another method must be used to induce the compressive residual stress in glass that is necessary to counteract the tensile stresses induced by the external bending loads that cause gross fracture of glass. Three techniques are used:

- a change of surface chemistry of the glass
- a thermal treatment called 'tempering'
- lamination

The first of these is achieved by ion exchange and is sometimes referred to as chemical strengthening. When a lithium glass is placed in a molten salt bath rich in sodium ions, larger sodium ions replace smaller lithium ions in the surface, resulting In residual compressive stress in the surface.

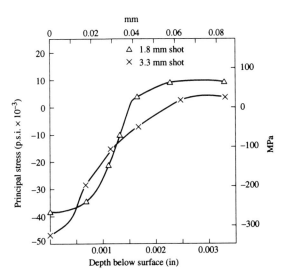

FIG. 10.23. The residual stress produced by shot peening with steel shot directed normal to the surface (after Ramanath 1986).

The tempering process consists of heating the glass above the softening temperature and then cooling it by blowing air on the surface to extract heat. Glass has an amorphous structure and gradually becomes rigid as its temperature is lowered. Above a particular temperature (T_G = glass transition temperature) the glass acts as a very viscous liquid, but below T_G it has the rigidity and behavior of a solid even though it is not crystalline. Tempering is performed by heating the glass above T_G and then extracting heat outward from the surface. The piece of glass contracts uniformly on cooling until T_G is first reached in the surface. As long as both surface and core are above T_G, residual stress will not develop, even though the surface will have contracted more than the core. Viscous flow will occur to accommodate the differential contraction. However, after T_G is reached in the core, the entire part will behave as an elastic solid on further cooling. The surface will then be at a lower temperature than the core and will tend to contract less than the core on cooling to room temperature. Since both surface and core must end up with the same length, the core will be stretched due to residual tensile stress while the surface is contracted by residual compressive stress. The resulting pattern will be as in Fig. 10.22. If only the surface is raised above the T_G temperature, as by rapid heating with a laser, then a tensile stress will result in the surface and a compressive stress in the core on cooling to room temperature. Eye glasses, side and rear automotive glass, and the vulnerable top edges of drinking glasses are strengthened by tempering.

The third method of obtaining residual compressive stress in a glass surface for strengthening purposes is by lamination. This involves using two glasses of different chemistries and hence different coefficients of linear expansion (α).

By fusing together a core glass of high α with two thin layers of skin glass of low α, upon cooling to room temperature the core will tend to contract more than the skin. Both the core and skin layers must end up having the same length, since they were fused together. This results in compressive residual stress in the skin and tensile residual stress in the core. Low-weight-high-strength dinnerware is made using this lamination method of providing compressive residual stresses in the surface to counteract tensile stresses induced by bending on impact.

This brief review of the role that residual stresses play in the strengthening of glass has been included since important counterparts are found in the grinding of metals.

Measurement

Residual stresses may be measured in a number of ways but only two of these have been widely used in grinding studies. Sachs (1927) measured residual stresses in cylindrical members by measuring the change in the outside diameter as material was progressively removed from the bore, or by measuring the diameter of the bore as the outside diameter was reduced progressively by turning. A related technique for nonsymmetrical parts is to drill holes at critical points and measure the adjacent change in strain due to the relaxation of stress. Measurement of Barkhausen noise intensity (Degnen 1976; Lauterlein 1978; Fix 1990) and changes in the velocity of an acoustic wave (Crecraft *et al.* 1967; Kino *et al.* 1979; Johnson 1981) are techniques that have been used for monitoring purposes. When a ferromagnetic material is subjected to an oscillating magnetic field each magnetic domain produces an electronic pulse. The intensity of these pulses is influenced by residual stress in the surface, and also by hardness and other structural changes in a secondary way. The velocity of an acoustic wave is not only sensitive to residual stress but also to other secondary effects that must be understood for intelligent interpretation.

These two monitoring techniques are somewhat akin to acoustic emission monitoring in wear analysis. They are useful monitoring methods that do not provide information concerning what is going on in basic terms, and may be used only after careful calibration over a limited range of interest.

The two methods most widely used in residual stress grinding studies are the plate deflection and X-ray diffraction methods.

In the deflection method, a relatively thin (0.2 in, or 5 mm) plate measuring 3-4 in (75-100 mm) on a side is ground under operating conditions of interest in a direction parallel to one of the sides. Before grinding, both surfaces of the specimen should be smooth, flat, and free of residual stresses (produced by gentle grinding conditions, by use of a superabrasive wheel, by lapping, or by elecropolishing). After test grinding, one side of the plate the curvature is measured on the unground face parallel and perpendicular to the grinding direction. The unground side is then masked, the ground face is etched for a short time, and the new thickness and the curvature of the unground surface are measured. This procedure is repeated until the total amount etched away corresponds to the depth of interest below the surface. Using a technique described by Treuting and Read (1951) a plot of residual stress versus depth below the ground surface is obtained.

The curvature may be measured in several ways:

- by measuring the depth below or above the edges of the specimen at its center (Treuting and Reed 1951)
- from interference fringes, obtained with monochromatic light when an optical flat is placed on the highly reflecting unground reference surface (Letner 1955)
- by use of a metallurgical microscope with a calibrated focusing movement and using a large-aperture short focal length lens (Knowles 1970, 1976)
- from strain gages mounted on the unground surface, which measure strain in the two principal directions (Brinksmeier et al. 1982)
- by means of a stylus tracing instrument (Ramanath 1986)

Letner (1953, 1955) used the Treuting and Read procedure to obtain residual stresses parallel and perpendicular to the grinding direction versus depth of cut. The residual stress curves for the two directions were similar, but those in the grinding direction were somewhat greater than those in the transverse direction.

The X-ray diffraction method determines the change in spacing of lattice planes by diffraction of X-rays of a given wavelength. From this and Bragg's Law an estimate of the corresponding residual stress may be derived. The X-ray diffraction measurements are made on the ground surface, rather than on the polished unground reference surface as in the deflection method. Hence, the surface roughness of the ground surface will have a greater influence on X-ray measurements than on deflection measurements. Also, the X-ray beam measures residual stress at a point, rather than a mean value for the entire surface, as in the deflection method. The X-ray beam penetrates an appreciable distance into the ground surface ($\cong 30 \mu$m for a Cu target but only about 3 μm for Cr).

CIRP (International Institution for Production Engineering Research) has conducted a round-robin investigation in which the principal residual stress (in the grinding direction) for identical ground steel surfaces was measured in several laboratories. Essentially, the same qualitative variation of stress with depth below the surface was obtained by the two methods. However, the magnitude of the peak stresses given by the two methods differed by as much as 30 percent. Figure 10.24 shows mean stress versus depth curves obtained by the two methods. Further experiments failed to reveal an acceptable reason for this lack of agreement between the two carefully performed sets of experiments.

One possible cause for the difference between deflection and X-ray results is that X-rays penetrate a significant distance below the surface and hence

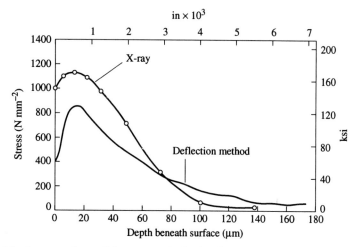

FIG. 10.24. A comparison of the mean principal residual stress distribution obtained by deflection and X-ray diffraction techniques for AISI 52 100 ball bearing steel ($R_c = 62$). Identical specimens measuring 40 × 40 × 4 mm (1.57 × 1.57 × 0.157 in) were surface ground under the following identical conditions and tested in several laboratories throughout the world. Wheel, A46-J8-V; wheel speed, 30 m s^{-1} (5900 f.p.m.); work speed, 400 mm s^{-1} (79 f.p.m.); wheel depth of cut, 7.5 μm (300 μin); fluid; 2 percent emulsion. (after Brinksmeier et al. (1982)

measure an average residual stress over the depth of penetration. If the stress gradient at the surface is steep—as it usually is—this can give rise to an appreciable effect which requires a correction. Koistenen and Marberger (1959) have discussed this problem and have suggested how the correction should be made. The deflection method takes care of this problem by use of etching increments that are small enough to allow the stress over the increment to be considered constant. If a correction for stress gradient were not required, the X-ray method could be considered a nondestructive one for measuring surface stress. However, since the correction for penetration involves the stress gradient at the surface, etching of the surface is required, as in the deflection method. An alternative method of correcting for X-ray penetration is to make measurements using two or more targets that give rays that penetrate to different depths.

When removing thin layers progressively to ascertain the change of stress with depth below the surface, it is important that the method of removal used not introduce stress. Letner (1955) showed that etching a ground steel surface with 5 w/o HNO$_3$ in ethanol satisfies this requirement.

Another difference between the X-ray and deflection techniques is that the X-ray method measures the residual stress over a relatively small area, whereas the deflection method gives an average value for the entire surface. Since, under certain circumstances, grinding conditions vary from point to point, this gives rise to variable residual stress over the surface and therefore, in such cases, X-ray measurements should be made at a number of points.

In applying the deflection method it is important that the grinding direction be the same over the entire specimen, since the principal residual stresses, which are parallel and perpendicular to the grinding direction, differ in magnitude.

The X-ray method is most accurate for single-phase materials, since inhomogeneities and second-phase particles in a heat-treated material cause diffuse patterns, leading to less accurate interpretation. Further details concerning X-ray diffraction stress measurement are available in Cullity (1978) and Macherauch and Hauk (1987).

Colwell (1955) used the X-ray technique to show that the residual stress at the surface of a conventionally ground surface was tensile.

It appears that the conclusion to be drawn is that either method may be used to obtain comparative results for different grinding conditions, but that absolute results obtained by the X-ray technique should not be compared with absolute results from the deflection technique. In the examples of residual stress distribution discussed below, the deflection method was used in all cases. Glickman was apparently the first to measure residual stresses in grinding in 1949.

Examples

Figure 10.25 gives (a) residual stress and (b) fatigue results for samples of AISI 4340 steel ($R_c 50$) ground under abusive, conventional, and gentle conditions. Both the conventional and abusive grinding conditions give very high tensile stresses at the surface (90 000–100 000 p.s.i., or 621–690 MPa), while gentle grinding conditions give a compressive stress at the surface. The principal difference between the conventional and abusive results is the depth of penetration of the residual tensile stress below the surface. The endurance limit in fatigue is seen to vary appreciably with the magnitude of the residual surface stress, which is always the case. The residual stress curves corresponding to the tests of Figs 10.19 and 10.20 were essentially the same as those for Fig. 10.25. Further experiments on hardened AISI 4340 steel (Field and Kahles 1985) have shown that the extent of the residual tensile stress below the surface increases with the depth of the UTM layer, but that the magnitude of the maximum tensile stress is approximately independent of the depth of the UTM layer. From Fig. 10.25(b) it appears that both the magnitude of the peak and the extent of the tensile residual stress below the surface are important in lowering the endurance limit. However, the magnitude of the peak residual tensile stress appears to be more important than the depth of penetration.

Residual stress patterns and fatigue results in reverse bending at room temperature are given in Fig. 10.26. for Inconel 718 and for a titanium alloy in Fig. 10.27. In these cases there was no visible change of surface structure for either abusive or gentle grinding conditions. A phase change is not involved in either case. For both materials, high peak tensile residual stress values and low endurance limits were obtained for abusive grinding, but low tensile or compressive residual stress and high fatigue values were obtained for gentle grinding.

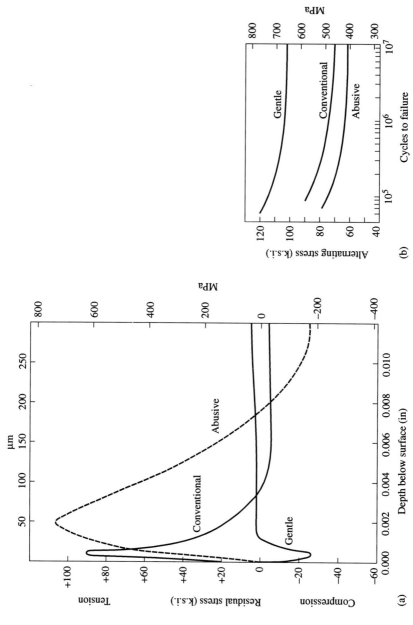

FIG. 10.25. Residual stress patterns (a) and S/N curves (b) for quenched and tempered AISI 4340 steel (R_c 50) surface ground under abusive, conventional, and gentle conditions. The grinding conditions pertaining in each case are given in Table 10.9 (after Field and Koster 1972).

TABLE 10.9 *Gentle, conventional and abusive grinding conditions for AISI 4340 steel ($R_c = 50$) shown in Fig. 10.25*

	Gentle	Conventional	Abusive
Wheel	A46-H-V	A46-K-V	A46-M-V
Wheel speed, f.p.m. (m s^{-1})	2000 (10.2)	6000 (30.5)	6000 (30.5)
Work speed, f.p.m. (m min^{-1})	40 (12.2)	40 (12.2)	40 (12.2)
Downfeed, in per pass (mm per pass)	†	0.001 (0.03)	0.002 (0.05)
Cross-feed, in per pass (mm per pass)	—	0.05 (1.27)	0.05 (1.27)
Fluid	Sulfur oil	1/20 Sol-oil	Dry

† 18 passes at 500 µin per pass (12.5 µm per pass); five passes at 200 µin per pass (5 µm per pass).

Origins of Residual Grinding Stresses

The previously discussed shot peening and glass tempering techniques for improving strength suggest an explanation for the results of Figs 10.26 and 10.27. In fine grinding of metals the removal mechanism is closely related to shot peening, but with small 'chips' extruded from the surface. A plastically deformed layer of appreciable depth is left behind in the finished surface. As long as the surface temperature is below an homologous value of 0.5, cold working should pertain and the resulting residual stress should be compressive, as in shot peening. However, if the surface temperature exceeds $\theta_H = 0.5$ there will be softening of the surface and then the glass 'tempering' mechanism will also be involved.

Heat will flow inward away from the surface and into the cooler substrate when grinding ceases. This will cause cooling of the surface to lag cooling of the core, which will give rise to a tensile component of residual stress in the surface. The net surface stress will be the resultant of the mechanical (peening) component and the thermal component. For gentle grinding the mechanical component is the only one present, except in the case of titanium, which is known for its high grinding temperatures. For abusive grinding the thermal component will be larger than the mechanical component, resulting in a net tensile stress in the surface.

For those cases in which a phase change is involved (hardened steel) there will be an additional effect. In the case of hardened 4340 steel, a layer of UTM is formed, which gives rise to an additional tensile component of stress. However, the peak stress is limited by the yield stress and the main influence that the UTM has relative to stress is in increasing the depth of the residual tensile stress: the greater the depth of UTM, the greater is the depth of the tensile stress. As previously mentioned in discussing Fig. 10.25, the magnitude of the peak residual tensile stress is more important relative to fatigue and fracture than the depth to which the tensile stress penetrates.

As the temperature of a material increases, its indentation hardness decreases, at first gradually and then more rapidly beyond the thermal softening temperature. Figure 10.28 shows values of hardness plotted against values

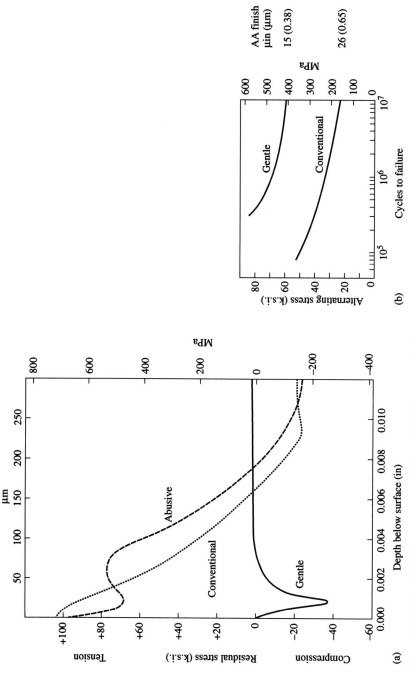

FIG. 10.26. Residual stress patterns (a) and S/N curves (b) for solution treated and aged nickel-based alloy (Inconel 718, $R_c = 44$) surface ground under abusive, conventional, and gentle conditions. The grinding conditions pertaining in each case are given in Table 10.9 (after Field and Koster 1972).

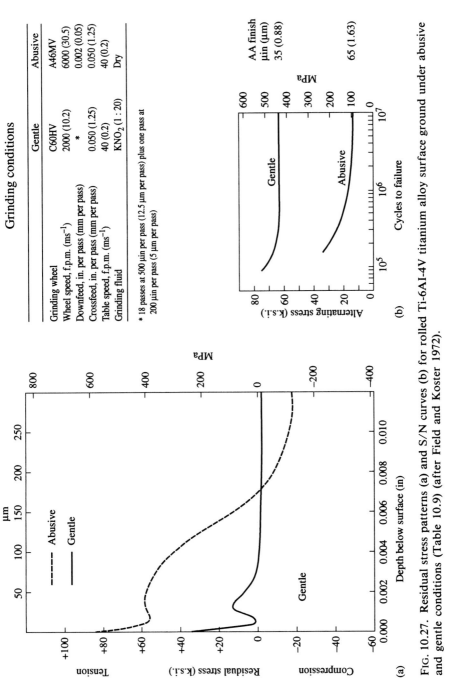

FIG. 10.27. Residual stress patterns (a) and S/N curves (b) for rolled Ti-6Al-4V titanium alloy surface ground under abusive and gentle conditions (Table 10.9) (after Field and Koster 1972).

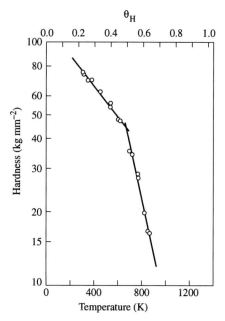

FIG. 10.28. The variation of indentation hardness with homologous temperature (θ_H) for copper (after Westbrook 1953).

of θ_H (homologous temperature) for Cu. It is evident that thermal softening occurs at $\theta_H \simeq 0.5$. Thus, to a first approximation, a thermal component of residual stress (tensile) should be expected at grinding temperatures above $\theta_H = 0.5$.

It is well established that when surface grinding steel under normal conditions with an aluminum oxide wheel a tensile residual stress will be obtained at the surface (Colwell 1955). However, when a cubic boron nitride (CBN) wheel is used under the same conditions, a compressive residual stress will be obtained at the surface (Brinksmeier et al. 1982). For example, Fig. 10.29 shows typical results for two abrasive types (Al_2O_3 and CBN). As discussed in Chapter 9, the surface temperature will be considerably lower with CBN than with Al_2O_3, due to the higher value of $(k\rho C)^{0.5}$ for CBN than for Al_2O_3.

Before CBN was commercially available it was found economical to grind highly alloyed tool steels with diamond wheels (Lindenbeck 1972), despite the fact that diamond wears relatively rapidly when grinding a ferrous material. Since diamond has a particularly high value of $(k\rho C)^{0.5}$, it is found that when steel is ground with a diamond wheel a compressive residual surface stress is obtained instead of the tensile residual stress obtained with Al_2O_3.

Dodd and Kumar (1985) found that the fatigue life of gears ground with CBN was 20 percent greater than for gears ground with Al_2O_3. Althaus (1985) found that the residual stress when internally grinding steel with Al_2O_3

was about $+30\,000$ p.s.i. ($+207$ MPa), but only about $-30\,000$ p.s.i. (-207 MPa) with CBN.

Ceramics

When grinding ceramics, residual surface stresses will tend to be compressive for the following reasons:

- low specific energy
- high $(k\rho C)_s^{0.5}$, particularly when using a diamond wheel
- the high melting point of ceramics, and hence a high θ_H

The first two of these give a relatively low surface temperature, while the last gives a high softening temperature.

Figure 10.30 shows representative results when a Ni–ZN ferrite memory disc material is ground with a resin-bonded diamond wheel. The residual stress at the surface is compressive. If this material were to be ground with an Al_2O_3 wheel, the surface stress would probably still be compressive since, despite the much higher temperature that would pertain, the thermal softening point would not be reached. Actually, this material cannot be successfully ground with Al_2O_3, since many surface cracks are obtained and there is considerable transfer of the material ground to the Al_2O_3 abrasive particles.

Residual surface stresses in Ni–Zn ferrite ceramic surfaces have been measured by Chandrasekar and Chaudhri (1993) using the deflection and X-ray techniques. Mechano-chemical polishing with a 7 v/o solution of orthophosphoric acid in water and 0.05 μm iron oxide abrasive powder was used to remove surface layers without introducing additional residual stress. The X-ray method showed substantial point-to-point variation in residual stress, and mean values of residual surface stress were about 20 percent higher by the X-ray method than by the deflection method. Residual stress measurements were also performed on annealed and tempered soda-lime glass surfaces by the deflection method, and by an indentation technique in which the radius of median half-penny cracks produced by a 90° apex angle indenter was used to estimate the residual stress.

The residual compressive stress (σ_c) in a brittle material may be estimated as follows for an indenter having a 90° apex angle (Lawn and Fuller 1975; Chandrasekar and Chaudhri 1993):

$$\sigma_c = (P^* - P)/\pi^{3/2} c^2 \tag{10.28}$$

where c is the radius of the half-penny median crack at the surface, P^* is the indenter load to produce a crack of radius c with residual compressive stress, and P is the indenter load to produce a crack of same radius c without compressive stress. It was found that values of residual compressive stress estimated by the indentation and deflection methods were in reasonably good agreement.

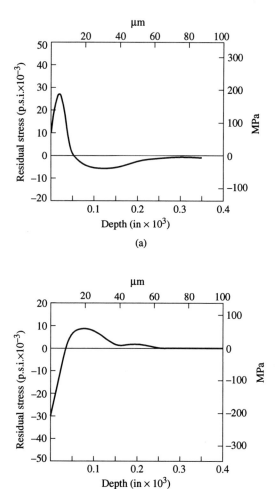

FIG. 10.29. The variation of residual stress parallel to the grinding direction with the depth below the ground surface (after Ramanath and Shaw 1986). (a) Case carburized steel ($R_c = 62$) ground in air with an Al_2O_3 wheel (A60 J8 VBE) at 6300 f.p.m. (32 m s^{-1}); wheel depth of cut, 0.0005 in (12 μm); table speed, 28 f.p.m. (0.14 m s^{-1}); cross-feed, 0.04 in per stroke (1 mm stroke). (b) The same grinding conditions with a resin-bonded CBN wheel (CB150-TBA-1/8).

Samuel and Chandrasekar (1989) have investigated the effect of residual stresses on the fracture strength of ground ceramics in four-point bending. Figure 10.31 shows residual stress patterns parallel and perpendicular to the grinding direction in plunge surface grinding for a silicon nitride ceramic. The residual stresses in the two orthogonal directions were about the same, unlike those for a metal, where the residual stress in the grinding direction is greater

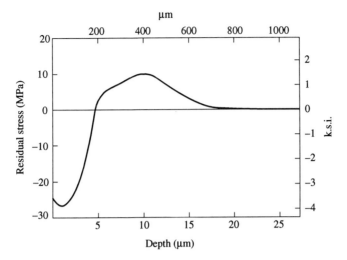

FIG. 10.30. The variation of the residual stress parallel to the grinding direction with the depth below the ground surface (after Ramanath and Shaw 1968). Work material Ni–Zn ferrite; wheel, resin-bonded diamond (400 grit size); wheel speed, 3280 f.p.m. (16.7 m s^{-1}); work speed 45 f.p.m. (0.023 m $^{-1}$); wheel depth of cut, 600 μin (15 μm).

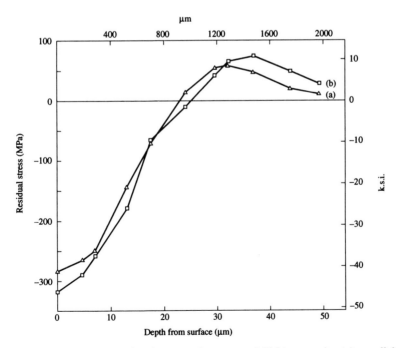

FIG. 10.31. Residual stresses in plunge surface ground Si_3N_4 ceramic, (a) parallel and (b) perpendicular to the grinding direction (after Samuel and Chandrasekar 1989). Wheel, 400 grit resin-bonded diamond; wheel speed, 37 m^{-1} (7282 f.p.m.); table speed, 23 mm s^{-1} (0.91 i.p.s.); depth of cut, 0.02 mm (800 μin).

than that perpendicular to the grinding direction. Similar patterns were found for other ceramics, including a tetragonal zirconia polycrystalline (TZP) ceramic. When TZP material is ground it undergoes a phase transformation from a tetragonal to a monoclinic structure, with an increase in volume. This gives rise to additional compressive stress which tends to close advancing cracks. It is found that in all cases the increase in compressive stress when ceramics are ground gives rise to an increase in fracture strength in four-point bending.

References

Alman, J. O. (1943). *SAE J. Trans.* **51**, 256.
Althaus, P. (1985). *Proc-Chicago Conf. on Superabrasives.*
Basuray, P. K., Sahay, B. V., and Lal, G. K. (1980). *Int. J. Mech. Tool Des. Res.* **20**, 265.
Basuray, P. K., Sahay, B. V., and Lal, G. K. (1981). *Int. J. Prod. Res.* **19**, 677, 689.
Brinksmeier, E., Cammett, J. T., Koenig, W., Leskovar, P., Peters, J., and Toenshoff, H. K. (1982). *Ann. CIRP* **31**(2), 491.
Chandrasekar, S. and Chaudhri, M. M. (1993). *Phil. Mag.* **A67**, 1187.
Colwell, L. V. (1955). *Trans. ASME* **77**, 149.
Crecraft, D.I. (1967). *J. Sound Vib.* **5**, 173.
Cullity, B. D. (1978). *Elements of X-ray diffraction.* Addison Wesley, Reading, Massachusetts.
Degnen, W. (1976). Dissertation, T. H. Karl Marx Stadt, DDR.
Dodd, H. D. and Kumar, K. V. (1985). *Proc. NAMRC*, 412.
Eda, H., Ohmura, E., and Yamauchi, S. (1993). *Ann. CIRP* **42**(1), 389.
Farmer, D. A., Brecker, J. N., and Shaw, M. C. (1967). *Proc. Instn Mech. Engrs* Part 3K, 182.
Field, M. and Kahles, J. F. (1964). *DMIC Report to US Air Force.*
Field, M. and Kahles, J. F. (1971). *Ann. CIRP* **20**(2), 153.
Field, M. and Kahles, J. F. (1985). *ASME PED* **16**, 175.
Field, M. and Koster, W. P. (1972). *Proc. Int. Grinding Conf.* Carnegie Press, Pittsburgh, Pennsylvania, p. 666.
Fix, R. M. (1990). *Abr. Engng Soc. Mag.*, Winter, 7.
Gatto, L. R. and Di Lullo, T. D. (1971). *SME Tech. Paper 1Q71*, 225.
Hahn, R. S. (1956). *Trans. ASME* **78**, 807.
Hahn, R. S. (1961). *J. Engng Ind.* **83**, 131.
Johnson, G. C. (1981). *J. appl. Mech.* **48**, 791.
Kino, G. S. *et al.* (1979). *J. appl. Phys.* **50**, 2609.
Knowles, J. E. (1970). *J. appl. Phys.* **D-3**, 1346.
Knowles, J. E. (1976). *Trans. IEEE* **MTT-24**, 117.
Koistenen, D. P. and Marburger, R. E. (1959). *Trans. ASM* **51**, 537.
Krauss, G. (1990). *Heat Treatment and Processing Principles.* ASM. Metals Park, OH.
Lauterlein, T. (1978). Dissertation, T. H. Karl Marx Stadt, DDR.
Lawn, B. R. and Fuller, E. R. (1975). *J. Mat. Sci.* **10**, 2016.
Letner, H. R. (1953). *Proc. Soc. Exp. Stress Anal.* **10**, 23.
Letner, H. R. (1955). *Trans. ASME* **77**, 149.

Lindenbeck, D. A. (1972). *New developments in grinding.* Carnegie Press, Pittsburgh, PA, p. 932.
Littman, W. E. (1967). *ASTME Preprint. MR67*, 593.
Macherauch, E. and Hauk, V. (1987). *Residual stresses in science and technology, vols I and II.* DGM Metallurgy Information, New York.
Malkin, S. (1984). *J appl. Metalworking*, **3**, 95.
Malkin, S. (1989). *Grinding technology.* Ellis Horwood, Chichester, U.K.
Montgomery, D. C. (1991). *Design and analysis of experiments.* John Wiley, New York.
Moyan, S. (1984). *J. Metals* **36**, 21.
Nakayama, K. and Shaw, M. C. (1968). *Proc. Instn Mech. Engrs*, 182. Part 3K.
Nishiyama, Z. (1978). *Martensitic transformation.* Academic Press, New York.
Orioka, T. (1957). *Rep. Fac. Engng, Yamanashi Univ*, Japan, No. 8.
Orioka, T. (1961). *J. Trans. Japan Soc. Grinding Engrs* **1**, 27.
Pahlitzsch, G. and Cuntze, E. O. (1966). *Ann. CIRP* **14**(1), 125.
Ramanath, S. (1986). PhD. dissertation, Arizona State University.
Ramanath, S. and Shaw, M. C. (1986). *Proc. NAMRC*, 636.
Sachs, G. (1927). *Z. Metallkunde* **19**, 352.
Samuel, R. and Chandrasekar, S. (1989). *J. Am. Ceram. Soc.* **71**, 1960.
Sato, K. (1950). *J. Soc. Prec. Mech.* **16**, 77.
Shen, C. H. (1983). Proc. NAMRC, 310, SME, Dearborn.
Shewmon, P. G. (1969). *Transformations in metals.* McGraw Hill, New York.
Snoeys, R. and Brown, D. (1969). *Proc IMTDR Conf.*, 325. Pergamon Press, Oxford.
Tarasov, L. P. (1950). *Machining theory and practice.* ASM, p. 409.
Tarasov, L. P. (1951). *Trans. ASME* **73**, 1144.
Tarasov, L. P. and Lundberg, C. O. (1946). *Trans. ASME* **68**, 389.
Tarasov, L. P., Hyler, S., and Letner, H. R. (1957). *Proc. ASTM* **57**, 601.
Treuting, R. G. and Read, W. T. (1951). *J. appl. Phys.* **22**, 130.
VanVlack, L. H. (1970). *Materials science for engineers.* Addison Wesley, Reading, Massachusetts.
Verkerk, J. and Pekelharing, A. J. (1978). *Ann. CIRP* **27**(1), 227.
Westbrook, J. M. (1953). *Trans. ASM* **45**, 221.
Yang, C. T. and Shaw, M. C. (1955). *Trans. ASME* **77**, 645.

11

WHEEL LIFE

Introduction

In FFG, wheel life depends upon when the wheel must be redressed due to wear. There are many reasons why a wheel needs to be redressed, including:
- excessive forces or power
- poor surface integrity (finish, phase transformation, and tensile residual stress)
- inadequate dimensional accuracy due to excessive temperature or wear
- instability, leading to waviness

In SRG periodic dressing is not required and wheel life is measured in terms of a grinding ratio (G = volume ground away/volume of wheel consumed).

Wheel wear involves the development of wear flats on abrasive particles, attritious wear, microchipping, and loss of entire grits due to bond fracture. The wear of individual abrasive particles without influence of bonding material is discussed in Chapter 3. This involves two types of test:
- rubbing without chip formation, which emphasizes attritious wear
- chip formation that gives greater emphasis to microchipping

In this chapter the wear and life of complete wheels in FFG is considered, which includes the influence of bond material. In FFG the rate of wheel wear is relatively low; a great deal of material must be ground away to obtain meaningful results and this involves considerable time. An accelerated test procedure that involves the expenditure of less material and time is highly desirable. Many unsuccessful attempts have been made to devise accelerated tool life test procedures in cutting and grinding by use of unusually severe test conditions or by extrapolating data taken over a relatively short time interval. The former has been unsuccessful, since exaggerated test conditions are not representative of those of practical interest, while the latter has been unsuccessful because the rate of wear is not linear and frequently involves cascading. This is largely responsible for the lack of success of radioactive testing. Although it does not appear possible to devise a meaningful procedure that conserves time, it is possible to conserve test material. This may be done by isolating a small area on the wheel face that is used to grind a groove in the work piece under conditions that gave the same mean undeformed chip thickness ($\bar{t}$) as that involved with the full-scale prototype. Since it has been found that grinding conditions are essentially the same for different grinding operations as long as $\bar{t}$ is the same, this should yield representative results.

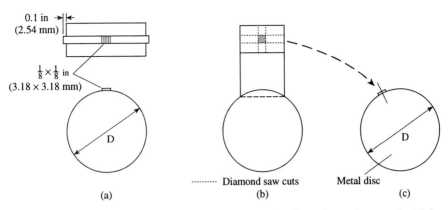

FIG. 11.1. Two methods of preparing clusters of abrasive grits to be tested: (a) by dressing away all but a small group of grits; (b) by sectioning a small specimen from the dressed wheel face and attaching it to the head of a screw mounted in the periphery of a metal disc, as shown at (c).

Cluster Overcut Fly Grinding (COFG)

A small cluster of grits may be used in two ways. The cluster may be formed directly on the surface of the grinding wheel by dressing away all but a $\frac{1}{8} \times \frac{1}{8}$ in (3.18 × 3.16 mm) area, as shown in Fig. 11.1(a), or it may be obtained by removing a small piece of material from the surface of the wheel using a diamond saw (Fig. 11.1(b)). In the latter case the small piece removed is cemented to the head of a screw that is mounted radially in the surface of a metal disc (Fig. 11.1(c)). In either case, the cluster is prepared after the wheel has been diamond dressed in the normal manner. The sectioning technique is of course advantageous when it is desired to examine a used cluster in the scanning electron microscope, or when it is desired to study wheel performance at speeds in excess of those that are safe for a complete grinding wheel. By preparing a cluster to be tested in this manner, the bond strength, the number of active grits per unit area, and all other pertinent items will be the same for the cluster as for the complete wheel.

The COFG technique is such a powerful one because each active grit in the cluster cuts its own groove, the profile of which is left behind in the final composite groove surface. This is so because the circumferential spacing of successive grits that make identical cuts (l_r) is greater than the length of the cluster ($\frac{1}{8}$ in = 3.18 mm), and may be demonstrated as follows. For a conventional grinding wheel of 60 grit size, the number of of active grits per square inch is about 600 (93 per cm²). In FFG, the maximum undeformed chip thickness (t) is characteristically about 100 μin (2.5 mm) and the corresponding scratch width (b') is about 20 times this value, or about 2000 μin (50 μm). The value of l_r may be estimated from the following relation (see eqn 9.1):

$$l_r b' C = 1.$$

Substituting values for b' and C,

$$l_r = 1/(600)(0.002) = 0.83 \text{ in } (21.1 \text{ mm}).$$

This is considerably greater than the circumferential length of the cluster (0.125 in, or 3.18 mm). Thus, the active grits in the surface of a cluster will be distributed so as to cut individual grooves in the work material and thus to leave behind a 'fingerprint' that may be subsequently analyzed. This provides, first of all, another convenient way in which to estimate the active number of cutting points per unit area (C). When a diamond stylus traces the geometry of the composite groove surface perpendicular to the cutting direction, a number of points (n) is observed. Then for a $\frac{1}{8} \times \frac{1}{8}$ in (3.18 × 3.18 mm) cluster, the number of active cutting points per square inch (C) will be

$$C = n/(0.125)^2 = 64n. \qquad (11.1)$$

By making successive transverse tracings at small intervals along the length of the groove, it is possible to identify the cuts made by the individual grits and to measure their corresponding depths of cut. Thus, the amount of wear corresponding to a precise number of cuts for each individual active grit, as well as the mean rate of wear for the cluster, may be found. If a complete wheel is tested having the same width ($\frac{1}{8}$ in = 3.18 mm) and the value of table speed (v) is adjusted to give the same values of $\bar{t}$, then the relation of length of work ground by the complete wheel (L_W) to that ground by the cluster (L_C) for the same amount of wear will be

$$L_W/L_C = \pi D/l_c \qquad (11.2)$$

where D is the wheel diameter and l_c is the circumferential cluster length (0.125 in = 3.18 mm). The foregoing is illustrated by the following example.

Example 1

It is desired to know the rate of wear of a 60 grit aluminum oxide wheel when surface grinding a steel specimen under the following conditions:

Wheel: A60-N8-V
Work material: AISI 52 100 steel ($R_c = 40$)
Spindle speed, N: 3450 r.p.m. (6323 f.p.m. = 32 m s^{-1})
Diameter of wheel, D: 7 in (178 mm)
Depth of cut, d: 250 μin (6.25 μm)
Mean undeformed chip thickness, $\bar{t}$ (eqn 4.5): 50 μin (1.25 μm)

For OCFG, the mean undeformed chip thickness will be

$$\bar{t} = (vl)/(ND). \qquad (11.3)$$

Substituting for l (eqn 9.1) and solving for v:

WHEEL LIFE   317

$$v = Nt'\sqrt{D/d},\qquad(11.4)$$

where N is the r.p.m. of the wheel.

The table speed (v_c) for COFG to give the same $\bar{t}$ as for the complete wheel will be:

$$v_c = \frac{(3450)(50 \times 10^{-6})}{12}\sqrt{\frac{7}{250 \times 10^{-6}}} = 2.41 \text{ f.p.m.}\ (0.012 \text{ m s}^{-1}).$$

A cluster of abrasive grits measuring $\frac{1}{8} \times \frac{1}{8}$ in (3.18 × 3.18 mm) was formed on the complete wheel as shown in Fig. 11.1 and a groove was cut in the work material at the above calculated table speed. Figures 11.2(a), (b) and (c) are traces of the composite groove made at distances $\frac{1}{16}$, $\frac{1}{2}$ and 1 in (2.19, 12.7 and 25.4 mm respectively from the start of the cut. It is seen that there are 12 active cutting points (marked 1, 2, 3, ..., 12 in Fig. 11.2) and that their cutting paths can be followed with grinding distance. Thus, it is possible to estimate C from eqn (11.1) as follows:

$$C = 64n = (64)(12) = 768 \text{ in}^{-2}\ (119 \text{ cm}^{-2}).$$

Table 11.1 gives the actual depths of cut of each active point and the mean depth of cut of the cluster shown in Fig. 11.2. It is seen that some points (1, 3, 4, etc.) show a decrease in d (positive wear), while some others (2, 5, 9) show an increase in d (negative wear). The negative wear observed for some points is due to the metal build-up on the tip of the abrasive grit formed during the cutting process. The mean depth of cut of the cluster in Fig. 11.2(a) is found to be 238.3 μin (5.96 μm), while in Fig. 11.2(c) it is 166.25 μin (4.16 μm). This corresponds to a mean wear of 72.08 μin (1.80 μm) in grinding a distance of $\frac{15}{16}$ in (23.81 mm) or after 112 cuts. Thus, the mean wear per cut of the cluster is found to be 0.64 μin/cut (0.016 μm/cut).

To obtain the same amount of wear as in COFG, the test with the complete wheel (subscript w) should be run with an active wheel width of $\frac{1}{8}$ inch (3.18 mm) and a table speed of

$$v_w = \bar{t}^2 VCr\sqrt{D/d},$$

in accordance with eqn (4.5).

From the COFG test we have $C = 768 \text{ in}^{-2}$ (119 cm^{-2}) and $r = 30$ (estimated mean value from Fig. 11.2) and, therefore

$$v_w = (50 \times 10^{-6})(6322)(768)(30)(7/250 \times 10^{-6})^{0.5}$$
$$= 60.93 \text{ f.p.m.}\ (0.31 \text{ m s}^{-1}).$$

The corresponding complete wheel–work length that will have to be ground is found from eqn (11.2):

$$L_W = L_C \pi D/l_C = \frac{(0.94)(7\pi)}{0.125} = 165 \text{ in}\ (0.42 \text{ m}).$$

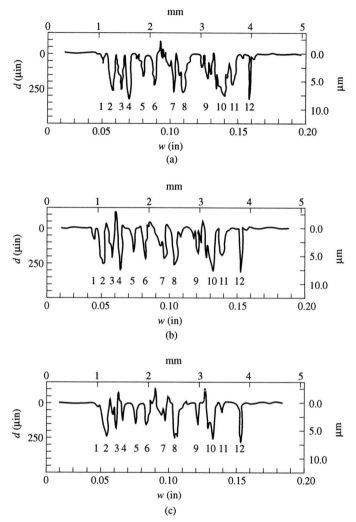

FIG. 11.2. Stylus traces transverse to the grinding direction of a groove produced under COFG conditions: (a) $\frac{1}{16}$ in (1.59 mm) from the start, after seven cuts; (b) $\frac{1}{2}$ in (12.7 mm) from the start, after 56 cuts; (c) 1 in (25.4 mm) from the start after 112 cuts (after Kumar *et al.* 1980).

Thus, only about 0.6 percent as much grinding is involved in the cluster test and, at the same time, statistical details of the wear of the individual grits are readily available for study, in addition to the mean rate of wear for an entire grinding wheel.

Example 2

A cluster of abrasive grits was formed on a conventional grinding wheel by dressing away all but a $\frac{1}{8}$ inch (8 mm) square area on the wheel face. Before

TABLE 11.1 Actual depth of cut (d) of active cutting points in COFG in Example 1

Distance from start (in)	Depth of cut of active points ((μ)in)												Average d ((μ)in)
	1	2	3	4	5	6	7	8	9	10	11	12	
1/16	75	250	250	325	165	225	270	270	175	310	220	325	238.33
1/2	65	255	210	295	170	225	225	255	190	300	195	310	224.58
1	30	250	170	125	150	145	125	240	150	255	75	280	166.25

forming the cluster the wheel was dressed with a single-point diamond tool. The dressing conditions and other pertinent information for this test are given below:

> Dressing conditions: four passes with 0.0005 in (12.5 μm) depth, in-feed = 2 f.p.m. (0.01 m s^{-1})
> Wheel: A60-J8-V
> Work: T-15 tool steel ($R_c = 60$)
> Spindle speed, N: 3450 r.p.m.
> Diameter of wheel, D: 6.0 in (152 mm)
> Wheel depth of cut, d: 300 μin (7.5 μm)
> Mean undeformed chip thickness, $\bar{t}$: 60 μin (1.5 μm) (from eqn 11.4)

The table speed (v) that will give $\bar{t} = 60$ μin (1.5 μm) is found from eqn (11.4) to be 2.44 f.p.m. (0.012 m s^{-1}).

Figure 11.3 gives traces of the grooves at different distances from the start of the cut. It is seen that there are 15 active cutting points in this case and hence the number of active cutting points per unit area is 960 per in^2 (1.49 mm^2).

Table 11.2 gives the actual depth of cut of each active grit and the mean depth of cut for the entire cluster for each trace. The depth of cut is different for each active point and is seen to vary with cutting distance. Figure 11.4 is a plot of the mean wear per cut of the cluster ($\triangle \bar{d}/n$) versus the number of cuts (n). It is seen that the rate of wear is initially high, after which it gradually decreases. The mean value of r (ratio of scratch width to scratch depth) from Fig. 11.3 is 30.

For comparable grinding conditions, values of t and l should be the same for COFG and complete wheel tests. If D and d are the same for the two tests, then l will be the same. The value of $\bar{t}$ is made the same by adjusting v_w to give $\bar{t} = 60$ μin (1.5 μm) and when this is done $v_w = 80$ f.p.m. (0.41 m s^{-1}).

The dimensions of the work material used were 6 × 4 in (152 × 102 mm). This allows 19 grooves to be ground for each test (each of length 6 in (152 mm) and width $\frac{1}{8}$ in (3 mm). The relation of the length of work ground by the complete wheel (L_W) to that ground by the cluster (L_C) that should give the same amount of wear is as follows:

$$L_W/L_C = 6\pi/0.125 = 151.$$

The value of L_W corresponding to 19 grooves ground with the complete wheel is 114 in (2900 mm) and hence $L_C = 0.75$ in (19 mm).

The COFG test involves intermittent cutting, while grinding with a complete wheel does not. It was therefore decided to run two sets of complete wheel tests to investigate the interruption effect. In the first complete wheel test, the number of interruptions was $114/6 = 19$, while in the second series, the workpiece was provided with transverse grooves that were 1.5 in (38 mm) apart, thus yielding $114/1.5 = 76$ interruptions. The number of interruptions for the COFG test with $L_C = 0.75$ was 84, which corresponds closely to the second test.

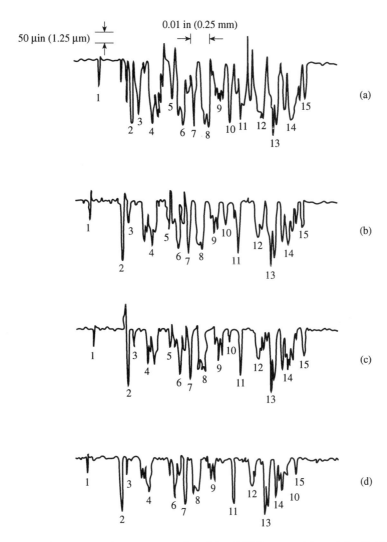

FIG. 11.3. Stylus traces of grooves produced in COFG tests: (a) at $\frac{1}{8}$ in (3.18 mm) from the beginning, after 14 cuts; (b) at $\frac{1}{2}$ in (12.7 mm) from the beginning, after 56 cuts; (c) at $\frac{3}{4}$ in (19.05 mm) from the beginning, after 84 cuts; (d) at 1 in (25.40 mm) from the beginning, after 112 cuts (after Kumar and Shaw 1979).

The two complete wheel tests were performed under conditions that were identical to those for the COFG test except for the table speed v, as explained above. Figure 11.5 shows the variations of wear ($\triangle d$) with length ground (L_W) both with and without transverse grooves. The two results are remarkably close, which suggests that interruption of grinding is of negligible importance. Figure 11.6 shows the variation in mean wheel wear ($\triangle \bar{d}$) versus

TABLE 11.2 Depth of cut of active points, d (μin) †

Distance from start (in)	Grit number															$\bar{d}$ (μ in)
	1	2	3	4	5	6	7	8	9	10	11	12	13	14	15	
1/8	115	300	250	260	170	310	310	315	190	300	285	270	360	280	170	259.0
1/2	100	300	110	210	140	230	255	235	135	120	255	180	310	210	130	194.7
3/4	80	270	80	170	90	215	250	200	150	50	220	150	305	195	140	171.0
1.0	75	260	85	160	0	200	225	170	105	20	205	145	280	195	80	147.0

† 40 μin = 1 μm.

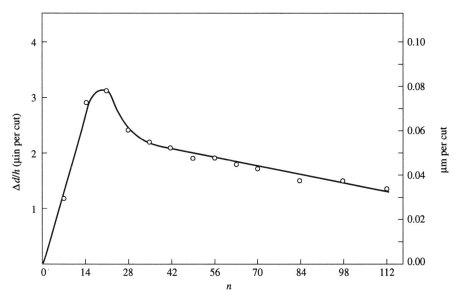

FIG. 11.4. The variation of the mean wear per cut $\triangle \bar{d}/n$ with the number of cuts n for COFG tests (after Kumar and Shaw 1979).

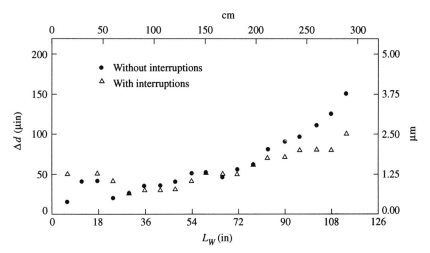

FIG. 11.5. The variation of cumulative wear $\triangle d$ with the length ground (L_W) for a complete wheel with and without a transverse groove (after Kumar and Shaw 1979).

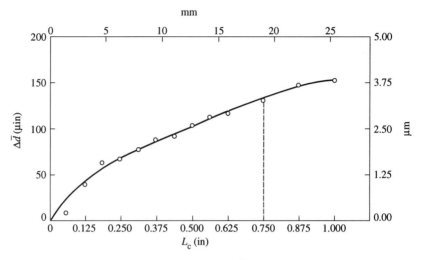

FIG. 11.6. The variation of cumulative wear $\Delta \bar{d}$ with length (L_C) for the COFG test of Fig. 11.4 (after Kumar and Shaw 1979).

distance ground (L_C) for the COFG test of Fig. 11.3. At the value of L_C corresponding to L_W = 114 in (2896 mm) (i.e. L_C = 0.75 in or 19 mm), the wear was found to be 129 μin (3.23 μm) in the COFG test. The corresponding wear in two complete wheel tests was found to be 150 μin (3.75 μm) without interruptions and 100 μin (2.50 μm) with interruptions. Thus, the amount of wear determined by COFG appears to be in good agreement with that obtained with a complete wheel.

Local Wheel Deflection

There is yet another type of test for which COFG is highly advantageous. This involves its use in conjunction with a quick-stop device. A quick-stop device is one normally used to stop the action of a cutting tool by moving it out of range so rapidly that the chip root that remains may be studied to reveal details of the chip formation process. It is extremely difficult to do this during grinding, since grinding speeds are so high that it becomes next to impossible to accelerate even a very small workpiece out of the cutting zone without the resulting chip root being influenced by the removal process. However, this is not the quick-stop test proposed here. Because the abrasive particles are in contact with the work for only a small fraction of each revolution, it is possible to remove a small workpiece during the time the abrasive cluster is out of contact with the work. The resulting specimen will not contain the usual chip root but, instead, will reveal the actual cutting path, which may be used to estimate local wheel–work deflection. The local wheel–work deflection will cause the effective wheel diameter D' to be greater than the actual wheel diameter D, and the ratio D'/D will be related to the grinding length ratio l'/l as follows:

$$D'/D = (l'/l)^{0.5}. \qquad (11.5)$$

The grinding length with deflection (l') may be measured using a stylus moving in the grinding direction on the COFG specimen.

COFG Summary

The COFG method of investigation of grinding performance provides a convenient means for studying the influence of a wide variety of grinding parameters (such as $\bar{t}$, C and l) on the grinding behavior of the abrasive materials used in different grinding systems. It is evident from the above discussion that the following items may be directly determined from a COFG test:

- the effective number of cutting points per unit area, corresponding to different values of undeformed chip thickness, dressing method, grinding fluid, wheel design, work material
- the mean value of the scratch width : scratch depth ratio (r) and the distribution of such values for different values of mean undeformed chip thickness ($\bar{t}$), wheel wear, etc.
- the wear per cut and the statistical distribution of such values for different grinding condition
- the actual wheel–work contact length (l'), including local wheel–work deflection for different grades of wheels and different operating conditions, using a much less critical quick-stop procedure than that required for chip root studies in conventional grinding
- the mean values of the grinding forces and the specific energy and the statistical distribution of these values for different values of mean undeformed chip thickness ($\bar{t}$) and amounts of wheel wear using the two-component piezoelectric dynamometer shown in Fig. 4.8

Materials that are Difficult to Grind

Two of the most difficult materials to grind are titanium alloy Ti-6AI-4V and high-speed steel with a high vanadium content (AISI T-15). The grinding characteristics of tool steel and titanium alloys have been studied by several researchers (Tarasov 1951, 1952; Yang and Shaw 1951; Shaw and Yang 1956; Ahmed and Dugdale 1976).

Titanium is a metal belonging to the tin group of the periodic table that resembles iron in many of its properties. The structural alloys of titanium are ductile, light in weight, and have good fatigue strength and corrosion resistance. The strength : weight ratio of titanium alloys is the highest of all structural materials.

The thermal conductivity and volume specific heat (and hence $(k\rho C)^{0.5}$) are unusually low, which accounts to a large extent for the difficulty in machining and grinding these alloys. Similarly to iron, titanium exhibits an allotropic transformation at 1625 °F (885 °C), changing from a hexagonal close-packed structure (α), stable at lower temperatures to a body-centered cubic structure (β) at the transformation temperature. While pure titanium cannot be hardened by heat treatment, it can be hardened by cold work. Since the strain hardening index for titanium is close to that for mild steel, the unusual difficulty in cutting and grinding titanium cannot be attributed to strain hardening.

Nitrogen, oxygen, carbon, and iron are soluble in titanium in the molten state and tend to make the metal harder, stronger, and less ductile. Titanium also has a strong affinity for compounds containing nitrogen, oxygen, carbon, and halogens, especially at elevated temperatures. This makes it difficult to refine titanium, since it has a strong tendency to react with the usual refractory materials, all of which have negative oxygen or carbon ions at their surfaces. At a temperature of 1475 °F (802 °C) titanium will burn with incandescence in an atmosphere of pure nitrogen to form TiN. At a temperature of 1125 °F (607 °C) titanium may be burned to TiO_2 in pure oxygen, or at 2200 °F (1204 °C) in air. At high temperatures oxygen reacts with titanium in preference to nitrogen.

Inasmuch as molten titanium reacts readily with refractory oxides, it should be expected to react with the glasses used as bonding materials in vitrified grinding wheels. Titanium also will form strong bonds with aluminum oxide or silicon carbide at temperatures as low as 1800 °F (982 °C) Titanium will decompose water at elevated temperatures, releasing nascent hydrogen. Titanium, as well as zirconium, has the property of absorbing hydrogen at high temperatures. This hydrogen forms interstitial hydrides, the lattice of the metal being expanded and, as a result, becoming more reactive.

The grinding of titanium alloys has been studied by Tarasov (1952) and by Yang and Shaw (1955). These studies revealed the importance of lowering the wheel speed when grinding titanium to approximately one-third of the conventional wheel speed to obtain good surface finish and a low rate of wheel wear. Al_2O_3 was found to have a lower rate of wear than SiC at low wheel speeds, while the opposite was true at the conventional wheel speed. While the above studies were useful in improving grinding performance, they do not consider the basic wear mechanisms involved.

Two types of experiments were conducted by Kumar and Shaw (1982) better to elucidate basic wear mechanisms for materials that are difficult to grind:

(1) COFG experiments as described above;
(2) Examination of ground surfaces, chips, wheel debris and metal built up on the tips of abrasive particles.

Figure 11.7 shows the mean cumulative wear of a cluster per cut ($\triangle \bar{d}/n$) versus the number of cuts (n) for Al_2O_3 and SiC abrasives when grinding the

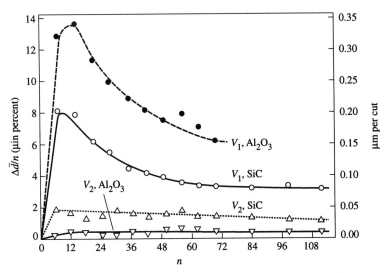

FIG. 11.7. The variation in the mean cumulative wear per cut ($\Delta \bar{d}/n$) with the number of cuts n for COFG tests on Ti-6Al-4V alloy. Wheel speeds, $V_1 = 6300$ f.p.m. (32 m s^{-1}) and $V_2 = 3300$ f.p.m. (16.8 m s^{-1}); mean undeformed chip thickness, $\bar{t} = 75$ μin (1.88 μm) (after Kumar and Shaw 1982).

titanium alloy Ti-16Al-4V ($R_c = 32$). The wheels used were A60 J8 V and C60 J8 V. SiC is seen to have a lower wear rate than Al$_2$O$_3$ but the difference in wear rate for these two abrasive grits decreases with decrease in wheel speed.

Figure 11.8 shows the mean cumulative wear per cut for a cluster ($\Delta \bar{d}/n$) versus the number of cuts (n) for Al$_2$O$_3$ and SiC when grinding T-15 tool steel (C, 1.57; W, 12.6; Cr, 4; V, 4.9; Co, 4.89).

Wear-resistant Tool Steel

Hardened AISI T-15 tool steel is one of the most difficult materials to finish. The constituents that give this steel unusual wear resistance, toughness, and hot hardness make it very difficult to grind. The hard complex carbides containing vanadium, chromium, and tungsten that are present approach the hardness of aluminum oxide abrasives normally used in grinding. In addition, the high percentage of cobalt found in the matrix provides an unusually strong support for these carbides. All of this results in very high rates of grinding wheel wear (the grinding ratio G, equal to volume of work ground/volume of wheel consumed, is often in the vicinity of one), which makes it difficult to grind this steel to required accuracy and surface integrity on an economic basis.

Three general approaches have been taken in the past to solve this problem:

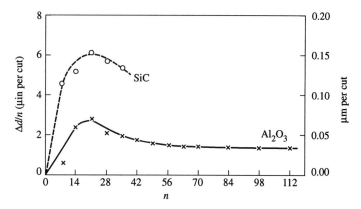

FIG. 11.8. The variation in the mean cumulative wear per cut $\Delta \bar{d}/n$ with the number of cuts n for COFG tests on T-15 steel. Wheel speed, 6300 f.p.m. (32 m s^{-1}); mean undeformed chip thickness $\bar{t}$, 60 μin (1.5 μm) (after Kumar and Shaw 1982).

(1) deviations from conventional grinding practice;
(2) improvements in the grinding wheel;
(3) alterations of workpiece chemistry and structure.

Relative to item (1), Tarasov (1951) found Al_2O_3 to be a suitable abrasive when used with a soluble oil containing high concentrations of sulfur and chlorine. However, the study conducted by Ahmed and Dugdale (1976) suggested that SiC performs better than Al_2O_3 when grinding tool steels with high carbon and with vanadium contents using a straight oil grinding fluid. This is an interesting result, since SiC and ferrous materials are generally considered to constitute a somewhat incompatible grinding system. The poor performance of SiC in grinding steel was explained by Geopfert and Williams (1959) by conducting high-temperature diffusion experiments between SiC and iron, resulting in the formation of pearlite at the interface. However, Ahmed and Dugdale (1976) explained the better performance of SiC when grinding high-carbon steel in terms of a lesser tendency for a material of higher carbon content to dissociate SiC.

Relative to item (2), Tarasov (1959a) demonstrated the economic advantage of using diamond grinding wheels having a vitrified bond when grinding T-15 tool steel. The substitution of cubic boron nitride (CBN) for diamond provides a further improvement in this direction. However, the high initial cost of diamond and CBN wheels and the consequent great cost of a mishap in the workshop have been deterrents to this approach to the problem.

Relative to item (3), Tarasov (1959a,b) demonstrated that additions of manganese sulfide to T-15 tool steel improved not only its machinability but also its grindability in the hardened state.

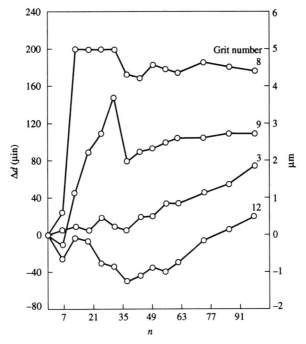

FIG. 11.9. The variation in the cumulative wear $\triangle d$ with the number of cuts n of T-15 tool steel ground with Al_2O_3 for a few active grits (grit numbers 8, 9, 3, and 12) within the cluster (after Kumar and Shaw 1982). The total number of active grits in an $\frac{1}{8} \times \frac{1}{8}$ in (3.18 × 3.18 mm) cluster is 16. Wheel speed V, 6300 f.p.m. (32 m s^{-1}; mean undeformed chip thickness $\bar{t}$, 60 μin (1.5 μm)

Figures 11.9 and 11.10 show individual rates of wear for a few of the active grits within a cluster when grinding T-15 steel. Al_2O_3 abrasives show both positive and negative wear, corresponding to microchipping and the formation of metal build-up at the tip of the abrasive respectively. The metal build-up causes chipping, resulting in further positive wear. The SiC abrasives, however, show only positive wear that is attritious in nature. In grinding titanium, it was also observed that both Al_2O_3 and SiC abrasives exhibited positive and negative wear, similar to that shown in Fig. 11.9, at both the high and low wheel speeds.

Auger electron spectroscopy (AES) conducted on T-15 specimens surface ground without cross-feed under conventional conditions using a complete wheel showed considerable oxidation of the ground surfaces and depletion of carbon, iron, and other constituents (Kumar and Shaw 1982). There was essentially no difference in the surface chemistry after grinding with Al_2O_3 or SiC, although there was a marked difference in the wear rates.

Similar AES results on ground titanium alloy surfaces showed no evidence of diffusion of Al or Si into the surface, although there was evidence of considerable oxidation and an increase in C concentration at the surfaces.

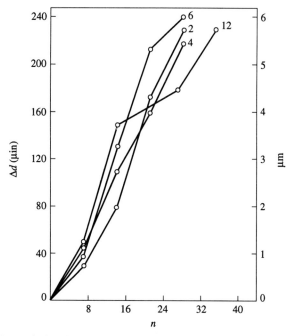

FIG. 11.10. The variation in the cumulative wear $\triangle d$ with the number n of cuts of T-15 tool steel ground with SiC for a few active grits (grit numbers 6, 2, 4, and 12) within the cluster (after Kumar and Shaw 1982). The total number of active grits in an $\frac{1}{8}$ in × $\frac{1}{8}$ in (3.18 × 3.18 mm) cluster is 15. Wheel speed V, 6300 f.p.m. (32 m s^{-1}); mean undeformed chip thickness $\bar{t}$, 60 μin (1.5 μm).

X-ray diffraction studies on T-15 grinding debris (Kumar and Shaw 1982) suggested the presence of complex carbide particles in the chips, similar to those reported earlier by Komanduri and Shaw (1975). An oxide phase could not be detected in the titanium chips by the diffractometer, presumably because the oxide layer was too thin.

Built-up material adhering to the tips of abrasive particles was examined using energy-dispersive X-ray analysis in conjunction with SEM examination. These results failed to reveal any evidence of solid state reaction between abrasives and work materials.

The foregoing results suggest that the chemical interaction between the abrasive and the work material is insignificant relative to the wear of the abrasive for the grinding systems considered here. The better performance of SiC when grinding titanium than when grinding steel can be explained as follows. The free energy for the first stable oxide to form may be calculated for an assumed interfacial temperature of 1000 °C using thermochemical tables (Barin and Knocke 1973) and this indicates the following reactions to be thermodynamically possible:

$$SiC(s) + O_2(g) \rightarrow SiO(g) + CO(g) + 89.69 \text{ kcal mol}^{-1},$$

$$Fe(s) + \tfrac{1}{2}O_2(g) \rightarrow FeO(s) + 44.89 \text{ kcal mol}^{-1},$$

$$Ti(s) + \tfrac{1}{2}O_2(g) \rightarrow TiO(s) + 99.73 \text{ kcal mol}^{-1}.$$

The reaction with the most negative free energy of oxide formation indicates the greatest affinity for oxygen. Thus, when grinding steel with SiC there should be a greater tendency for SiC to be oxidized than there is for the iron in steel to be oxidized. However, when grinding titanium there should be a greater tendency for the oxidation of titanium than for the oxidation of SiC. As a result the SiC abrasive should wear more rapidly when grinding steel than when grinding titanium under the same conditions.

In contrast, Al_2O_3 is thermally more stable than SiC and its wear in grinding steel and titanium will be accompanied by metal build-up. The build-up of material in such cases is primarily due to the high temperatures and pressures existing at the interface, resulting in the formation of strong adhesive bonds.

In grinding T-15 steel with SiC, it is found that the wear of SiC is primarily due to oxidation, which results in an attritious type of wear. In contrast, the wear of Al_2O_3 is predominantly due to metal build-up, which results in a microchipping type of wear. When grinding a titanium alloy it is found that both Al_2O_3 and SiC abrasives exhibit a microchipping type of wear that results in a lower rate of wear at low wheel speeds than at normal wheel speeds.

The ground surface and the chips formed in fine grinding T-15 steel and Ti-6Al-4V alloy showed no evidence of solid state chemical reaction between the abrasive and the work material. The material adhering at the tip of the abrasive is primarily a result of a welding action which results in increased grinding forces and subsequent microchipping of the grit.

Improved Grindability of T-15 Tool Steel

The grindability of T-15 tool steel may be improved by an alteration of the steel-making practice. Instead of following the conventional procedure of casting and hot working, the steel may be made by atomizing the liquid alloy in an inert atmosphere and consolidating the resulting powder by sintering into hydrostatically compressed billets (Dulis and Neumeyer 1970). A valuable treatise on the atomization of molten materials has been produced by Yule and Dunkley (1994).

Kasak *et al.* (1972) have reported an improvement in wheel life of from 60 to 350 percent for this particle metallurgy steel in transverse cylindrical grinding. Komanduri and Shaw (1975) have conducted a study to explain the improved grindability of this steel in fundamental terms. In this investigation, the grinding performance of the following four work materials was compared:

(1) quenched and tempered T-15 tool steel produced by the consolidation of metal particles, designated (CPM);

(2) the annealed version of item (1) (CPM-A);
(3) conventionally produced T-15 tool steel in the quenched and tempered state (CON);
(4) the annealed version of item (3) (CON-A).

The atomized particles cooled at a very high rate, resulting in particles ranging in size from 0.0005 to 0.005 in (12.5–125 μm). The individual carbides in each particle were extremely small (1.3–3.5 μm or 52–140 μin) and uniformly distributed. The individual carbides in the conventional steel ranged in size from 6.2 to 34 μm (248–1360 μin) and were not uniformly distributed. Figure 11.11 shows photomicrographs of carbides in the conventional and particle metallurgy steels.

Fly milling tests were performed as shown in Fig. 11.12, where the abrasive particle was mounted in the periphery of a 20 in (508 mm) aluminum disc and the small ($\frac{1}{2} \times \frac{1}{4} \times 0.040$ in, or $12.7 \times 6.4 \times 1$ mm) specimen was mounted on the piezoelectric dynamometer described in Chapter 5. The specimen was inclined in the axial direction so that the depth of cut (d) increased progressively. Two types of abrasive grits of 36 screen size were tested with all four work materials.

Figure 11.13 shows the specific energy, based upon the maximum circumferential force (F_P) and the maximum depth of cut (d) versus the maximum depth of cut (d). No significant difference in specific energy was found between the CPM and CON samples of T-15 for either the quenched and tempered or annealed conditions. The specific energy was found to decrease from a value that approached 10×10^6 p.s.i. (69 GPa) at a value of d of about 50 μin (1.25 μm) to a value approaching 2.5×10^6 p.s.i. (17 GPa) at a value of d of about 1000 μin (25 μm). The large test-to-test scatter is probably due to variation in grain tip geometry.

It was next decided to investigate the rate of wheel wear and the surface finish produced by a complete grinding wheel used to remove material from CPM and CON samples. The 10 in diameter by 1 in face width wheel (254 × 25.4 mm) had 60 screen size white aluminum oxide grits, a vitrified bond, and was operated at a speed (V) of 6000 f.p.m. (30.5 m s^{-1}). The table speed (v) was 75 f.p.m. (0.38 m s^{-1}), the downfeed (d) was 0.001 in per pass (25 μm per pass) and a cross-feed (b) of 0.050 in per stroke (1.25 mm per stroke) were used. A 5.5 percent concentration of a water-based fluid was applied and the total downfeed was 20 × 0.001 in (20 × 25 μm) in each test. The grinding ratios (G) were as follows:

CON 1.8, CPM, 11.9.

Values of surface finish produced without spark-out were essentially the same for the CPM and CON materials (5 μin AA (0.13 μm) for CPM and 6 μin AA (0.15 μm) for a 0.03 in (0.76 mm) cut-off). However, the ground CON surface was wavy, while that for the CPM steel was less wavy.

Plunge surface grinding (no cross-feed) of the quenched and tempered CPM

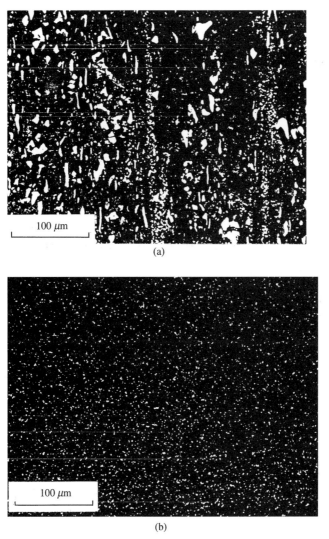

FIG. 11.11. Photomicrographs (picral etch) of the structure of quenched and tempered AISI T-15 steel; (a) conventional steel; (b) steel produced by particle metallurgy (courtesy of Crucible Steel Co.).

and CON steels was investigated next, using the same wheel and grinding conditions as for the cross-feed grinding tests; except for the downfeed per pass (d), which was 0.0002 in per pass (50 μm per pass), and the width ground, which was 0.813 in (20.7 mm). The results obtained are given in Table 11.3.

SEM photographs of chips revealed many semicircular ring cracks around the carbides in the CPM steel, but not for the CON steel. These cracks provide more flexible support for the CPM carbides than for the CON carbides.

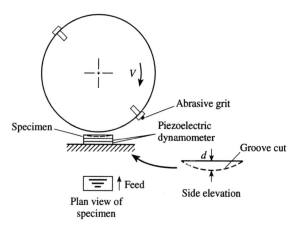

FIG. 11.12. A schematic of the plain fly milling test.

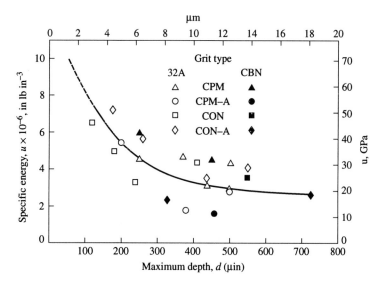

FIG. 11.13. The variation of the specific energy u with the maximum depth of cut d for four work materials and two grit types. Wheel speed, 7500 f.p.m. (38.1 m s^{-1}) (after Komanduri and Shaw 1975).

In the fly milling tests no difference in grinding forces or in specific grinding energy was found between the CPM and CON varieties of steel. The reason for this lies in the fact that in this operation the material is cut only once. Material that flows to the side is not removed by subsequent grits, as in an actual grinding operation involving a complete grinding wheel. Therefore, the advantage of the presence of the soft support of carbides provided by the ring

TABLE 11.3 *Mean grinding results for plunge surface grinding of quenched and tempered steels using a 48A60 L5 V wheel at a wheel depth of cut (d) of 200 μin (5 μm) and a wheel speed of 6000 f.p.m. (30.5 m s^{-1})*

Work material	Grinding Ratio, G	Specific energy, p.s.i. (GPa)	AA finish, μin (μm)
CON	2.03	2.6×10^6 (17.9)	25 (0.63)
CPM	3.48	9.5×10^6 (65.5)	36.5 (0.91)

cracks that develop in CPM is not realized, and the CON and CPM materials appear to have similar characteristics in the plain milling test.

The lack of a distinct difference in specific grinding energy for annealed and hardened specimens has been noted before. This is due to the fact that the specific energy is the product of a mean stress and a mean strain. While the flow stress for a hard material will be relatively high, the strain at fracture will be relatively low. The reverse will be true for the soft material. The net effect is that the product of the stress and strain associated with grinding chip formation will tend to be constant regardless of the hardness of the material.

In the tests without cross-feed, the difference in grinding ratio was only 1.7-fold in favor of the CPM material. Apparently, the tool is damaged less when small carbides are encountered than when very much larger and harder carbides are encountered. This is to be expected, just as a ship striking a small block of ice will be damaged less than when an iceberg is struck.

In the tests with cross-feed, a 6.6-fold difference in grinding ratio G was observed. This large difference is believed to be due to the fact that in complete wheel grinding there is a great deal of material that is ground more than once before it is ejected from the surface as free chips. This in turn provides ample opportunity for the elastic mounting of carbides due to ring cracks to reduce the wear of abrasive grits.

The specific energy for CON is seen to be much larger than for CPM in Table 11.3. At first this does not appear to be consistent with the plain fly milling results, where no difference in specific energy was observed for the two steels. However, for plunge grinding there is considerable regrinding of material (although less than in cross-feed surface grinding). For plain fly milling there is no regrinding. This difference in regrinding is responsible for the observed difference in the effectiveness of CPM relative to the specific energy and forces in the two cases.

The surface finish in plunge grinding (without spark-out) was better for CON than for CPM. This is due to the greater amount of burnishing of the surface that occurs with the nonuniform carbide distribution of CON. The greater burnishing action with CON arises due to the higher specific energy and lower G values pertaining.

Even though better finish may occasionally be obtained with CON due to burnishing, as indicated in Table 11.3, the uneven carbide distribution gives rise to considerable waviness in the finished surface. However, this waviness will not be apparent when a surface finish measurement is made with an instrument having the usual cut-off of 0.03 in (0.75 mm).

The most important characteristic of T-15 steel produced by particle metallurgy (CPM) is the presence of ring cracks around the carbides in steel that has been heavily plastically deformed. These ring cracks provide an elastic support for carbides in the surface, which allows them to recede from a mating surface rather than causing a scoring action that would lead to abrasive wear. It appears that the relative advantage of CPM steel will increase as the amount of regrinding increases. This should be large for grinding operations with cross-feed and with spark-out.

References

Ahmed, O. I. and Dugdale, D. S. (1976). *Proc. 17th IMTDRC*, Macmillan, London.
Barin, I. and Knocke, O. (1973). *Thermochemical properties of inorganic substances*, Springer, New York.
Dulis, E. J. and Neumeyer, T. A. (1970). *Proc. Conf. Materials for Metal Cutting*, p. 126. The Iron and Steel Institute, London.
Geopfert, G. J. and Williams, J. L. (1959). *Mech. Engr* **81**, 69.
Kasak, A., Steven, J., and Neumeyer, T. A. (1972). *Proc. Auto Engng Congr.*, Detroit.
Komanduri, R. and Shaw, M. C. (1975). *Proc. 3rd NAMRC*. Carnegie Press, Pittsburgh, Pennsylvania, p. 481.
Kumar, K. V. and Shaw M. C. (1979). *Ann. CIRP* **28**(1), 205.
Kumar, K. V. and Shaw, M. C. (1982). *Wear* **82**, 57.
Kumar, K. V., Cozminca, M., Tanaka, Y., and Shaw, M. C. (1980). *J. Engng Ind.* **102**, 80.
Shaw, M. C. and Yang, C. T. (1956). *J. Engn Ind.* **78**, 86.
Tarasov, L. P. (1951). *Trans. ASM* **43**, 1144.
Tarasov, L. P. (1952). *Am. Mach.* **96**, 135.
Tarasov, L. P. (1959a). *Tool Engr* **43**, 109.
Tarasov, L. P. (1959b). *Tool Engr* **42**, 77.
Yang, C. T. and Shaw, M. C. (1955). *J. Engng Ind.* **77**, 645.
Yule, A. J. and Dunkley, J. J. (1994). *Atomization of metals*. Oxford University Press, Oxford.

12

THE GRINDING ENVIRONMENT

Introduction

Grinding is influenced by the environment in which it occurs, whether in the presence of vacuum, a vapor, a liquid, or an adsorbed solid. The temperature of the environment is important, as well as its chemistry. The method of applying and distributing a fluid is also of importance, both with regard to cooling and also in keeping the wheel surface clean of adhering metal that is being ground.

Gases

In order to gain some insight into the importance of the atmosphere surrounding the wheel and workpiece during grinding, forces were measured when grinding in nitrogen (Outwater and Shaw 1952). A Neoprene observation balloon was fitted around the wheel, workpiece, and dynamometer and the air displaced by nitrogen. The observed effect of the nitrogen depended on its purity and it was only after repeatedly filling and emptying the bag and grinding for several passes that the residual oxygen was consumed and a significant change in grinding was observed. When the oxygen concentration was very low, the grinding forces were unusually high — in some instances being 25 times as high as the corresponding values in air. In order to be sure that the inertness of the atmosphere was responsible for these results and not the nitrogen itself, the tests were repeated using helium. It is conceivable that under the unusual conditions of high pressure and temperature and a clean nascent surface, the nitrogen molecule might be split and a metal nitride formed. However, the helium results were the same as those with nitrogen. When a trace of air was admitted, the grinding forces dropped to atmospheric values. Representative force values are given in Table 12.1.

These observations collectively indicate that a trace of oxygen plays a very important role in the grinding process. Were it not for the oxygen present in the air, the rate at which metal is removed in finish grinding would have to be greatly reduced, due to the danger of damaging the finished surface by the development of excessively high temperatures.

The important role that a trace of oxygen plays in grinding is thought to be due to the decreased tendency for metals to weld if a thin layer of oxide is present. A nascent heated iron surface will oxidize extremely rapidly in air to form a primary film of iron oxide from 10 to 20 Å thick. However, subsequent increase in oxide thickness requires an increasingly longer time. In atmospheric grinding such a primary oxide forms practically instantaneously

TABLE 12.1 *Grinding data in air and inert gases. Wheel speed, 6280 f.p.m. (31.9 m s^{-1}); work speed, 2 f.p.m. (0.01 m s^{-1}); depth of cut, 0.0005 in (13 μm) (after Outwater and Shaw 1952)*

Atmosphere	F_Q		F_P	
	lb	N	lb	N
Air	2.70	12	1.3	6
N_2	60.0	267	35.0	156
He	62.5	278	37.5	167

FIG. 12.1. Welded chips attached to the end of an AISI 52100 specimen ground in nitrogen (after Outwater and Shaw 1952).

and thus prevents the chip from rewelding to the workpiece. In the absence of oxygen, chips will reweld to the finished surface, and the same metal may have to be sheared several times before it finally leaves the system as a free chip. This mechanism accounts for the large increase in grinding energy when grinding in an inert atmosphere.

The tendency for one chip to weld to the next in an inert atmosphere is illustrated in Fig. 12.1. Here a great many individual chips are shown welded into one solid mass, which is attached to the end of a specimen ground in nitrogen.

Courter (1949) has reported some interesting orientation results when metal is ground in vacuum. Using electron-diffraction he found that metals ground in air showed no preferred orientation in the surface, whereas metals ground in vacuum were strongly oriented. In the case of steel ground in vacuum, the [111] planes were oriented parallel to the ground surface, whereas with magnesium, the basal [001] planes were aligned along the ground surface. Surfaces

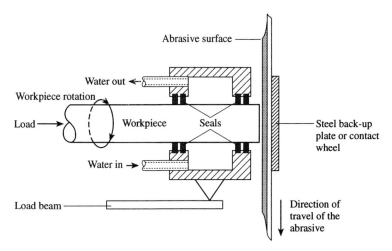

FIG. 12.2. An apparatus for studying the effect of gaseous and liquid fluids in belt grinding.

ground in air merely showed oxide rings but no orientation. It would appear that the much greater energy associated with grinding in an inert atmosphere could be responsible for the observed orientation.

Tanaka and Ueguchi (1971) confirmed the importance of atmospheric oxygen in preventing chip–grit adhesion by grinding over a very wide range of atmospheric pressure (10^{-5} to 8 atm). Stock removal in fine grinding became practically impossible at pressures below 10^{-1} atm.

Duwell *et al.* (1966) have studied the role of gases in plunge belt grinding using the apparatus shown in Fig. 12.2. A water-cooled tube (1 in OD and $\frac{1}{2}$ in ID, or 25 × 12.5 mm) was fed axially into a belt supported by a steel back-up plate. The gas or fluid was pumped through the tube and a load beam was used to measure the grinding force (F_p) as the slowly rotating tube was fed axially at a constant rate (v) against the belt. Figure 12.3 shows plots of specific energy (u) versus the ratio of volume removed to area of belt involved (cut per path, $in^3 in^{-2}$ or $mm^3 mm^{-2}$) for different abrasives and when argon and and air were pumped through the tube. The specific energy is initially low, rises rapidly and then at a lower rate, and finally rises extremely rapidly. The belt life is taken to be the value of cut per path at which the specific energy begins to rise rapidly. Air is seen to increase belt life for both abrasives but more effectively for Al_2O_3.

Figure 12.3 is for AISI 1020 steel. Similar results were obtained for a 304 stainless steel, where a much thinner oxide layer was found to be equally important (Fig. 12.4). A water mist was no more effective than air but a heavy duty water-based liquid lubricant containing compounds of sulfur and chlorine was much more effective than air or a water mist.

It was also found that reactive gases such as H_2S, HCl, and Cl_2 were more

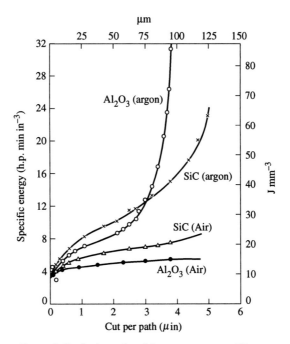

FIG. 12.3. The effect of displacing air with argon on specific energy when grinding AISI 1020 steel with Al_2O_3 and SiC coated abrasive belts (after Duwell et al. 1966). Belt speed, 2750 f.p.m. (14 m s^{-1}); removal rate, 0.1 in^3 min^{-1} (1639 mm^3 min^{-1}); belt length 132 in (3.35 m); grit size, 60.

effective than air in reducing specific energy and in increasing belt life, while unreactive gases such as N_2 and CO_2 were much less effective than air (Duwell and McDonald 1961).

The role of gases in improving belt life can be negative or positive relative to grinding in air. Inert gases that exclude air which forms an effective oxide that decreases the tendency for adhesion between metal and grits give a negative result. Reactive gases can be even more effective than air in preventing adhesion. Neither inert nor reactive gases have a significant influence on surface temperature. A water mist is an exception, since it does have a cooling effect. However, this is not a gas but a two-phase material, consisting of finely divided liquid water particles in a matrix of air. The water particles extract heat of vaporization as water is converted to steam.

Liquids

In FFG liquids have two major roles:

(1) to prevent build-up of metal on active grit tips by contamination (lower μ);
(2) to lower the mean interface temperature (θ) by cooling action.

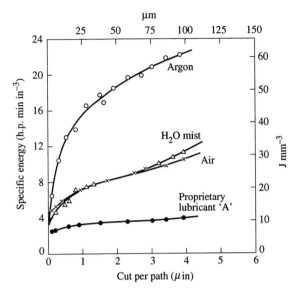

FIG. 12.4. The effect of an argon atmosphere and water-based mist on specific energy when grinding a 304 stainless steel tube with a 60 grit Al_2O_3 belt (after Duwell et al. 1966). Belt speed, 5610 f.p.m. (28.5 m s^{-1}); removal rate, 0.1 in^3 min^{-1} (1639 mm^3 min^{-1}); belt length, 132 in (3.35 m).

Grinding liquids are either water-based materials or are oil based. In either case, additives are generally employed to lower the specific energy, prevent corrosion and foaming, increase wetting, or retard the growth of bacteria. Water- or oil-based liquids are generally used depending on whether excess temperature or metal transfer is the major concern. Practically all grinding fluids are proprietary materials, so that the production engineer is forced to seek the advice of the manufacturer's representative. These liquids are marketed as much on the basis of the advisory service provided as on the performance of the liquid. In many cases, operating conditions are improved at the same time as the fluid is changed and, if the net result is positive, the fluid is credited with the improvement.

In fine grinding flats gradually form on the active grits and weld areas which develop an increase in size until a critical value is reached that leads to grit pullout. Periodic dressing of the wheel is required to keep the average flat size below the critical value. Active lubricants tend to prevent metal transfer to the grit flats by contaminating the surfaces. This can occur by chemical or physical action. Chlorine, sulfur, and phosphorous additives tend to react with a freshly ground surface to form inorganic compounds which tend to prevent adhesion. Fatty additives in oil adsorb on clean metal surfaces to screen them against welding.

Patterson (1986) suggests that the nature of wear flats that develop during wheel wear plays an important role in grinding performance. When using a heavy duty 10 percent emulsion to plunge cylindrical grind AISI 52100 steel ($R_c = 60$) with $V = 9830$ f.p.m. (50 m s^{-1}), $v = 60$ f.p.m. (0.3 m s^{-1}) and in feed = 0.033 i.p.m. (0.84 mm s^{-1}), the grinding forces were found to cycle between high and low values due to periodic metal build-up and removal by grit fracture. The use of a grinding oil and a reduction of wheel speed from 9830 f.p.m. (50 m s^{-1}) to 5600 f.p.m. (29 m s^{-1}) gave a substantially reduced wheel wear rate and the disappearance of cyclic force behavior. SEM examination of wear flats on active grit tips revealed a microscopic roughness obtained with grinding oil but a smooth surface with a water emulsion. It was suggested that the microscopic peak-to-valley roughness with oil results in a microcutting action by the wear flats instead of periodic metal build-up and relatively coarse grit fracture. The peak-to-valley roughness R_t was found to be about 100 μin (2.5 μm) on wear flats that produced a noncyclic wear rate, which is greater than the mean grit depth of cut. It has been found (Patterson 1986) that when abrasive wear flats are striated due to microchipping instead of cyclic relatively gross chipping, the following beneficial results are obtained:

- greater removal rate
- longer dressing interval
- lower tensile residual stress in ground surface

Stainless steels require more ductile grits since they do not form protective films as readily as plain carbon steels. Stainless steels are also in greater need for screening action, since they are stronger and more refractory than carbon steels. High-temperature alloys (Ni- or Co-based) and refractory materials such as Mo are still more stable and even more in need of help from active ingredients in the grinding liquid than are stainless steels.

Titanium alloys (discussed in Chapter 11) are the most difficult to screen against strong weld areas due to their affinity for oxides, carbides, and nitrides and their unusually low value of β, resulting in high mean values of surface temperature. The titanium alloys are so difficult to grind that they provide a good system to study from the point of view of grinding fluid action, and this will be done in a later section of this chapter.

In Chapter 9, data were presented (Fig. 9.24) that show that water-based fluids limit mean surface temperature primarily by heat extraction, while oil-based fluids limit temperature primarily by reduction in specific grinding energy (u).

In the application of grinding liquids there are many practical considerations that may prevent a material that looks attractive in the laboratory from being used in practice. In addition to the surface integrity, wheel life, and surface temperatures normally studied in the laboratory, the following items must not be overlooked in application (Shaw 1970):

- adverse physiological action on the operator
- atmospheric pollution (misting and fuming tendency)

- adverse action on machinery (rust, deposits, paint removal, and reaction with nonmetallic machine elements)
- bactericidal growth and odor development
- ease of separation from swarf and from parts
- stability and life of active ingredients
- ease of disposal when spent
- inconvenience associated with use of a fluid of any type

An important consideration with water-based fluids is the quality of water used to dilute the concentrate. Hard water that contains large amounts of calcium, magnesium, aluminum, iron, or zinc ions will tend to force the fluid concentrate out of solution. In such cases, use of deionized water becomes necessary. Corrosion inhibitors and other materials that adsorb on metal surfaces are removed relatively rapidly with the swarf. This requires careful monitoring and make-up of active ingredients. In some cases, make-up may amount to 25 percent of the concentrate per day. When diluting a fluid concentrate to form an emulsion, it is important to add the concentrate to water in order to prevent an inverted emulsion (water in oil) from being produced.

When a large amount of heat is involved, a large sump, or the use of a heat exchanger in the sump, is important. Use of a high-pressure jet to improve fluid penetration to the wheel–work contact or for wheel cleaning will tend to cause foaming of the fluid. This may require use of a larger baffled sump to allow more time for air bubbles to be dissipated, or use of an effective antifoaming agent.

Another practical problem is prevention of fine Swarf from being entrained in the fluid recirculated to the grinding zone. This may require more time in the sump for the swarf to settle out, or filtering in the case of very fine swarf.

An interesting concept introduced by Pahlitzsch and Appun (1954) was a dual liquid system involving a water-insoluble oil and a water-based solution (*Zweistoff-Zweiweg-Kuhlung*). The oil was pumped through the wheel, while the water-based coolant was conventionally applied. The oil floated to the top of the liquid in the sump and was available for recirculation. This arrangement required a wheel of unusually high porosity and careful filtering (~ 3 μm capability) to prevent clogging and to keep the oil pressure at a reasonable level.

Salje and Riefenstahl (1949) successfully employed through-the-wheel cooling (*Innenkuhlung*) in internal grinding, where it is particularly difficult to get the fluid to penetrate the wheel-work interface. This required a dual filter system. The first stage was a paper filter with a 15–35 μm range, while the second stage employed a micro-paper filter with a 2–20 μm range.

With ever increasing environmental concerns, the disposal of spent fluids is becoming an important issue, and companies that specialize in this activity are beginning to emerge. For smaller operations it is frequently less costly to engage such services than to go it alone.

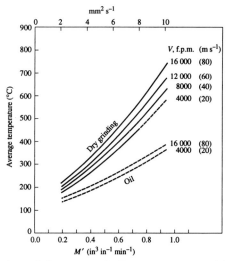

FIG. 12.5. The variation of the mean grinding temperature with the rate of removal (M') for different wheel speeds (V) and fluids (after Opitz and Gruehring 1968). Wheel, EK80-J7 VX; work, AISI 1045 steel; work speed v, 90 f.p.m. (0.46 m s^{-1}).

High-speed Grinding

Starting in the late 1960s, there has been an interest in higher wheel speeds in FFG in order to increase removal rate without adverse effects on surface integrity. Speeds as high as 16 000 f.p.m. (81.3 m s^{-1}) have been used in plunge cylindrical and form grinding.

From eqn (9.27) it is evident that the mean grinding temperature increases in proportion to $V^{0.5}$. Thus, the main problem in high-speed grinding is to prevent overheating of the workpiece. This may be done most effectively by use of a grinding liquid, making sure that it gets to the wheel and work surfaces.

Since high temperature is the item of concern in high-speed grinding, we might assume that a water-based fluid would be most effective. However, in practice, it turns out that oil-based materials give better results in this type of grinding. Figure 12.5 shows the variation in mean grinding temperature with metal removal rate M' (in^{-3} in^{-1} min^{-1} or mm^2 min^{-1}) in cylindrical plunge grinding for wheels operated at different speeds (V), both dry and with an active grinding oil. The temperature is seen to increase significantly with removal rate (M') but to a lesser extent with increase in wheel speed (V). The mean grinding temperature is very much lower when a grinding oil is used than when grinding dry. A water-based fluid was found to give results that are intermediate between the dry and oil cases in this instance.

At high wheel speeds it is difficult to get the fluid to penetrate the boundary layer of air which clings to the wheel face. If the fluid is applied externally it must either have sufficient velocity to penetrate this boundary layer or the boundary layer must be scraped from the wheel face before the fluid is

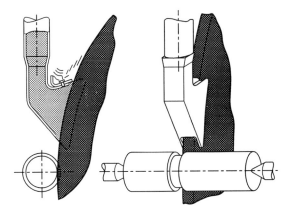

FIG. 12.6. A special nozzle for use in high-speed grinding (after Gruehring 1967).

applied. To penetrate the boundary layer requires a fluid jet accelerated through a nozzle by a pressure of the order of 250 p.s.i. (1.72 MPa). A successful boundary layer removal technique is shown in Fig. 12.6. Just before the wheel reaches the work, a scraper removes the air boundary layer and then fluid is immediately applied to the wheel face by means of a close fitting nozzle box. This type of system requires a large fluid flow rate at moderate pressure (~50 p.s.i. = 0.34 MPa).

In high-speed grinding with a fluid, considerable heat goes into the fluid. The quantity of fluid in the sump must be sufficient to handle this heat without causing too high a temperature rise, and normally a heat exchanger must be used to remove heat from the sump.

High-speed grinding involves a number of special problems, and it is important to be sure that the advantages outweigh the disadvantages before adopting such a system. The main advantages, of course, are a reduction in the production time and hence an increase in productivity and a decrease in labor cost. High-speed grinding is of greatest interest when the actual chip-forming time represents a large percentage of the cycle time. Items that may be involved in high-speed grinding include the following:

- more expensive, higher power, more stable machine tools
- greater risk, calling for more elaborate safety precautions
- higher maintenance costs and a shorter machine tool life
- a higher level of skill required and hence a higher unit labor cost
- sufficient production to keep the machine and operator fully occupied

Inorganic Fluids for Titanium Alloys

The main difficulty when grinding titanium alloys is the tendency for strong bonds to form between titanium chips and grits that are oxides or carbides

TABLE 12.2 *Preliminary tests—cylindrical grinding. Work, Ti 150A; wheel, A46-K5-V; Wheel speed, 2200 f.p.m. (11 m s^{-1}); work speed, 30 f.p.m. (0.15 m s^{-1}); wheel depth of cut, 0.0005 in (12.5 μm); cross-feed, 0.05 in per stroke (1.27 mm per stroke) G = grinding ratio = wear volume/volume removed (after Shaw and Yang 1956)*

Fluid	G
Water	0.5
Air	0.9
Soluble oil	1.1
5 percent $AlCl_3$	1.4
5 percent NH_4Cl	1.5
5 percent $NaNO_2$	1.7
10 percent $NaNO_2$	2.0
10 percent $NaNO_2$	2.5
10 percent KNO_2	3.3

(practically all common abrasives). This bonding tendency increases with temperature and the low $\beta = (k\rho C)^{0.5}$ for titanium gives unusually high grinding temperatures. When metal builds up on the tips of grits this gives higher values of specific energy and increases the tendency for loss of grit material by chipping. An effective way of reducing chip–grit adhesion when grinding titanium alloys is to decrease the mean grinding temperature by reducing wheel speed to about one-third of the usual value. Another way is to have material adsorb on the contacting surfaces that inhibit bonding.

In grinding titanium a water-based fluid is preferred, due to the danger of fire. If titanium is ground at 6000 f.p.m. (30.5 m s^{-1}) the sparks are brilliant and there is danger of the grinding swarf becoming ignited if a combustible fluid such as oil is used. At the recommended speed for titanium, 2000 f.p.m. (10.2 m s^{-1}), the chips do not glow and there is little chance of fire. However, it is thought best to use only noncombustible fluids in grinding titanium, since a workman may inadvertently grind with too high a speed and thus set a fire.

Tarasov (1951, 1952) was the first to show that rust-inhibiting additives such as sodium nitrite are effective in water as grinding fluids for titanium alloys. Shaw and Yang (1956) undertook a study to explain and optimize the performance of dilute solutions of inorganic salts in water as grinding fluids for titanium alloys. A few preliminary tests (Table 12.2) showed that:

- ordinary soluble oils are ineffective grinding fluids for titanium alloys
- distilled water is inferior to air
- inorganic salts of heavy alkali metals are most effective
- both the anion (negatively charged ion) as well as the cation (positively charged ion) are important in determining the effectiveness of a given salt

The additive sodium nitrite, of course, ionizes in water as follows:

TABLE 12.3 *Grinding conditions for surface grinding with cross-feed*

Wheel: 7 in (178 mm) dia, A60-L5-V
Work: Ti RC 130B
Wheel speed, V: 2000 f.p.m. (10 m s^{-1})
Work Speed, v: 33.3 f.p.m. (0.17 m s^{-1})
Wheel depth of cut, d: 0.001 in (25 μm)
Cross-feed, b': 0.05 in per stroke (1.27 mm per stroke)
Dress: four passes of 0.001 in (25 μm) across pyramidal diamond

TABLE 12.4 *Grinding results with 0.1 M solutions of inorganic salts in water. For cutting conditions, see Table 12.3. G = grinding ratio relative to water (=1.00) (after Shaw and Yang 1956)*

Salt	G	Salt	G
Series 1: cation varied		*Series 2: anion varied*	
LiNO$_3$	1.58	Na$_2$PO$_4$	3.20
NaNO$_3$	2.14	Na$_2$S	3.00
KNO$_3$	3.06	Na$_2$SO$_4$	2.95
CsNO$_3$	2.78	Na$_2$B$_4$O$_7$	2.69
AgNO$_3$	0.99	Na$_2$CrO$_4$	2.68
NH$_4$NO$_3$	1.05	NaOH	2.68
Mg(NO$_3$)$_2$	1.31	NaCO$_3$	2.56
Ca(NO$_3$)$_2$	2.08	NaNO$_2$	2.44
Sr(NO$_3$)$_2$	2.83	NaCl	2.22
Ba(NO$_3$)$_2$	3.55	NaNO$_3$	2.14
Co(NO$_3$)$_2$	0.96	*Series 3: miscellaneous*	
Cu(NO$_3$)$_2$	0.98	Air	1.33
Cd(NO$_3$)$_2$	1.04	Distilled water	1.00
La(NO$_3$)$_3$	2.24	K$_3$PO$_4$	3.60
Ce(NO$_3$)$_3$	2.14	Ba(HO)$_2$	3.15
Al(NO$_3$)$_3$	1.12	BaS	2.56
Fe(NO$_3$)$_3$	0.93	BaH PO$_4$	1.11
Th(NO$_3$)$_4$	1.36	10 percent water-based fluid	1.27
Pb(NO$_3$)$_4$	1.17	100 percent grinding oil	1.97

$$NaNO_2 \underset{H_2O}{\Rightarrow} [Na^+] + [NO_2^-].$$

Since many nitrates are available it was decided first to keep the anion constant [NO$_3^-$] as the cation was varied and then to keep the cation constant [Na$^+$] as the anion was varied. A one-tenth molar solution was used in all cases. All tests were performed in cross-feed grinding under the conditions of Table 12.3, and the results are summarized in Table 12.4.

Concentrations of 0.1 M are much lower than would be used in practice and the G values for more practical concentrations would be several times those given in Table 12.4.

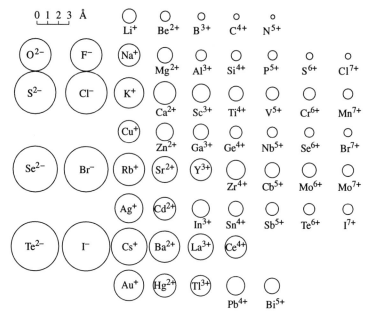

FIG. 12.7. The variation of ion size with position in the periodic table (after Pauling 1948).

Examining the variation of the relative G values with respect to many properties when the cation was varied (series 1 of Table 12.4) it was found that ion size gave the best correlation. Figure 12.7 shows ion sizes to scale for the ions involved in this study.

Aluminum oxide used in grinding wheels is α corundum (Al_2O_3), which consists of a hexagonal close-packed array of oxygen ions with aluminum ions interspersed in two-thirds of the cusps between oxygen ions. The oxygen ions [O^{-2}] are very large compared to the aluminum ions [Al^{3+}], as may be seen by reference to Fig. 12.8. Two views of an aluminum oxide surface are shown to scale in Fig. 12.8. Since the components of aluminum oxide are true ions, having been formed by the transfer of electrons from oxygen to aluminum atoms, the bonding responsible for the strength of this material is electrostatic. This type of bond is referred to as an ionic bond. The aluminum ions may be ignored, since they are very small and are buried between oxygen ions.

Also shown in Fig. 12.8 is a titanium surface. Titanium is held together by metallic bonds. The building blocks in a metal are positive ions produced from neutral metal elements by removal of the valence electrons. Having four valence electrons titanium gives rise to positive ions having a quadruple charge when the valence electrons are removed. In the case of a metal, valence electrons do not transfer to other elements as in the case of an ionic crystal, but distribute themselves throughout the metal much as the molecules are distributed in a gas. The metallic bonding force arises from the interaction of

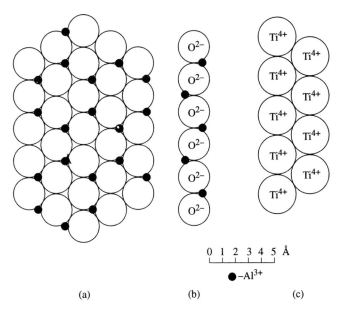

FIG. 12.8. Aluminum oxide and titanium surfaces shown to scale: (a) plan view of Al_2O_3 surface; (b) side view of Al_2O_3 surface; (c) side view of titanium–metal surface (free electron cloud not shown) (after Shaw and Yang 1956).

the cloud of valence electrons and the positively charged ions. The result is a structure that is close-packed and strong. The size of the ions in a metal lattice differ considerably from those in an ionic crystal, since each is subject to entirely different forces. The titanium surface shown in Fig. 12.8 corresponds to the size of the titanium ion in a metal lattice as determined by X-ray diffraction studies (Wells 1950).

Comparison of the titanium ion with that of the oxygen ion in Al_2O_3 shows that the two are about the same size. This provides the good fit essential for the formation of strong electrostatic bonding forces. The large charge difference between titanium (plus four) and oxygen (minus two) ions adds to the tendency for strong bonding between a clean (freshly ground) titanium surface and an aluminum oxide surface. Thus, titanium has a strong affinity for refractory oxides, which is why it is so difficult to grind titanium alloys with refractory oxide abrasives.

Just as it is possible for titanium ions in a metal lattice to form strong electrostatic bonds with the surface ions in Al_2O_3, it is equally possible for other positively charged ions (cations) to be attached to the oxide surface and become strongly attached or adsorbed. The sodium ions in a sodium nitrite solution can attach themselves electrostatically to the oxygen ions of Al_2O_3 and thus prevent titanium from getting close enough to the surface to form a strong bond. The effectiveness of cations used in this way might be expected to vary with their size and charge. Size is important in that the ion adsorbed

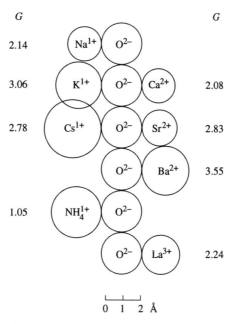

FIG. 12.9. Several cations shown to scale with the oxygen ions of an Al_2O_3 surface for comparison (after Shaw and Yang 1956).

must fit the oxide surface and completely shield it, and charge is important in that the layer must be strongly bonded to the oxide surface.

A few of the cations are shown to scale in Fig. 12.9, together with the G values observed. The improvement observed in proceeding from sodium to potassium is due to a better match between the sodium and oxygen ions. Cesium is not as good as potassium, since it is too large. Similarly, in proceeding down the second column of the periodic table, we observe an increase in G value in going from calcium to strontium to barium. The potassium ion which gave the best result in column (1) of the periodic table is seen to have the same size as the barium ion which gave the best result in column (2). The superiority of barium over potassium is attributed to the double charge on the barium ion.

When trivalent and tetravalent metals were examined it was found there are none large enough to do an outstanding job (i.e. La, Ce, Th, etc.).

From the tests that have been run on cations it is clear that none is as good as the barium ion, since this ion has the optimum charge-size combination of all the elements.

The ammonium cation $[NH_4^{+1}]$ is seen to give a G value that is unusually low for its size (Fig. 12.9). This anomalous result will be discussed later.

Just as material adsorbed on the Al_2O_3 surface should be expected to

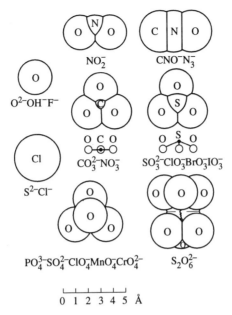

FIG. 12.10. The size and configuration of representative negative radicals according to Bragg.

prevent strong bonding between titanium and the abrasive, material adsorbed on the titanium metal surface should be effective. Those anions most closely matching titanium in size and having the greatest charge should be the most effective. As before, the comparison of results with regard to ion size is quite simple, but the matter of effective ion charge requires further discussion.

The anionic parts of the salts of Table 12.4 (series 2) consist of radicals rather than being simple elements. A typical radical is $[NO_2^{-1}]$. This group of elements is held together by a covalent bond—the type of bond found in practically all organic compounds. Ordinarily, it is a relatively weak type of bonding force and accounts for the low strengths and decomposition temperatures of organic materials such as lubricating oils and waxes. The ordinary covalent bond arises from a sharing of electrons between two elements such as carbon and hydrogen in an organic compound.

The effective charges on any of the oxygen atoms of the negative radicals of Table 12.4 (series 2) can be obtained as follows. The phosphate radical is $[PO_4^{-3}]$. However, there are four oxygen atoms on which three charges are distributed equally in time and hence the effective charge on a single oxygen atom is $-\frac{3}{4}$.

The size and configuration of a number of negative radicals is shown in Fig. 12.10. In those radicals involving oxygen (i.e. the angular $[NO_2^{-1}]$, the planar $[NO_3^{-1}]$, or tetragonal $[SO_4^{-2}]$ radicals, etc.) and another element, the oxygen is large while the other element (metal) is relatively small. Thus, with

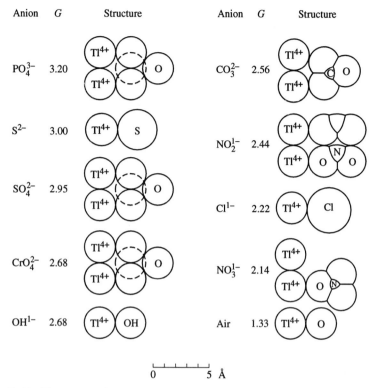

FIG. 12.11. The expected arrangement of anions adsorbed on a titanium surface.

regard to size, oxygen ions of equal size arranged into different configurations are involved in the case of all of the negative radicals of Table 12.4.

In Fig. 12.11 the anions tested are shown to scale in conjunction with the ions in a metallic titanium surface. The corresponding values of G are also listed. Since practically all ions correspond approximately in size to the oxygen ion, the size of the individual atoms constituting the cations need not be considered. The expected arrangements of the several anions on the surface are as shown in Fig. 12.11.

In Table 12.5 the mean charge per atom contacting the titanium surface (A) is given together with the number of atoms per anion making contact (B) and the product of these two quantities (AB). We should expect the effectiveness of a given anion to vary directly with this product AB, since it is a measure of the firmness with which the anion becomes attached to the titanium surface. Excellent correlation with G values is observed for all cases except that for [Cl^{-1}], which is a very large ill-fitting particle.

Now, we return to the cation [NH_4^{+1}] which was observed to give an anomalous result in Fig. 12.9. From size considerations alone, the ammonium radical should be expected to give a G value between that for potassium and

TABLE 12.5 *The relation between the effective bonding charge per ion and the grinding ratio (G) (after Shaw and Yang 1956)*

Anion	A Mean charge per atom	B Number of bonding atoms per anion	Product AB = effective bonding charge	G
$[PO_4^{-3}]$	$-3/4$	3	2.25	3.20
$[S^{-2}]$	-2	1	2.00	3.00
$[SO_4^{-2}]$	$-1/2$	3	1.50	2.95
$[CrO_4^{-2}]$	$-1/2$	3	1.50	2.68
$[OH^{-1}]$	-1	1	1.00	2.68
$[CO_3^{-1}]$	$-1/3$	2	0.67	2.56
$[NO_2^{-1}]$	$-1/2$	1	0.50	2.44
$[Cl^{-1}]$	-1	1	1.00	2.22
$[NO_3^{-1}]$	$-1/3$	1	0.33	2.14

cesium (Fig. 12.9) or to yield a G value of about 3. However, the observed value was closer to 1. The reason for this lies in the resonating structure of this radical. The [NH4$^+$] group consists of a large nitrogen atom surrounded by four small hydrogen atoms in the form of a tetrahedron. A single positive charge resonates between the four hydrogen atoms and thus the effective charge on a single hydrogen atom is only $+\frac{1}{4}$. Hence the [NH^{4+}] group behaves as though it were a potassium ion with a positive charge of $\frac{1}{4}$ instead of 1.

The values of G given in Table 12.4 (series 3) are in general agreement with what might be expected from the series 1 and 2 tests, but indicate that the anionic and cationic effects are not strictly separable and independent. While all solutions were 0.1 M the change in ion concentration due to valence differences must of course be taken into account in making comparisons.

Distilled water is significantly poorer than air and there are several reasons why this might be so:

1. Titanium has a strong tendency to oxidize and the oxygen in the air would provide a thin oxide coating on freshly produced titanium surfaces, thus providing a screening layer between the titanium and aluminum oxide surfaces. We should not expect this action to be too effective, however, since in most instances the nascent titanium surfaces will be in contact with aluminum oxide immediately upon being formed and hence will form a strong bond with the oxygen surface of the Al_2O_3 rather than with the oxygen in the air. The presence of water would exclude atmospheric oxygen and thus prevent the limited beneficial effect of its presence in dry grinding.

2. Water vapor is known to decompose into its elements at elevated temperatures in the presence of titanium. The nascent hydrogen atoms that would thus be formed are known to be very small and would rapidly diffuse interstitially between titanium ions to form a titanium hydride. The presence of

the hydrogen atoms in the titanium surface would enhance the positive charge associated with this surface and tend to cause even stronger bonds with the negative oxygen ions of the Al_2O_3 surface. Nascent hydrogen is also a positive fluxing agent.

3. At high temperatures, water will react with aluminum oxide to form the hydroxide. This material is very weak and such a reaction should increase the frequency with which particles of Al_2O_3 would be plucked from the surface following the formation of welds between the titanium and the abrasive.

While it is not possible to say which of the foregoing actions is most likely, it is expected that they may all contribute to the poor performance of pure water as a grinding fluid for titanium alloys. In this connection it is instructive to consider the grinding of steel. When steel is ground in dry air the rate of wheel wear is relatively low, since the freshly ground metal is coated with a thin layer of iron oxide. The strength of titanium oxide is greater than iron oxide and is comparable to that of Al_2O_3. Consequently, titanium oxide is not nearly as effective as iron oxide in preventing grinding-wheel wear. If a trace of water vapor is introduced when steel is ground in air, the forces rise significantly. This is probably due to the partial reduction of the protective oxide coating on the metal surface by the nascent hydrogen produced when the water is decomposed, along with action 3 previously mentioned. When the wheel is flooded with water there is no further decrease in the forces, which indicates that oxygen exclusion action (item 1) is probably not the major role of water, at least in the case of steel.

When a salt goes into solution part of it will ionize and the remainder will not. The fraction of the total salt that ionizes is known as its activity coefficient. The activities of the salts of monovalent electrolytes are normally about 0.8, while those for similar divalent salts are closer to 0.5. It is found that G values are dependent upon the concentration of the salt used. However, examination of the results of Table 12.4 with regard to activity coefficients reveals that this quantity is without effect. Thus, we have a process that depends upon the adsorption of ions but which is independent of the ion concentration, being dependent only upon the total salt content. The explanation of this paradox would seem to lie in the fact that the relaxation time required for ionization is small compared with the time for which a particle is in close contact with the surface. Thus, while the total salt concentration would be important in influencing the probability of a particle contacting the surface at a point where needed, the question as to whether the contacting particle was ionized or undissociated upon striking the surface would be unimportant.

Corrosion Inhibitors

While there is considerable similarity between the action of corrosion inhibitors and the action of the inorganic salts of this investigation, there are also important differences. For example, it is well known that whereas a very small concentration (0.1 percent or less) of sodium nitrite ($NaNO_2$) will inhibit the corrosion of iron or steel immersed in aerated water, sodium nitrate ($NaNO_3$) is not effective in this regard. Yet, it will be observed in Table 12.4 (series 2) that sodium nitrate and sodium nitrite are almost equally effective in grinding. To understand this, the action of a corrosion inhibitor such as sodium nitrite should be considered.

When an iron specimen is exposed to dry air the rate of oxidation of its surface is very low. This rate is almost equally low in deaerated water, but is greatly accelerated when oxygen or air is entrained in the water. Corrosion is particularly rapid at the point at which the specimen protrudes through a fluctuating surface. If the iron surface was generated in air or exposed to air before being immersed, it will be covered by a layer of adsorbed oxygen ions. However, these oxygen ions are a poor fit on the iron ions of the surface (atomic radius of oxygen = 1.40 Å; atomic radius of iron ion in metal lattice = 1.26 Å) and hence holes will exist through which small particles can migrate.

The small particles that migrate to the surface in this case are oxygen molecules and hydrogen ions. Upon reaching the surface each hydrogen ion takes up an electron from the free electron gas of the metal and an equivalent number of metal ions are freed from the surface. The nascent hydrogen then reacts with an adjacent oxygen ion to form an hydroxide ion [OH^{-1}]. This particle fits the iron surface no better than the original oxygen ion and further holes are provided for the migration of oxygen molecules. These in turn accept electrons from the metal surface and the resulting oxygen ions are adsorbed. The foregoing cycle is repeated over and over. A thick membrane of 'rust', consisting of alternate layers of hydroxide and iron ions, eventually develops on the iron surface.

The foregoing procedure will occur only with those metals that have ill-fitting oxides and happen to be above hydrogen in the electrochemical series. For example, aluminum does not behave in this manner even though it is more electronegative than iron, since it has a closely fitting oxide (the atomic radius of oxygen ion = 1.40 Å, while the atomic radius for the aluminum ion in a metal lattice = 1.43 Å). On the other hand, copper does not rust in water even though it has a poorly fitting oxide (atomic radius of copper ion in a metallic lattice = 1.28 Å), since it is electropositive relative to hydrogen.

The corrosion of iron in aerated water can be prevented if a close-fitting film that is impervious to molecular oxygen and hydrogen ions is deposited on the surface. Sodium nitrite provides such a coating (Fig. 12.12). Such a layer can be adsorbed from a dilute aqueous solution of a nitrite (such as sodium nitrite) or by pretreating the iron with concentrated nitric acid. In the latter case the iron becomes passive. Since concentrated nitric acid is known to be reduced by iron to form nitrite ions, it would seem that iron is rendered passive in the same way when either treated by concentrated nitric acid or a dilute solution of an ionizable nitrite.

The mutual repulsive forces of oxygen atoms in the [NO_2^{-1}] ions, each of

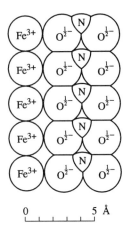

FIG. 12.12. An absorbed layer of nitrite ions on an iron surface.

which has an effective resonant charge of $-\frac{1}{2}$, cause an extremely close-packed arrangement and a close-knit protective layer. The size of the oxygen atoms will be less in this close-packed array than in the free state, i.e. as in Fig. 12.10, and hence a good fit will occur on the metal surface. The close packing will not only insure a completely filled layer but will augment the attractive electrostatic force between the oxygen atom and the metal surface in anchoring the coating in place.

When sodium nitrate is used in place of sodium nitrite an ill-fitting array, as shown in Fig. 12.13, results which is not capable of preventing the migration of oxygen molecules and hydrogen ions to the metal surface. The nitrate ion is effective, however, in raising the G value, since diffusion through the adsorbed film is not present in the grinding problem, as in the case of corrosion. The nitrate ion is not quite as effective in titanium grinding as the nitrite ion, since the former is not as firmly attached to the titanium surface.

The effectiveness of the nitrite ion in inhibiting iron against corrosion is enhanced when an organic chain is attached to one end of the nitrite ion. This can be accomplished by reacting sodium nitrite with an organic amine to form a nitrite amine:

$$\underset{\text{alkyl amine}}{R - NH_2} + \underset{\text{sodium nitrite}}{NaNO_2} \Rightarrow \underset{\text{sodium alkyl ammonium nitrite}}{R - NH_2 - NO_2 - Na}$$

where R represents an organic chain. Sometimes all three positions on the amino nitrogen are substituted by organic groups, as with the well-known vapor phase inhibitor tricyclohexylammonium nitrite. The ammonium nitrite ionizes in water to give an ammonium nitrite anion $[R - NH_2 - NO_2^{-1}]$ that adsorbs on an iron surface as in Fig. 12.12, except that R-chains extend outward from the outer oxygen atoms. These chains offer a further barrier to the passage of oxygen molecules or hydrogen ions to the metal surface and help

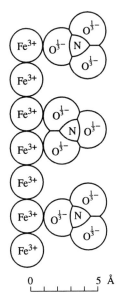

FIG. 12.13. An adsorbed layer of nitrate ions on an iron surface.

hold the protective layer on the surface. Actually, the nitrite group is soluble in water and if fresh water should be substituted for the inhibited nitrite solution, rusting could occur. The organic chain attached to the nitrite group renders the ammmonium nitrite complex hydrophobic (water repellent), which in turn keeps the protective layer in place on the surface even in the presence of uninhibited water.

This development in the field of corrosion inhibition suggests that cations for grinding titanium should also have long chains attached.

Summary: Inorganic Solutions For Titanium

The systematic study of dilute inorganic aqueous solutions has revealed that these solutions are effective grinding fluids for titanium alloys due to the formation of adsorbed layers on both the aluminum oxide and freshly ground titanium surfaces. These adsorbed layers screen the titanium and abrasive surfaces and prevent strong bonds from being established between them which, when ruptured, result in rapid loss of abrasive material.

Three considerations govern the effectiveness of the adsorbed layer:

(1) The ability of the adsorbing ion to fit the ions of the surface closely;

(2) The effective charge per ion tending to anchor in place the screening ions adsorbed on the surface;

(3) The length of the adsorbed ion extending from the surface.

TABLE 12.6 *The effect of tripotassium phosphate solution on abrasive belt performance. Work material, Ti-6A1-4V; abrasive, 60 grit size SiC; belt length, 132 in (335 cm); axial load, 40 p.s.i. (0.28 Pa) (after Cadwell et al. 1958)*

Experiment number	Lubricant	Total time (s)†	Total cut (cm^3)
1	Air	27	1.85
2	Water	48	2.30
3	0.3 percent K_3PO_4	192	8.20
4	2.8 percent K_3PO_4	810	35.3
5	5.0 percent K_3PO_4	1500	62.7
6	15.5 percent K_3PO_4	2130	97.3
7	30.0 percent K_3PO_4	2145	101.3

All experiments were terminated when a grinding rate of 0.03 cm^3 s^{-1} was reached.

The barium ion [Ba^{+2}] was found to be the most effective cation for protecting aluminum oxide surfaces, while the phosphate ion [PO_4^{-3}] was found to be the most effective anion for protecting titanium surfaces. Unfortunately, barium phosphate cannot be used, since it is insoluble in water. Barium hydroxide solutions have been found to be quite effective and such solutions are noncorrosive on steel parts.

The concentration of the active ingredient was found to be important, best results being obtained with about one-half molar concentrations.

Belt Grinding of Titanium Alloys

Cadwell *et al.* (1958) verified the improved performance obtained with aqueous solutions of tripotassium phosphate in belt grinding of titanium alloys. Table 12.6 indicates that in belt grinding water gives slightly better performance than air and that the belt life increases with the concentration of K_3PO_4. Hong *et al.* (1971) have also successfully used phosphate solutions to belt grind titanium alloys.

Since K_3PO_4 solutions are sufficiently basic to injure human skin and remove paint and grease from machinery, it should be buffered by addition of an acidic phosphate salt to bring the pH down to an acceptable level (pH ~ 9.0). Hong (1972) employed a 10 w/o solution of K_3PO_4 in water buffered with an equimolar mixture of K_3PO_4 and NaH_2PO_4 (wt K_3PO_4/wt NaH_2PO_4 = 212/120). Tests were performed on a 10 hp machine at a belt speed of 3100 f.p.m. (15.75 m s^{-1}) and a grit size of 60, using a smooth contact wheel of 8 in (203 mm) diameter (Fig. 12.14). The work material was Ti-6A1-4V, which was ground from a diameter of 1.05 in (26.67 mm) to 0.95 in (24.13 mm). Silicon carbide abrasive was found to out-perform Al_2O_3 abrasive for this unusually severe belt grinding operation (centerless grinding). The relative performance of the grinding fluids tested with SiC and Al_2O_3 abrasives are given in Table 12.7 relative to air.

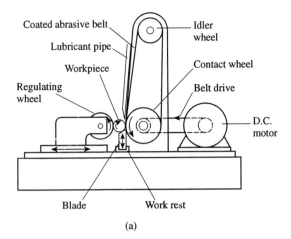

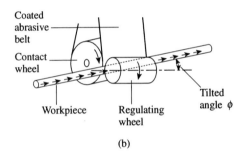

FIG. 12.14. A belt centerless grinding operation: (a) end view; (b) side view (after Hong 1972).

Subsequent belt grinding studies (Duwell *et al.* 1985b) have confirmed the superiority of SiC abrasive over Al_2O_3 when belt grinding a titanium alloy using an inorganic solution.

Upon evaporation of a phosphate solution, a residue is deposited which may have an adverse influence on machine tool performance. Also, even buffered phosphate solutions attack paint slowly and can cause skin irritation. Therefore, care must be taken that these fluids be used under conditions in which the machine and operator are not adversely affected.

Mittal *et al.* (1991) have compared the performance of 5 percent soluble oils and a 10 percent buffered phosphate solution when plane surface grinding a titanium alloy with vitrified SiC and Al_2O_3 wheels. It was found that the phosphate solution gave lower grinding forces and specific energy than the soluble oils tested with SiC and Al_2O_3 wheels. The surface finish was also better with the phosphate solution than with the soluble oils, as was the grinding ratio G. SEM photomicrographs of grit tips showed less metal build-up with the phosphate solution than with the soluble oils. SEM photomicrogaphs

TABLE 12.7 *Performance of SiC an Al_2O_3 abrasive belts when grinding Ti-6A1-4V alloys with different fluids relative to air. These relative G values are for the heavy removal rate of $5.1 \times 10^{-3} in^3 min^{-1}$ ($83.6 mm^3 min^{-1}$) (after Hong 1972)*

	Abrasive	
	SiC	Al_2O_3
Air	1.00	1.00
Water	1.38	1.62
10 percent heavy duty lubricant	2.00	2.00
10 percent buffered K_3PO_4 solution	1.58	3.29

of ground surfaces also showed less side flow (better finish) with the buffered solution than with the soluble oils.

It has been found that a buffered phosphate solution is also effective when Al_2O_3 abrasive is used in belt grinding of ferrous alloys. However, the improvement over dry grinding is not as dramatic as when grinding a titanium alloy.

Diamond Grinding of Titanium Alloys

Under ordinary surface grinding conditions, diamond grinding of titanium alloys gives very poor results. The strong affinity of titanium for carbon gives strong bonds, leading to metal build-up on the grit tips. This build-up leads to high temperatures, despite the unusually high value of β for diamond, and then more build-up with rising temperature until chipping occurs. One way of making it possible for diamond to be used to grind titanium is to employ cryogenic cooling, but this is awkward and may not be cost-effective. Chattopadhyay *et al.* (1985) employed a stream of liquid nitrogen to the point at which the wheel enters the work and found a sizeable reduction in forces and an improvement in surface integrity.

Kumagai and Kiyoshi (1984) found that Ti-6A1-4V could be successfully ground at a conventional wheel speed ($32 m s^{-1}$) if a high-pressure jet of coolant was applied to the wheel at the point of entry of the wheel into the work. Best results (high G) were obtained with a tough form of diamond. Lower G values were obtained with SiC, Al_2O_3, and CBN abrasives than for D, even with a high-pressure jet. The authors attribute this to the following chemical reactions, suggested by analysis of ESCA traces:

for SiC, $SiC + O_2 \Rightarrow SiO + C$,
$SiC + 2O_2 \Rightarrow SiO_2 + CO_2$,
$Ti + SiO_2 + O_2 \Rightarrow TiO_2 + SiO_2$,
$Ti + SiC + O_2 \Rightarrow TiC + SiO_2$;
for Al_2O_3, $2Al_2O_3 + Ti \Rightarrow 3TiO_2 + 4Al$;

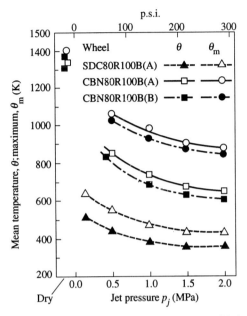

FIG. 12.15. The variation in the mean grinding temperature with jet pressure: (a) 80R, tough grit D wheel; (b) 80R, CBN grit wheel (after Kumar 1990). Wheel speed, 32 m/s (6300 f.p.m.); work speed, 0.1 m s^{-1} (19.7 f.p.m.); wheel depth of cut, 20 μm (800 μin); work, Ti.

for CBN, $2B_4N_3 + 8O_2 \Rightarrow 8BO_2 + 3N_2$,
$B_4N_3 + 7O_2 \Rightarrow 4BO_2 + 3NO_2$,
$BO_2 + Ti + O_2 \Rightarrow TiO_2 + BO_2$.

When grinding temperatures were measured by the buried thermocouple technique, very low values were obtained for diamond relative to the other abrasives when a high-pressure jet was employed. Figure 12.15 shows variation of the mean grinding temperature at the wheel face with jet pressure (p_j) for diamond and CBN wheels. At a pressure of 2 MPa (290 p.s.i.) the mean temperature was only 373 K, but with dry grinding the mean temperature was above the transformation temperature (~1350 K) for all wheels. A high jet pressure (2 MPa) gave a high compressive residual stress for both SiC and D wheels, but a high tensile residual stress was obtained with dry grinding for all wheels. Surface finish was significantly better for all wheels with an increase in jet pressure.

In studying the Kumagai and Kiyoshi paper (1984), it appears that diamond grinding of titanium alloys is possible when a high-pressure jet is employed for two reasons:

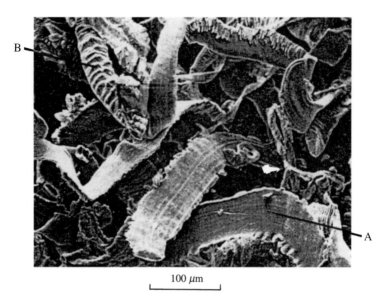

FIG. 12.16. SEM of Ti 6Al 4V chips from surface grinding operation with a resin-bonded moderate toughness diamond wheel (after Kumar 1990).

(1) the superior cooling action;
(2) the prevention of build-up by wheel cleaning associated with a high-pressure jet.

Kumar (1990) performed tests on Ti-6Al-4V ($R_c = 35$) using diamond and CBN resin-bonded wheels (100–120 grit size, 100 concentration) in conventional surface (pendulum) grinding with cross-feed. Five percent heavy duty grinding coolant was applied using two nozzles, as Kumagai and Kiyoshi (1984) had done, but up grinding was used instead of down grinding. A vitrified SiC wheel (GC 60 K 8VBE) was also used for comparison. Wear, specific energy, and finish (without spark-out) were measured. The grinding conditions were as follows:

Wheel speed, V: $30\,\text{m s}^{-1}$ (5900 f.p.m.)
Work speed, v: $0.25\,\text{m s}^{-1}$ (49 f.p.m.)
Wheel depth of cut, d: 0.025 mm (0.001 in)
Cross-feed per stroke (b'): 1.25 mm (0.050 in)

Representative results obtained are given in Table 12.8. The force ratio F_Q/F_P was close to 2.0 in all cases.

These results confirm those of Kumagai and Kiyoshi (1984) and indicate that with sufficient cooling and wheel cleaning, diamond may be very successfully used to grind titanium alloys.

Figure 12.16 is an SEM of chips produced by the medium tough grade of

TABLE 12.8 *Representative surface grinding results for superabrasive grinding of titanium alloy Ti-6Al-4V, for grinding conditions givens in the text (after Kumar 1990)*

Abrasive	Grinding ratio, G	Specific energy, u		Finish, R_a	
		p.s.i.	GPa	μin	μm
Diamond†	160	4.44×10^6	30.6	60	1.5
CBN†	35	7.40×10^6	51.0	60	1.5
SiC	1	6.90×10^6	47.6	28	0.7

† Medium tough grade.

diamond referred to in Table 12.8. One side of these chips is relatively smooth (A, for example), while the other side (B) has a saw-tooth pattern, having a small pitch of the order of 5 μm (200 μin). The saw-tooth type of chip is also found when titanium is machined in single-point cutting, and the mechanics of this type of chip formation, which involves periodic fracture planes extending part way across the chip, is discussed in Shaw and Vyas (1993). The smooth sides of the chips are against the tool face in cutting but in contact with the plastic region in FFG (Fig. 4.1), while the free surface is where the periodic fracture planes emanate.

The mean chip width in Fig. 12.16 is about 50 μm (0.002 μin). The mean radius of curvature (ρ) of the grit tip may be estimated from eqn (4.8) (for a 100–120 grit size this is about $g/4 = 0.0064/4 = 0.0016\,in = 40\,\mu m$).

Superabrasives used to grind difficult to grind materials in the aerospace industry have been discussed by Kumar (1992). These materials include nickel-based alloys (Ni > 50 w/o), titanium alloys, and nickel and titanium aluminides. All of these materials tend to form strong bonds with both conventional and superabrasive materials. This leads to build-up of work material, leading to increasing forces and temperatures, and resulting in excessive wear.

Kumar (1992) used the coolant/cleaning arrangement shown in Fig. 12.17, which made it possible successfully to grind the above-mentioned aerospace materials with superabrasives (CBN and D). The two cooling jets each delivered a heavy duty water-based coolant (5 percent concentration) at a flow rate of 4.75 g.p.m. (18 l min^{-1}) and a pressure of 43 p.s.i. (0.3 MPa). The cleaning jet operated at 2.19 g.p.m. (8 l min^{-1}) and 430 p.s.i. (3 MPa). Up grinding as shown in Fig. 12.17 was more effective than down grinding.

Tests were performed on nickel-based alloys using vitrified superabrasive wheels (100–120 grit size, 175 concentration, 30 percent porosity, 6.9 in (175 mm) in diameter) in a creep feed surface grinding mode (Chapter 13). Grinding conditions were as follows:

Wheel speed: 5510 f.p.m. (28 m s^{-1})
Work speed: 1.31 f.p.m. (0.4 m min^{-1})
Wheel depth of cut: 0.06 in (0.152 mm)
Width ground: 0.107 in (2.7 mm)

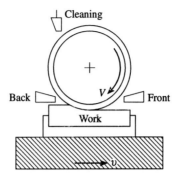

FIG. 12.17. An arrangement used to grind aerospace materials with superabrasives (after Kumar 1992).

A tough grade of CBN gave better results than D. Mean results were as follows for a typical nickel-based alloy (52.8 w/o Ni, 19 w/o Cr, 18.5 w/o Fe, 5 w/o Cb, and 3 w/o Mo):

> Grinding ratio, G: 250
> Specific energy, u: 13×10^6 p.s.i. (90 GPa)
> Surface finish, R_a: 20 µin (0.5 µm)

For a work speed of 0.65 f.p.m. (0.2 m min^{-1}) the mean G value was about 50 percent higher, the u value was about 70 percent higher, and the mean R_a value was about 30 percent lower than the above values for the higher work speed of 1.31 f.p.m. (0.4 m min^{-1}).

Conventional reciprocating surface grinding tests were performed on titanium alloy Ti 6A1 4V ($R_c = 35$) with cross-feed using resin-bonded D and CBN wheels. Grinding conditions were as follows:

> Wheel speed: 5900 f.p.m. (30 m s^{-1})
> Work speed: 49.9 f.p.m. (15.2 m min^{-1})
> Cross-feed: 0.06 in per stroke (1.5 mm per stroke)
> Wheel: 100–120 grit size,
> 100 concentration, resin bond.

Tough grades of D and CBN gave the best results. Mean results were as follows:

Grinding ratio, G: 165 (D), 30 (CBN)
Specific energy, u: 4.7×10^6 p.s.i. (32.4 GPa) for D and
7.9×10^6 p.s.i. (54.6 GPa) for CBN

The heavy duty water-based coolants used in these tests were typical of those used in the aerospace industry and the grinding ratio varied by about 20 percent between brands. The grinding ratio of a SiC wheel under comparable grinding conditions was only about one.

Even though titanium alloys have a very strong affinity for carbon and nitrogen, adhesion and build-up can be controlled by use of effective cooling and wheel cleaning. With ordinary coolant application, results when grinding titanium alloys with D or CBN wheels are very much poorer.

Nickel and titanium aluminides used in the aerospace industry because of their high-temperature strength and low specific weight have also been successfully ground by Kumar (1992) using diamond wheels in conjunction with the special coolant system shown in Fig. 12.17. These materials are essentially ungrindable with conventional abrasives. For example, when a material having the composition 65 w/o Ni, 12 w/o Al, and 23 w/o Ta was ground with a resin-bonded diamond wheel (100–120 grit size, 75 concentration, friable type diamond), the following results were obtained:

Grinding ratio, G: 25
Specific energy, u: 3.7×10^6 p.s.i. (25.2 GPa)

Grinding conditions were as follows:

Wheel speed: 4900 f.p.m. (25 m s^{-1})
Work speed: 49.9 f.p.m. (15.2 m min^{-1})
Cross feed, 0.120 in/str (3 mm/str)
Wheel depth of cut, 0.002 in (50 µm)

The foregoing examples show that superabrasives may be used to grind materials used in the aerospace industry that are very difficult to grind provided that proper cooling and wheel cleaning precautions are taken. The abrasive of choice for nickel-based alloys is a tough form of CBN, but it is diamond for titanium alloys and nickel alumide. Semi-tough diamond gave surfaces free of cracks and burn for titanium alloys, but friable diamond giving relative low G values was required for nickel and titanium aluminides for good surface integrity.

Fluid Evaluation by COFG

In Chapter 10, metal transfer and wear of SiC and Al_2O_3 abrasives when dry grinding AISI T15 tool steel and a titanium alloy (Ti-6Al-4V) were discussed in terms of COFG results. While Al_2O_3 gave less wear when grinding T15 tool steel, SiC gave less wear when grinding the Ti-6Al-4V alloy. SiC was found to oxidize excessively when grinding T15 tool steel, giving rise to a relatively high attritional wear rate. By contrast, Al_2O_3 tended to acquire a layer of built-up metal when grinding both T15 and the Ti-6Al-4V alloy, which gave rise to a microchipping mode of wear. While SiC also tended to form a built-up layer and to microchip when grinding the Ti-6Al-4V alloy, this action was not as pronounced as for the Al_2O_3/(Ti-6Al-4V) grit/work combination. In no case was there any evidence of a solid state reaction between work and abrasive.

As described in Chapter 10, the COFG method employs a small cluster of

TABLE 12.9 *Nominal compositions (w/o) of work materials used in COFG fluid evaluation tests (after Chandraeskar and Shaw 1983)*

Material	Composition								
	C	O	Fe	W	Cr	V	Co	Al	Ti
T15 tool steel	1.57	–	72.04	12.6	4.0	4.9	4.89	–	–
Ti-6Al-4V	0.1	0.2	–	–	–	4.0	–	6.0	89.7

grits measuring $\frac{1}{8} \times \frac{1}{8}$ in (3.2 × 3.2 mm) on the dressed wheel surface to surface grind a groove $\frac{1}{8}$ in (3.2 mm) wide in the workpiece surface. The wheel speed V and wheel depth of cut d are the same in the COFG test as in the corresponding complete wheel situation. The table speed v_c in the COFG test is reduced from the speed v_w in the full wheel case, so that the undeformed chip thickness t of individual chips remains the same. This is done since practically all grinding results are functions of t and it is desired that all conditions be as identical as possible in the COFG and full wheel cases.

Wheel wear should be the same for COFG and complete wheel tests when the length L_w of work ground by the complete wheel is related to the length L_c of work ground by the cluster as follows:

(1) when microchipping wear is dominant,

$$L_w/L_c = \pi D/l_c; \qquad (12.1a)$$

(2) when attritional wear is dominant,

$$L_w/L_c = v_w/v_c; \qquad (12.1b)$$

where D is the wheel diameter, and l_c is the circumferential length ($\frac{1}{8}$ in = 3.2 mm) of the cluster.

The nominal chemical compositions of the work materials used are given in Table 12.9. The T-15 steel was a sintered material with a median carbide size of 0.3 μm (12 μin) and was heat-treated to a hardness of $R_c = 65$. The Ti-6Al-4V alloy was an annealed material of hardness $R_c = 32$. The grinding wheels used were A60 J8 V and C60 J8 V. The wheel speed (V) was 6300 ft min^{-1} (32 m s^{-1}) and the undeformed chip thickness (t) was 150 μin (3.75 μm) in all cases. This corresponds to a wheel depth of cut (d) of 500 μin (12.5 μm) and a table speed (v) of 75 f.p.m. (0.38 m s^{-1}) for the complete wheel. The rate of fluid flow was 0.5 U.S. gallons per minute (3.15 × 10^{-5} m^3 s^{-1}).

Except as noted elsewhere, the following fluids were used: (a) air; (b) 10 w/o NaNO$_2$ in water; (c) 10 w/o K$_3$PO$_4$ in water, buffered with an equimolar concentration of NaH$_2$PO$_4$; (d) a 10 percent concentration of a heavy duty (S + Cl) water-based cutting fluid in water; and (e) undiluted grinding oil.

Fluids (b) and (c) are inorganic salt solutions that have been found to be effective in grinding titanium alloys in industry. The undiluted grinding oil was not used on the Ti-6Al-4V alloy because this may constitute a fire hazard.

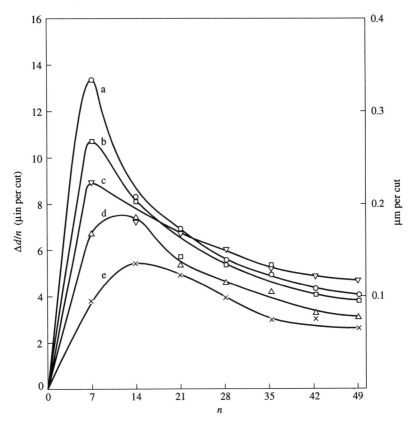

FIG. 12.18. The variation in the mean cumulative wear per cut ($\Delta d/n$) with the number of cuts n for Al_2O_3 in COFG of T15 tool steel for various fluids: (a) air; (b) 10 w/o $NaNO_2$; (c) 10 w/o K_3PO_4; (d) 10 w/o heavy duty grinding fluid; (e) undiluted grinding oil (after Chandrasekar and Shaw 1983).

Figures 12.18 and 12.19 show the mean cumulative wear per cut ($\Delta d/n$) plotted against the number of cuts (n) for Al_2O_3 and SiC grits respectively when cutting in the COFG mode using different fluids. Al_2O_3 is seen to wear less rapidly in all cases and the best fluid of the group when grinding T15 tool steel with Al_2O_3 was the undiluted grinding oil. The best fluid for the SiC–T15 combination was the heavy duty water-based fluid, although these results were not nearly as good as those for the best Al_2O_3 combination.

Figure 12.20(a) shows the mean cumulative wear for several individual Al_2O_3 grits when grinding the T15 tool steel using the undiluted grinding oil, while Fig. 12.20(b) shows comparable results when grinding in air. The grinding oil is seen to be relatively effective in reducing adhesion and build-up and hence in reducing microchipping wear. Grit number 12 of Fig. 12.20(a) is the only one that showed a slight indication toward build-up, while grit number

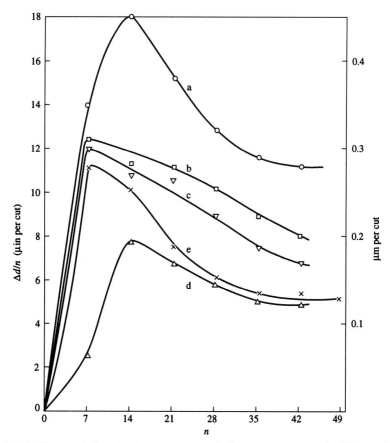

FIG. 12.19. The variation in the mean cumulative wear per cut ($\Delta d/n$) with the number of cuts n for SiC in COFG of T15 tool steel for various fluids: curves (a)–(e) as for Fig. 12.18 (after Chandrasekar and Shaw 1983).

12 of Fig. 12.20(b) shows considerable evidence of build-up in the form of 'negative wear'.

Figure 12.21 shows plots of mean cumulative wear Δd against the length L_w ground for Al_2O_3 and SiC abrasive grinding of T15 tool steel using grinding fluid (curve a) with complete wheels and for the COFG technique (L_c converted to L_w). These results are representative of a number of plots, all of which show excellent agreement between full wheel and COFG tests when interpreted in terms of eqn (12.1).

Scanning electron microscope (SEM) pictures of used wheel surfaces were helpful in verifying the presence of metal build-up on abrasive surfaces.

Figures 12.22 and 12.23 show the mean cumulative wear per cut ($\Delta d/n$) plotted against the number of cuts (n) for Al_2O_3 and SiC grains respectively when cutting in the COFG mode using different fluids. SiC is seen to wear less rapidly in all cases and the best fluid for the SiC–(Ti-6A1-4V) system was

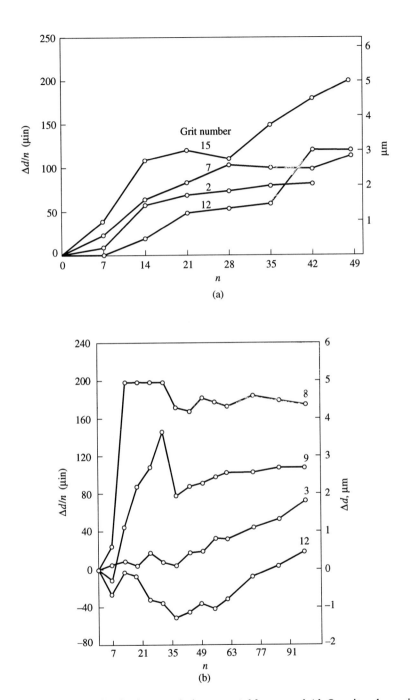

FIG. 12.20. The variation in the cumulative wear Δd for several Al_2O_3 grits when grinding T15 tool steel: (a) with undiluted grinding oil; (b) dry in air (after Chandrasekar and Shaw 1983).

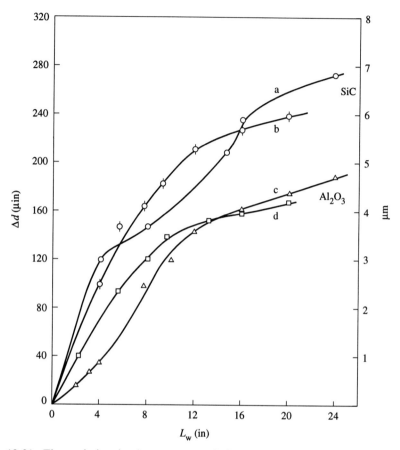

FIG. 12.21. The variation in the mean cumulative wear Δd with the length ground (L_W) for Al_2O_3 and SiC, grinding T15 tool steel using a 10 w/o heavy duty solution: (a) full wheel with SiC; (b) COFG with SiC; (c) full wheel with Al_2O_3; (d) COFG with Al_2O_3 (after Chandrasekar and Shaw 1983).

the heavy duty solution. The best fluid for the Al_2O_3–(Ti-6Al-4V) combination was the $NaNO_2$ solution.

Figure 12.24 shows the mean cumulative wear Δd for several individual SiC grits when grinding the Ti-6Al-4V alloy using the heavy duty solution. Considerable negative wear (metal build-up, MBU) is seen to occur on grit 15 even when the most effective fluid of the group is used. The extent of the MBU was greater for the titanium alloy than for the T15 steel, particularly with the less effective fluids.

Figure 12.25 shows representative wear results of COFG and full wheel tests. This is for the Al_2O_3–(Ti-6Al-4V), heavy duty solution system. Other combinations gave equally good correlation between COFG and full wheel tests.

When the Ti-6Al-4V alloy was ground, chips were found to adhere to the

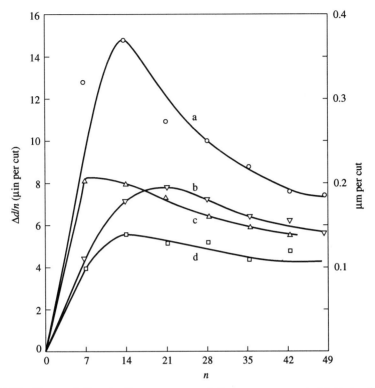

FIG. 12.22. The variation in the mean cumulative wear per cut ($\Delta d/n$) with the number of cuts n for Al_2O_3 in COFG of Ti-6Al-4V alloy for various fluids: (a) air; (b) 10 w/o equimolar K_3PO_4 and NaH_2PO_4; (c) 10 w/o heavy duty solution; (d) 10 w/o $NaNO_2$ (after Chandrasekar and Shaw 1983).

wheel when less effective fluid was used, which led to chipping and a high rate of wheel wear.

Figure 12.26 shows the curves of mean cumulative wear per cut ($\Delta d/n$) versus number of cuts (n) for the Al_2O_3-(Ti-6Al-4V) system when using $NaNO_2$ solutions of various concentrations. Results for higher concentrations were about the same as for the 10 wt percent concentration. Since the residue left behind on the machine when the fluid evaporates becomes more troublesome as the salt content increases, the 10 wt percent concentration of inorganic salt appears to be about optimum. The same conclusion was reached when using different concentrations of $NaNO_2$ with the SiC-(Ti-6Al-4V) system.

The COFG technique gives wear results that are similar to those obtained with a complete wheel when different grinding fluids are evaluated, and hence provides a useful method of evaluating different fluids in less time, with the use of less ground material, and with a smaller sample of the fluid being

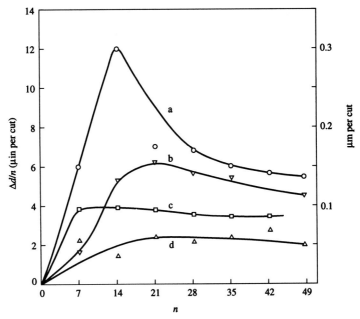

FIG. 12.23. The variation in the mean cumulative wear per cut ($\Delta d/n$) with the number of cuts n for SiC in COFG of Ti-6Al-4V alloy for various fluids: curves (a)–(d) as for Fig. 12.22 (after Chandrasekar and Shaw 1983).

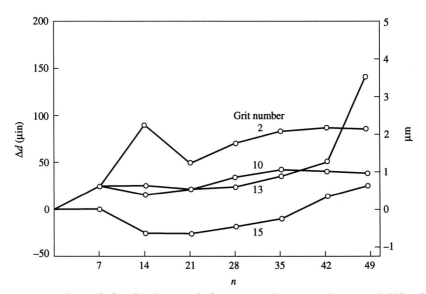

FIG. 12.24. The variation in the cumulative wear Δd versus n for several SiC grits when grinding Ti-6Al-4V alloy using 10 w/o heavy duty solution (after Chandrasekar and Shaw 1983).

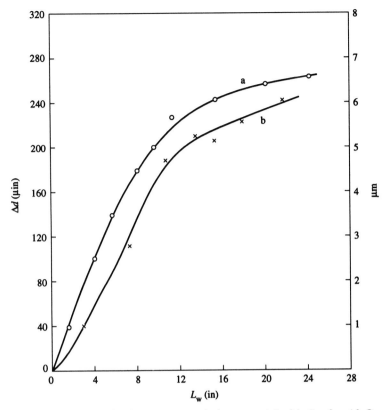

FIG. 12.25. The variation in the mean cumulative wear Δd with L_W for Al_2O_3 grinding Ti-6Al-4V alloy using 10 w/o heavy duty solution: (a) COFG; (b) full wheel (after Chandrasekar and Shaw 1983).

evaluated. The tendency for chips to adhere to the wheel face, causing chipping, is essentially the same for COFG and when grinding with a complete wheel. When T15 tool steel is ground at a conventional wheel speed (6300 f.p.m., or 32 m s^{-1}), Al_2O_3 gives better results than SiC and the undiluted oil is the most effective fluid tested. When Ti-6Al-4V alloy is ground at conventional wheel speed (6300 f.p.m., or 32 m s^{-1}), the SiC abrasive gives better results than Al_2O_3 and the heavy duty solution is the most effective fluid.

Grinding Aids

Grinding aids are solid materials impregnated into the pores of grinding wheels to improve performance. Barash (1979) has abstracted a Russian report that described work with impregnated Al_2O_3 wheels used to grind hardened ball bearing steel. The following types of impregnation were employed:

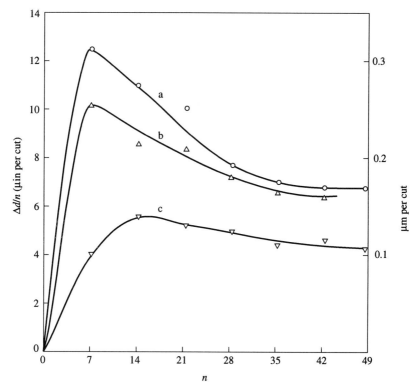

FIG. 12.26. The variation in the mean cumulative wear per cut ($\Delta d/n$) with the number of cuts n for Al_2O_3 grinding Ti-6Al-4V alloy in the COFG mode: (a) 1 w/o $NaNO_2$; (b) 5 w/o $NaNO_2$; (c) 10 w/o $NaNO_2$ (after Chandrasekar and Shaw 1983).

- sulfur
- 60 percent S, 30 percent stearin, 10 percent aluminum stearate
- 50 percent S, 30 percent stearin, 10 percent aluminum stearite, 10 percent potassium permanganate

The wheels were impregnated by soaking in the molten substance at 125 °C. Taking the life of an unimpregnated wheel as 1, the relative wheel lives resulting in grinding burn for the three impregnating materials were (a) 2.2, (b) 2.9, and (c) 3.3.

X-ray diffraction revealed that iron sulfide forms on the grits, and then oxidizes to form the solid lubricant iron sulfate. The important role of oxidation was verified by the fact that sulfur alone (case (a)) was relatively ineffective when grinding in an inert atmosphere, but that when the oxidizing agent (potassium permanganate) was present (case (c)) improved performance was obtained in air or an inert atmosphere.

Yamamoto and Ueda (1978) studied the action of sulfur and other materials

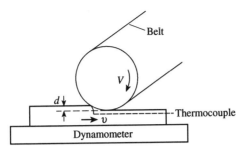

FIG. 12.27. Belt surface grinding with an 'erosion' thermocouple penetrating into ground surface (after Cosmano *et al*. 1984).

(paraffin, stearic acid, and carnuba wax in chlorinated paraffin) in the pores of honing sticks. It was found that grinding performance (grinding ratio or removal rate) was best for impregnations with sulfur. The role of impregnation depends on whether the operating temperature is above or below the melting point of the impregnating material. If the material in the pores remains solid (as is usually the case in honing) then structural reinforcements are obtained. If the material in the pores is molten, as in FFG, then the principal influence is in preventing chip-grit adhesion.

Cosmano *et al*. (1984) have ground 304 stainless steels, several carbon steels of different carbon contents, ranging from 0.18 to 0.95 w/o C, and a gray cast iron using 3 × 80 in (75 × 2000 mm) regular Al_2O_3 (60 grit size) belts and an 18 in (457 mm) aluminum contact wheel, as shown in Fig. 12.27. Tests were performed at a wheel speed (V) of 5600 f.p.m. (28.5 m s^{-1}), wheel depths of cut d of 0.001 and 0.002 in (25 and 50 μm), and table speeds (v) of 10, 20, 30, and 40 f.p.m. (0.05–0.20 m s^{-1}) with and without a proprietary grinding oil (presumably containing S and Cl) in the pores of the belt. No grinding fluid was used. A $\frac{1}{4}$ in (6.35 mm) diameter 'eroding' thermocouple was mounted flush with the surface at the center of each 7 in long by 1 in wide (127 × 25 mm) work piece to measure the mean grinding temperature (θ).

When grinding 304 SS without a grinding aid, the specific grinding energy (u) varied from about 5.9×10^6 to 3.4×10^6 p.s.i. (40–23 GPa) as the undeformed chip thickness ($t \sim v^{0.5} d^{0.25}$) increased by a factor of two. With the grinding aid all values of u were approximately half as great as without the grinding aid. Values of the grinding force ratio F_Q/F_P were close to two in all cases. The measured mean temperature rise (θ) went from 460 °C to 425 °C as u went from 5.9×10^6 to 3.4×10^6 p.s.i. (40 GPa to 23 GPa) without the grinding aid, and from 195 °C to 155 °C as u went from 3.1×10^6 to 1.8×10^6 p.s.i. (21 GPa to 12 GPa) with the grinding aid. The specific grinding energy has a predominant influence on θ, as theory suggests it should have, and the grinding aid had a marked influence in decreasing both u and θ when grinding 304 SS.

When several unhardened plain carbon steels were ground at a removal rate of 1 in³ min⁻¹ (16.4 cm³ min⁻¹) the specific energy was only about 1.4×10^6 p.s.i. (9.65 GPa) with and without a grinding aid. When gray cast iron was ground at 1 in³ min⁻¹ (16.4 cm³ min⁻¹) the specific grinding energy was about 0.93×10^6 p.s.i. (6.4 GPa) with and without the grinding aid. The fact that the grinding aid is beneficial in lowering the specific energy for stainless steel but not for carbon steel and cast iron suggests that rapid oxidation decreases adhesion in the case of steel, and that graphite has a similar effect in the case of cast iron, thus making chemical action by the grinding aid less important.

It is interesting to note that the values of u for belt grinding under SRG conditions (1 in³ min⁻¹, or 16.4 cm³ min⁻¹) approach those for metal cutting.

While grinding aids are usually deposited in the pores of a wheel or belt, it is possible that solid additives to the work would have a similar role in decreasing adhesion. A naturally occurring example of this is the graphite flakes in gray cast iron. In sintered products, it is possible to include solid-phase grinding aids in the work material. It has been found (Yamaguchi *et al.* 1993) that additions of glass to sintered steel improve the machinability of the steel without adversely influencing its structural properties. Apparently, the molten glass covers the tool face and decreases adhesion. By selecting a glass that wets abrasive surfaces, it is conceivable that a similar result might be obtained in grinding.

Wheel Loading

Wheel loading occurs when chips fill the pores between active grits in the surface of a grinding wheel. This is particularly troublesome when grinding relatively soft ductile materials at a high removal rate. When the removal rate exceeds the rate of chip storage available, chips will tend to become lodged in the chip storage space. This in turn causes an increase in grinding temperature and an increase in wedging pressure, tending to cause loss of grit material by chipping or whole grit removal.

Solutions to this problem include the following:

- a decrease in the removal rate
- an increase in the wheel speed
- the use of an effective fluid or grinding aid to decrease chip–grit adhesion
- an increase in the hardness and brittleness of the work material
- the use of a high-pressure cleaning jet
- the use of a more open structure wheel and/or a coarse dress
- the use of a wire or fiber brush or rubber wheel to clean the surface and prevent pore clogging

The last of these (the rubber wheel) is an interesting possibility when combined with the transmission of power from the motor to the wheel. Rubber

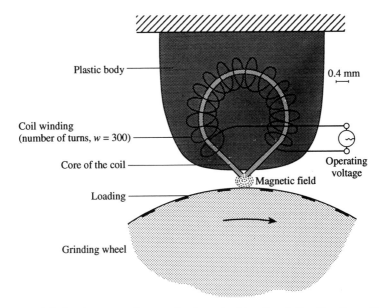

FIG. 12.28. A inductive sensor for monitoring wheel loading (after Koenig and Lauer-Schmaltz 1978).

is an extremely forgiving material when subjected to abrasive action. If this were not so, automotive tire life would be a fraction of the relatively long life obtained when a 100 hp or more is transmitted by two tires in contact with a very abrasive road surface. The use of a motor-driven rubber wheel in contact with a grinding wheel 180° from the grinding zone to drive the wheel, as wheel deflection is reduced and continuous wheel cleaning is achieved, offers an interesting machine design possibility.

Radhakrishnan (1993) has suggested use of a high-powered laser to remove material continuously from the pores of a grinding wheel by evaporation. The challenge in this case is to remove the clogging chip material without adversely affecting wheel life due to bombardment of the grits.

Koenig and Lauer-Schmaltz (1978) have described a sensor (Fig. 12.28) for continuously measuring the degree of wheel loading. When ferritic chip material enters the magnetic field the impedance of the coil winding changes and is measured by a bridge circuit. The operating voltage is 0.2 V at a frequency of 300 kHz. In a second paper (Lauer-Schmaltz and Koenig 1980) the wheel-loading mechanism is discussed further.

References

Barash, M. (1979). *Manufact. Engr* **82**, 41.
Cadwell, D. E., Weisbecker, H. L., and McDonald, W. J. (1958). *ASME Paper 58-SA44*.

Chandrasekar, S. and Shaw, M. C., (1983). *Wear* **86**, 139.
Chattopadhyay, A. B., Bose, A., and Chattodadhyay, A. K. (1985). *Prec. Engng* **7**(2), 93.
Cosmano, R. J., Abrahamson, G. R., and Duwell, E. J. (1984). *Proc. Int. Grinding Conf.*, SME, Dearborn, Michigan.
Courtel, R. (1949). *Rev. Metall.* **46**, 24.
Duwell, E. J. and McDonald, W. J. (1961). *Wear* **4**, 384.
Duwell, E. J., Hong, I. S., and McDonald, W. J. (1966). *Wear* **9**, 417.
Duwell, E. J., Cadwell, E. J., Cosmano, R. J., Abrahamson, G. R., and Gagliardi, J. J. (1985a). *J. Engng Ind.* **110**, 19.
Duwell, E. J., Cosmano, R. J., Abrahamson, G. R., and Gagliardi, J. J. (1985b). *ASME PED* **16**, 251.
Gruehring, K. (1967). D.Ing. dissertation, Aachen T.H.
Hong, I. S. (1972). *Proc. Pittsburgh Int. Grind. Conf.* Carnegie Press, Pittsburgh, Pennsylvania, p. 860.
Hong, I. S., Duwell, E. J., McDonald, W. J., and Mereness, E. (1971). *ASLE Trans.* **14**, 8.
Koenig, W. and Lauer-Schmaltz, H. (1978). *Ann. CIRP*, **27**(1), 217.
Kumagai, N. and Kiyoshi, K. (1984). *Proc. 5th Int. Conf. Titanium*, Deutsche Gesellesehaft für Metallkunde, Oberursel, Germany.
Kumar, K. V. (1990). *SME 4th Int. Grinding Conf.*, MR90505, SME.
Kumar, K. V. (1992). *Proc. 3rd Int. Conf. High Tech.*, Chiba, Japan.
Lauer-Schmaltz, H. and Koenig, W. (1980). *Ann. CIRP*, **29**(1), 201.
Mittall, B., Barber, G. C., and Malkin, S. (1991). *ASME Ped 54/TRIB2*, 15.
Opitz, H. and Gruehring, K. (1968). *Ann. CIRP*, **26**(1), 61.
Outwater, J. O. and Shaw, M. C. (1952). *Trans. ASME* **74**, 73.
Pahlitzsch, G. and Appun, J. (1954). *VDI Z.*, No. 11/12, 96.
Patterson, H. B. (1986). *SME Grinding Conf.*, MR86-630.
Pauling, L. (1948). *Nature of the chemical bond* (2nd edn) Cornell University Press, Ithaca, NY.
Radhakrishnan (1993) Private Communication.
Salje, E. and Riefenstahl, J. (1949). *HGF* 182/49 1322, Girardet, Essen.
Shaw, M. C. (1970). *SME Publ. MR 70 277*.
Shaw, M. C. and Vyas, A. (1993). *Ann. CIRP* **42**(1), 29.
Shaw, M. C. and Yang, C. T. (1956). *Trans. ASME* **78**, 861.
Tanaka, Y. and Ueguchi, T. (1971). *Ann. CIRP* **19**(1), 449.
Tarasov, L. (1951). *Trans. ASM* **43**, 1144.
Tarasov, L. (1952). *Am. Mach.* **96**, 135.
Wells, A. F. (1950). *Structural inorganic chemistry* (2nd edn). Oxford University Press, Oxford.
Yamaguchi, K., Nakamoto, T., Kitano, M., Suzuki, M., and Abbay, P. A. (1993). *ASME PED 67/TRIB 4.*
Yamamoto, A. and Ueda, T. (1978). *Bull. Japan Soc. Proc. Engrs* **12**, 27, 201.

13

SPECIAL PROCESSES

Introduction

In this chapter the following special grinding operations will be discussed:

- internal grinding
- creep feed grinding
- cylindrical grinding
- centerless grinding
- belt grinding
- nontraditional grinding
- deburring, lapping, and polishing

Up to this point, surface grinding has been emphasized in discussing FFG, since it is the simplest grinding process and may be used as a basis for understanding other operations having more complex kinematics. Such transitions involve the following basic concepts:

- the equivalent wheel diameter, D_e
- the mean undeformed chip thickness, $\bar{t}$
- the specific grinding energy, u
- the mean forces acting on a single grit, F_P'' and F_Q''

Internal Grinding

Introduction

Internal grinding is most widely used for finishing bores of relatively small diameter (Fig. 13.1(a)), for grinding internal surfaces of complex shape (Fig. 13.1(b)), and for grinding the inner races of ball bearings (Fig. 13.1(c)). These are typical internal grinding situations, all of which use wheels of relatively small diameter. A common characteristic is that the wheel-work contact is of high conformity. This gives rise to a relatively long length (l) of wheel-work contact and an angle of approach (θ_a) that is unusually small (Fig. 13.2).

It has been shown previously (Chapter 4) that the equations for wheel-work contact length l and mean undeformed chip thickness $\bar{t}$ for internal grinding are the same as those for surface grinding, provided that the actual wheel diameter D_s is replaced by the following equivalent diameter D_e, which takes the difference in wheel-work conformity into account (see eqn 4.11)

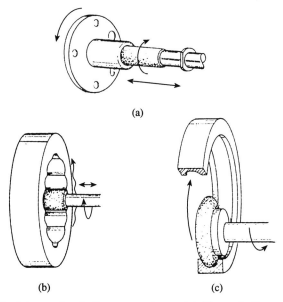

FIG. 13.1. Typical internal grinding operations. (a) The grinding of sintered iron bushing with a CBN wheel: (140-170, 150 conc., V); $D_s = 12.0$ mm (0.47 in); $D_w = 14.3$ mm (0.563 in); $V = 45$ m/s^{-1} (9000 f.p.m.); $v = 45$ m min^{-1} (147 f.p.m.); fluid $= 100:1$, G ratio $= 2000$. (b) The grinding of the injector pump cam ring with a vitrified CBN wheel. (c) The grinding of a ball bearing inner race with a vitrified bond microcrystalline CBN wheel (courtesy of GE Superabrasives).

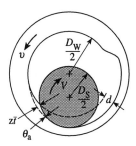

FIG. 13.2. An internal grinding operation indicating the length of wheel–work contact (l) and the grit approach angle (θ_a) in up grinding operation.

TABLE 13.1 *A comparison of internal grinding operation and equivalent surface grinding operation*

Variable	Internal	Surface
Wheel diameter (D_S), in (mm)	1 (25.4)	1 (25.4)
Work diameter (D_W), in (mm)	1.25 (31.8)	∞
Equivalent wheel diameter (D_e), in (mm)	5 (127)	1 (25.4)
Wheel depth of cut (d), in (μm)	0.0002 (5)	0.0002 (5)
Wheel–work Contact length (l), in (mm)	0.032 (0.813)	0.014 (0.356)
Wheel speed (V), f.p.m. (m s^{-1})	6000 (30.5)	6000 (30.5)
Work speed (v), f.p.m. (m s^{-1})	60 (0.3)	60 (0.3)
Active grits per area (C),	1000	1000
Scratch width/scratch depth (r),	15	15
Mean underformed chip thickness ($\bar{t}$), μin (μm)	65 (1.62)	97 (2.43)
Grit engagement angle (θ), min	13.96	47.8

$$D_e = \frac{D_w D_s}{D_w - D_s},$$

where D_w is the work diameter.

The mean undeformed chip thickness $\bar{t}$ for internal grinding then becomes (see eqn 4.13):

$$\bar{t} = \sqrt{\frac{v}{VCr}} \sqrt{\frac{d}{D_e}}.$$

The important role that D_e plays with regard to l and θ is illustrated in Table 13.1, in which an internal grinding operation is compared with the equivalent surface grinding operation. The value of θ_a shown in Fig. 13.2 is as follows, to a good approximation:

$$\theta_a = \sin^{-1}(2\bar{t}/l). \tag{13.1}$$

From Table 13.1 it is evident that the value of l is over twice as large for the internal grinding operation as the value for the equivalent surface grinding operation. Also, $\bar{t}$ is only about two-thirds as large for the internal grinding operation as for the equivalent surface grinding operation, and θ_a is only about 30 percent as large for internal grinding as for the surface grinding example.

These differences in θ_a and $\bar{t}$ due to conformity make it more difficult for the grit to penetrate the work, and this results in a greater amount of rubbing before cutting begins and a higher value of specific energy for internal grinding than for the equivalent surface grinding operation.

Hahn has conducted many fundamental grinding studies, mostly from an internal grinding perspective. He was the first to point out that chip formation is preceded by rubbing and plowing as the wheel enters the work (Hahn 1962).

For chip formation to begin, the radial force on the wheel (F_Q) must reach a critical value (Hahn calls this the threshold value) which increases with the indentation hardness of the work, improved wheel–work lubrication, and dulling of the active grits, and with an increase in conformity of the wheel–work contact (a decrease in θ_a). The threshold value is greater for down grinding than for up grinding. Hahn (1962) attributes this to a greater effective clearance at the trailing edge of a grit in up grinding than in down grinding. Because the normal component of the grinding force plays such an important role in internal grinding, Hahn suggests that this should be considered as the main input parameter in grinding.

Constant-force Grinding

The removal rate (vdb) is normally considered to be the input and forces, finish, wear rate, etc. are taken to be outputs. Adoption of normal force as input has led Hahn (1964) to suggest that the control of a grinding machine should be in terms of F_Q rather than feed rate (v) or removal rate (vbd) (i.e. constant-force grinding as opposed to more conventional constant-removal-rate grinding). Constant-force grinding has the advantage of automatically reducing removal rate as wheel wear occurs, which makes subsurface damage less likely. This is particularly important in internal grinding where undeformed chip thickness is usually low, giving rise to high specific energy and hence high surface temperature. High conformity between the wheel and the work in internal grinding gives rise to larger values of the contact length, which further aggravates the surface temperature problem.

Constant-force grinding is an attractive alternative for those grinding operations in which grinding burn is a problem. The use of a hydraulic feed in place of a lead screw provides an automatic reduction in the removal rate as wheel wear occurs, thus enabling the initial removal rate for a sharp wheel to be higher than would be possible if the removal rate were not automatically reduced as wheel wear and hence F_Q increased. Higher productivity without subsurface damage results with constant-force (F_Q) grinding.

Hahn and Lindsay (1971) present grinding data in the form of charts, where the radial force on the wheel per unit width ground ($F'_Q = F_Q/b$) is the main independent variable (abscissa). Figure 13.3 is such a chart, where the following quantities are dependent variables (ordinates):

$\dot{M}'_w$ = volume removal rate per unit width ground
$\dot{M}'_s$ = volume rate of wheel loss per unit width ground
R_a = AA surface roughness
horsepower

Three zones are identified:

(1) a rubbing zone in which F'_Q is too small to displace metal;
(2) a rubbing zone in which F'_Q is sufficient to plow material to the side without any removal; and

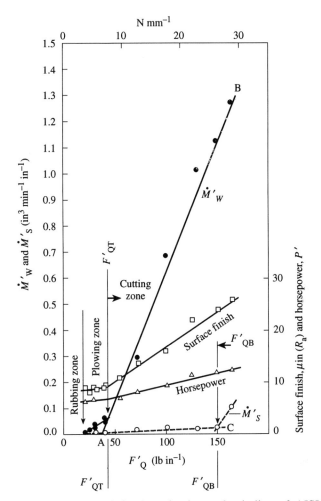

FIG. 13.3. A grinding characteristic chart for internal grinding of AISI 52100 steel ($R_c = 60$) (after King and Hahn 1986). Wheel, A80K4V; dress, 0.003 i.p.r. (75 μm rev^{-1}) lead and 0.0002 in (5 μm) depth; wheel speed V, 12 000 f.p.m. (60 m s^{-1}); work speed v, 250 f.p.m. (1.27 m s^{-1}); $D_e = 2$ in (50 mm); coolant, water-based.

(3) a cutting zone in which F'_Q is sufficient to cause removal.

The slope of line AB is

$$\Lambda_w = \Delta \dot{M}_w / \Delta F'_Q. \tag{13.2}$$

This is 0.010 in^3 min^{-1} lb^{-1} (36.8 mm^3 min^{-1} N^{-1}) in this case, and is called the work removal parameter (WRP)

The slope of line AC is

$$\Lambda_s = \Delta \dot{M}_s / \Delta F'_Q \tag{13.3}$$

This is 0.0002 in³ min⁻¹ lb⁻¹ (0.74 mm³ min⁻¹ N⁻¹) and is the corresponding wheel wear parameter.

The WRP (Λ_w) is a convenient measure of wheel sharpness S, where

$$S = \Lambda_w / V \quad [L^2 F^{-1}]. \tag{13.4}$$

The quantity S may be considered to be the cross-sectional area (bd) of the ribbon of material (the chip sheet area) removed per unit F'_Q and will have a higher value for a sharply dressed wheel and for grits of high friability.

The nondimensional grinding ratio G will be:

$$G = \text{volume removed/volume wheel wear} = \Lambda_w / \Lambda_s. \tag{13.5}$$

The work removal parameter is the quantity used by Hahn and Lindsay to correlate forces, specific energy, wear rate, and surface integrity in place of the mean undeformed chip thickness ($\bar{t}$) or the chip equivalent (t_e) used in the constant-feed-rate approach.

The surface finish improves with increase in F'_Q in the cutting zone but will have a minimum value in the plowing zone. With spark out F'_Q will drop to a value in the plowing zone as wheel–work elastic relaxation occurs, and hence a good final finish will be obtained.

Force F'_{QT} is the threshold force, which is the radial force per unit width ground at which cutting actually begins. This is also a measure of wheel sharpness and of the grindability of a work material.

Force F'_{QB} is the breakdown force per unit width ground, beyond which rapid breakdown of the wheel occurs. Precision grinding should be performed between F'_{QT} and F'_{QD}.

Specific power (P') is the power required per unit grinding width (h.p./b or kW/b). This is related to the specific energy (u, in lb in⁻³) as follows:

$$u = (h.p.)(33\,000)/(vbd) = 33\,000 P'/(vd), \tag{13.6}$$

where v is the work speed in f.p.m., b is the width ground, and P' is the total power required including threshold power.

Charts such as those of Fig. 13.3 are very useful for process planning when the control variable is F'_Q. However, since most grinding machines are designed with workpiece speed v as the control (independent) variable, these charts are somewhat less convenient. A disadvantage of the constant-force grinding approach is that the very important variable mean undeformed chip thickness ($\bar{t}$) that plays such an important role relative to specific grinding energy is not highlighted, but the metal removal parameter Λ_w is emphasized instead. Nevertheless the constant-force approach represents a useful alternative to the constant-removal-rate approach adopted in this monograph. The purpose of the foregoing discussion is to emphasize the similarities of the two approaches to grinding performance and to demonstrate the equivalence of the key aspects of each.

Figure 13.3 gives representative internal grinding results for a hardened ball

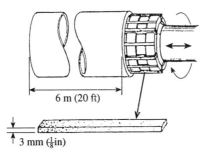

FIG. 13.4. The honing of a long 50 mm (2 in) dia. AISI 4140 steel ($H_{RB} = 90$) cylinder with 8 stick CBN (80–100, 35 conc.) hone (courtesy of GE Superabrasives). Hone speed, 260 f.p.m. (80 m min^{-1}); axial speed, 45 f.p.m. (14 m min^{-1}); removal rate, 0.35 in^3 min^{-1} (5.7 cm^3 min^{-1}); finish, 60 μin AA (1.5 μm); grinding ratio, 600.

bearing steel specimen ground with an 80 grit Al$_2$O$_3$ wheel operating at high speed (12 000 f.p.m. = 61 m s^{-1}).

Cubic boron nitride wheels are used for internal grinding, particularly when the required geometry involves a wheel of very small diameter, as in Fig. 13.1(b). Grinding conditions for the examples of Fig. 13.1 using CBN wheels are given in the figure caption.

One of the problems associated with internal grinding involving small-diameter wheels is that the spindle must have a small diameter. Spindle deflection will then limit the removal rate for precision grinding. Vibration of the wheel and spindle may also occur. Hahn has addressed the stability problem in considerable detail (Hahn 1978; King and Hahn 1986).

Honing

Honing is an alternative to internal grinding for finishing large-diameter bores of internal combustion engine cylinders or long hydraulic cylinders. In honing, abrasive sticks are loaded against the bore and rotated and oscillated as shown in Fig. 13.4.

The kinematics are such that a criss-cross scratch pattern is produced. This gives small chips even though the length of cut may be several feet. Only the tops of the ridges made during one stroke are removed in the next stroke, similar to the action in vertical spindle surface grinding (Chapter 8).

Honing is a classical example of the constant-force grinding discussed in the previous section. Outward pressure is applied to the four or more honing sticks (50–250 p.s.i., or 0.35–1.72 MPa) mounted in the mandrel. Higher pressures are used for roughing ($R_a \simeq 50$–100 μin, or 1.3–2.6 μm) and for softer materials. Since the abrasive is in continuous balanced contact with the work in honing, there is less deflection of the work.

The abrasive used may be either Al$_2$O$_3$ or CBN. When CBN is used the rotary speed should be higher (up to 700 f.p.m., or 3.6 m s^{-1}) than for Al$_2$O$_3$. The grinding ratio with a typical CBN hone (B 60/80. 50 conc. M) may be as high as 2000 for hard materials but will be much less when honing soft materials (as low as 200). A honing oil should always be used. The axial velocity should be about 20 percent of the rotary velocity.

Ueda and Yamamoto (1984) have presented a valuable analytical and experimental study of the honing process. It was found that the specific energy increased with a decrease in mean undeformed chip thickness $\bar{t}$, as in ordinary grinding. However, in honing the density of active grits (C) changes markedly as the process proceeds, requiring a semi-empirical approach to determine $\bar{t}$. It was found that at the outset of a honing operation when $\bar{t}$ is relatively large ($\sim$250 μin, or 6.25 μm) the specific energy (u) for carbon steel was comparable to values in ordinary SRG ($\sim$11 GPa, or 1.6 × 10^6 p.s.i.). However, at the end of the operation, for values of $\bar{t} \sim 5 \mu$in, or 0.13 μm, u was extremely high ($\sim$5000 GPa = 725 × 10^6 p.s.i.). In constant-pressure honing the specific energy has a low value at the beginning of the process but rises exponentially to extremely high values as the process proceeds, due to an increase in rubbing without removal as the number of active grits with flats increases.

Creep Feed Grinding

Introduction

Creep feed grinding was inadvertently discovered in Germany in the early 1960s. It was found that if horizontal spindle surface grinding is performed with an unusually large wheel depth of cut (d) and a correspondingly low work speed (v), some unusual performance characteristics are obtained. Most applications have been in a surface grinding or profile surface grinding mode and hence surface grinding will be considered in the analysis presented here.

Figure 13.5 illustrates the difference between conventional and creep feed surface grinding. In the conventional process d is very small and v is relatively large. The wheel is cycled back and forth across the work, removing a thin layer with each pass, as shown in Fig. 13.5(a). In the creep feed process (Fig. 13.5(b)) all of the material is removed with a single pass of the wheel. The conventional process is usually called pendulum grinding, while that in Fig. 13.5(b) is usually called creep feed grinding. Other terms sometimes used are swing grinding for Fig. 13.5(a) and deep grinding for Fig. 13.5(b). Creep feed grinding has been likened to milling with a grinding wheel. The entire field of creep feed grinding may be divided into the following two regimes, which have quite different characteristics:

(1) with common abrasives such as Al$_2$O$_3$;
(2) with superabrasives such as CBN.

These will be considered in that order.

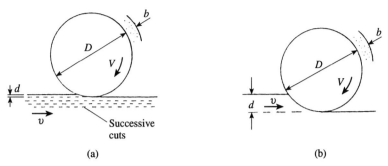

FIG. 13.5. A comparison of (a) conventional (pendulum) surface grinding and (b) creep feed surface grinding.

Creep feed grinding has been applied mainly to materials that are difficult to grind in the surface grinding or profile surface grinding modes. A typical example of the first type is the production of deep slots in hydraulic pump production, while an example of the second type is the production of the fir tree pattern at the base of a gas turbine blade.

There are several important differences between the two processes shown in Fig. 13.5 and these may best be discussed in terms of a representative example. Table 13.2 compares the processes when performed with the same wheel, wheel speed, and removal rate ($\dot{M} = vbd$). The wheel depth of cut ratio adopted here is 100, but applications may involve values as high as 1000. The numbers of active grits per unit area (C) and scratch width/scratch depth (r) are based on values given in Chapter 4 for a 60 grit Al_2O_3 wheel. Equations (4.10) and (4.13) were used to compute the values of the wheel–work contact (l) and the mean undeformed chip thickness ($\bar{t}$) respectively. The specific energy values are based on the reciprocal relationship between u and $\bar{t}$ for fine grinding. The approach angle values (θ_a) were obtained from eqn (13.1).

The remaining values in the table were obtained from the following relationships:

$$F'_T \simeq \text{tangential force per width ground} = uvd/V, \qquad (13.7)$$

$$F'_R \simeq \text{radial force per width ground} = 2F'_T, \qquad (13.8)$$

$$F''_T = \text{mean tangential force per grit} = F'_T/lC, \qquad (13.9)$$

$$F''_R = \text{mean radial force per grit} = 2F''_T. \qquad (13.10)$$

When the two grinding modes are compared, the following important differences are evident where the subscript p is used for the pendulum case while subscript c is used for the creep feed case:

- l_c is ten times l_p
- t_c is about one-third of t_p

TABLE 13.2 *A comparison of pendulum and creep feed surface grinding at the same moderate removal rate*

Variable	Pendulum grind	Creep feed grind
Wheel	A 60H 8V	Same
Diameter (D), in (mm)	10(254)	Same
Speed (V), f.p.m. (m s^{-1})	6000 (30.5)	Same
Depth of cut (d), in (μm)	0.002 (5.1)	0.02 (5100)
Work speed (v), f.p.m. (m s^{-1})	60 (0.31)	0.60 (0.0031)
Active grit per area (C), in^{-2} (cm^{-2})	1200 (186)	600 (93)
Scratch width/scratch depth, (r)	20	30
Wheel–work contact length (l), in (mm)	0.045 (1.14)	0.45 (11.4)
Mean underformed chip thickness ($\bar{t}$), μin (μm)	4.3 (1.08)	15.8 (0.40)
Specific energy (u), p.s.i. (MPa)	8×10^6 (55.2)	24×10^6 (165)
Approach angle (θ_a), min	6.6	0.24
Tangential force/width (F'_T), lb in^{-1} (N cm^{-1})	16(28)	48(84.1)
Radial force/width (F'_R), lb in^{-1} (N cm^{-1})	32(56)	96(168.1)
Tangential force/grit (F''_T), lb, (N)	0.3 (1.33)	0.18 (0.80)
Radial force/grit (F''_R), lb (N)	0.6 (2.66)	0.36 (1.60)
β_w (steel), J m^{-2} s$^{-0.5}$ °C^{-1}	10^4	Same
β_s (Al$_2$O$_3$), J m^{-2} s$^{-0.5}$ °C^{-1}	12.5×10^3	Same
β_c (water), J m^{-2} s$^{-0.5}$ °C^{-1}	10^3	Same
β_s (CBN), J m^{-2} s$^{-0.5}$ °C^{-1}	48.4×10^3	Same

- the time of wheel–work contact (l/v) is 10^4 times as long for the creep feed case as for the pendulum case
- the creep feed forces are about three times the pendulum values
- the mean forces per grit for the creep feed case are 60 percent of the pendulum values
- the approach angle θ_a for the creep feed case is only 3.6 percent of the pendulum value

In addition to these differences, the ratio of the mean creep feed temperature to the mean pendulum temperature may be estimated by use of eqn (9.23):

$$\theta \cong \frac{Ruvd}{\beta_w \sqrt{vl}},$$

where R is the fraction of the energy going to the work, which is given to a good approximation by eqn (9.28) when grinding with a water-based coolant:

$$R = \left(1 + \frac{\beta_s}{\beta_w}\sqrt{\frac{VA_R}{vA}} + \frac{\beta_c}{\beta_w}\sqrt{\frac{V}{v}}\right)^{-1}.$$

Table 13.3 gives values of thermal properties required to apply eqns (9.23)

TABLE 13.3 *Thermal analysis for comparison of pendulum and creep grinding with Al_2O_3 and water*

Steel: $\beta_w = 10^4 \, J \, m^{-2} \, s^{-0.5} \, °C^{-1}$
Al_2O_3: $\beta_s = 12.5 \times 10^3 \, J \, m^{-2} \, s^{-0.5} \, °C^{-1}$
Water: $\beta_c = 10^3 \, J \, m^{-2} \, s^{-0.5} \, °C^{-1}$

From eqn (9.28): $R_p = 1 \bigg/ \left[1 + \left(\dfrac{12.5}{10}\right) + 1 \right] = 0.307$

$$R_c = 1 \bigg/ \left[1 + \dfrac{12.5}{10} \sqrt{10^4(10^{-2})} + 0.1(10^4)^{0.5} \right] = 0.043$$

---

and (9.18) to the comparisons already made in Table 13.2. For the same removal rate and work material, eqn (9.23) gives the following ratio for creep feed temperature (subscript c) to pendulum temperature (subscript p):

$$\underset{(a)}{\theta_c/\theta_p} = \underset{(a)}{(R_c/R_p)} \, \underset{(b)}{(u_c/u_p)} \, \underset{(c)}{[(vd)_c/(vd)_p]} \, \underset{(d)}{(\beta_{wp}/\beta_{wc})} \, \underset{(e)}{[(vl)_p/(vl)_c]^{0.5}}. \tag{13.11}$$

When appropriate values are substituted from Tables 13.2 and 13.3 the following result is obtained for θ_c/θ_p when grinding steel with an Al_2O_3 abrasive at the same removal rate. Of the five nondimensional factors (a) to (e), some will have a value of unity. For example, (c) and (d) will be equal to 1 if the creep feed and pendulum cases have the same removal rate and abrasive material.

$$\underset{(a)}{\theta_c/\theta_p} = \underset{(a)}{(0.140)} \, \underset{(b)}{(3)} \, \underset{(e)}{(3.16)} = 1.32.$$

This analysis shows the following:

- R_c (0.043) is considerably less than R_p (0.307)
- θ_c/θ_p is greater than one due to u_c/u_p and $[(vl)_p/((vl)_c]^{0.5}$ both being considerably greater than one

The reason for the first of these results is due to the effect that a large value of $(V/v)^{0.5}$ (=100 in the creep feed case in this example) has in decreasing R_c.

From the above analysis, it is evident that creep feed grinding involves unusually long thin chips, high forces, appreciable rubbing (low θ_a), high temperatures but relatively low values of force per chip. The low force per chip means that softer wheels that are more porous to accommodate the long thin chips should be used, provided that wheel safety requirements are not violated. In this connection, the much safer segmental wheel design of Figs 6.9 and 6.10 should be considered for creep feed grinding with vitrified wheels.

The high temperature for creep feed grinding means that wheel life and surface integrity may be problematic unless wheel speed is reduced to lower the temperature (according to eqn (9.27), $\theta \sim V$), and very effective cooling is employed or a more refractory abrasive with better thermal properties (such as CBN) is used.

Wheel Speed with Al_2O_3

Shimamune *et al.* (1990, 1991) have suggested that better wheel life may be obtained in creep feed grinding with Al_2O_3 by the use of unusually low wheel speeds (700–1400 m min^{-1}, or 2300–4600 f.p.m.). The use of low wheel speeds will tend to offset the effect on $\bar{t}$ (and hence on u and θ) that small values of v pertaining in creep feed grinding will have. The optimum $(V/v)^{0.5}$ value was found to vary from about 140 to 200, when creep feed grinding hard ball bearing steel, the higher value pertaining for larger values of l or v. This influence of l and v is probably a consequence of the need for larger values of V with an increase in either l or v from the chip accommodation point of view. A value of $(V/v)^{0.5}$ of about 170 is in contrast to a value of about 10 for pendulum grinding.

Up Versus Down Grinding

A detail concerning creep feed grinding that has been debated without a logical conclusion concerns whether up or down grinding is preferable. Figure 13.5(b) shows up grinding. For down grinding the work velocity would be reversed. The confusion comes from the incorrect impression that the energy required in down grinding is substantially less than in up grinding.

Shiozaki *et al.* (1977) have measured the grinding forces F_P and F_Q for ball bearing steel ($H_{RC} = 61$) using vitrified 60 grit Al_2O_3 wheels having different grades under up and down creep feed grinding conditions. In all cases, the F_Q (vertical dynamometer force) values were about the same but the F_P (horizontal component force) values were about 30 percent lower in down grinding than in up grinding (Fig. 14.10). From this it might be incorrectly assumed that the power consumed by the wheel is 30 percent lower in down grinding than in up grinding under otherwise identical conditions. However, as shown in Chapter 14, when d/D is relatively large, a distinction must be made between the tangential component of force F_T, which is proportional to the power consumed at the wheel face, and the horizontal dynamometer force F_P, which is not. When this is done the energy consumed in down grinding is found to be only about 14 percent less than in up grinding, as the analysis in Table 14.4 clearly shows. In the case of CBN, the down grinding energy is found to be only about 8 percent less than for up grinding when dynamometer data is properly interpreted (Table 14.5). However, when improperly interpreted (not distinguishing between F_P and F_T), the down grinding energy is found to be 48 percent less than the up grinding energy for CBN creep feed grinding.

Another point to consider in comparing up and down creep feed grinding is that there is more shock involved in down grinding as the wheel enters the work with maximum undeformed chip thickness, whereas in up grinding the wheel enters with minimum undeformed chip thickness. This shock at impact undoubtedly has an influence on wheel wear, and will require a more expensive machine with greater rigidity and stiffness, as in the case of a milling machine used in the down milling mode.

Still another point to consider in comparing up and down creep feed grinding is a thermal aspect. With up grinding, unheated fluid and abrasive particles enter the work at the critical point at which the final surface is being generated. In down grinding the final surface is ground with fluid and grits that have been heated throughout the wheel-work passage. The point of maximum temperature along the wheel–work contact will be near the point at which the wheel exits the work in up grinding. It is important that a second jet of fluid be directed at this point, so that heat may be extracted outward across the original surface and not be permitted to flow inward to the new surface, where it may cause subsurface damage, including residual tensile stress.

A valid point frequently made in support of down creep feed grinding is that there will be less rubbing when grits exit the point of minimum undeformed chip thickness than when they enter the work with minimum undeformed chip thickness.

From the above discussion it is evident that the up versus down creep feed grinding question is complex, and is best solved empirically for each case on the basis of the performance criterion that happens to be most important (wear, finish, power, stability, removal rate, etc.).

Cooling

The most important detail concerning the successful use of the creep feed process has to do with cooling. It is important that a high-pressure jet of water-based fluid be introduced at the point at which the wheel enters the work. The location and orientation of the nozzle is as important as the pressure. It is essential that the fluid be carried uniformly over the wheel–work contact area, so that it may reduce the unusually high temperatures involved in creep feed grinding. It is also important that a scraper be used to remove the air boundary layer from the wheel surface, which will prevent the coolant from reaching the wheel face. An open-structure wheel is important in creep feed grinding to provide fluid reservoirs. The use of grooves in the surface of the wheel offers an additional means for providing sufficient cooling. In addition to providing better cooling, an open-structure wheel is important in helping to solve the chip storage problem in creep feed grinding. Creep feed chips are unusually thin and long, and it is difficult to pack them densely in the space available without excessive rubbing pressure being developed in the chip–tool interface. Even when the removal rate is the same, the chip storage problem becomes more difficult the longer and thinner the chips are. With

CBN creep feed grinding, to be discussed presently, the removal rate is frequently greater than in the corresponding grinding operation and thin, long chips are then very difficult to accommodate in the wheel–work contact zone.

Temperature Instability

Andrew (1979) has described an interesting coolant-related phenomenon associated with creep feed surface grinding at high removal rates. At a critical rate of removal the grinding power was observed to surge cyclically, with corresponding cyclic evidence of surface damage on the ground surface.

Safto (1975) compared the energy dissipated at the wheel face to that going to the workpiece in creep feed grinding with coolant by a combination of subsurface thermocouple measurements and finite element analysis, and found that less than 5 percent of the energy dissipated at the wheel face was conducted into the workpiece. This is in good agreement with the 4.3 percent value obtained in the example of Table 13.2. Safto concluded that the bulk of the energy was going to the coolant and that the surge phenomenon was related to the burn-out problem in boiler technology, in which an abrupt increase in surface temperature occurs at a critical heat flux when boiling changes from nucleate to film boiling.

As long as isolated bubbles form and leave the surface, they enhance heat transfer to the water by agitation and the surface temperature remains close to the boiling point. However, when the heat flux is sufficient to cause these bubbles to coalesce and form a continuous insulating blanket of steam, the surface temperature will suddenly rise to a very high value. Figure 13.6 shows the change in temperature of a nickel wire in water with rise in heat flux. In this case, when the heat flux reaches about $1.2\,\mathrm{MW\,m^{-2}}$ the surface temperature of the wire suddenly goes from about 110 °C to 1000 °C.

It is of interest to note that in pendulum grinding a greater percentage of the energy dissipated at the wheel face goes to the workpiece and hence less to the fluid, since the time available for heat transfer to the coolant is much less for pendulum grinding.

Safto (1975) carried out creep feed grinding tests for a range of conditions and measured the heat flux at which temperature surges and force fluctuations occurred. This was found to be $32\text{-}36\,\mathrm{MW\,m^{-2}}$, which is somewhat higher than the critical value of Fig. 13.6.

Powell (1979) simulated creep feed grinding thermal conditions by having a grinding wheel operating at conventional speed contact a thin metal strip supported by asbestos. An electric current was passed through the metal resistor and the thermal flux measured at which a sudden temperature increase (burn-out) occurred under different cooling conditions. Burn-out occurred at a higher temperature by:

- an improved fluid nozzle location
- a lower coolant inlet temperature

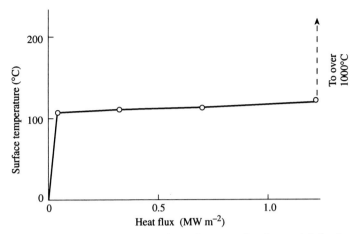

FIG. 13.6. Surface temperature versus heat flux transfer from nickel wire to water (after Farber and Scorah 1948).

- a shorter simulated wheel–work arc length
- a more porous wheel for better fluid distribution
- a slotted wheel face for better fluid distribution
- an increase in wheel speed to reduce the duration of wheel–work contact

All of these indirectly support the burn-out theory of temperature surge.

Safto (1975) suggested that the cyclic behavior was due to cyclic self-dressing of the wheel. At the critical heat flux the high temperature caused rapid wheel wear (dressing), which resulted in a sharper wheel with lower specific energy and a temporary reduction in grinding force.

Creep Feed Machines

Machines for creep feed grinding require greater rigidity, stability, and power (three to five times) than machines used for pendulum grinding. Special attention must be directed to the coolant system and to wheel dressing. It has been found, when using Al_2O_3 wheels, that continuous dressing using a diamond roller dressing tool is helpful to keep the specific energy at a lower level, which helps to solve the severe thermal problem associated with creep feed grinding. With continuous dressing it is important that the abrasive particles removed be washed clear of the wheel before the fluid reaches the inlet cooling jet.

Surface Finish

Surface finish is normally not a problem with creep feed grinding, because surface finish improves with decrease in undeformed chip thickness and creep

TABLE 13.4 *Thermal analysis for comparison of pendulum grinding and creep feed grinding with CBN and water*

Steel: $\beta_w = 10 \times 10^3 \, \text{J m}^{-2} \text{s}^{-0.5} \, °\text{C}^{-1}$
CBN: $\beta_s = 48.4 \times 10^3 \, \text{J m}^{-2} \text{s}^{-0.5} \, °\text{C}^{-1}$
Water: $\beta_c = 10^3 \, \text{J m}^{-2} \text{s}^{-0.5} \, °\text{C}^{-1}$

From eqn (9.28): $\quad R_p = 1 \bigg/ \left[1 + \left(\dfrac{48.4}{10}\right) + 1 \right] = 0.146$

$$R_c = 1 \bigg/ \left[1 + \dfrac{48.4}{10} \sqrt{10^4(10^{-2})} + 0.1(10^4)^{0.5} \right] = 0.017$$

From eqn (13.11): $\quad \dfrac{\theta_c}{\theta_p} = \dfrac{(0.017)}{(0.146)} (3)(10)^{0.5} = 1.10$

---

feed grinding chips are normally thin (Table 13.2). With continuous dressing the surface roughness will be slightly greater because there is less burnishing action with a sharp, freshly dressed wheel.

Cubic Boron Nitride

Creep feed grinding with CBN is quite different than with common abrasives such as Al_2O_3. The examples of Tables 13.2 and 13.3 will be extended to illustrate some of these differences, particularly with regard to grinding temperatures. First, the example of Table 13.2 will be repeated with all items the same except that CBN replaces Al_2O_3 as the abrasive. Table 13.4 gives the thermal properties then involved and the resulting values of R for pendulum (R_p) and creep feed (R_c) grinding, together with $\theta_c/\theta_p = 1.09$. This is seen to be much less than the corresponding value for $Al_2O_3 = 1.32$ due to the higher value of β_s for CBN and hence the lower value of factor a in eqn (13.11).

Since CBN is much more refractory than Al_2O_3 and has better thermal properties, it has been found possible to increase the wheel speed V with CBN instead of decreasing it when going to creep feed grinding as with Al_2O_3. The increased wheel speed is beneficial to the chip storage problem. The effect of doubling the wheel speed to 12 000 f.p.m. (61 m s^{-1}) for the creep feed case only in the CBN example just considered will lead to a substantial reduction in θ_c/θ_p. The following values then obtain:

$R_p = 0.146, \quad R_c = 0.012, \quad \theta_c/\theta_p = 0.780$ (compared with 1.08 before V was doubled).

Since increasing V for creep feed grinding with CBN has such a dramatic effect in reducing θ_c/θ_p, this suggests that it should be possible to increase the removal rate (productivity) along with wheel speed V_c and still have a reasonable creep feed temperature. The removal rate per unit width ground

(vd) may be increased by increasing either v or d, or by a combination of the two.

If both the creep feed wheel speed (V_c) is doubled and v_c is increased by 1.25 with CBN, while the pendulum conditions and all others remain the same as in Table 13.2, then the following values are obtained:

$$R_p = 0.146, \quad R_c = 0.0134, \quad \theta_c/\theta_p \simeq 0.973.$$

Similarly, the following values for CBN are found when V_c is doubled, and d_c is increased by 1.25, but v_c remains the same, as do the pendulum conditions given in Table 13.2:

$$R_p = 0.146, \quad R_c = 0.012, \quad \theta_c/\theta_p \simeq 0.922.$$

It is thus seen that when CBN is used in creep feed grinding the wheel speed V should be increased as well as the removal rate, and that the temperature ratio (θ_c/θ_p) will then be less than it would be with Al_2O_3 without an increase in V_c or v_c.

The small difference in θ_c/θ_p between the last two cases, in which v_c and d_c are doubled respectively, will probably disappear when the fact that an increase in v_c will cause an increase in $\bar{t}$ and hence a decrease in u_c which has not been taken into account in the calculations. Doubling d_c will have a much smaller effect on u_c than when v_c is doubled.

Creep feed grinding coupled with CBN is one attractive possibility for increasing productivity, improving surface integrity, and decreasing cost, despite the high cost of CBN wheels. The use of smaller-diameter wheels, single grit layer electrolytic or brazed wheels, and wheels of lower concentration are possibilities for lowering the wheel cost. Figure 13.7 shows three examples of creep feed grinding with CBN wheels.

High-efficiency Deep Grinding

When CBN is used at high wheel speed and at an increased removal rate in creep feed grinding with a copious supply of fluid, a new grinding concept called high-efficiency deep grinding (HEDG) emerges. The foregoing analysis explains why, when CBN is used with increased wheel speed, it is possible to increase removal rate (productivity).

Brinksmeier and Minke (1993) have suggested that at high wheel speeds and copious fluid flow in HEDG a hydrodynamic component of force develops that adds to the force required for material removal. This hydrodynamic effect shows up as an increase in spindle power with an increase in wheel depth of cut d (i.e. with increase in l) or wheel speed.

Creep Feed Cylindrical Grinding

While creep feed grinding has been primarily applied to surface and profile surface grinding applications, it was demonstrated by Liverton (1979) to be applicable also to plunge cylindrical grinding. Figure 13.8 shows the test

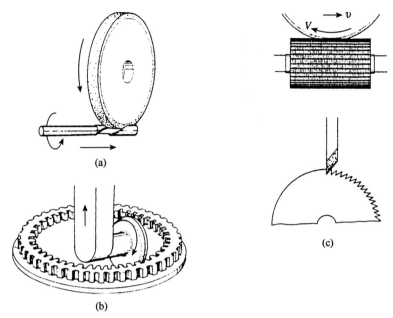

FIG. 13.7. Examples of creep feed grinding with CBN (courtesy of GE Superabrasives). (a) Flute grinding from solid using CBN with a single layer of electrolytically attached 100–120 CBN grits, with oil as the grinding fluid. (b) Creep feed grinding of saw teeth in a 3 in (75 mm) stack of thin discs using a 140–170, 100 concentration CBN wheel, with oil as the grinding fluid. In this case there will be an axial component of force as well as tangential and radial components (see Fig. 13.18). (c) Finish profile grinding of hardened gear teeth by creep feed grinding, using a resin-bonded CBN wheel.

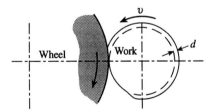

FIG. 13.8. A cylindrical creep feed down grinding operation.

arrangement used, where the work was plunged into the wheel to full depth and then rotated at a slow speed (1 r.p.m.) without further in feed. The machine performed satisfactorily at a wheel depth of cut of 2 mm (0.080 in) in the down grinding mode with a copious supply of oil, a wheel speed of 30 m s^{-1} (5900 f.p.m.) and continuous dressing of a vitrified Al_2O_3 wheel. The target application was thread grinding. While full thread depth was produced in one revolution, a second spark-out revolution was required to provide the desired surface finish of 0.25 μm R_a (10 μin).

FIG. 13.9. A specimen ground in cylindrical creep feed up grinding mode with a wheel depth of cut d of 7 mm (0.28 in. The fluid was suddenly cut off and the wheel quickly removed (after Zhou and Shaw 1981).

Zhou and Shaw (1981) have presented plunge cylindrical creep feed grinding results obtained on a special 55 kW (74 h.p.) controlled force creep feed grinding machine having high-pressure jet cleaning, hydrostatic bearings, and a high-capacity coolant system with a special supply nozzle. Creep feed tests have been performed with this machine with a wheel depth of cut as high as 7 mm (0.28 in) and a work speed of 0.25 r.p.m. Figure 13.9 shows a steel specimen up ground under the above conditions. Fluid was suddenly shut off and the wheel quickly removed from the work. Burn is evident only at the point at which the wheel exits the work, which is where the temperature is a maximum.

Table 13.5 compares creep feed and conventional cylindrical grinding results when a steel specimen was up ground with Al_2O_3 wheels at approximately the same removal rate per width ground (6 mm^2 min^{-1} = 0.95 in^2 min^{-1}). The work material was AISI 1045 steel (H_B = 230 kgm mm^{-2}). The wheel depth of cut in the creep feed case was 5 mm (0.2 in) and in the conventional case 0.0172 mm (0.0007 in). This corresponds to a d ratio of 5/0.0172 = 291. The corresponding work speeds to give approximately the same removal rate were 122 mm min^{-1} (0.4 f.p.m.) and 34 532 mm min^{-1} (113 f.p.m.). The fluid used in both cases was a 2 w/o soluble oil in water. Both of the grinding wheels used had a 60 grit size, but the creep feed wheel had grit of greater friability and a more open structure than the wheel used under conventional conditions. Also, the wheel speed in the creep feed case was twice as high as in the conventional case.

From Table 13.5 it is evident that the creep feed specific energy was about 1.6 times the conventional grinding value. This is to be expected when the mean undeformed chip thickness values are compared (creep speed $\bar{t}$/conventional $\bar{t}$ = 0.18). The grinding ratio was nearly five times as high in the creep feed case as in the conventional cylindrical grinding case.

The fact that creep feed cylindrical grinding may be performed with a lower

TABLE 13.5 *Comparison of creep feed and conventional cylindrical grinding operations*

Item	Creep feed	Conventional
Wheel	38 A 60 I V	19A 60 K/L V
Wheel diameter (D), mm (in)	600 (23.6)	Same
Wheel speed (V), m s^{-1} (f.p.m.)	60 (11 800)	35 (6890)
Wheel depth of cut (d), mm (in)	5 (0.2)	0.0172 (0.0007)
Work	AISI 1045 (230 kg mm^{-2})	Same
Work diameter (D), mm (in)	155 (6.1)	115 (4.5)
Work speed (v), mm min^{-1} (f.p.m.)	122 (0.4)	34 350 (113)
Width of cut (b), mm (in)	30 (1.18)	50 (1.97)
Specific energy (u), J mm^{-3} (p.s.i.)†	50.0 (7.25 × 10^6)	31.0 (4.50 × 10^6)
Grinding ratio, G	113	24
Effective wheel diameter (D_e), mm (in)	123 (4.84)	97 (3.80)
Wheel-work Contact length (l), mm (in)	24.8 (0.98)	1.29 (0.05)
Mean underformed chip thickness ($\bar{t}$), μm (μin)	0.55 (22)	3.1 (124)

† Based on $C = 1.5$ mm^{-2} (968 in^{-2}); $r = 15$.

rate of wheel wear than conventional grinding suggests that the main advantage of the creep feed technique does not lie in the reduction of the number of impacts at the beginning of successive wheel passes, as originally suggested in Europe. In conventional cylindrical grinding there is but one impact at the start of the cut, as with creep feed surface grinding. The main advantage appears to lie, instead, in the fact that more of the heat is convected away by the wheel (a low value of R) and that the mean force per grit is less in creep feed grinding than in conventional grinding.

Creep feed cylindrical grinding appears to hold promise not only for easily ground materials but also for materials that are difficult to grind (hard facing alloys, high-temperature alloys for turbines, prehardened parts, etc.). Creep feed cylindrical grinding should be useful in the production of many cylindrical parts containing deep grooves or a substantial variation in radial depth, such as ball bearing outer races, surface cams, and camshafts.

Speed Stroke Grinding

One of the main advantages of creep feed grinding over pendulum grinding is a reduction in the time during which the wheel is out of contact with the work. This increase in productivity is very significant when grinding work of short length in the grinding direction, and when many passes are required to remove the desired depth in pendulum grinding. A technology competing with creep feed grinding in decreasing nongrinding cycle time is called speed stroke grinding. This involves pendulum grinding at a very high frequency (up to 400 strokes per minute), the capability of very short strokes (as small as 10 mm, or 0.040 in), and very short nongrinding overtravel (as small as 1 mm, or 0.040 in). Speed stroke grinding appears to be establishing a niche, despite the fact that such grinders are more complex and expensive than conventional pendulum grinders. The work speed (v) in this process will be much greater than in conventional pendulum grinding. However, as mentioned in Chapter 9, the grinding temperature will be approximately independent of v, since the increase in v in eqn (9.23) ($\theta \sim v^{0.5}$) will be offset by a reduction in u ($u \sim v^{-0.5}$). At the same time, the surface finish will be poorer.

It is important to note that machines required to take full advantage of creep feed grinding with CBN will also cost much more than conventional pendulum machines. Also, CBN wheels will have an initially greater cost than common abrasive wheels. Furthermore, it is a general rule that, as the level of technology advances, a higher level of worker capability with a higher labor cost is necessary to protect greater capital investment.

The decision of whether to adopt new technology is a very complex one, involving many technical as well as business-related factors that exist now, as well as changes in the future. Such decisions should be made after careful study of all carefully weighted factors that can be identified. In no case should such important decisions be based merely on a desire to appear modern and progressive.

Cylindrical Grinding

Introduction

There are two types of cylindrical grinding:

(1) with axial feed (v_a);

(2) plunge grinding (no axial feed).

The counterparts of these exist in horizontal spindle surface grinding where the cross-feed per stroke plays the same role as axial feed per revolution of the work does in plunge cylindrical grinding. Grinding with axial feed is much more complex than without and hence most basic studies involve plunge grinding. The basic differences between plunge grinding and grinding with axial feed have been discussed in Chapter 5 in connection with surface grinding. Axial feed grinding may be characterized as a type leading to a relatively fine finish, small chips, and relatively high specific energy. Where there is less interest in finish and generation of a profile is involved, plunge grinding is employed. In plunge cylindrical grinding spark-out is usually used to improve the finish produced.

The basic input variables in surface grinding are: wheel speed (V), work speed (v), wheel depth of cut (d), and width ground (b). In cylindrical plunge grinding the same variables pertain, except for d, which is replaced by radial feed per revolution of the work:

$$d = v_r/N_w, \qquad (13.12)$$

where v_r is the axial feed per unit time and N_w is the r.p.s. of the work. The equation for the removal rate will be:

$$\dot{M} = v_r b d. \qquad (13.13)$$

The equations for mean undeformed chip thickness ($\bar{t}$), wheel-work contact length (l), and mean surface temperature (θ) are then the same as for plunge surface grinding if the wheel diameter D is replaced by the equivalent wheel diameter (D_e) where, for cylindrical grinding:

$$D_e = \frac{D_s D_w}{D_s + D_w} \qquad (13.14)$$

In cylindrical grinding, wheels are generally large, spindles stiff, and spark-out time is relatively short. The in-feed velocity is usually controlled rather than the specific normal force ($F'_Q = F_Q$ per width ground) since this is simpler.

The wheel–work contact length is relatively short in cylindrical grinding for the same value of d, due to the low conformity pertaining. For example, if $D_s = 10 \, \text{in}$ (250 mm) and $D_w = 2 \, in$ (50 mm), then $D_e = 1.67 \, \text{in}$ (42.3 mm). For $d = 0.001$ in (0.025 mm), values of l for cylindrical grinding and surface grinding will be as follows:

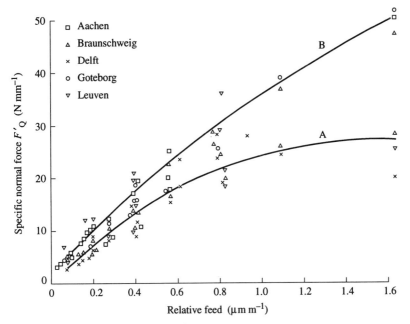

FIG. 13.10. A plot of the specific normal force (F_Q') versus the relative feed (v_r/V) for two grinding wheels (A and B) operating under a wide range of conditions (after Snoeys and Decneut 1971) Wheels, A80 J7 V (A) and A60 L7 V (B); work, AISI 52100 steel (R_c = 62–63); wheel dia. (D_s), 500–750 mm (19.7–29.5 in); wheel speeds (V), 30, 45, and 60 m s^{-1} (5900, 8850, and 11 800 f.p.m.); speed ratio ($q = V/v$) 60; radial feed (v_r), 5, 12, 25, and 50 μm s^{-1} (200–2000 μin sec^{-1}); fluid, 3 percent sol oil of at 40 l min^{-1} (10.6 g.p.m.); dressing, plane of diamonds, depth = 0.05 mm (0.002 in), lead = 0.20 mm rev^{-1} (0.008 i.p.r.); spark-out, none.

cylindrical, $l = 0.041$ in (1.04 mm),
surface, $l = 0.100$ in (2.54 mm).

CIRP Study

The International Institution for Production Engineering Research (CIRP) has conducted an extensive cooperative study of cylindrical plunge grinding. The first phase of the study was to determine whether the same results would be obtained in different research laboratories using different instrumentation and machine tools but identical materials (steel, wheels, dressing tool, and fluid). The materials involved were carefully produced in the same plant and at the same time and then distributed to the participating laboratories.

Snoeys and Decneut (1971) have presented a summary of this calibration stage of the study. Representative results are shown plotted in Figs 13.10 and 13.11 for two Al$_2$O$_3$ wheels of different grade and grit size (wheel A = 80J

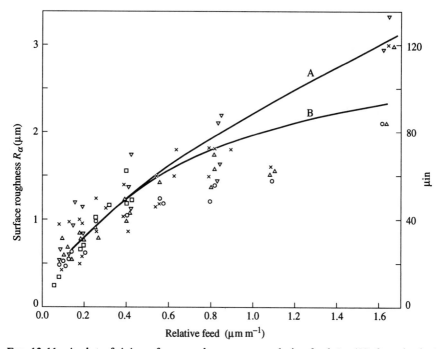

FIG. 13.11. A plot of AA surface roughness versus relative feed (v_r/V) for wheels A and B. Grinding conditions as in Fig. 13.10.

and wheel B = 60L). Both were vitrified and had a structure number of 7. The data points shown represent a wide range of grinding conditions (different values of V, v, and d). After a good deal of trial and error, the abscissa employed (v_r/V, called the relative feed) was found to give the correlation shown.

Some time later, it was discovered that v_r/V, which has the units of length, is closely related to the mean undeformed chip thickness ($\bar{t}$) in these experiments.

$$v_r/V \approx v_w d/V. \tag{13.15}$$

Since the work diameter was always the same in these tests,

$$v_r/V \sim (v_w d)/V. \tag{13.16}$$

For plunge cylindrical grinding,

$$\bar{t} = v_w d/VCl\bar{b}'. \tag{13.17}$$

The quantity $v_w d/V$ consists of machine variables, while the remaining quantities consist of variables associated with the fine geometry of individual cuts and will be essentially constant for these tests. The first quantity is also the equivalent undeformed chip thickness (t_e, eqn 4.15) which is a two-dimensional approximation for eqn (13.17).

Thus, for these experiments, $v_r/V \sim \bar{t}$.

Since it is well established that specific forces (F'_P and F'_Q), specific energy (u), and surface finish (R_a) are directly related to $\bar{t}$ in surface grinding, it is to be expected that the same will hold for cylindrical grinding as well as for other types of grinding in which the basic parameters V, v, and d are properly evaluated in terms of the kinematics.

The curves of Fig. 13.10 and 13.11 show that:

- the specific normal force F'_Q increases with an increase in the undeformed chip thickness, and more rapidly as the wheel hardness increases
- the surface roughness increases with an increase in the undeformed chip thickness, and more rapidly as the wheel hardness decreases

There is considerable scatter of data produced in different laboratories. Snoeys and Decneut (1971) suggest that this is no greater than might be expected as test-to-test variation in the same laboratory. This appears to be an overly optimistic interpretation. Since in many other tests forces and finish correlate quite well with undeformed chip thickness ($\bar{t}$), the following items are suggested as possible causes for some of the scatter of results obtained:

1. Control of the experiments was not complete, despite the great care taken with the materials involved. Major differences in machine tool capacity (6–25 kW) and wheel diameter (450–750 mm, or 18–30 in) still existed.
2. The relative feed (v_r/V) is too poor an approximation to the undeformed chip thickness $\bar{t}$.

In a continuation of the CIRP study, summarized in Snoeys and Peters (1974), considerable data is presented concerning forces, specific energy, surface finish, and wheel wear. All of these quantities were found to be closely related to the equivalent undeformed chip thickness (t_e), which is a better approximation to $\bar{t}$ than is v_r/V.

Wheel Wear

There are three main mechanisms by which grinding wheels wear:

(1) bond fracture, leading to the loss of complete abrasive grits;
(2) grit fracture, part of the grit fractures, leaving a reasonably sharp portion of grit held in the wheel;
(3) attritious grit wear, the process by which a wear flat is formed on the surface of the grit.

Grinding wheel wear is frequently measured by the decrease in radius of the wheel after grinding for a set time interval. The wear may than be expressed in terms of the grinding ratio (the ratio of the wheel wear volume to the volume of metal removed). This procedure is quite satisfactory for processes in which the wheel is not frequently dressed — such as cut-off grinding or snagging (SRG operations). In these operations a steady state condition is approached in which grits fracture or are knocked out of the bond as attritious wear causes large flats to form on their clearance faces However, in FFG such as surface grinding or cylindrical grinding, unacceptable grinding burn, resulting from the formation of wear flats, occurs and the wheel has to be periodically dressed. As Brueckner (1960) and Malkin (1968) have pointed out, the radial wheel wear between redressing operations is only about one-tenth of a grain diameter for finish grinding operations. Thus, in operations of this type, grinding ratio is not the best measure of wheel wear.

Attritious wear is undesirable in finish grinding. The other wear processes (grit fracture and bond fracture) are generally beneficial, since they cause the removal of dull grits. Because of the practical importance of attritious wear, there is a need to study this wear mechanism independently from the other mechanisms. About the only method of doing this for a complete grinding wheel is to examine the wheel surface with a microscope. Brueckner (1960), Grisbrook (1962), Malkin (1968), and others have used this procedure, but it is rather unsatisfactory because subjective judgements are involved in deciding just what constitutes a flat.

Malkin (1968) measured the size of flats developed on grits in a surface grinding study, and at the same time collected all the grinding debris by mounting the wheel in a plastic box coated with petroleum grease on the inside. The debris was then separated from the grease by means of a solvent and the metal chips removed by dissolving them in acid.

Measurement of the size distribution of abrasive debris collected during grinding showed that harder wheel grades gave finer debris than softer grades. This is consistent with the concept that grits are more firmly bonded in hard wheels. The size distribution obtained from both hard and soft wheels was not very different from the size distribution of abrasive before it was bonded into the wheel. This suggests that the major wear process contributing to loss of wheel volume is bond fracture, grit fracture and attritious wear contributing only a small proportion of the total volume lost from the wheel. However, since the formation of wear flats determines when the wheel must be redressed, attritious wear is the process of most importance in FFG. An analogy to this result is that there is very little difference in the weight of a worn-out automotive engine and a new one.

Malkin (1968) also collected and analyzed the material removed during dressing. The debris produced in dressing a hard wheel was finer than that for a soft wheel. The stronger bond of the hard wheel reduces the tendency for complete grits or large fragments of a grit to be removed during dressing. This should result in a smaller deviation in height of grits below the wheel surface for hard wheels than for soft. A fine dressing lead tends to give larger initial flat areas. This results in more rapid growth of flats to a size

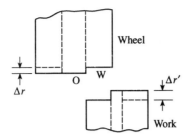

FIG. 13.12. A method of measuring radial wear.

at which wheel burn occurs. However, fine dressing gives a better surface finish.

Figure 13.12 shows the procedure used to measure wheel wear in plunge cylindrical grinding. Periodically, the change in wheel radius is measured by grinding a thin (0.002 in, or 0.051 mm) piece of steel shim stock to obtain the distance Δr between the original and worn wheel surface. An optical comparator is used to measure Δr. The corresponding distance on the work ($\Delta r'$) is measured to obtain the volume of material ground away per unit width ground (M'). Figure 13.12 shows how two tests may be performed per dressing operation. The solid lines correspond to Δr and $\Delta r'$ for a completed test, while the dashed lines are for the next test to be run on the other side of the wheel.

Figure 13.13(a) is a plot of radial wheel wear versus the volume of material ground away per unit wheel width (M'). This curve shows three regions: a region of rapid wear, a region of low wear and, finally, a second region of rapid wear ending with chatter or workpiece burn.

Figure 13.13(b) shows curves for grinding conditions similar to those of Fig. 13.14(a) for wheels of different grade. The softer H and K wheels are seen to wear more rapidly than the L grade wheel, but chatter or burn requiring dressing is postponed due to a self-dressing action for the softer wheels.

Figure 13.13(c) shows similar results, with all operating conditions the same except for wheel speed (V). This shows little change in wheel wear with V, which puts in question the oft cited claim of much less wheel wear with increased wheel speed.

From Fig. 13.13(a) the slope of the curve, which is related to the grinding ratio, is seen to vary continuously. Hence, the use of a single value of G to characterize the wear of a wheel in plunge cylindrical grinding is not very realistic.

The cutting forces (F_P and F_Q) were measured by mounting solid state strain gages on the centers, as shown in Fig. 13.14(a). Figure 13.15(a) shows that specific energy increases as wheel wear progresses. These values of specific energy are clearly in the region between SRG (2×10^6–3×10^6 p.s.i., or 13.8–20.7 J mm^{-3}) and FFG (approaching 10^7 p.s.i., or 69.0 J mm^{-3}).

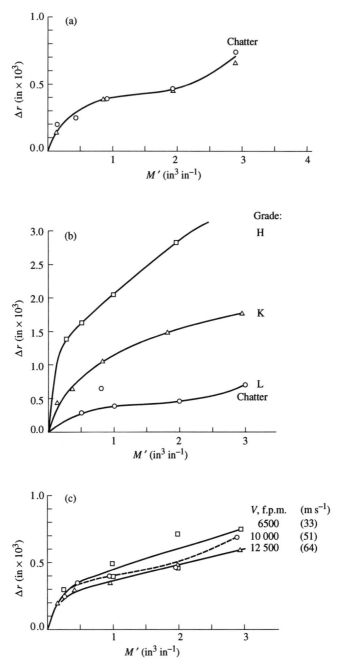

FIG. 13.13. (a) The variation of radial wheel wear (Δr) with the volume of material removed (M'). Wheel, 32A 54L V, 30 in. dia (760 mm); wheel speed, V: 6500 f.p.m. (33 m s^{-1}); $V/v_w = q = 35$; $v_r = 0.0011$ i.p.s. (0.027 mm s^{-1}); work, AISI 52100 steel ($R_c = 63$), 4 in (102 mm) dia.; machine, 10 × 24 plain cylindrical grinder,

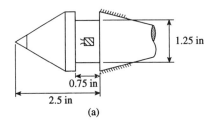

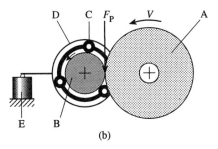

FIG. 13.14. (a) Solid state strain gages mounted on a turned down center to measure forces F_P and F_Q. (b) A method of measuring cylindrical grinding torque by Salje: A, wheel; B, work; C, roller; D, outer ring; E, damper. (After Opitz 1952.)

Figure 13.14(b) is of historical interest. It shows the method of measuring grinding torque in cylindrical grinding used by Professor Salje in his doctoral dissertation. This was abstracted by Professor Opitz in the first volume of the *Annals of CIRP* (Opitz 1952). The most practical method of obtaining spindle power is of course by measuring the power consumed by the motor directly (using a d.c. motor or making a power factor correction by some convenient means) and subtracting windage and friction losses with the wheel close to the work with fluid flowing, but without grinding. While this may appear overly approximate, it is sufficiently accurate when one considers the other approximations involved in subsequent use of the result. Also, by virtue of the Galileo Principel (Chapter 9), relative values will be more representative of performance than absolute ones. Direct measurement of spindle power also eliminates the need to correct workpiece dynamometer values when grinding with a large wheel depth of cut (d), as discussed in Chapter 14.

Figure 13.15(b) shows that surface roughness without spark-out is essentially constant at about $R_a = 50 \mu$in (1.25 μm), dropping only slightly as wheel life is approached, probably due to a greater burnishing action as wear flats develop on the active grits.

FIG. 13.13. *cont* 15 h.p. (11.2 kW); fluid, 3 percent water-based. (b) same as (a) but $V = 10\,000$ f.p.m. (51 m s^{-1}) and two additional wheels used: 32A 54 H V and 32A 54K V. (c) same as (a) except three wheel speeds employed.

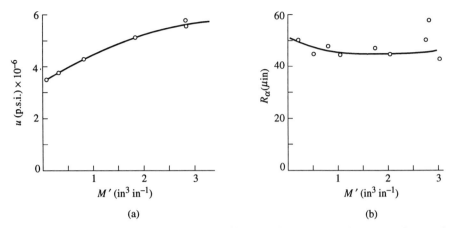

FIG. 13.15. (a) The variation of the specific grinding energy with the volume of material removed (M'). (b) The variation of surface roughness (R_a) with the volume of material removed (M'). Grinding conditions: same as (a) except wheel speed (V) = 10 000 f.p.m. (51 m s^{-1}).

Wheel Life

Wheel life (i.e. when dressing is required) is usually associated with excessive surface temperature (burn) or chatter. The problems associated with burn and what can be done to avoid it have been discussed in Chapter 9. A brief discussion of chatter follows. Chatter is a complex subject that has received a good deal of attention. A comprehensive discussion of this topic may be found in King and Hahn (1986).

There are some situations in which vibration may have desirable effects on the grinding process. In operations involving a large wheel–work area, the introduction of a controlled vibration can effectively reduce the instantaneous contact area and may improve the surface finish and wheel life. However, in most cylindrical grinding situations, vibration of the machine will produce an undesirable chatter pattern on the workpiece.

A grinding machine is a complex structure capable of several modes of vibration. Before a troublesome vibration can be corrected, its type and cause must be identified. There are two main categories of vibration in grinding— forced and self-excited. The forced vibration arises from some periodically varying phenomenon either in the machine tool drive or from some external source (for example, some other machine tool nearby). The most common cause of forced vibration is unbalance of the grinding wheel. The resultant chatter marks on the workpiece will then have a frequency corresponding to the wheel speed. Unbalance of some drive shaft or misalignment of bearings in the grinding machine head may also cause chatter marks at a frequency equal to the wheel speed or some integer multiple of V.

Self-excited vibration involves a feedback of energy from the grinding process to the frame of the machine. It may arise whenever some deflection of the machine frame alters the grinding conditions so that the grinding forces fluctuate out of phase with the deflection. Depending on the phase relationship between the deflection and force fluctuation and the damping characteristics of the machine, it is possible for a large chatter vibration to grow from a small transient force variation. This type of vibration will have a frequency equal to the free natural frequency of the part of the machine frame which is deflecting cyclically. There are several different mechanisms by which self-excited vibrations may develop in grinding.

Speed Ratio

The ratio of wheel speed (V) to work speed (v) is sometimes called the speed ratio, and assigned the symbol q'. In the discussion of creep feed grinding, the quantity $(q')^{0.5}$ was found to play a very important role relative to the fraction of energy going to the work (eqn 9.28). The quantity $(q')^{0.5}$ also plays an important role relative to undeformed chip thickness ($\bar{t} \sim (q')^{0.5}$ in equation 4.13)) and, as a consequence of the size effect, it also has an important influence on the specific energy (u). From eqn (9.23) the grinding temperature θ is proportional to u and hence is also influenced in a major way by $(q')^{0.5}$.

The normal range of $(q')^{0.5}$ in pendulum grinding is 7–10 ($q' \simeq 50 \simeq 100$). However, in creep feed grinding values of $(q')^{0.5}$ may be as high as 150 ($q' \simeq 25\,000$). *Saije et al.* (1983) have pointed out that whereas values of q' greater than 50 have been used in creep feed grinding, the region below 50 remained unexplored. They therefore examined grinding behavior for values of q' below 50 in plunge cylindrical grinding. This was done by comparing the following important grinding parameters with conventional values over a wide range of q' values at the same wheel speed (V) and removal rate (vbd):

- wheel work contact length, l
- grinding forces, F_P and F_Q
- specific energy, u
- workpiece surface temperature, θ
- surface roughness, R_t

Plunge cylindrical grinding is a convenient process to use to study performance at low values of $(q')^{0.5}$ and conventional wheel speed because it simply involves using work speeds that are higher than usual.

In considering wheel–work contact length at low values of q', it is important to take into account the fact that the motion of a grit relative to the work is cycloidal rather than circular, and that for very small values of wheel depth of cut the peak-to-valley roughness (R_t) increases the effective value of d. When these two factors are included, the following result for l pertains for up grinding:

$$l = \sqrt{D_e(d + R_t)} \left(\frac{V + v}{V}\right) = \sqrt{D_e(d + R_t)} \left(1 + \frac{1}{q'}\right). \quad (13.18)$$

For values of q' in the usual range ($q' > 50$), eqn (13.18) reduces to eqn (4.12) $(D_e d)^{0.5}$). Figure 13.16(a) shows the variation of l with q' for a representative grinding case at constant wheel speed V and specific removal rate $\dot{M}'$. The wheel–work contact length is seen to be a minimum in the pendulum grinding range ($q' \simeq 50$–100), to rise appreciably in the creep feed range ($q' > 100$), but to rise only slightly in the low q' range ($q' < 50$). While chip accommodation can be troublesome in the creep feed range (long, thin chips) this is less likely to be the case in the low q' regime.

The total specific energy (u) consists of two parts in the low q' regime:

$$u_s = \text{specific energy to the wheel} = F_P V/\dot{M}, \quad (13.19)$$

$$u_w = \text{specific energy to the work} = F_P v_w/\dot{M}; \quad (13.20)$$

and

$$u = u_s + u_w. \quad (13.21)$$

In the ordinary pendulum and creep feed regimes ($q' > 50$), the second term is negligible and $u = u_s$ to a good approximation. Figure 13.16(c) shows values of u (dashed lines) and u_w (solid lines). The second term in eqn (13.21) is seen to be negligible for values of $q' > 50$. The total specific energy is approximately constant in the low q' regime, but rises appreciably in the creep feed regime. The surface temperature versus q' curve will have a shape similar to that of the u versus q' curve of Fig. 13.16(c) — low in the low q' regime, medium in the pendulum regime, and high in the creep feed regime.

Figure 13.16(b) shows the observed variation in tangential (F'_P) and normal (F'_Q) forces per unit width ground for the same example as in Fig. 13.16(a). The normal force is about twice the tangential force, and both rise with q' continuously, particularly in the creep feed regime ($q' > 100$).

Figure 13.16(d) shows the variation of peak-to-valley roughness (R_t) with q'. The surface roughness is seen to decrease as q' increases or decreases from the pendulum regime shown shaded, approaching spark-out roughness (R_t) at very large and very small values of q'.

Figure 13.17(a) shows the path taken by successive grits cutting in the same groove in creep feed grinding. These successive grit paths are actually circular arcs, but are shown here with a much greater magnification in the vertical direction than in the horizontal direction, so that both d and R_t may be shown ($d \gg R_t$). Figure 13.17(b) shows the pattern left on the finished surface having a pitch p and a chordal rise R_t. This value of R_t is the major component of surface roughness due to feed marks only, and does not include the influence of metal build-up, chatter, burnishing action, or other secondary effects. The relation between p and R_t will be:

$$p = 2\sqrt{D_e R_t}. \quad (13.22)$$

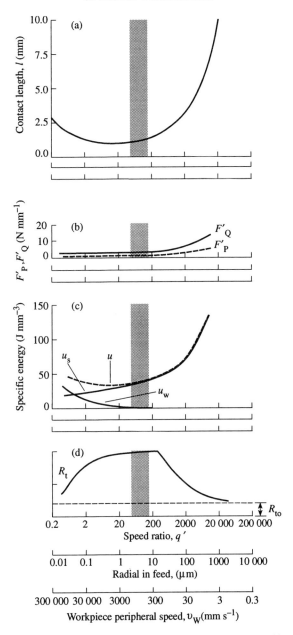

FIG. 13.16. The variation of grinding characteristics over a very wide range of speed ratios q' for a constant wheel speed and removal rate: (a) wheel–work contact length versus q'; (b) the specific cutting forces F_P' and F_Q' versus q'; (c) variation of the specific energy at the wheel (u_s), at the work (u_w), and the total (u) with q'; (d) variation of R_t with q' (after Salje *et al.* 1983). Grinding conditions: mode, up grinding; wheel, A60 K8 V; work, AISI 1045 steel; equivalent wheel diameter (D_e), 100 mm (3.94 in); wheel speed (V), 60 m s^{-1} (11 800 f.p.m.); specific removal rate ($\dot{M}'$), 180 mm^2 min^{-1} (0.28 in^2 min^{-1}). The shaded area is the usual pendulum grinding range of q'.

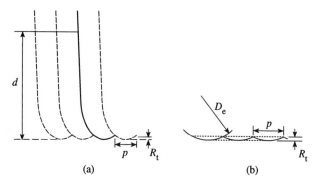

FIG. 13.17. (a) Successive grit paths for creep feed and pendulum grinding ($d \gg R_t$) with much greater magnification in the vertical direction than in the horizontal direction. (b) The feed pattern left on the finished surface in low speed ratio grinding when $d = R_t$.

The corresponding spacing of successive grits in the wheel face cutting in the same groove will be (from eqn 9.1):

$$l_r = (Cb')^{-1} = (Cp)^{-1}. \qquad (13.23)$$

The quantities p and l_r are related as follows:

$$p = (v/V)l_r = l_r/q'. \qquad (13.24)$$

Combining these three equations and solving for R_t,

$$R_t = (4D_e Cq')^{-1}. \qquad (13.25)$$

Thus, for the same wheel R_t associated with feed, marks should vary inversely with q' and approach zero as q' becomes very large. However, when secondary effects are included the spark-out value of R_t will be approached instead of zero. The inverse proportionality relating R_t and q' will appear as shown in Fig. 13.16(d), since semi-logarithmic coordinates are employed.

In the region in which q' is small, R_t will approach d as q' decreases. Figure 13.17(b) shows the feed pattern left behind on the finished surface in relation to the wheel depth of cut d for a low value of q'. Here $R_t = d$. When d reaches a value equal to R_t in pendulum grinding ($R_t \simeq 1\ \mu$m), as q' is reduced then R_t should equal d with further reduction in q', and the R_t curve should drop with a decrease in q' (and hence d for a constant removal rate), as shown in the left-hand region of Fig. 13.16(d).

Salje summarizes the comparison of up grinding characteristics for a constant wheel speed and removal rate over a wide range of speed ratio q' in Table 13.6.

Machine tools for low-speed ratio grinding should have a higher work speed and power capability and be equipped with live centers. High wheel speed requires better wheel balance. To enable work changing downtime to be reason-

TABLE 13.6 *Comparison of up grinding characteristics for a constant wheel speed and removal rate over a wide range of speed ratio (q') (after Salje et al. 1983)*

	$q' = 1$	$q' = 100$	$q' = 10\,000$
Wheel–work contact length, l	Medium	Low	High
Forces F_P and F_Q	Low	Medium	High
Total specific energy, u	Low	Low	High
Temperature, θ	Low	Medium	High
Surface roughness, R_a	Low	High	Low
Wheel wear	Medium	High	Low

able, rapid acceleration and deceleration of the work becomes important. An alternative to a use of higher work speeds with a conventional wheel speed to obtain low values of q is to reduce the wheel speed so that the work speed need not be as high.

An interesting possibility is to use low-speed ratio grinding in conjunction with speed stroke grinding. For a 100 mm (3.94 in) stroke at 400 strokes per minute, the mean work speed would be about $4\,\text{m min}^{-1}$ (13 f.p.m.). At wheel speeds of $30\,\text{m s}^{-1}$, the values of q would be 7.5 and 3.8 respectively.

Axial Force In Profile Grinding

Salje and Damlos (1991) have experimentally studied the relation of forces in plunge surface grinding (Fig. 13.18(a)) and profile (form) grinding involving an inclined surface (Fig. 13.18(b)). In the latter case there will be an axial component of force F_A, and the following force relations were found to hold:

- force components F_T and F_R are independent of angle α for the same removal rate
- force component $F_A = F_R \tan \alpha$
- the resultant of F_T and F_R will be normal to the inclined surface

These relationships may be used to find the net axial force across the wheel width (b) for any complex profile by integration. The principle is illustrated in Fig. 13.18(c) in which an inclined surface of width b_1 with positive slope and an inclined surface of width b_2 with a negative slope are shown.

The net axial force (F_A) acting to the left ($-$) in this case will be:

$$F_A/F_R = (b_1/b)\sin\alpha_1 + (b_2/b)\sin\alpha_2, \qquad (13.26)$$

where values of α are positive with positive slope, and vice versa.

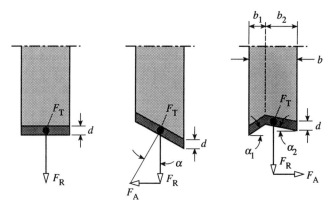

FIG. 13.18. A comparison of forces pertaining for (a) ordinary surface grinding, (b) profile grinding of an inclined surface, and (c) profile grinding of axisymmetric inclined surfaces.

Centerless Grinding

Introduction

Centerless grinding is very similar to cylindrical grinding except for the manner in which the workpiece is supported and driven. In cylindrical grinding the work is supported on centers and driven by a member connecting the headstock and the work. In centerless grinding the work is supported in part by a work blade and is friction-driven by a regulating wheel (Fig. 13.19). The weight of the part and the vertical components of force at the wheel–work and regulating wheel–work contacts provide positive contact between the work and the work blade. The up grinding mode is used as shown in Fig. 13.19 to provide more positive contact between the work plate and the work, and to keep the tangential grinding force from lifting the work out of contact with the grinding wheel and regulating wheel (a dangerous possibility with down grinding).

As in cylindrical grinding, both plunge and cross-feed modes are employed. In the latter case the process is called through-feed centerless grinding, since the work is then fed axially past the two wheels. This is accomplished by adjusting the axis of the regulating wheel to have a small angle ϕ (a few degrees) to the grinding wheel axis. This causes the regulating wheel to provide not only rotation of the work but also axial feed motion. In plunge centerless grinding the axes of the grinding wheel and the regulating wheel are parallel and the regulating wheel is fed inward at a fixed rate. Peets (1925) first discussed the centerless grinding process in qualitative terms.

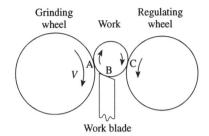

FIG. 13.19. A centerless grinding operation.

Applications

The centerless grinding process is essentially a mass production operation. It is widely used in the rolling element bearing and automotive industries over a wide range of workpiece diameters, from as small as 2 mm (0.080 in) to several inches (several hundred mm) in diameter. It Is particularly well suited to long slender workpieces where workpiece deflection is a problem in cylindrical grinding. In centerless grinding the workpiece finds it own center and is completely supported in the grinding zone.

Centerless grinding with ordinary equipment is capable of producing parts with a diameter tolerance of 0.0002 in (0.005 mm), a roundness of 0.0008 in (0.02 mm) and an R_a finish of about 12 μin (0.3 μm). In through-feed grinding axial speeds of up to 400 f.p.m. (122 m min^{-1}) have been used.

Centerless grinding wheels are usually large in diameter and are often quite wide (up to 20 in = 500 mm), particularly for through-feed grinding. Vitrified Al_2O_3 wheels having grit sizes and structure numbers similar to those used in cylindrical grinding are used. For rough grinding open-structure wheels are used. Wheel speeds range from about 6000 to 12 000 f.p.m. (30–60 m s^{-1}). The regulating wheel usually has a rubber bond which gives long life, good frictional characteristics for driving the work with little slip, and good damping to suppress chatter.

The decision to use centerless grinding instead of cylindrical grinding rests largely on differences in lot size, set-up time, and the production rates possible. Centerless grinding may be used successfully to grind a wide range of materials, including metals, ceramics, glasses, plastics, and even cork.

Mechanics

Figure 13.20 shows the workpiece with the three points of contact at A, B, and C. Since the work is supported on a surface being ground there is a tendency for feedback leading to lobes on the ground surface. Dall (1946) first investigated theoretically the rate at which a defect on the surface would

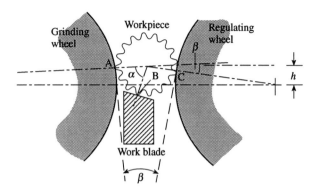

FIG. 13.20. A centerless grinding operation showing the work blade setting angle (α) and the workpiece setting angle (β) that determines the amount by which h, the work center, is above the line of centers (after Rowe et al. 1989).

disappear or grow with a perfectly stiff machine. This was found to depend on the height of the workpiece center above the line of centers (h) and the work blade angle (α) shown in Fig. 13.20. It is possible to produce a perfectly round specimen by proper selection of α and β in the absence of dynamic instability.

Furukawa et al. (1971) derived relations for the development of lobes as shown in Fig. 13.20:

$$\alpha = n_e'\beta, \qquad (\pi - \beta) = (n_e - 1)\beta, \qquad (13.27, 13.28)$$

where n_e and n_e' are even numbers.

For rounding, these conditions are to be avoided and for optimum rounding β should be 7 or 8 degrees and α should be $(n_e' - 1)\beta$. In practice, relations such as these are used. These are generally programmed into a data bank and used to select values of α and β to avoid geometric instability. In addition to geometric instability there is dynamic instability that also influences the ability for rounding. To avoid dynamic instability, a stiff machine is required.

Rowe et al. (1965) have used a simple experimental technique to check setting angles that provide a strong corrective action to remove errors initially present on the workpiece. The workpiece shown in Fig. 13.21 is ground and the number of revolutions of the work required to remove the flat is determined. As a matter of safety, the chordal rise must be a small percentage of the work diameter (18 percent in Fig. 13.21). This technique confirmed that $\beta = 7°$ is about right and that a work blade angle (α) of about 30° should be used.

Reeka (1967) has demonstrated that the roundness error is a function of the number of workpiece revolutions (U_w) and the depth of cut d, as shown in Fig. 13.22.

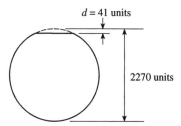

FIG. 13.21. A specimen with a small flat for use in checking rounding capability in centerless grinding (after Rowe *et al.* 1965).

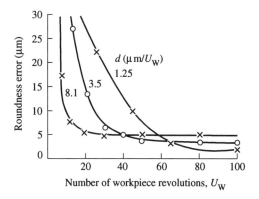

FIG. 13.22. The variation of roundness with number of workpiece revolutions (U_w) for different values of in-feed per revolution (*d*) in plunge centerless grinding (after Reeka 1967).

Gurney (1964) and Rowe (1979) have discussed the dynamic stability aspects of centerless grinding in considerable detail, and Frost and Fursdon (1985) have discussed the use of stability maps.

If the value of *h* is too great and the grinding force ratio F_Q/F_P is too high, as with a glazed wheel in need of dressing, there will be a tendency for the work to jump up and down as it periodically leaves contact with the work blade.

Udupa *et al.* (1988) have studied the possibility of removing instability by applying pressure to the top of the workpiece. This was done by use of a close fitting (0.5 mm = 0.020 in clearance) air bearing containing several 0.6 mm (0.024 in) diameter holes, as shown in Fig. 13.23. It was found that when the supply pressure was 10^4 N m^{-2} (1.45 p.s.i.) essentially all dynamic instability was removed. The finish with the stabilizing air bearing was improved by

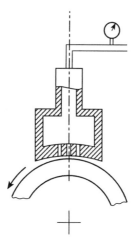

FIG. 13.23. An air bearing applying force to a workpiece in centerless grinding to remove dynamic instability (after Udupa *et al*. 1988).

about 15 percent and the circularity was improved. by about 30 percent at a work speed of 30 m min^{-1} (98 f.p.m.) and a setting angle β of 8°.

Udupa *et al*. (1986) studied the influence of the type of cutting fluid on through-feed centerless grinding. It was found that when oil is used regulating wheel slip increases and surface finish improves, but at the expense of circularity. When circularity is the major concern, it is best to use a fluid that does not contain oil. Miyashita *et al*. (1982) have published a diagram for selecting Chatter-free conditions in centerless grinding.

Machines

Machines of higher power and stiffness, larger grinding wheel diameter and width (particularly in through-feed grinding), and higher wheel speeds are required in order to realize the full potential of plunge and through-feed centerless grinding. This is particularly true when CBN wheels are economically justified.

Rowe *et al*. (1987) have proposed a new machine layout designed to increase the structural stiffness and stability of centerless grinding machines. One of the items incorporated is use of the centerless grinding wheel itself to dress the regulating wheel, instead of using a single-point diamond dressing tool. This has been shown by Hashimato *et al*. (1983) to improve the accuracy of the process by reducing run-out of the regulating wheel. It was found that this dressing technique was capable of reducing surface roughness from $R_a = 0.32\,\mu$m (12.8 μin) to 0.12 μm (4.8 μin) and roundness error from 1.7 μm (68 μin) to 0.2 μm (8 μin).

Optimization

Centerless grinding is usually employed to obtain high circularity and finish. The objective is to obtain the required quality in the least production time. High production rates are possible due to the ease of loading and unloading the work from the machine. The production time may be further reduced if the rate of removal (vbd) can be increased without adversely affecting quality. However, when the production rate is increased wheel wear will increase, and the total unit grinding cost will rise if the saving in reduced labor and machine cost is more than offset by an increase in the unit cost associated with dressing. Most companies engaged in centerless grinding will process a variety of parts on a given machine. Only in a few cases, involving very large production of a single product, will a machine be dedicated to a single product. Most large companies will have a data bank, based on previous experience, programmed into their computer and used for process planning. This data bank will provide a reasonable set of operating conditions for start-up. However, it is not practical to include all variables that will have an influence on the unit grinding cost for a given application. In fact, some of these variables, such as workpiece hardness, shop temperature, operator fatigue, etc., will change during a production run. It is therefore important to monitor the process continuously and to adjust some easily changed variable to insure operation under conditions that are close to optimum.

The ideal way of doing this is by automatic adaptive control. However, this would involve a huge set of sensors, an equally large set of interfaces between sensors and machine controls, and a complex fuzzy logic computer program. While all of this is being worked on it appears that it will be a very long time before such a complex control system is available at reasonable cost. The alternative for the foreseeable future is to provide the machine operator with a simple tool to make it possible for him to act as the set of sensors, interfaces, and decision-making programs. The manual adaptive control (MAC) procedure, briefly discussed in Chapter 6, is a useful monitoring tool, and Stelson *et al.* (1977) have applied MAC to monitor a through-feed centerless grinding operation, as discussed below. Since there is essentially no slip between the regulating wheel and the work, the axial velocity v_a of the work will be

$$v_a = v_R \tan \phi \simeq v_R \phi, \qquad (13.29)$$

where v_R is the circumferential speed of the regulating wheel and ϕ is the small inclination angle between the grinding wheel and the regulating wheel axes. Assuming zero slip at the regulating wheel, the material removal rate ($\dot{M}$) will be

$$\dot{M} = v_a (\pi D_w) b. \qquad (13.30)$$

For the same work diameter (D_w) and wheel width (b),

$$\dot{M} \sim v_a \sim v_R \phi \sim N_R \phi \qquad (13.31)$$

where N_R is the regulating wheel r.p.m. Thus, $\dot{M}$ may be varied by changing either v_R or ϕ. However, it is more convenient to change v_R than ϕ.

When using a centerless grinding machine in the continuous mode, the problem is to select the cost optimum regulating wheel speed N_R after the wheel width b and inclination angle ϕ have been decided upon. The desired optimum usually corresponds to minimum grinding cost per part. At this operating point, the best trade-off is the lowest sum of the cost associated with dressing the wheel and the costs related to labor, machine, and overheads. If the machine is operated at too low a metal removal rate, the labor, machine, and overhead cost will be too great. If the process is operated at too high a metal removal rate, the grinding wheel will become dull prematurely and the wheel dressing cost will be too high.

The general equation for grinding cost per unit volume per part (¢) is:

$$¢ = x/\dot{M} + y/G + ¢_D/(\Delta V), \tag{13.32}$$

where x is the value of the machine operator and overheads (¢ min^{-1}), y is the mean value of the grinding wheel (¢ in^{-3}), G is the grinding ratio = volume of metal removed/volume of wheel consumed, $¢_D$ is the cost of dressing the wheel, including machine down time, the value of wheel removed and the dressing diamond cost, and ΔV is the volume of metal removed between dressing operations.

For stock removal grinding (SRG), the third term on the right is normally zero since the wheel is never dressed. For form and finish grinding (FFG), the second term on the right will usually be negligible and hence, for axial feed centerless grinding (a FFG operation):

$$¢ = x/\dot{M} + ¢_D/(\Delta V). \tag{13.33}$$

More convenient variables in the workshop than $\dot{M}$ and ΔV are:

P = number of parts/dress = $(\dot{M}T)/B$, and
T = operating time between dressing operations.

Then, $\Delta V = BP$, where B is the volume ground away per part.

When these substitutions are made, eqn (13.33) becomes:

$$¢ = (xT + ¢_D)/BP. \tag{13.34}$$

This equation may be used to program a hand-held calculator to give the value of ¢ for constants x, $¢_D$, and B, and variables T and P. The procedure to be used is as follows:

1. The machine is set up with values of α, β, ϕ, and $\dot{M}$ from the data bank for the part being ground.
2. The operator begins making useful parts, starting with these data bank recommendations.
3. The operator periodically inspects the parts and when the wheel needs to be redressed for whatever reason (unacceptable finish or circularity,

SPECIAL PROCESSES

chatter, excess power, workpiece burn, noise, etc.) two quantities, P and T, are inserted Into the calculator.

4. A specified key (A) is pressed which enters the value of ¢ from eqn (13.34) into storage.
5. The regulator wheel speed N_R is then increased by a small increment and production is continued at the new removal rate.
6. When the wheel again requires dressing, the new values of T and P are inserted, and when key A is pressed the new value of ¢ is given and put into storage. When key B is pressed the old value is taken from storage and the new value is subtracted. If the difference is positive, this indicates that the new condition gives a lower cost per part than the old value.
7. If a positive value is obtained with key B, then the value of N_R is again increased by a small increment; otherwise, N_R is simliarly reduced.
8. After a few such cycles, the value will cycle between a positive and a negative value, indicating that the operating conditions are then very close to the cost optimum value.

Example

In order to check the foregoing procedure, tests were run under production conditions in a plant manufacturing earth-moving equipment. The grinding wheel employed was 24 × 20 × 12 in (610 × 508 × 305 mm) with the designation A54 L8 V. Two steel parts shown in Fig. 13.24 were studied. The data bank recommendations for the first part (a) were $N_R = 30$ r.p.m. and $\phi = 3°$. Cost values for three successive sets of parts are given in Table 13.7. Beyond the last point the operator could not load and unload parts at the production rate required and two operators had to be used to continue toward cost optimum conditions. Eventually the optimum cost per part was found to be 1.82 ¢ per part, which is 42 percent of the data bank value (4.31 ¢ per part). The machine was operating satisfactorily and producing parts within tolerance at the optimum condition.

A second steel part (Fig. 13.24(b)) was produced, again starting with conditions recommended by the data bank ($N_w = 56$ r.p.m., $\phi = 3°$) and the values given in Table 13.8 were obtained. Since the initial parts per dress (P) was low, it was decided to begin lowering N_R, which turned out to be the correct thing to do. At $N_R = 33$ r.p.m. the cost per part was only 73 percent of the initial data bank value.

It is to be expected that, *on the average*, the axial feed rate from the data bank (v_a) will be close to the cost optimum value. However, individual parts may be produced under conditions that are far removed from the cost optimum conditions unless some continuous monitoring method such as MAC is used to fine tune data bank recommendations.

There are several constraints which may prevent the economic optimum from being obtained. These include power limitation, surface finish and size

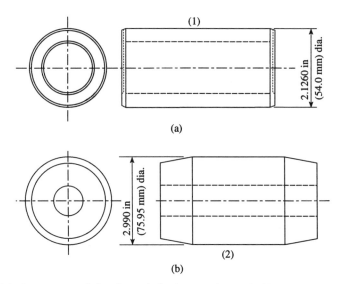

FIG. 13.24. Parts ground in through-feed centerless grinding optimization experiments: (a) surface (1) ground in first experiment; (b) surface (2) ground in second experiment.

TABLE 13.7 *Input data (first three columns) and output data (last three columns) for the first part (Fig. 13.24(a))*

N_R (r.p.m.)	T (min)	Parts per dress, P	Cost per part		
			¢ tool	¢ labor	¢ total
30	31.2	468	0.78	3.53	4.31
40	16.3	408	0.90	2.12	3.02
50	13.5	391	0.94	1.83	2.77

required, and the rate at which the machine may be loaded. If the machine is underpowered and the economic optimum is at a production rate greater than that corresponding to the available power, then the constrained optimum will be at the point of maximum power.

It is also possible that the surface finish at the economic optimum is unacceptable. This may be overcome by reducing the angle between the wheels (β) or decreasing the regulating wheel r.p.m. Another constraint is the ability of the operator to load the machine rapidly enough. When the operator is loading the machine as rapidly as possible, this corresponds to the constrained

SPECIAL PROCESSES

TABLE 13.8 *Input data to calculator (first three columns) and output data (last three columns) for the second part (Fig. 13.24(b))*

N_R (r.p.m.)	T (min)	Parts per dress, P	Cost per part		
			¢ tool	¢ labor	¢ total
56	1.65	47	7.81	1.89	9.67
50	3.38	73	5.03	2.45	7.48
33	6.90	104	3.53	3.52	7.05
25	9.97	125	2.94	4.23	7.17

TABLE 13.9 *The use of data from Table 13.8 to determine parts produced per hour when the removal rate (N_R) is increased. Here the total time to redress the wheel is assumed to be 10 minutes*

N_R (r.p.m.)	T (min)	Parts/dress, P	Time per part (min)	Parts per hour
56	1.65	47	0.250	240
50	3.38	73	0.183	329
33	6.90	104	0.163	368
25	9.97	125	0.160	375

optimum rate of production. However, if this occurs frequently it suggests that the economics of installing a high-speed automatic loading device should be considered.

Production Rate Optimum

Under some conditions it is important that the machine be operated to produce the maximum number of parts per hour regardless of cost. This would be the case if the operation considered is a bottleneck or in case of a national emergency. The optimization variable then becomes the total grinding time per part (T_P):

$$T_P = (T + T_D)/P, \qquad (13.35)$$

where T_D is the total time required for dressing.

This equation may be similarly preprogrammed into a calculator by a production engineer for use at the machine by the operator.

If the total wheel dressing time is assumed to be 10 minutes, then the data in Table 13.8 will yield the values given in Table 13.9. Here it is evident that the maximum production rate will occur at a removal rate greater than the removal rate for minimum cost per part. This is always the case. In these examples where the time between dressing operations is relatively low and the dressing time is relatively high, it is evident that it would be advantageous if

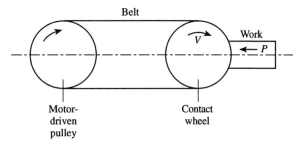

FIG. 13.25. Constant-force (*P*) belt grinding with a contact wheel.

the time between dressing operations could be increased by a better selection of the wheel and operating conditions.

Belt Grinding

Introduction

Belt grinding involves coated products mentioned at the end of Chapter 2. These consist of a single layer of oriented grit attached to a substrate in the form of a disc or belt. Figure 2.27 shows the cross-section of typical product. Abrasive belts are continuous ribbons that operate between pulleys. Figure 13.25 shows a belt operating between driven and idler pulleys, with the work being fed against one of the pulleys, called the contact wheel. The contact wheel provides resistance to the work as it is fed into the belt. The configuration shown in Fig. 13.25 is the simplest type of belt grinding, which is plunge grinding under constant axial force. This type of belt grinding is widely practiced as an SRG operation on castings. Abrasive belts are not dressed but are replaced by new ones when flats develop on active grits, resulting in burn or excessive feed force or power.

Operations such as that in Fig. 13.25 may involve manual feed at essentially constant pressure or mechanical feed at a constant rate. Belt grinding is well adapted to SRG because the active grits are relatively far apart, thus providing good chip storage even when the wheel–work contact length is large. Due to the elongated shape and the electrostatically induced orientation of grits in belts, the maximum force per grit before grit loss is large, making it possible to grind under conditions resulting in large mean undeformed chip thickness ($\bar{t}$) and hence low specific energy. This results in relatively low grinding temperatures at high rates of removal, but relatively poor finish. While the operation shown in Fig. 13.25 is essentially an SRG one, other configurations are used where the objective is to produce surfaces of high precision and good finish.

Belts are used to grind a wide variety of materials, ranging from wood, plastic, and leather to metals, ceramics, and glass. The grit material employed

FIG. 13.26. A grooved rubber contact wheel.

depends upon the material being finished. For wood, relatively inexpensive silica sand or garnet may be used, while for ductile metals and castings, Al_2O_3 and SiC grits are used, including the zirconia alloys, sol-gels, and agglomerates discussed in Chapter 2. For high-temperature alloys, CBN grits may be used, while for ceramics and glass, diamond grits are often economically justified. For SRG operations belt grinding can consume up to 30 h.p. per inch of belt width ($8.8\,\text{kW}\,\text{cm}^{-1}$) at a removal rate as high as $10\,\text{in}^3\,\text{min}^{-1}$ per inch of belt width ($65\,\text{cm}^3\,\text{min}^{-1}\,\text{cm}^{-1}$).

Contact Wheels

The contact wheel is usually a rubber-coated metal pulley. The rubber provides good traction and resilience to enable more grits to be active and to provide damping to prevent instabilitly. The hardness of the rubber is important and is specified in terms of an empirical durometer reading (resistance to penetration of a standard indenter). The practical range of durometer hardness (H_D) is 20 (soft) to 90 (hard). A hard backup wheel will give a high removal rate and relatively poor finish, while a soft backup wheel gives a better finish at a lower removal rate. The main function of rubber hardness is to change the belt–work contact length. The softer the backup wheel, the greater is the contact length for a given feed force, and hence the smaller the force per active grit.

Rubber backup wheels are frequently serrated, with slots in the face of the wheel at an angle to the grinding direction ranging from 90 to about 45 degrees (Fig. 13.26). The ratio of groove width to land width is important, as well as the size of the groove width. While a groove angle of 90° (grooves parallel to wheel axis) is most aggressive, causing the active grits to penetrate deeper for higher removal rates, this is rarely used because it leads to short belt life and a high siren-like noise level. The main action of the grooves in a backup wheel is to change the mean force per grit for a given feed force. As the ratio of groove width to land width increases, the mean force per grit decreases and hence the removal rate decreases as the surface finish improves. Thus, surface

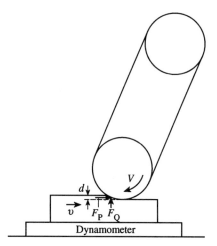

FIG. 13.27. Plunge surface grinding with an abrasive belt.

finish may be improved at the expense of removal rate by using a soft contact wheel having inclined grooves of large width.

Matsui and Syoji (1978) studied the influence of the stiffness of ungrooved rubber contact wheels on the wheel–work contact length and grinding forces. A contact wheel and belt arrangement replaced the grinding wheel in a plunge surface grinding experiment, as shown in Fig. 13.27. Three rubber-coated contact wheels with durometer hardnesses (H_D) of 35, 50, and 80 were used. A quick-stop device (Sauer and Shaw 1974) was used to disengage the work from the belt quickly during a cut. From the scratches remaining on the workpiece, the length of wheel–work contact (l) for the different values of contact wheel hardness was determined for otherwise identical grinding conditions, as shown in Fig. 13.28. Also shown in Fig. 13.28 is the contact length versus d curve for a conventional grinding wheel.

An elastic contact wheel will also have a major influence on the number of grits cutting per unit time. This was measured by Matsui and Syoji (1978) by mounting a workpiece of low mass on a piezoelectric crystal and recording the normal force pulses when grinding under identical conditions with the three contact wheels of different hardness. The results shown in Fig. 13.29 were obtained. Here it is evident that the spacing of active grits increases with a decrease in contact wheel hardness, as does the force and hence depth of cut for each active grit.

The decrease in active grit density (C) with a decrease in contact wheel hardness will give an increase in $\bar{t}$ and, hence a decrease in u that more than offsets the opposite effect that an increase in l has on $\bar{t}$. Otherwise, a decrease in backup wheel hardness would not give a decrease in the mean force per active grit, as shown in Fig. 13.29. The reduction in mean force per grit accounts in part for the excellent wheel life performance of belt grinding at high

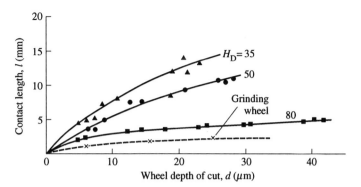

FIG. 13.28. The variation of the wheel-work contact length (l) with wheel depth of cut (d) for three ungrooved contact wheels of different durometer hardnesses (H_D). The dashed curve is for a conventional vitrified grinding wheel (WA 80H V) (after Matsui and Syoji 1978).

removal rates with a resilient backup wheel. An additional reason is that a coated abrasive system gives a larger spacing between active grits cutting in the same groove, which reduces the chip storage problem that often becomes troublesome at high removal rates.

Namba and Tsuwa (1967) studied the grinding performance of abrasive belts with a 45° serrated rubber contact wheel that had 12.7 mm (0.5 in) lands and grooves. Fixed load plunge surface grinding of cold rolled steel was performed as shown in Fig. 13.30. Measurements were made of wheel depth of cut for a wide range of grinding conditions, and it was found that the wheel depth of cut (d) varied as indicated by the following proportionality:

$$d \simeq (F_Q V)/v, \tag{13.36}$$

where the proportionality constant is a function of the workpiece cutting characteristics and the grit shape pertaining. The proportionality constant will also be a function of the backup wheel durometer hardness and geometry, in accordance with the experiments of Matsui and Syoji (1978).

Belt Life

In belt grinding the change in belt thickness when the belt is replaced is relatively small and hence the grinding ratio (G) is not a convenient measure of belt life. Instead the quantity

cut per path = volume removed per unit area of belt used

is a better measure of belt life. As a belt wears, flats develop on the abrasive grits and this causes an increase in the power required in fixed feed grinding. Figure 13.31 shows representative curves of h.p. in^{-1} versus cut per path for

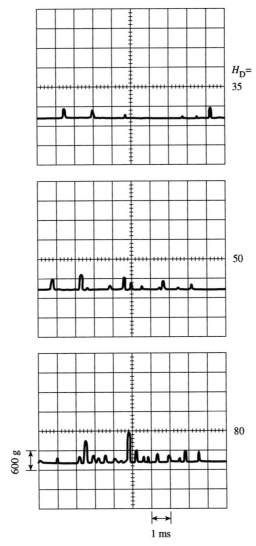

FIG. 13.29. Oscilloscope traces of belt grinding with ungrooved contact wheels of different hardnesses (after Matsui and Syoji 1978).

belts having two different abrasive types. As the volume cut per unit area of belt traversed increases, the power per unit belt width is seen to increase. If belt life in this case is limited to 25 h.p. in^{-1}, then the belt life in terms of cut per path is seen to be about 3.2 in^3 in^{-2} for the fused Al_2O_3 abrasive, but 13.8 in for the ceramic sol-gel abrasive (ratio of 4.3:1).

The two abrasives in Fig. 13.31 are present on the belt as a single layer of monolithic particles. As mentioned in Chapter 2, abrasive particles that consist

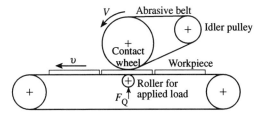

FIG. 13.30. The arrangement used by Namba and Tsuwa (1967) to study the influence of a serrated contact wheel in fixed load surface grinding with an abrasive belt.

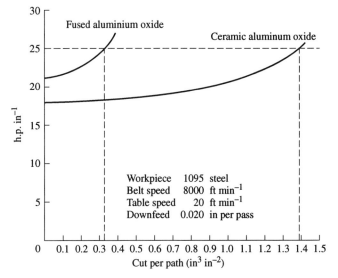

FIG. 13.31. The variation of the power per unit belt width with (volume cut)/(belt area) in belt surface grinding with fused Al_2O_3 grit and sol-gel ceramic grit (after Gagliardi *et al.* 1988). Work AISI 1095 steel; belt speed, 8000 f.p.m. (40.7 m s^{-1}); work speed, 20 f.p.m. (6.1 m min^{-1}); wheel depth of cut (d), 0.020 in (0.30 mm).

of an agglomeration of smaller particles are also used. Figure 13.32 shows an abrasive belt coated with a single layer of agglomerate particles. Typically, the individual abrasive particles in an agglomerate are about 20 percent as large as the final particle, which has a 100–300 mesh size. This type of abrasive is unusually tough and wear resistant and gives the belt a self-sharpening action as wear flats develop. This gives the belt a more consistent action from beginning to end than the more usual single-layer design.

Figure 13.33 shows the variation of stock removal rate with time for a single-layer belt and one with agglomerate grits of the same individual particle

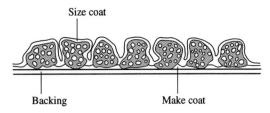

FIG. 13.32. An abrasive belt with grits consisting of agglomerations of smaller particles (after Gagliardi *et al.* 1988).

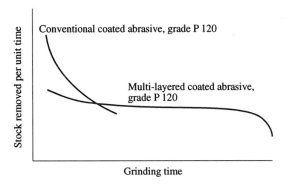

FIG. 13.33. Representative curves showing the variation of the stock removal rate versus grinding time for constant-force belt grinding with conventional and multilayered agglomerate grits of the same size (after Gagliardi *et al.* 1988).

size (120 mesh). The belt with agglomerate particles is seen to have an initially lower rate of removal but a much longer belt life.

Surface Finish

Figure 13.34 shows relative surface finish values versus time for single-layer and agglomerate grits having the same individual particle size (100 mesh). While the single grit layer gives a better surface finish, the agglomerate design gives a sustained more constant surface finish after initial break-in. As the single grit layer wears, the surface finish improves due to more burnishing action as flats develop. In general, a better finish is obtained with decreased grit size.

Typical surface finish values for a belt grinding with a contact wheel and monolithic grits range from about $R_a = 120$ for a screen size of 50 to about $R_a = 25$ for a screen size of 180. Surface finish and removal rate increase with belt speed, but the power required and the surface temperature also increase with the increased belt speed. When good finish and flatness are

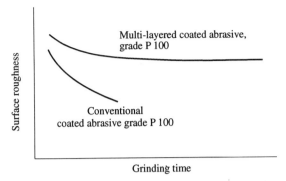

FIG. 13.34. Representative curves showing the variation of the surface roughness with grinding time for conventional and multilayered agglomerate grits of the same size (after Gagliardi *et al.* 1988).

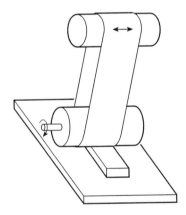

FIG. 13.35. Platen belt grinding with a solid metal contact wheel and axial belt oscillation to provide a good finish (after Gagliardi *et al.* 1988).

required, stock removal grinding is sometimes followed by platen grinding with a solid metal contact wheel (Fig. 13.35).

While belt grinding with a contact wheel is usually performed dry, a fluid is sometimes used with platen grinding. Fluids have been found useful when belt grinding certain metals, notably titanium and high-temperature alloys. This was discussed in Chapter 12.

Special Operations

In addition to SRG, for which belt grinding is well adapted because of the unusually wide spacing of active grits cutting in the same groove, belt grinding

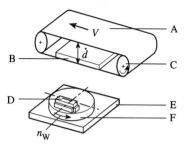

FIG. 13.36. Constant down-feed belt surface grinding using a metal backing plate and a rotary work table (after Spur and Becker 1990). A, Coated abrasive belt; B, backing plate; C, driven pulley; D, work; E, sliding table; F, rotary table.

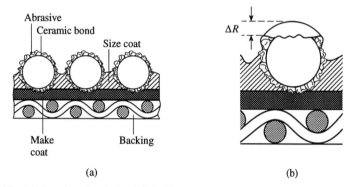

FIG. 13.37. (a) An abrasive belt with hollow agglomerate grits before grinding. (b) A single worn hollow agglomerate sphere. (After Spur and Becker 1990.)

is also used where surface finish and precision are major concerns. Examples of the latter are platen grinding (Fig. 13.35) and centerless belt grinding of rods and tubes. In centerless belt grinding, the contact wheel replaces the grinding wheel, but the arrangement is otherwise the same as for conventional through-feed centerless grinding. By use of a belt of fine grit size and a small removal rate, a surface roughness as low as $R_a = 2\,\mu\text{in}$ (80 μm) may be obtained.

An interesting application of precision belt grinding to produce flat surfaces is shown in Fig. 13.36. Here a metal backing plate is used in place of a contact wheel in a constant-feed rotary surface grinding operation. The belt (Fig. 13.37) has a polyester backing and hollow SiC agglomerate grits. These hollow agglomerate grits consist of a single layer of 60 screen size grits, held together by a ceramic bond. The hollow spheres are about ten times the diameter of the individual grits and are bonded to the backing by a resin make-coat and a second resin size coat, to further anchor them to the backing.

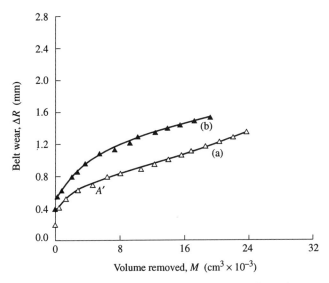

FIG. 13.38. The variation of linear belt wear ($\triangle R$) versus the volume removed (M) for (a) 60 screen size Al_2O_3 hollow spheres and (b) 60 screen size SiC hollow spheres in precision platen belt grinding of gray cast iron surfaces (after Spur and Becker 1990). Belt speed (V), 25 m s^{-1} (4920 f.p.m.); downfeed rate (d), 0.05 mm s^{-1} (0.002 i.p.s.); table speed, 72 r.p.m.; fluid, 1.5 percent emulsion.

Cast iron parts were surface ground using a 1.5 percent emulsion. After break-in the hollow abrasive particles developed flat tops containing a ring of fine abrasive particles, as shown in Fig. 13.37(b). From this point on, surface finish and removal rate proceeded uniformly, as a self-sharpening action of individual grits in the surface of the hollow spheres took place.

Figure 13.38 shows the linear wear (ΔR) versus the volume ground away for hollow agglomerate grits of SiC and Al_2O_3 for the same individual grit size (60). The Al_2O_3 belt is seen to give about one-quarter to one-third less wear than the SiC belt for a given amount ground away. The wheel life M (about $18 \times 10^3\text{ cm}^3$, or 1100 in^3) for SiC is also less than that for Al_2O_3 (about $20 \times 10^3\text{ cm}^3$, or 1220 in^3). This is because the grits are more friable than the Al_2O_3 ones, leading to a higher rate of microchipping wear.

It is also seen in Fig. 13.38 that above an initial break-in period corresponding to ΔR of about 0.7 mm (0.028 in) for Al_2O_3, ΔR varies linearly with the volume removed (M). Figure 13.39 shows a hollow sphere with individual 60 screen size grits that are about 0.3 mm (0.012 in) in diameter. Also shown are lines corresponding to $\Delta R = 0.7$ mm (end of break-in) and $\Delta R = 1.35$ mm (belt life) for the Al_2O_3 curve of Fig. 13.38. A ratio of hollow agglomerate size to grit size of 10 has been assumed in constructing Fig. 13.39. This suggests that break-in ends and the lower linear wear rate begins when about

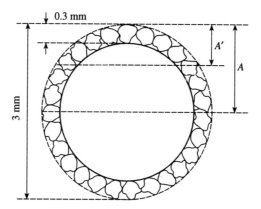

FIG. 13.39. A diagrammatic sketch of the hollow 60 screen size Al_2O_3 involved in Fig. 13.38, where A' is the value of $\triangle R$ at the end of break-in (0.7 mm = 0.028 in) and A is the value of $\triangle R$ at wheel life (1.35 mm = 0.053 in).

one-quarter of the hollow sphere has worn away, while belt life corresponds to the point at which about 50 percent of the hollow sphere has worn away.

Figure 13.40 shows the variation of tangential (F_P) and normal (F_Q) grinding forces with the rate of downfeed, $\dot{d}$. Both forces are seen to vary linearly with $\dot{d}$ and F_Q/F_P is about 3. The specific energy for this case will be

$$u = (F_P V)/(A\dot{d}) \qquad (13.37)$$

where A is the area being ground.

Solving for F_P,

$$F_P = (uA/V)\dot{d}. \qquad (13.38)$$

The slopes of the dashed lines in Fig. 13.40 ($=F_P/d$) are seen to increase as $\dot{d}$ decreases. This is to be expected, since u will not be constant as $\dot{d}$ decreases but will increase due to the size effect ($u \simeq 1/t_e^n$), where t_e is the chip equivalent discussed in Chapter 4 (also called chip sheet thickness).

For this operation the chip sheet thickness will be

$$t_e = (A\dot{d})/V \qquad (13.39)$$

and hence t_e will decrease in proportion to the decrease in $\dot{d}$, since A and V are constant for Fig. 13.40. Thus, as $\dot{d}$ decreases, u will increase, which accounts for ratio $F_P/\dot{d}$ for point A being about three times the value for point B in Fig. 13.40.

Since the area ground in Fig. 13.40 was 9011 mm^2 (Spur 1994), from eqn (13.38) the value of specific energy at point A is found to be 97.1 J min^{-3} (14.1 × 10^6 p.s.i.), while at point B, $u = 30.52$ J mm^{-3} (4.43 × 10^6 p.s.i.).

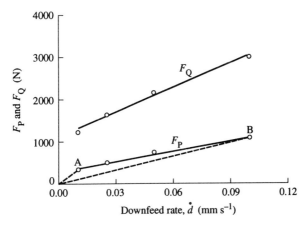

FIG. 13.40. The variation of the tangential (F_P) and normal (F_Q) grinding forces versus the downfeed rate ($\dot{d}$) for precision platen belt grinding of gray cast iron surfaces with hollow agglomerate spheres containing 60 grit size SiC abrasive particles (after Spur and Becker 1990). Belt speed (V), 25 m s^{-1} (4920 f.p.m.); table speed, 72 r.p.m.; fluid 1.5 percent emulsion.

The Future

Belt grinding has progressed in the last quarter of a century from an SRG process for finishing castings and other roughing operations to a process that is now being used in precision operations. In addition to the use of more than one layer of abrasive particles in the form of solid and hollow agglomerate spheres, patches of multilayered abrasive particles of various sizes and shapes are being used in abrasive belts. These multilayer patch designs, with considerable spaces between the particles, are being used primarily with superabrasive particles for grinding glass and other hard brittle materials. These thin patches of multilayered abrasive particles have a flat surface, and hence have an advantage over the spherical agglomerates in that there will be a shorter break-in period.

An advantage of the hollow agglomerate approach is that there is greater chip storage space between active grits cutting in the same groove. This makes it possible to grind with long uninterrupted grinding paths without overloading the chip storage space, which leads to excess temperatures, rapid wear, and poor finish. The grinding process shown in Fig. 13.30 would not be possible due to the long uninterrupted grinding paths involved were it not for the large chip storage space provided by the hollow particles.

The agglomerate patch design shown in Fig. 13.41 provides the chip storage advantage of hollow spheres without the disadvantage of break-in. It also has abrasive particles only at the leading and trailing edges of the thin patches and this conserves abrasive. Abrasive conservation is particularly important in the use of superabrasives.

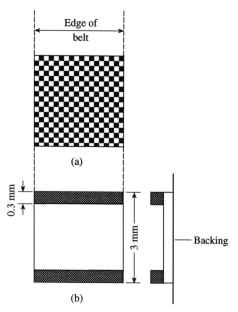

FIG. 13.41. (a) A plan view of a proposed arrangement of abrasive patches attached to an abrasive belt. (b) A plan view and side elevation of an enlarged single patch, showing the region containing a multilayer of abrasive particles in black. The recessed area provides chip clearance similar to the empty space in the hollow sphere design.

The use of rubber contact wheels of different hardness values makes it possible to change the performance of a given belt over a wide range. When a hard contact wheel is used, the belt performs in an SRG mode, but when a soft contact wheel is used it will perform in a FFG mode. Figure 13.29 shows this very clearly. If the effective hardness of contact wheels and platens could be changed instantaneously without interrupting the grinding process, a dream of the grinding community could be realized. This involves use of the same tool for roughing and finishing purposes. At present this is possible only to a very limited extent with grinding wheels, by adjusting conditions to alter the mean undeformed chip thickness ($\bar{t}$ or t_e). The unique properties of rubber make it possible to change its effective hardness reversibly and instantaneously by adjusting its environment. By the use of a layer of rubber of adjustable hardness on contact wheels and platens, a degree of control of belt grinding performance not possible with rigid grinding wheels is possible.

It appears that use of belt grinding has a very bright future and that, with imagination and further development, it is possible for it to seriously challenge grinding with rigid wheels, not only in the stock removal regime as in the past but also in the area of ultra-precision grinding.

SPECIAL PROCESSES 437

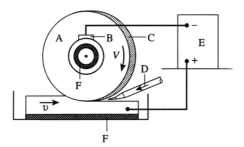

FIG. 13.42. An electrochemical grinding operation using an Al_2O_3 wheel with an electrically conducting resin bond. A, Wheel; B, brush, C, protruding nonconducting grit; D, electrolyte; E, d.c. power supply; F, insulation.

Nontraditional Operations

Introduction

After World War II a number of nontraditional removal operations emerged to deal with special problems in new industries associated with jet engines, nuclear engineering, electronics, and the exploration of outer space. These include the following:

- chemical milling, engraving, and blanking (CHM)
*• electrochemical grinding (ECG)
- electrochemical machining (ECM)
*• abrasive jet machining (AJM)
*• abrasive water jet cutting (AWJ)
*• ultrasonic grinding (USG)
- electrodischarge machining (EDM)
- wire EDM
- electron beam machining (EBM)
- laser beam machining (LBM)
- plasma beam machining (PBM)
- ion beam machining (IBM)

Only those processes marked by an asterisk involve abrasives and hence are the only ones considered here. The ultrasonic grinding (USG) process is discussed in Chapter 14, since it is used primarily for very hard brittle materials such as glasses and ceramics.

Electrochemical Grinding (ECG)

In ECG, material is removed by a reverse electroplating process. Figure 13.42 illustrates the action. The tool and work are flooded with an electrolyte and attached to a low voltage (0–15 V) high-current (300–1000 A) d.c. power supply. The negative terminal (cathode) of the power supply is attached to the tool and the positive terminal (anode) is attached to the work. The workpiece and grinding wheel must be electrically conductive, but the abrasive grits must have low conductivity. The function of the grinding wheel is:

- to provide space without contact between the conductive bond and the work through which the electrolyte can flow
- to keep the surface of the work in a clean, nascent condition
- to control the shape and finish of the surface produced

The bulk of the material is removed electrolytically as metal ions and only a small amount is removed by mechanical grinding action (less than 10 percent). Grit protrusion must be from 0.0005 to 0.005 in (10–100 μm) to prevent short-circuits between wheel bond and work. Electrons flow from the work through the external circuit and power supply to the wheel.

At the surface of the work, metal leaves as positively charged ions which migrate through the electrolyte to the negatively charged wheel, where they accept electrons and precipitate as an oxide. If these oxides are allowed to build up on the work surface, the resistance increases and the removal rate decreases. An important function of the abrasive particles is to prevent this from happening, by a scouring action.

The removal rate in ECG depends primarily on Faraday's Law, which states that in order for one atom to leave the work surface as an ion, a number of electrons equal to the valence of the ion must be removed. In order for a mole of material (atomic weight in grams) to be removed, 6.022×10^{23} n electrons must flow from the work surface to the power supply (where 6.022×10^{23} is Avogadro's number and n is the valence of the ion involved). The charge in coulombs (A s) required to remove one mole of monovalent ions in 1 s will be

$$(1.6 \times 10^{-19})(6.022 \times 10^{23}) = 96\,500 \text{ coulombs} = 1 \text{ Faraday},$$

where 1.6×10^{-19} coulombs is the charge on an electron.

The following example illustrates how the rate of removal of iron may be estimated by use of Faraday's Law. The removal of one ion of iron [Fe^{+2}] will require the removal of two electrons. The number of coulombs required to remove one gram equivalent of iron (56 g) will be $2 \times 96\,500$. Since iron has a density of 7.86 g cm^3, the rate of removal in cm^3 min^{-1} for a current flow of 1000 A will be

$$(56 \times 10^3) \times 60/(2 \times 96\,500 \times 7.86) = 2.21 \text{ cm}^3 \text{ min}^{-1} (0.135 \text{ in}^3 \text{ min}^{-1}).$$

This is a theoretical value that assumes that none of the energy is consumed as heat due to resistance to current flow at the work surface. The actual value would be about 80 percent of this.

TABLE 13.10 *Theoretical removal rates in ECG for several metals at a current flow of 1000 A (after Phillips 1986)*

Metal	Valence	Density lb in^{-3}	Density g cm^{-3}	Metal removal rate at 1000 A lb h^{-1}	Metal removal rate at 1000 A in^3 min^{-1}	Metal removal rate at 1000 A cm^3 min^{-1}
Aluminum	3	0.098	2.67	0.74	0.126	2.06
Berylium	2	0.067	1.85	0.37	0.092	1.50
Chromium	2	0.260	7.19	2.14	0.137	2.25
	3			1.43	0.092	1.51
	6			0.71	0.046	0.75
Cobalt	2	0.322	8.85	2.42	0.125	2.05
	3			1.62	0.084	1.38
Niobium	3	0.310	8.57	2.55	0.132	2.16
(Columbium)	4			1.92	0.103	1.69
	5			1.53	0.082	1.34
Copper	1	0.324	8.96	5.22	0.268	4.39
	2			2.61	0.134	2.20
Iron	2	0.284	7.86	2.30	0.135	2.21
	3			1.53	0.090	1.47
Magnesium	2	0.063	1.74	1.00	0.265	4.34
Manganese	2	0.270	7.43	2.26	0.139	2.28
	4			1.13	0.070	1.15
	7			0.65	0.040	0.66
Molybdenum	3	0.369	10.22	2.63	0.119	1.95
	4			1.97	0.090	1.47
	6			1.32	0.060	0.98
Nickel	2	0.322	8.90	2.41	0.129	2.11
	3			1.61	0.083	1.36
Silicon	4	0.084	2.33	0.58	0.114	1.87
Silver	1	0.379	10.49	8.87	0.390	6.39
Tin	2	0.264	7.30	4.88	0.308	5.05
	4			2.44	0.154	2.52
Titanium	3	0.163	4.51	1.31	0.134	2.19
	4			0.99	0.101	1.65
Tungsten	6	0.697	19.3	2.52	0.060	0.98
	8			1.89	0.045	0.74
Uranium	4	0.689	19.1	4.90	0.117	1.92
	6			3.27	0.078	1.29
Vanadium	3	0.220	6.1	1.40	0.106	1.74
	5			0.84	0.064	1.05
Zinc	2	0.258	7.13	2.69	0.174	2.85

Phillips (1986) has presented a valuable discussion of the practical aspects of ECG, including Tables 13.10 and 13.11. The first of these gives theoretical values of removal rate for a current flow of 1000 A for a wide variety of metals. The actual values of the removal rate will be less than these values by an efficiency factor that depends upon the conductivity of the work material. This will be about 90 percent for a good conductor such as copper,

TABLE 13.11 *Achievable current densities for various metals in ECG (after Phillips 1986)*

Material	Current density (A in^{-2})†
Cast iron cutting grade of carbide	500–800
Steel cutting grade of carbide	800–1000
Low-carbon steels	3000–4000
High-carbon steels	2000–3000
Stainless steels	3000–4000
Stellite	2000–3000
Udimet 500	3000–4000

† For A cm^{-2} divide by 77.5.

but appreciably less for poorer conductors such as tungsten carbide and stainless steel. Table 13.11 gives ranges of current density for different materials, beyond which there will be sparking and pitting of the workpiece. These tables are useful for planning purposes and for estimating a good starting voltage that may be raised to increase removal rate until the surface finish or quality of cut no longer meets requirements.

While elections flow in the external circuit, current flow in the eletrolyte is of course by ion diffusion, which is a much slower process. The small distance between wheel and work in ECG is advantageous from the point of view of current flow rate and hence removal rate. However, when grit protrusion is reduced by wear to the the point at which the removal rate falls off substantially due to shorting, protrusion must be reestablished. This is normally done by reversing the polarity and grinding a good conductor such as copper. This causes the bond to be deplated, thus reestablishing grit protrusion.

The removal rate increases as the gap between wheel and work decreases. This tends to yield a constant rate of removal even though the feed rate may fluctuate slightly. Nevertheless, a smooth constant low-speed feed mechanism is required in EGG, as in creep feed grinding.

Electrolytes commonly used in ECG are aqueous solutions of NaCl, KCl, or NaNO$_3$. These are not very machine friendly and corrosion resistance is an important consideration in ECG machine design. The debris that accumulates in the electrolyte is very fine (~ 1 μm), requiring expensive filtering equipment in production applications. For short-run applications, used low-cost electrolytes are discarded after use. For optimum performance the power supply in ECG should automatically maintain the current density at a constant value as the area being ground changes. Since material is removed atom by atom in ECG, it is very inefficient relative to energy consumption. The specific energy in ECG is an order of magnitude greater than in conventional grinding and two orders of magnitude greater than in metal cutting.

One of the important advantages of ECG is the ability to produce surfaces that are burr and burn free, and also free of residual stresses and subsurface damage. Since there is little mechanical removal requiring high forces, ECG is useful in finishing delicate parts where deflection of the work would otherwise be a problem.

ECG is well suited for materials that are difficult to grind, such as sintered tungsten carbide and high-temperature alloys. The hardness of the workpiece is unimportant relative to grindability in ECG. Since most of the material is removed electrolytically in ECG, tool life will be up to an order of magnitude greater than in conventional grinding.

The roughness of ECG surfaces range from $R_a = 5\,\mu m$ to $R_a = 40\,\mu m$ (0.05–1 μm) depending on the current density employed. Removal rates may be as high as $60\,in^3\,hr^{-1}$ ($983\,cm^3\,hr^{-1}$). When a very good finish is required a roughing pass at high current density (high removal rate) may be followed by one at low removal rate (low current density).

ECG has been used to perform a wide variety of operations, including those replacing milling and conventional grinding of all types, including cut-off.

Electrochemical machining (ECM) is a process that evolved from ECG, in which the scouring action of the work surface by abrasive particles is replaced by a very high rate of fluid flow. In ECM an hyrodynamic film of electrolyte separates the tool from the work, in place of the similar role played by nonconducting abrasive particles in ECG. In ECM there is no grinding action and 100 percent of the unwanted material is removed by electrolytic action.

This nonabrasive operation has been used to drill holes and for die sinking operations in materials that are difficult to machine. In the case of hole drilling, a hollow tube usually constitutes the tool, with the electrolyte pumped at very high pressure through the tube and out through the space between the electrically conducting tube and the surface of the hole being drilled. To prevent overcutting, leading to a tapered hole, the sides of the tube are insulated by a very thin layer of tightly adhering oxide on the outside diameter of the tube. An interesting application of the ECG process is production of long very small diameter cooling holes in gas turbine blades. In this case, a very thin titanium tube is the tool and H_2SO_4 is the electrolyte employed. In ECM, fluid pressures up to 200 p.s.i. (1.38 MPa) are employed.

Abrasive Jet Machining (AJM)

In AJM a high velocity stream of air (500–1000 f.p.s., or $150–300\,m\,s^{-1}$) containing very small abrasive particles is directed on to a workpiece surface. Each impinging particle removes a small amount of material. The nozzle employed is usually tungsten carbide or sapphire, and has a diameter of 0.003–0.018 in (75–450 μm). The abrasive particles usually Al_2O_3 are 15–40 μm (600–1600 μm) in size. The velocity of the impinging particles is controlled by the pressure of air supplied (30–120 p.s.i., or 0.21–0.83 MPa).

The removal rate is very small (measured in $mg\,min^{-1}$) and depends on the abrasive size, shape, and velocity, as well as the angle of impact, the distance between the nozzle and the work ($\sim$0.32 in, or 0.81 mm), and the brittleness of the work. This process is mainly used for deburring, cleaning, and surface texturing, such as the production of frosted glass surfaces, where 10 μm (0.25 μin) diameter abrasive powder is used to give about a 7 μin (0.18 μm) AA finish.

The nozzle life is rather short being about 15 hr for a tungsten carbide nozzle, but about 300 hr for a sapphire nozzle. For delicate cleaning purposes, very soft material such as sodium bicarbonate is used.

The AJM process is not widely used for precisely altering the shape of parts. It is mentioned only because it was a precursor of the more promising water jet cutting process discussed next.

Abrasive Water Jet (AWJ) Cutting

In about 1975, slitting of relatively thin materials by a high-speed water jet was introduced. Filtered water was raised to pressures as high as 60 000 p.s.i. (414 MPa) and discharged through a 0.010 in (250 μm) diameter sapphire nozzle to give a fine coherent jet with supersonic velocities as high as 3000 f.p.s. (915 m s^{-1}). A wide range of soft materials may be cut by this method, including paper, cardboard, cloth, rubber, plywood, polymers, lead, and thin sheets of aluminum. Relatively shallow grooves may also be produced in hard brittle materials (glass to depths up to 0.125 in, or 3.2 mm and granite to depths up to 0.25 in, or 6.4 mm).

The advantages of this method of cutting include the possibility of cutting in all directions with little airborne dust and temperature rise, low kerf loss and forces, and essentially no damage or contamination of the material cut (of importance to food). Jet traverse speeds (v) range from 1 i.p.s. (25 mm s^{-1}) for glass to 120 i.p.s. (3 m s^{-1}) for cardboard. The main disadvantage of the process is the relatively high cost of the equipment required.

The high pressure required is generated in an intensifier (Fig. 13.43) where pressure is increased in the ratio of the oil piston area to the water piston area. For an intensifier rate of 20:1, an oil pressure of 3 k.s.i. will give a water pressure of 60 k.s.i. Since water at 60 k.s.i. is 12 percent compressible, an accumulator is required to insure an essentially constant rate of flow. The water should be filtered and softened to prevent clogging of the small orifices (0.010–0.020 in, or 250–500 μm diameter). Figure 13.44 shows a schematic of a water jet cutting system. The function of the catcher is to receive the jet and dissipate the remaining energy at a reasonable noise level (80–90 dB).

The jet velocity (V) may be approximated by Bernouilli's equation as follows:

$$V = \sqrt{2p/\rho}, \qquad (13.40)$$

where p is the pressure and ρ is the mean density of the water.

The rate of flow (Q) will be

$$Q = C_D A V \qquad (13.41)$$

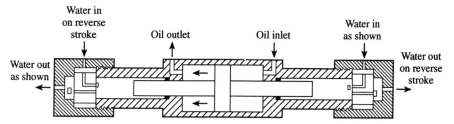

FIG. 13.43. An intensifier for increasing the oil pressure as the large piston moves to the left against water pressure acting on the small piston. At the end of the stroke, the oil flow is reversed (after Olsen 1980).

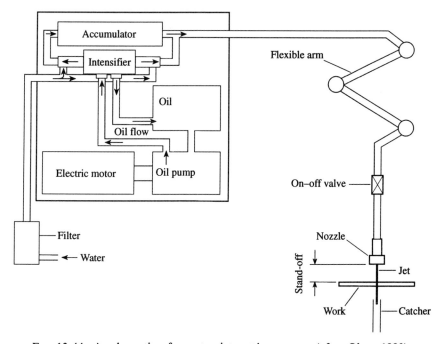

FIG. 13.44. A schematic of a water jet cutting system (after Olsen 1980).

and the fluid power required (P) will be

$$P = pQ = C_D A \sqrt{\frac{2}{\rho}} p^{1.5} \tag{13.42}$$

where C_D is the discharge coefficient of the nozzle (≈ 0.7), and A is the area of the jet. The mechanical efficiency of the pumping system will be about 70 percent.

From these equations, for a 60 k.s.i. water pressure and 0.008 in (200 μm)

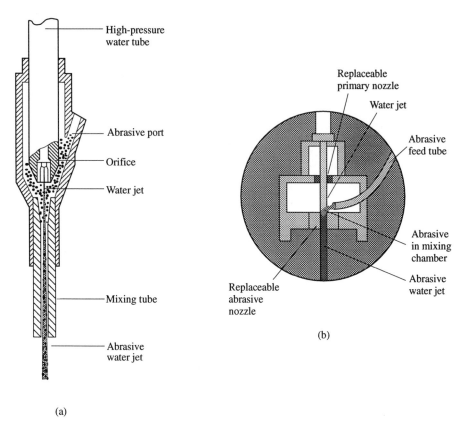

FIG. 13.45. Abrasive mixing chambers. (a) Mixing tube (after Hashish 1988a). (b) Mixing chamber with abrasive nozzle (after Hitchcock 1986).

nozzle diameter, the rate of flow would be 0.34 g.p.m. (1.29 l min^{-1}), the fluid power would be 17 h.p. (12.7 kW), and the motor power required would be about 25 h.p. (18.1 kW).

The stand-off distance (Fig. 13.44) is relatively unimportant as long as it is less than 1 in (25 mm). The depth of penetration of the jet depends upon the material cut and the transverse speed of the jet (v), increasing as the energy per length of traverse increases. The surface roughness of the cut improves as the energy per length of traverse increases. Thus, finish improves with decrease in speed of traverse of the jet when other variables are held constant. Verma and Lal (1984) have presented an experimental study of water jet cutting.

In the late 1970s, inexpensive abrasive material was incorporated into the high-velocity jet stream. This enables much deeper cuts in more difficult materials at fluid pressures comparable to those used in water jet cutting without abrasive. The abrasive is introduced down-stream from the primary nozzle as shown diagrammatically in Fig. 13.45. Here two mixing arrangements are shown. In Fig. 13.45(a) a long tungsten carbide mixing tube (length to diameter ratio = 50-100) is used, while in Fig. 13.45(b) a second sapphire or B_4C nozzle is used. In both cases, a vacuum is produced in the mixing chamber by the high-velocity jet that traverses the chamber. The life of either type of mixing arrangement is much shorter than that for the primary water jet nozzle, due to the presence of the abrasive. The abrasive particles metered into the mixing chamber at rates up to 3 lb min^{-1} (1.36 kg min^{-1}) are drawn into the water jet and accelerated to a velocity approaching that of the water.

AWJ cutting makes it possible to slit steel. The depth of cut increases linearly with the abrasive flow rate and the pressure. For deeper cuts a lower traverse rate is used with a lower stand-off distance. According to Burnham (1990), carbon steel up to 0.75 in (19 mm) thickness may be slit at jet traverse rates of 4-8 i.p.m. (10-20 cm min^{-1}). Also, nickel and titanium alloys and stainless steels may be slit in thicknesses of 1 in (25 mm) at a rate of 1 i.p.m. (2.5 cm min^{-1}). Thin titanium sheets may be machined at much higher rates. Hitchcock (1986) reports that titanium 0.063 in (1.6 mm) thick can be slit at a speed of 12 i.p.m. (30 cm min^{-1}) using a 1.5 lb min^{-1} (0.68 kg min^{-1}) of 60 garnet abrasive at a pressure of 50 k.s.i. (345 MPa) and a water flow rate of 1 g.p.m. (3.78 l min^{-1}). Gray cast iron having a thickness of 0.825 in (22 mm) may be cut at a traverse rate of 4 i.p.m. $\simeq$10 cm min^{-1} (Hitchcock 1986). AWJ cutting is used in the foundry for removing fins, gates, and risers. The surface roughness for such cuts is an acceptable 150-250 μin (2.5-6.3 μm).

The abrasives normally used in AWJ cutting are garnet or silica sand. While garnet costs about 30 percent more than the silica, garnet sand cuts about 30 percent faster than silica sand. Other abrasives are normally too expensive since, once used, grits are normally discarded.

Important operating variables influencing the depth of cut in a single slitting pass are pressure, traverse rate, abrasive flow rate, grit type and size, and water jet diameter. Figure 13.46 shows how the depth of cut varies with fluid pressure for different abrasive rates when slitting mild steel at a consant traverse rate and garnet grit size. The depth of cut is essentially linear with increase in pressure to 40 k.s.i. (0.28 MPa) and then falls off. Hence, AWJ cutting is usually conducted at a pressure of from 30 to 40 k.s.i. (0.21-0.28 MPa). Figure 13.47 shows the variation of slitting depth with traverse speed when cutting Ti-6Al-4V at a constant pressure and garnet grit size, but with different values of jet diameter and abrasive flow rate. Figure 13.48 shows similar results when slitting 304 stainless steel at variable pressures, but constant orifice diameter and garnet flow rate. Hashish (1985) has found that the best results are obtained in AWJ cutting with a grit size in the 60-80 range.

When greater depths of cut than those shown in Fig. 13.46-13.48 are desired, multiple-pass cutting is an option. However, the depth of cut per pass falls off with subsequent passes, due to an increase in the stand-off distance and interaction of the sides of the previous cut with the jet.

Hashish (1988a) has visually studied the action of abrasive water jets by

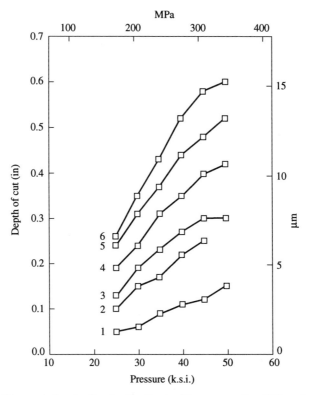

FIG. 13.46. The variation in the depth of cut with pressure for different abrasive flow rates (after Hashish 1985). 1, 0.07 lb min^{-1} (32 g min^{-1}); 2, 0.15 lb min^{-1} (68 g min^{-1}); 3, 0.17 lb min^{-1} (77.2 g min^{-1}); 4, 0.23 lb min^{-1} (104 g min^{-1}); 5, 0.32 lb min^{-1} (145 g min^{-1}); 6, 0.41 lb min^{-1}). Mild steel slit at 6 i.p.m. (15.2 cm min^{-1}) with a 0.010 in (0.254 mm) diameter nozzle and 80 grit garnet.

cutting transparent plastics and glass and recording the results cinematically. It was found that cutting near the surface is steady, but as the jet proceeds below the surface cutting becomes cyclic, producing striations that have a pitch corresponding to the distance traveled by the jet in the time it takes the jet to traverse a distance equal to the diameter of the jet. In nonthrough cutting, the bottom of the cut shows similar waviness, and this is one of the reasons why AWJ cutting is not a viable substitute for end milling (Hashish 1989).

Figure 13.49 shows the arrangement used by Hashish (1987) for AWJ turning. While removal rates could approach those for conventional turning, instability of the jet leading to waviness is a problem. This usually requires a finishing cut with fine abrasive particles and a low traverse speed to meet

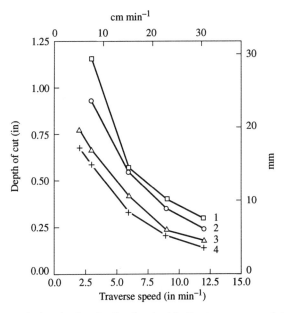

FIG. 13.47. The variation in the depth of cut with the traverse speed for different jet diameters and abrasive flow rates when slitting Ti-6Al-4V with 80 grit garnet at a pressure of 30 k.s.i. (207 MPa) (after Hashish 1985). 1, 0.018 in (0.46 mm) jet diameter and 0.54 lb min^{-1} (245 g min^{-1}); 2, 0.014 in (0.36 mm) jet diameter and 0.54 lb min^{-1} (245 g min^{-1}); 3, 0.010 in (0.25 mm) jet diameter and 0.54 lb min^{-1} (245 g min^{-1}); 4, 0.007 in (0.18 mm) jet diameter and 0.32 lb min^{-1} (145 g min^{-1}).

finish requirements. However, materials that are very difficult to machine, such as a Mg/B$_4$C metal matrix composite, are about as easy to turn by AWJ as ordinary steel. Thus, rough turning of materials that are very difficult to machine by AWJ, followed by a conventional finishing operation where needed, may be a useful option.

A Mg/B$_4$C metal matrix composite with 15 percent B$_4$C was turned by AWJ from a diameter of 25 mm (1 in) to 6 mm (0.24 in) in one pass at a feed rate of 7.2 mm min^{-1} (0.28 i.p.m.). This required a jet power of 13 kW (17.4 h.p.) and a flow rate of garnet abrasive of 4.8 gm min^{-1} (0.01 lb min^{-1}) (Hashish 1987). In AWJ turning the force on the work is negligible, making it possible to turn parts to very small diameter without deflection. The rotational speed of the work is not important in AWJ turning and the work may be rotated in the positive or negative directions indicated in Fig. 13.49. Hashish (1987) found that the finished diameter showed somewhat less waviness when the work was rotated in the positive direction. The debris in AWJ turning consists of fine powder.

Hashish (1989*b*) found that AWJ could drill very small holes in glass to

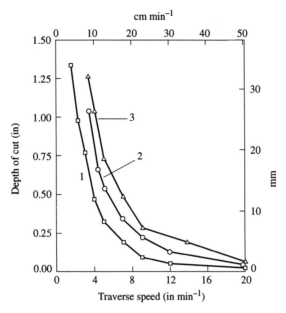

FIG. 13.48. The variation in the depth of cut with the traverse speed for different pressures when slitting 304 stainless steel with garnet at a flow rate of 2 lb min^{-1} (908 g min^{-1}) and a jet diameter of 0.020 in (0.51 mm) (after Hashish 1985). 1, 20 k.s.i. (0.138 MPa); 2, 30 k.s.i. (0.207 MPa); 3, 40 k.s.i. (0.276 MPa).

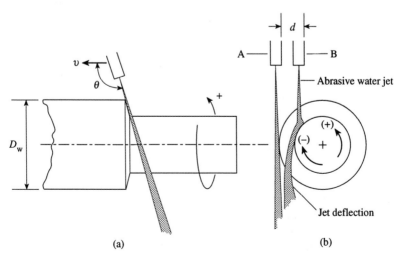

FIG. 13.49. A schematic of an AWJ turning operation. A and B in (b) show the jet progressing from the outside diameter toward the center as the depth of cut d is generated at the beginning of a cut (after Hashish 1987).

TABLE 13.12 *Relative values of cost per hour for a representative AWJ foundry cutting operation (based on Metcut 1986)*

Item	Percentage
Operator and overheads	35
Machine†	25
Abrasive	15
Water nozzle	Negligible
Abrasive nozzle	10
Pump maintenance	10
Power	5
	100

† Amortized over 5 yr (10^4 hr).

considerable depth rather quickly. For example, a 0.5 mm (0.020 in) diameter hole was drilled through a piece of glass 1 in (25 mm) thick in 10 sec.

In summary, it would appear that AWJ cutting is well suited to through-slitting of materials that are difficult to machine involving straight cuts or those of complex shape. It is particularly useful for expensive materials where kerf loss is important and where edges of high surface integrity that are burr-free are required. AWJ also appears to be a good way to drill deep holes of small diameter in hard, brittle materials.

Metcut (1986) has presented representative cost figures for a typical AWJ foundry cutting operation. However, rather than present absolute values here, relative values are presented in Table 13.12, since these are more apt to be meaningful over a longer time period. The abrasive, abrasive nozzle, and pump maintenance costs are substantial, amounting to about 35 percent collectively.

A competing technology for slitting sheet metal and for foundry use is plasma beam machining (PBM). Metcut (1986) has presented Table 13.13, which compares the principal characteristics of the PBM and AWJ cutting methods.

Deburring, Lapping, and Polishing

Burr Technology

Burrs that form on edges of cast, blanked, machined or ground parts often play an important role in the failure of a part due to fatigue, or gross fracture. Difficulty may also be involved with debris between surfaces associated with small clearance, making accurate measurements, and in the assembly of parts. Beginning in the early 1970s, the technology of burr formation, prevention, and removal began to be studied and, since then, considerable progress has been made relating to the production of quality edges.

TABLE 13.13 *A comparison of the characteristics of plasma beam machining (PBM) and abrasive water jet machining (AWJ) for sheet metal and foundry use (after Metcut 1986)*

Condition	PBM	AWJ
Cutting rate	+	
Kerf	=	=
Heat-affected zone		+
Surface roughness	+	
Noise	=	=
Cleanliness/dust		+
Radiation/safety		+
Equipment cost	+	
Reoccurring costs	+	
Size/weight of unit		+
Work thicknesses feasible	=	=
Auxiliary systems/floor space	+	
Hand-held units not as accurate	=	=

+, Advantage; =, equality.

The first priority is to produce edges that meet specifications relative to soundness and roundness or sharpness directly by proper design and production procedures. When this cannot be achieved, a deburring operation is called for. Since deburring often involves some form of abrasive action, it is briefly discussed here. Electrochemical grinding is an example of a removal operation that is burr-free.

A common method of deburring edges, particularly for large parts such as castings, is by grinding with a wheel or belt. Smaller parts are often deburred by agitating a mixture of parts and small bonded abrasive pieces that resemble cutting tool inserts in size and shape in a ball mill, in a vibrating container, or on a vibrating belt. The weak burr material is removed by a combination of mild impact and abrasive action. This is a complex process that can be very effective but is highly empirical.

Wire brushing is another method used for deburring edges. For very coarse deburring, as with castings, wheels containing twisted groups of heat-treated medium carbon steel wires are used. Burrs are removed in this case by considerable impact and some wiping action. For edges that require a less drastic deburring action, single-strand, crimped, or straight steel wires of smaller diameter are used. Wires of a wide variety of other materials may be used, including stainless steel, titanium, aluminum, brass or even natural fibers (tampico), and polypropylene. Nylon co-extruded with fine ($\sim$ 120 grit size)

SPECIAL PROCESSES 451

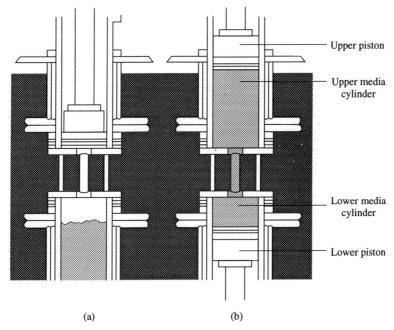

FIG. 13.50. A schematic of an AFM machining operation. the abrasive medium is shown in the lower cylinder (a) and after being transferred to the upper cylinder (b). The abrasive medium is cycled back and forth until the desired sizing, radiusing, deburring, or polishing is obtained (after Rhoades 1988).

SiC or Al_2O_3 abrasive particles is used in fine deburring brushing operations. For a grit size of 120 the filament diameter would be about 0.050 in (1.25 mm) and the unsupported filament length up to 8 in (200 mm). With so-called long string brush deburring, the action is quite gentle. In mass production brush deburring operations it is advantageous to use a numerically controlled automated system. The rate of deburring increases with brush speed, and brush life increases as the applied pressure decreases. A good burr removal rate and good life is obtained with a combination of high brush speed and light pressure.

Abrasive Flow Machining (AFM)

Abrasive flow machining is a light finishing operation that employs a thixotropic putty-like silicone material containing fine abrasive particles (usually SiC) as the abrasive medium. The basic principle is illustrated in Fig. 13.50. The equipment consists of upper and lower pumps and fixturing to guide and restrict flow of the pseudo-solid abrasive medium through and across the surfaces to be sized and deburred or polished.

The apparatus is usually highly automated to enable the required number of back and forth cycles to be accomplished without attention. The apparatus is capable of reducing surface finish to 10 percent of the initial value in a short time. The removal rate is controlled by the volume rate of flow of the abrasive medium, which is influenced by the viscoelastic properties of the grade of silicone employed, the size and concentration of abrasive particles, the pressure employed, and the clearance provided between fixturing and part. All of this is a highly empirical selection process.

This technique is particularly useful for deburring and polishing regions that are hard to reach, such as small passages in intricate valving designs. The process is capable of machining very thin sections without deflection by arranging balanced pressure on opposing surfaces of the part. The process can accurately produce tangential radii from 0.025 to 1500 mm (0.001–0.060 in).

For deburring and radiusing an abrasive medium of relatively low viscosity is used, while for polishing a more viscous material is used to provide a uniform rate of flow. The viscosity of the abrasive medium depends upon the mean l/d ratio of the passages involved. For $l/d < 2$, a more viscous medium is used than otherwise. The abrasive used is normally SiC or B_4C, ranging in size from 0.0002 to 0.060 in diameter (5–1500 μm). Finer abrasive particles give a lower rate of removal but a better finish. Pressures involved vary from 100 to 3200 p.s.i. (0.69–22 MPa). Usually, simple tooling is required for dies, but carefully designed tooling is required to direct and monitor the flow for external surfaces. Surface finish improves with decrease in grit size and pressure and exponentially with the number of stokes employed.

Ultrasonic Cleaning

Lindbeck (1952) first described the use of ultrasonics for cleaning and burr removal for small high-precision parts. It was suggested that the oscillating energy should be supplied to the fluid rather than directly to the part. Frequencies from 20 kHz to 50 kHz, provided by a magnetostrictive generator consisting of a stack of nickel plates subjected to an oscillating magnetic field, were recommended. Cavitation impact forces which require a radiation density of about $7\,kW\,cm^{-2}$ at the cavitation threshold constitute the removal mechanism involved. This process is widely used today for cleaning small parts.

Superfinishing

Superfinishing is related to honing, in that a shaped bonded abrasive stick is used, as illustrated in Fig. 13.51. This process differs from conventional honing in that the applied pressure and oscillating stroke are smaller. The part is rotated at a rate that, when combined with the oscillatory velocity, yields a criss-cross scratch pattern. This process is not for size control but rather for fuzz removal and the refinement of surface finish. When a carefully filtered

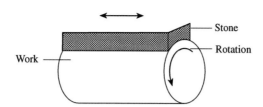

FIG. 13.51. A superfinishing operation.

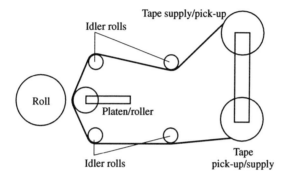

FIG. 13.52. An arrangement used to superfinish rolling mill rolls using film-backed coated abrasive tape (after Visser and Dinberg 1988).

water-based coolant is used, an R_a finish of 2–3 μin (0.05–0.075 μm) is possible. In order to reduce processing time, a sequence of abrasive grit sizes is often used. Oscillating frequencies up to 1650 min^{-1} and rotational speeds up to 400 f.p.m. (122 m min^{-1}) are used.

This process was first developed in the 1930s to prevent front-end automotive roller bearings from being damaged when cars were shipped by rail across the country. It was found that superfinishing the rollers solved the problem.

Special film-backed (0.005 in = 125 μm polyester) abrasive belts having micron-sized abrasive particles (60–9 μm) of Al_2O_3 are used for the final finishing of rolling mill rolls by superfinishing, using an arrangement such as that shown in Fig. 13.52.

Lapping

Lapping is a process involving loose abrasive particles that become embedded in a substrate, often from a slurry. Frequently, the substrate is cast iron, but other materials such as copper, pitch, tin, or even polymers and cloth may be used. Lapping machines usually involve complex motion between lap and work, so that a criss-cross scratch pattern is obtained (Fig. 13.53). Depending

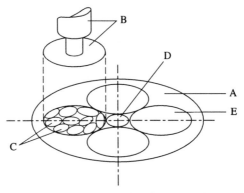

FIG. 13.53. A schematic of a lapping machine for producing flat surfaces: A, lap; B, pressure plate and air cylinder; C, workpieces; D, driven sun gear; E, geared retaining ring.

on the hardness of the workpiece, lapping pressures may be as high as 150 k.s.i. (1.0 MPa). When flat surfaces are involved, opposing surfaces are often lapped simultaneously using upper and lower lapping plates. When lapping curved surfaces as with glass lenses, shaped laps are used.

Fine lapping, as in the production of gage blocks, usually involves a progression of grit sizes, and in such cases it is important that all traces of a coarse abrasive be removed before going to the next finer stage. It is also important that lapping pressures and surface speeds (temperatures) are not too high. Otherwise, fine abrasives are apt to be sintered into much larger ones, giving rise to scratches that are unexplained by the initial grit size. In one attempt to increase production rate in the lapping of gage blocks, surface speed and lapping pressure were increased. When deeper than expected surface scratches were encountered, microscopic examination of the fluid revealed particles an order of magnitude larger than the initial abrasive size. These large particles of Al_2O_3 were found to be magnetic and were observed to consist of an assembly of smaller particles separated from each other by a thin black network. The problem was that, under the higher temperature and pressure conditions, the material lapped away was oxidized to black (magnetic Fe_3O_4) iron oxide, which acted as a sintering aid for the Al_2O_3 particles. Reluctantly, the removal rate had to be reduced to prevent the sintering action. In this case a low-melting eutectic spinel, produced by reaction of Al_2O_3 and Fe_3O_4, acted as the bonding agent in the sintering process. It was found that the addition of a surface active agent (on active sulfur compound) to the lapping oil helped to prevent sintering and enabled sinter-free performance at an intermediate lapping rate.

Loladze et al. (1982) have performed a finishing operation on very hard materials (cemented carbides, ceramics, quartz, silicon, etc.) using the arrangement shown in Fig. 13.54. This is essentially a lapping operation, with the face

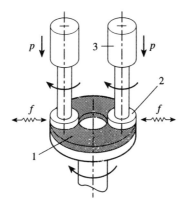

FIG. 13.54. A schematic of an LPG finishing method in which a bonded grinding wheel is used in place of a lap: 1, grinding wheel; 2, cassette retaining work materials; 3, feed device.

of a bonded grinding wheel replacing the lap. Workpieces are retained in cassettes that are oscillated radially at a frequency of 500–4000 c.p.m. Wheel speed was very low ($\sim 1\,\text{m/s}^{-1} \simeq 200\,\text{f.p.m.}$), as was the normal pressure (~ 0.2–$9\,\text{kg/cm}^{-2}$, or 2.8–1.25 p.s.i.). Surface integrity was unusually good even at higher removal rates ($\sim 200\,\mu\text{m}\,\text{min}^{-1}$ for WC, $\sim 125\,\mu\text{m}\,\text{min}^{-1}$ for ceramics, and $500\,\mu\text{m}\,\text{min}^{-1}$ for monocrystalline silicon). The very good suface integrity is due to the low surface temperatures pertaining (approximately room temperature) and the authors refer to this as low-temperature precision grinding (LPG).

Funck (1991) has also used a bonded vitrified grinding wheel (grit size about 180 and J grade) and grinding oil to replace the lapping disc while maintaining the geometry and kinematics of the lapping process. Lapping speeds of 100–200 m min^{-1} (328–656 f.p.m.) were used, with contact pressures from 10 to 100 kPa (1.5–15 MPa). A criss-cross pattern typical of lapping was obtained, which is advantageous relative to the chip storage problem characteristics of processes involving long grit–work contact arcs. Surface finish and flatness values comparable to those obtained in conventional lapping were obtained.

Free Abrasive Grinding

Like lapping, free abrasive grinding invoves unbonded abrasive particles. However, instead of the abrasive becoming embedded in a soft lap, a hard ($R_c = 60$–62) backup wheel is used, so that the particles remain free to roll between the plate and the work. This provides a longer abrasive life, since all sides of the abrasive are used, and the possibility of a high removal rate with good finish, accuracy, and flatness is achieved. The process is capable of producing flat surfaces up to 6 ft (1.83 m) in diameter, parallel to 0.0001 in

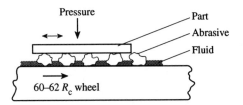

FIG. 13.55. A detail of rolling abrasive particles in slurry between work and a hard surface in a free abrasive grinding operation.

(25 μm), a flatness of 80 μin (2 μm), and an R_a finish of 45 μin (1 μm) for Al and 60 μin (1.5 μm) for hard steel. The machine arrangement used is the same as that in Fig. 13.53 for a surface lapping machine, except that a hard backup plate replaces the lap. Figure 13.55 shows a single workpiece with loose abrasive particles in a slurry separating wheel and work. A sun gear rotates in alternate directions and the hard wheel rotates continuously in one direction. This improves flatness. The lower plate is water-cooled to prevent thermal distortion.

A very wide range of materials may be finished by this method, including tungsten carbide, steel, titanium, brass, aluminum, glass, rubber, nylon, and Teflon. Like surface lapping, no special fixtures or waxing are required and low surface temperatures are involved, giving rise to surfaces of high integrity. For free abrasive grinding of steel, the pressure required on the workpiece is about 3 p.s.i. (2.07 N cm^{-2}). A more detailed description of the Free Abrasive Grinding process is to be found in Zorn (1983).

Magnetic Field Assisted Finishing

Traditional fine finishing operations such as grinding, lapping, or honing employ a rigid tool that subjects the workpiece to substantial normal stress. When finishing hard brittle materials such as Al_2O_3, Si_3N_4, or glass, the normal force on individual abrasive particles is high and often causes subsurface cracks to develop at relatively low removal rates. Magnetic field assisted grinding, which was first used in the 1940s but not studied seriously until the 1980s (Fox *et al.* 1994), provides a means of polishing and finishing at a greatly reduced force per grit.

Two methods of application have been employed:

- magnetic abrasive finishing
- magnetic fluid finishing

In the first method, relatively large particles of iron (~ 40 grit size) are sintered with much finer Al_2O_3 or SiC abrasive particles (~ 400–1200 grit

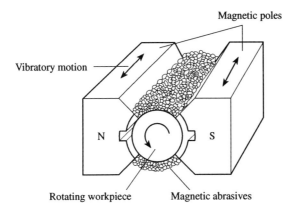

FIG. 13.56. A schematic of a magnetic abrasive process (after Fox *et al.* 1994).

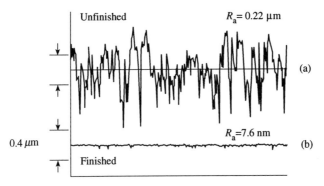

FIG. 13.57. The surface finish of (a) as-received stainless steel rod, and (b) after magnetic field assisted finishing (after Fox *et al.* 1994).

size). Figure 13.56 shows a typical application for finishing a cylindrical workpiece. The composite magnetic abrasive particles surround the workpiece, which is suspended between two magnetic poles either with or without a solid lubricant. The workpiece is rotated and the magnetic poles are given an oscillating motion (15–25 Hz). The electromagnets consist of low-carbon steel wound with copper wire, through which a current of 0.5–5 A flows. The workpiece may be either a ferromagnetic or nonmagnetic material. In the case of a nonmagnetic material, the magnetic field density will be about one-quarter that for a magnetic workpiece and hence a higher current flow is required for the same rate of removal. Figure 13.57 shows the improvement in surface finish that may be achieved for a nonmagnetic stainless steel specimen by the magnetic abrasive finishing technique in going from a ground

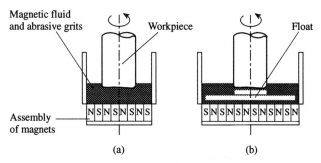

FIG. 13.58. Magnetic fluid grinding: (a) without a float; (b) with a float (after Umehara 1994).

to a polished condition in about 3 h. The use of a solid lubricant (~5 w/o zinc stearate) mixed with the sintered magnetic abrasive particles gives a better finish. Increasing the magnetic flux up to the saturation level causes a more rapid approach to the best finish obtainable and lowers the magnitude of this finish. The removal rate increases with the rotational speed of the work and the oscillating frequency should be adjusted to give a criss-cross pattern corresponding to an included angle of 30–70°.

Anzai *et al.* (1993a) have described a procedure for making iron-bonded diamond magnetic pellets. Iron and graphite powders are ball miled with stainless steel balls for 10 minutes; then 45 v/o 6.7 μm diamond and 1 w/o zinc stearate is added, followed by 10 minutes more ball milling. The material is then pressed to 490 MPa (3380 k.s.i.) and sintered for 0.5 h at 1423 °C in an NH_3 atmosphere. The finished slug is then crushed and screened. Tests were conducted using 212–300 μm (0.008–0.012 in) diameter pellets with 45 w/o of 6.7 μm (268 μin) diamond (Anzai *et al.* 1993b).

In the second method, unbonded abrasive is mixed with colloidal magnetic (iron) powder to form a magnetic fluid (Imanaka *et al.* 1981). Figure 13.58(a) shows the method of applying this technique for polishing the end of a rotating nonmagnetic rod. The assembly of permanent magnets shown interacts with the iron particles, creating a small upward force of abrasive particles against the work. However, this force is insufficient to give a practical rate of removal. It was found that use of a nonmagnetic 'float', as shown in Fig. 13.58(b), causes a larger pressure (about two orders of magnitude greater) to be obtained at the grit–work interface. The pressure obtained depends upon the total thickness of the float, which controls the distance between the float and the magnets. Umehara (1994) has discussed the mechanics of magnetic fluid polishing with a float. Since there is no contact between the float and the work, very high work speeds (up to 10^4 r.p.m.) may be used without the danger of vibration-induced impact. Since the abrasive is free to rotate, all sides are exposed to the work, giving a longer grit life than with a fixed abrasive system. The cooling action of the water-based fluid further increases grit life. Figure 13.59 shows the application of magnetic fluid finishing with

SPECIAL PROCESSES 459

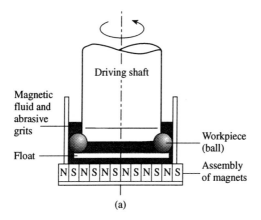

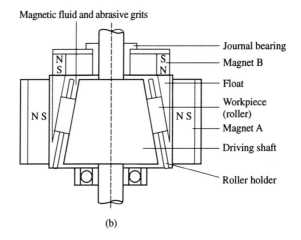

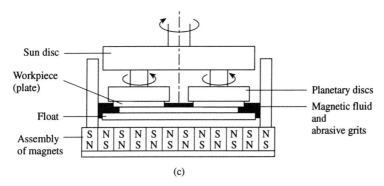

FIG. 13.59. Magnetic fluid grinding of Si_3N_4 parts with a float: (a) balls; (b) rollers; (c) plate (after Umehara 1994).

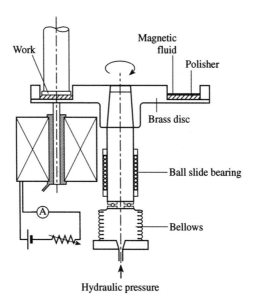

FIG. 13.60. A magnetic fluid polishing set-up (after Kurobe *et al.* 1986).

a float to silicon nitride balls, rollers, and plates. For Si_3N_4 balls, sphericity was reduced from 50 μm (2000 μin) to 0.14 μm (5.6 μin) in about 3 hr, at which time the peak-to-valley roughness (R_t) was 0.1 μm (4 μin).

Kurobe *et al.* (1986) have described a process of magnetic field assisted 'lapping'. The lap (polisher) was a 3 mm (0.12 in) thick polyurethane sheet, containing a close-packed array of 5 mm (0.20 in) diameter holes. A magnetic field was developed between a stationary electromagnet and a rotating iron spindle to which the workpiece is attached, as shown in Fig. 13.60. The magnetic fluid consisted of a liquid suspension of fine (< 150 Å) Fe_3O_4 particles and the abrasive was 40 μm Al_2O_3 in a concentration of 11 volume percent. The number of active grits was found to increase with the current (field strength), and this resulted in up to a three-fold increase in removal rate when the work material was soda lime glass.

Anzai *et al.* (1993b) have used their magnetic abrasive in a face milling mode to deburr small milled parts on an NC milling machine. A sintered magnetic grit size of 212–300 μm (0.0085–0.012 in) and one pass was satisfactory for deburring milled parts; but a better finish was obtained with a 425–600 μm (0.017–0.024 in) grit size, although up to six passes were then required for complete burr removal. In these tests a gap between electromagnet and work of 1.6 mm (0.064 in) was used. The optimum feed rate was 5 mm/min^{-1} (0.2 i.p.m.), with a high electromagnet rotational speed.

FIG. 13.61. A cushioned coated abrasive product (after Archer 1992).

Polishing

In the technical processing literature, a distinction is made between polishing and buffing. Polishing is used to describe a process in which improved performance is the objective, while buffing is more related to improved esthetics and hence marketability. Polishing, as generally understood, employs abrasive bonded to a backing and used as a belt or flap wheel, or the abrasive is distributed throughout the web of cloth discs used in flap wheels for three-dimensional polishing. A scratch pattern remains and the finished surface is not reflective. An important objective of polishing is to produce a finished surface with desired geometry, with the finest scratch pattern to be removed by further processing. Cushioned abrasive belts and discs are useful in this connection. Figure 13.61 shows the structure of a cushioned coated product. The resilient layer allows abrasive particles initially at different levels to recede to a constant level under pressure. This produces a more uniform scratch pattern of lower maximum depth for the same removal rate.

Buffing generally refers to finishing with more flexible cloth wheels impregnated with finer and sometimes softer (rouge = Fe_3O_4, tripoli, pumice, etc.) abrasives operated at lower r.p.m. to make the rotating wheels less rigid. A visible scratch pattern is not evident after buffing, and the resulting surface will normally be highly reflective. Such surfaces are sometimes said to have a mirror finish.

The traditional production of surfaces for metallographic examination illustrates the steps normally taken to produce a flat specimen with mirror finish. After rough grinding, the surface to be examined is polished by use of a coated product having progressively finer grit sizes, usually in four steps. After each step the specimen is polished until all previous scratches are removed, and then rotated 90° before the next step. Polishing is followed by one or more buffing steps, using a very fine loose abrasive applied to the surface of a flat layer of cloth or felt, attached to a metal disc. The abrasive is usually applied as an aqueous slurry.

There has been considerable speculation concerning the removal mechanism involved in buffing. Some have believed that it is merely a leveling of the surface without removal by melting or plastic flow (for example, Beilby 1903), while others (Aghan and Samuels 1970) suggest that buffing is the same as polishing but with finer chips. A third possibility, suggested by Rabinowicz (1971), is that the final stage of polishing involves removal of material by adhesive wear on a very fine scale. The first of these may be ruled out by a measurable weight loss during buffing, and the second by the presence of scratches in polishing but their absence in buffing. This suggests an entirely different mechanism for polishing and buffing, which is consistent with the proposed mechanism of Rabinowicz. The precision finishing of glass has received a great deal of scientific attention over the years because of its importance to optics, and results that have been obtained lend considerable support to an adhesive wear mechanism for buffing.

Precision Polishing of Glass

In the glass industry, polishing refers to the production of lustrous surfaces with negligible change of dimension or shape. This is called buffing in the metals industries. To avoid confusion, the term 'precision polishing' will be used when discussing the buffing of glass.

Cook (1990) has presented a review of the precision polishing of glass in which the case is made for a chemico-mechanical approach. Traditionally, the precision polishing of glass employs, a viscoelastic pitch lap and an aqueous abrasive slurry. The presence of water is essential, even diamond having a low polishing rate in the absence of water. Water diffuses into the surface of the glass, converting SiO_2 to $Si(OH)_4$, which bonds to the abrasive and is removed. The basic mechanism involves elastic impact of the surface, bonding of the abrasive, and subsequent removal of the hydrated surface material on unloading. This is essentially the mechanism proposed by Rabinowicz, except for the surface hydration aspect.

Different abrasive materials yield different polishing rates due to their adhesive bond to their hydrated SiO_2 surface strength. Cerium oxide gives the highest removal rate, followed by ThO_2 ($\sim$ 80 percent of CeO_2), TiO_2 ($\sim$ 60 percent and ZrO_2 ($\sim$ 50 percent of the CeO_2 rate). SiO_2 will even polish hydrated glass, but at a low rate except at a high pH value (> 10). However, SiO_2 is used to polish Si in the semiconductor industry. In the precision polishing of glass, water enters the surface, depolymerizes silica, and repolymerizes the hydrated silica on the surface, which must be removed by adhesive transfer to the abrasive particle.

The precision polishing rate of glass also depends on the chemistry of the glass. The polishing rate decreases with an increase in chemical durability which, in turn, decreases with the alkali content of the glass. The polishing rate also depends on the rate of diffusion of molecular water into the surface of a given glass composition and also the rate of SiO_2 dissolution under the pressure and temperature conditions pertaining. The precision polishing rate of glass also depends on the rate of repolymerization of removed silicate ions leaving the trailing edge of the abrasive back on to the polished surface. Chelating agents such as 2 w/o ammonium molybdate added to an aqueous solution tend to form colloids of the removed silicate, thus decreasing redeposition. Very fine wood flour added to a pitch lap also tends to enhance collold formation, thus decreasing redeposition.

Parks and Evans (1994) have investigated the influence of backing material

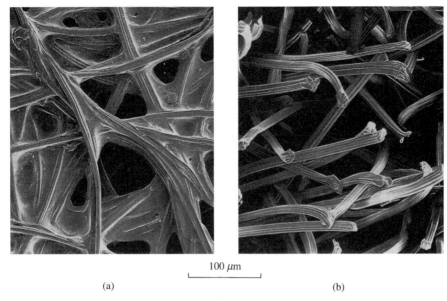

FIG. 13.62. SEMs of nonwoven material supporting free abrasive: (a) synthetic fabric; (b) green felt (after Parks and Evans 1994).

on the rate and finish obtained in precision finishing of diamond-turned electroless nickel on an aluminum substrate. Several cloth and felt materials with and without a resilient foam backing were used. A nonwoven synthetic fabric with foam backing was found to give the best precision polishing rate of all combinations investigated, including green and red felt and foamed polyurethane and the unwoven fabric. Figure 13.62 shows SEMs of the nonwoven synthetic fabric and the green felt. While a marked difference in structure is evident, no reason was suggested for the former giving a better removal rate than the latter. Two additional observations were made:

1. The polishing rate was greater when precision polishing was carried out parallel to the grinding lay than when carried out in the orthogonal direction.
2. Finish improved as previous grinding scratches disappeared, but then decreased slightly with further removal.

The first of these is surprising and remains unexplained, while the second is probably due to redeposition of material already removed.

Elastic Emission Machining (EEM)

Tsuwa *et al.* (1979) devised an ingenious method of precision polishing with numerical control (NC) of geometry. The principle is illustrated in Fig. 13.63.

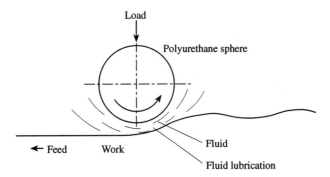

FIG. 13.63. The principle of the elastic emission machining (EEM) process (after Tsuwa *et al.* 1979).

A polyurethane rubber sphere rotating at high speed progresses over the surface to be polished without contact, guided to maintain a constant clearance of a fraction of a millimeter (a few thousandths of an inch) by an NC machine. This constitutes an hydrodynamic bearing with a slurry of ultrafine abrasive particles of Al_2O_3 or ZrO_2 (0.02–20 μm, or 0.8–800 μin) in water as the lubricant. The Reynolds number is sufficiently large (> 2000 based on the clearance; Shaw and Macks 1949) to give turbulence. This causes abrasive particles to be projected against the surface being ground with sufficient energy to give *elastic* penetration.

Operating conditions are such that the volume elastically deformed is sufficiently small that no dislocations or dislocation sources are encountered and thus the material remains perfectly elastic. If the work material is an oxide it will adhere to an oxide abrasive rather strongly, and when unloaded and the stress reverses a small particle will be removed due to *transverse* crack propagation provided that the abrasive-to-work bond strength is greater than the work-to-work bond strength. Since no plastic flow is involved, there will be no subsurface deformation or residual stress; and no subsurface cracks, since only transverse cracking is involved on unloading in this micro-removal process. The individual particles removed will be of atomic size, resulting in a surface finish that is close to the maximum achievable value (about $R_a = 5$ Å).

It is interesting to note that this method of precision polishing without surface damage depends on the 'size effect', whereby the yield stress of a material increases with a decrease in the loaded specimen size due to a decrease in the probability of encountering a dislocation or a dislocation generating source. If the loaded specimen value becomes small enough there can be no plastic flow, even for a material such as copper (which is very ductile when tested in the usual size range).

It was found by accident (Herring and Galt 1952) that very small samples of metals free of dislocations may be produced by elecrodeposition at very low

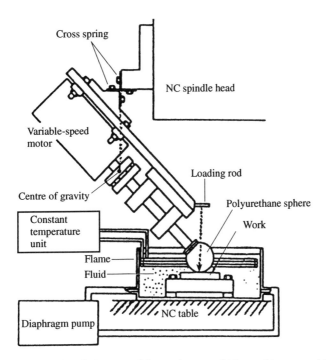

FIG. 13.64. A schematic of the assembly used on an NC machine to guide an EEM tool over the surface being precision polished without solid contact (after Tsuwa *et al.* 1979).

values of potential. These 'whiskers' are produced so slowly that there is plenty of time for each atom to find its proper place and there are no accidents of growth (dislocations). When such whiskers are tested in bending they:

- are elastic to the point of fracture (no plastic flow)
- are nonlinearly elastic, the elastic stress–strain curve resembling the first quarter cycle of a sine wave
- have a very large strain at fracture, approaching $\gamma = 0.25$

Figure 13.64 shows the arrangement for applying the tool to an NC machine to gulde the rotating ball over the surface previously produced on the same machine. The control variables are the speed of the sphere and the load applied to the loading rod to counterbalance the upward hydrodynamic force component.

Namba (see McClure 1985) has developed the version of EEM precision polishing shown in Fig. 13.65 at the High Energy Physics Laboratory at Tsukuba Science City, Japan. This is designed to produce ultra-smooth flat surfaces for large-scale integrated circuit chip production. In this case the

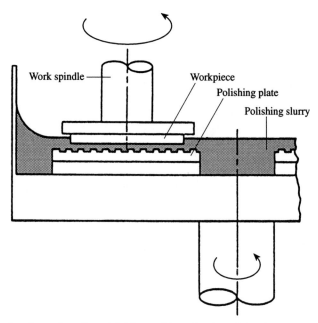

FIG. 13.65. A schematic of an EEM machine developed by Namba for producing ultra-flat and smooth precision polished surfaces (after McClure 1985).

work floats on a hydrodynamic film and an NC machine is not required for guidance of the work over the tool.

The abrasive used by Namba was ultra fine (~ 60 Å) SiO_2. The rotating tool is a tin disc with concentric grooves to produce turbulence in the clearance space between tool and work. Both tool and work rotate at 50–100 r.p.m. The machine has produced glass samples that have an R_a surface finish that is estimated to be 2 Å (0.0008 μin). In all applications of EEM precision polishing, it is important to maintain the temperature of the slurry constant.

If EEM is to be applied to precision polishing of metals, it may be necessary to change the fluid to increase adhesion between the oxide abrasive and the metal. It appears that a fluid (liquid or gas) that would oxidize the freshly polished surface between passes might be useful to provide a stronger oxide-to-oxide bond than a metal-to-oxide bond would give.

Large Optical Components

Since about 1980 there has been considerable interest in producing large terrestrial and space optics for surveillance and astronomical research applications. Many of these have been in the 1–4 m (3–15 ft) size range with low f number (~ 1.5) aspheric surfaces. A contour accuracy as low as 50 nm (2 μin) and a surface roughness approaching an R_a value of 1 nm are being

TABLE 13.14 *Surface Generating Processes Compared by Stowers* et al. (1988)

1. Conventional lens production employing fixed abrasive diamond wheels followed by free abrasive pitch laps employing modest pressure (1–3 kPa and speeds (<0.25 m s^{-1})
2. Computer numerical control (CNC) of grinding and polishing in (1) above
3. Magnetic field assisted finishing using a magnetic fluid and float (Fig. 13.58)
4. Elastic emission machining (Figs 13.63 and 13.64)
5. Abrasive water jet machining (AWJ)
6. Chemical hydrodynamic polishing
7. Magnetic field assisted fine finishing using fine (<15 nm) Fe_3O_4 suspended in a magnetic fluid in conjunction with a rubber pad that presses the abrasive particles against the work
8. Electrolytic abrasive mirror finishing. Removal is primarily by electrolytic action with fine abrasive suspended in the electrolyte to provide a mild scouring action as the fluid moves across the surface at high velocity
9. Neutral ion beam figuring. A 1500 eV source of [Ar^+] or [Kr^+] ions removes material from the work by a sputtering action
10. Plasma assisted etching. A chemical reaction between a gas (CF_4, O_2, or Cl_2) is caused to occur with the surface at an energy level below that causing sputering. The volatile reaction product is swept away by the gas
11. Single-point diamond turning (SPDT)
12. Ultra-precision diamond grinding (UPDG)

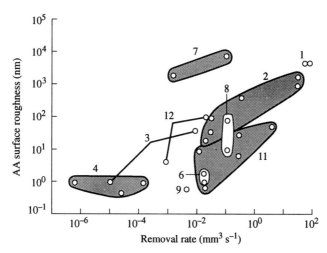

FIG. 13.66. A comparison of the surface roughness (R_a) and the removal rate achievable for the processes listed in Table 13.14 (after Stowers *et al.* 1988).

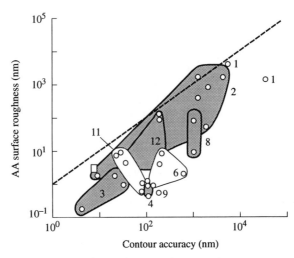

FIG. 13.67. A comparison of the surface roughness and contour accuracy achievable for the processes listed in Table 13.14 (after Stowers *et al.* 1988).

specified. Removal rates in excess of $10 \text{ mm}^3 \text{ s}^{-1}$ are required in order to keep production costs within bounds. Stowers *et al.* (1988) have compared the precision surface generating capability of the several processes listed in Table 13.14.

Figure 13.66 shows the relation between the AA surface roughness (R_a) and removal rate for the several processes listed in Table 13.14, while Fig. 13.67 shows the corresponding relationship between surface roughness (R_a) and contour accuracy. A target combination of 50 nm roughness at a removal rate of $10 \text{ mm}^3 \text{ s}^{-1}$ can be achieved by SPDT, which explains why this process has received so much attention for the production of large high-quality optical components. Figure 13.67 shows that, in general, the countour accuracy achievable is poorer than the corresponding surface finish.

Erosive jet machining results are not shown in Fig. 13.66 and 13.67, since a rippled surface is produced due to instability of the jet as it moves across the surface at a grazing angle. Plasma-assisted chemical etching results are also not shown in Fig. 13.66 and 13.67.

References

Aghan, R. L. and Samuels, L. E. (1970). *Wear* **16**, 293.
Andrew, C. (1979). *Int. Conf. Creep Feed Grinding*, University of Bristol.
Anzai, M., Otaki, H., Kawashima, E., and Nakagawa, T. (1993*a*). *Int. Japan Soc. Prec. Engrs* **27**, 223.
Anzai, M., Otaki, H., Sudo, T., and Nakagawa, T. (1993*b*). *J. Japan Soc. Prec. Engrs* **27**, 223.

Archer, J. F. (1992). *AES Mag.* **31**, 17.
Beilby, G. T. (1903). *Proc. R. Soc.* **72**, 217.
Brinksmeier, E. and Minke, H. (1993). *Ann. CIRP* **42**(1), 367.
Brueckner, D. K. (1960). *Ind. Anzeiger* **82**, 23.
Burnham, C. (1990). *Abr. Engng Soc. Mag.* **30**, 8.
Cook, L. M. (1990). *J. Noncrystall Solids* **120**, 152.
Dall, A. H. (1946). *Mech. Engng* **68**, 325.
Farber, E. A. and Scorah, R. L. (1948). *Trans. ASME* **70**, 369.
Frost, M. and Fursdon, P. M. T. (1985). *ASME PED* **16**, 313.
Funck, A. (1991). *ZwF* **88**(10), 454.
Fox, M., Agrawal, K., Shinmura, T., and Komanduri, R. (1994). *Ann. CIRP* **43**(1), 181.
Furukawa, Y., Miyashita, M., and Shiozaki, S. (1971). *Int. J. Mach. Tool Des. Res.* **11**, 145.
Gagliardi, J., Hong, I. S., and Duwell, E. (1988). *Abr. Engng Soc. Newslett.* July, 14.
Grisbrook, H. (1962). *Adv. Mach. Tool Des. Res.*, 155. Pergamon Press, Oxford.
Gurney, J. P. (1965). *J. Engng Ind.* **87**, 163.
Hahn, R. S. (1962). *3rd MTDR Conf.*, Birmingham, UK.
Hahn, R. S. (1964). *J. Engng Ind.* **86**, 287.
Hahn, R. S. (1978). *SME Tech. Paper MR78331*.
Hahn, R. S. and Lindsay, R. (1971). Five articles in *Mach. Mag.*, July–Nov.
Hashimoto, F., Kanai, A., and Miyashita, M. (1983). *Ann. CIRP* **32**(1), 237.
Hashish, M. (1985). *Proc. Nontraditional Machining Conf.* ASM.
Hashish, M. (1987). *J. Engng Ind.* **109**, 281.
Hashish, M. (1988a). *Exp. Mech.*, June, 159.
Hashish, M. (1988b). *9th Int. Conf. on Jet Cutting Technol.* Sendai. Japan.
Hashish, M. (1989). *Manufacturing Rev.* **2**, 142.
Herring, C. and Galt, J. (1952). *Phys. Rev.* **85**, 1060.
Hitchcock, A. L. (1986). *Metal Prog.* **86**, 33.
Imanaka, O., Kurobe, T., and Matsushima, K. (1981). *Proc. Japan Soc. Prec. Engrs Spring Conf.*, 777.
King, R. I. and Hahn, R. S. (1986). *Handbook of modern grinding technology*. Chapman and Hall, New York.
Kurobe, T., Imanaka, O., and Shimeno, K. (1986). *Bull. Japan Soc. Prec. Engrs* **20**, 49.
Lindbeck, B. (1952). *Ann. CIRP* **1**, 71.
Liverton, J. W. (1979). *Int. Conf. Creep Feed Grinding*, University of Bristol, U.K., p. 207.
Loladze, T. N., Batiashvili, B. I., Butskhrididze, D. S., Mamulashvili, G. L., Grdzelishivili, G. L., and Markevich, J. I. (1982). *Ann. CIRP* **31**(1), 205.
Malkin, S. (1968). *Sc.D. Dissertation*, M.I.T., Cambridge, Massachusetts.
McClure, R. (1985). *ONRFE Sci. Bull.* 10, No. 4.
Matsui, S. and Syoji, K. (1978). *Tech. Rep. Tohoku Uni.* **13**, 255.
Metcut (1986). *News Letter* No. 2. Metcut Research Associates Cincinnati, Ohio.
Miyashita, M., Hashimoto, F., and Kanai, A. (1982). *Ann. CIRP* **31**(1), 221.
Namba, Y. and Tsuwa, H. (1967). *Tech. Rep. Osaka Univ.* **17**, No. 796.
Olsen, J. H. (1980). *Cutting by waterjet*. Flow Industries, Inc., Kent, Washington.
Opitz, H. (1952). *Ann. CIRP* **1**, 23.
Parks, R. E. and Evans, J. C. (1994). *Prec. Engng* **16**, 223.

Peets, W. J. (1925). *Mech. Engng* **47**, 695.
Phillips, R. E. (1986). *Carbide Tool J.* **18**, 12.
Powell, J. (1979). Ph.D. dissertation, University of Bristol, U.K.
Rabinowicz, E. (1971). *Scient. Am.* **218**, 91.
Reeka, D. (1967). D. Ing. dissertation, Aachen T. H.
Rhoades, L. J. (1988). *Mfg. Engng* Nov.
Rowe, W. B. (1979). *Prec. Engng* **1**, 75.
Rowe, W. B., Barash, M. M., and Koenigsberger, F. (1965). *Int J. Mach. Tool Des. Res.* **5**, 203.
Rowe, W. B., Miyashita, M., and Koenig, W. (1989). *Ann. CIRP* **38**(2), 1.
Rowe, W. B., Spreggett, S., and Gill, R. (1987). *Ann. CIRP* **36**(1), 207.
Safto, G. R. (1975). Ph.D. dissertation, University of Bristol, U.K.
Salje, E. and Damlos, H. H. (1981). *NAMWR Conf.* SME, Dearborn, Michigan, p. 240.
Salje, E., Teiwes, H., and Heidenfelder, H. (1983). *Ann. CIRP*, **32**(1), 241.
Sauer, W. S. and Shaw, M. C. (1974). *Proc. Int. Conf. Prod. Engng* Tokyo, p. 645.
Shaw, M. C. and Macks, E. F. (1949). *The analysis and lubrication of bearings*. McGraw-Hill, New York, p. 355.
Shimmamune, T., Mochida, M., and Ono, K. (1990). *Bull. Japan Soc. Prec. Engrs* **24**, 206.
Shimamune, T., Mochida, M., and Ono, K. (1991). *Int. J. Japan Soc. Prec. Engrs* **25**, 11.
Shiozaki, S., Furukawa, Y., and Ohishi, S. (1977). *Bull. J. Soc. Prec. Engrs* **11**, 97.
Snoeys, R. and Decneut, A. (1971). *Ann. CIRP* **19**(1), 507.
Snoeys, R. and Peters, J. (1974). *Ann. CIRP* **23**(2), 227.
Spur, G. (1994). Private communication.
Spur, G. and Becker, K. (1990). *ZwF* **85**, 397.
Stelson, T. S., Komanduri, R., and Shaw, M. C. (1977). *4th Int., Conf. Prod. Res., Tokyo*. Taylor and Francis, London.
Stowers, I. F., Komanduri, R., and Baird, E. D. (1988). *Lawrence Livermore National Laboratory*, UCRL 99735.
Tsuwa, H., Ikawa, N., Mori, Y., and Sugiyama, K. (1979). *Ann. CIRP* **28**(1), 193.
Udupa, N. G., Shunmugam, M. S., and Radhakrishnan, V. (1986). *All India MTDR Conf.*, p. 12.
Udupa, N. G., Shunmugam, M. S., and Radhakrishnan, V. (1988). *J. Engng Ind.* **110**, 179.
Ueda, T. and Yamamoto, A. (1984). *J. Eng. Ind.* **106**, 237.
Umehara, N. (1994). *Ann. CIRP* **43**(1), 185.
Verma, A. P. and Lal, G. K. (1984). *Int. J. Mech. Tool Des. Res.* **24**, 19.
Visser, R. G. and Dimberg, A. P. (1988). *AES Mag.*, Mar./Apr., 16.
Zhou, Q. E. and Shaw, M. C. (1981). *Proc. 9th NAMWR Conf.*, SME, Dearborn, Michigan, p. 243.
Zorn, F. J. (1983). *AES Mag.*, Mar./Apr., 10.

14

HARD WORK MATERIALS

Introduction

In this chapter grinding characteristics of hard brittle materials are considered. These materials include rock (marble, granite, basalt, etc.), concrete, ceramics, glasses, and very hard metals.

Rock and Concrete

Diamond Wire Sawing

Diamond tools are widely used for grinding rock and concrete. Diamond wire sawing was developed in the 1970s for quarrying granite and marble. The modern wire saw for quarrying consists of a steel cable on which diamond-impregnated beads are threaded. The beads are separated by springs or regions of injection-molded plastic. Figure 14.1 shows a typical wire sawing operation, while Fig. 14.2 shows a typical diamond cutting wire. In making a cut the first step is to drill two orthogonal intersecting holes — one vertical and one horizontal. The wire is then threaded through the holes passed over the drive and guide pulleys and the ends joined to form a continuous loop. The drive wheel drives the wire at speeds up to $45\,\mathrm{m\,s^{-1}}$ (8850 f.p.m.), while the drive unit forces the wire against the stone. Forty U.S. mesh size diamonds are bonded to a steel core in a single layer by electrodeposited nickel, or in the form of a sintered multilayered structure. Steel springs are used in sawing marble, but are replaced by impervious plastic molded spacers when sawing more abrasive granites, which would rapidly wear the wire rope if springs were used. Typical dimensions involve 10 mm diameter (0.4 in) beads mounted on a 5 mm (0.2 in) cable for marble, but a somewhat smaller size for sawing granite (Hawkins *et al.* 1990). Typically, there are about 30 beads per meter (9 beads per foot). Marble is generally sawed using friable self-sharpening diamonds, while more blocky, tougher diamonds are used for sawing granites. For sawing granite a typical wire speed is $20\,\mathrm{m\,s^{-1}}$ (3940 f.p.m.). Higher speeds are generally used for marble. There is an optimum rate of removal, depending upon the type of granite being cut. This is controlled by the force with which the wire is forced against the rock and the area being sawed. A typical cutting rate for a relatively easily cut granite would be about $4\,\mathrm{m^2\,h^{-1}}$ ($43\,\mathrm{ft^2\,h^{-1}}$) to give a wire life of about $6\,\mathrm{m^2\,m^{-1}}$ ($19.7\,\mathrm{ft^2\,ft^{-1}}$). For a very difficult granite, both of these values might be half as high.

Other methods of extracting blocks of stone in quarrying are flame burning and percussion drilling. However, these techniques are being displaced because

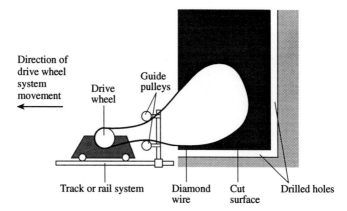

FIG. 14.1. A typical wire sawing arrangement (after Hawkins *et al*. 1990).

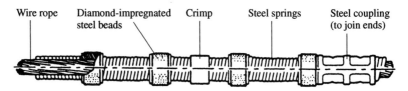

FIG. 14.2. A typical diamond sawing wire. The crimps are about every five beads along the cable, to prevent bead sliding (after Hawkins *et al*. 1990).

diamond sawing is more productive (faster, less noisy and arduous, and less labor-intensive) and is more gentle, resulting in less scrap loss. On wires employing springs as spacers (for sawing marble) the beads can rotate to provide uniform circumferential wear. This is not the case for wires with welded plastic spacers. Alternatively, uniform wear can be achieved by giving the wire about one turn about its centerline per meter of length as it is being threaded.

Wire sawing is also used in the construction industry for demolition and alteration of concrete and other structures (Hawkins and Brauninger, 1991).

Fine Wire Sawing

Wire sawing is also performed using a fine wire coated with diamond over its entire length. In this case a steel wire about 0.008 in (200 μm) in diameter having a single layer of copper bonded grits on its surface is used. Figure 14.3 shows a typical fine diamond sawing arrangement. In this case the ends of the wire are not joined but are attached to a capstan. The wire is unwound from one end of the capstan and rewound on the other. When the wire is completely unwound, the rotational direction is reversed and the wire traverses the work in the opposite direction. Figure 14.4 shows two types of wire used in fine wire

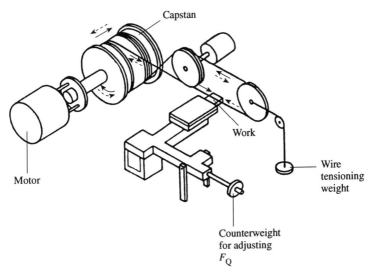

FIG. 14.3. An arrangement for fine wire sawing using a reversing capstan (after Ito and Murata 1987).

sawing. The first, in Fig. 14.4(a), shows a single strand of diamond coated wire, while a multistranded wire is shown in Fig. 14.4(b).

Ito and Murata (1987) have investigated fine wire sawing, as shown in Fig. 14.3, and have found the ratio of normal force F_Q to tangential force F_P to be used depends on the ductility of the material being cut:

Work material	F_Q/F_P
Al_2O_3 ceramic	2.0
Glass	1.8
Brass	1.4
Aluminum	1.2

These values are for fresh wire, and they increase as diamond wear proceeds. It was also found by Ito and Murata (1987) that cutting ability was improved significantly when a coolant was used.

Fine wire sawing is sometimes performed, using a fine abrasive ($\sim 600\ \mu m$ mesh) as a slurry in a liquid that is somewhat viscous to keep the abrasive in solution. In this case, the wire is uncoated. Fine stranded wire has also been used in fine abrasive sawing using an abrasive slurry.

FIG. 14.4. Two types of bonded abrasive wire used in fine diamond sawing (after Ito and Murata 1987).

Grindability of Granite

Stone occurs in many chemical compositions, which are determined by the origins of the stone. Three main categories exist: igneous, sedimentary, and metamorphic. Igneous rocks are formed by the solidification of molten rock (magma): examples are granite, quartz, and basalt. Sedimentary rocks are formed by deposition of the wear debris of weathered igneous rock or inorganic material, the debris being transported into land depressions by wind or water: sandstone, coal, bauxite, and limestone are examples. Metamorphic rocks result from alterations to igneous and sedimentary rocks, these alterations being brought, about by shifting of the Earth's tectonic plates, as in earthquakes—slate, marble, and schist are metamorphic rocks.

Stone is steadily gaining importance as an engineering material in the construction and monument industries. This is because of the emergence of modern abrasive materials and techniques that make stone more economic to cut, size, and polish. Of all the rock categories, granitic igneous rocks offer the best resistance to environmental weathering and air pollution. Granites are available in a wide range of colors and patterns. They can be shaped accurately and polished to a lustrous finish: hence they are a popular choice with architects for external cladding of buildings and in the monument industry. An outstanding example of the latter is the Vietnam Veterans Memorial in Washington, D.C.

Birle and Ratterman (1986) have studied the grindability of granite. Factors influencing the sawability of granite include composition, grain size, residual stress, and the degree of weathering. No correlation of sawability with bulk properties has been found and sawability is best evaluated by standardized sawing tests typical of conditions pertaining in industry.

Birle and Ratterman (1986) used the mean ratio of area sawed to the depth of radial wheel wear (Y, in $m^2 mm^{-1}$) in a standardized cylindrical grinding test as the criterion for sawability. It was found that only two of seven factors considered had a major influence on Y (the w/o of quartz (B) and the abrasive wear resistance (C, in $m^2 mm^{-2}$)). The abrasive wear resistance (C) was measured in terms of an empirical ASTM test (C 241-51, 1958) that measures the resistance of stone to foot traffic. This test consists of sliding 2 in (5.08 cm)

TABLE 14.1 *A comparison of measured and predicted grindabilities (Y, $m^2\,mm^{-1}$) for ten samples of granite studied by Birle and Ratterman (1986)*

Granite type	w/o quartz	Y (m^2 mm^{-2}) Measured	Predicted
Academy Black	9.6	12.3	10.5
Lake Placid Blue	0	10.7	12.1
Sierra White	24.9	9.3	8.7
Barre	22.0	6.8	4.3
Charcoal Black	20.5	4.6	4.5
Rockville	24.6	4.3	5.2
Inada	35.0	4.2	5.4
Agate	27.0	4.1	5.2
Sunset Red	34.3	4.0	3.2
Bright Red	20.0	1.8	1.9

square specimens over an iron lap at a low constant speed and a load of 2 kg (4.4 lb) with a constant flow of dry 60 grit Al_2O_3. The abrasion resistance is expressed in terms of the volume of material lost for a given sliding distance.

The wear performance (Y) was found to vary from about 12 to 2 for the wide range of granites considered in Table 14.1. The predicted value of Y was determined from the following empirical equation, obtained by regression analysis,

$$Y = 18 - 0.13B - 0.17C, \qquad (14.1)$$

and was estimated to be reliable to within ±15 percent.

Equation (14.1) was used to predict the grindability (Y) of a large number of samples of granite from around the world. Table 14.2 gives the resulting values.

Birle and Ratterman (1986) then divided these materials into the following four classes based on a range of Y values:

Class	Y
I (easy to grind)	>7
II	4–7
III	3–4
IV (difficult to grind)	<3

In addition to quartz content (B) and abrasive resistance (C), quartz grain size, which varies from about 0.5 to 3.0 mm (0.020 to 0.120 in) is also found to be important, a large quartz grain size being more difficult to grind than a small grain size.

TABLE 14.2 *Predicted values of grindability (Y, $m^2\,mm^{-1}$) for a wide variety of rock types plus their assigned class of grindability (I high–IV low) (after Birle and Ratterman 1986)*

Rock type	Predicted grindability	Class†
Honkomatsu	12.3	I
Lake Placid Blue	12.1	I
Basalt Lava — Rhon	10.9	I
Academy Black	10.5	I
Dyorit Metaphyr	9.9	I
Veined Ebony	9.9	I
Coarse Black	9.5	I
Angola Granite	9.2	I
Africa F.G.	8.7	I
Labrador	8.7	I
Sierra White	8.7	I
Texas Pearl	8.2	I
Impala	7.8	I
Labrador Al. Syenita	6.6	II
Tarn Royal Granite	6.6	II
Injou	6.4	II
Vermont Blue Pearl	6.0	II
Sardinian Gray	5.8	II
Opalescent	5.7	II
Diamond Pink	5.5	II
Makabe	5.5	II
Africa Belfast	5.4	II
Carnelian	5.4	II
Inada	5.4	II
Agate	5.2	II
Rockville	5.2	II
Outo	4.9	II
South African Black	4.7	II
Iridian	4.6	II
Bansai	4.5	II
Charcoal Black	4.5	II
Deer Island Pink	4.4	II
Barre	4.3	II
Lac Dubonnet	4.3	II
Epprechtsteiner	3.8	III
Koesseine	3.8	III
Orienta	3.8	III
Bethel White	3.5	III
Tokuyama	3.5	III
Rosso Yuca	3.4	III
Baveno	3.2	III
Sunset Red	3.2	III
India Granite	3.0	III

TABLE 14.2 *Continued*

Rock type	Predicted grindability	Class†
Bright Red	2.9	IV
Aji	2.8	IV
Eishu	1.8	IV
Rosa Sardegnia	1.6	IV
Oshima	1.5	IV
Laurentian Rose	1.1	IV
Vanga	0.4	IV
Korall	−0.2	IV
Desert Pink	−0.7	IV
Colombo	−0.9	IV
Tranas	−1.1	IV
Switzerland	−2.2	IV

† Class I, $Y > 7$; Class II, $Y = 4$–7; Class III, $Y = 3$–4; Class IV, $Y < 3$.

Diamond Circular Sawing of Granite

Circular sawing is also extensively used to process granite. Vastly superior wear performance and hence a long tool life make diamond the preferred abrasive over other abrasives such as silicon carbide and aluminum oxide. Diamond-impregnated metal segments are attached to the periphery of a metal wheel at regular intervals separated by slots. The saw blade is held firmly with large flanges and made to spin at high speeds while it is fed across the work material with a large depth of cut. Although the process is used for the purposes of stock removal and cut-off, this sawing process resembles the creep feed grinding process (Chapter 13) more than it does the abrasive cut-off process used on metals (Fig. 6.1). Saw diameters vary from 8 in (200 mm) to as large as 12 ft (3.7 m). Wheel peripheral speeds (V) are around 6000 f.p.m. (30 m s^{-1}) and the traverse rate (v) is of the order of 3–15 f.p.m. (1–5 m min^{-1}). The depth of cut achievable depends on the nature of the work material, machine rigidity and, of course, the machine horsepower. Under the right conditions a large blade can cut a depth of 10–12 in (250–300 mm) in a single pass.

Figure 14.5 illustrates a typical application for grinding granite using a 100 kW (134 h.p.) heavy duty bridge saw with an infinitely variable d.c. motor.

Figure 14.6 shows directly measured values of spindle power when grinding Orienta (Class III) granite. Each cut was 1.3 m (4.25 ft) long. Twelve sets of cuts, each at a wheel depth of cut of 10 mm (0.40 in), were made — six in the up grinding mode and six in down grinding. Figure 14.7(a) shows dynamometer values F_H and F_V for a sequence of up grinding cuts, while Fig. 14.7(b) shows corresponding results for down grinding for the same grinding conditions as for Fig. 14.6.

The spindle power values were obtained using a wattmeter, nulled to zero

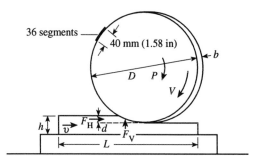

FIG. 14.5. Grinding granite on 100 kW (134 h.p.) machine using a large wheel with diamond segments (after Brach *et al*. 1988). Wheel D, 650 mm (25.6 in), $b = 4.5$ mm (0.17 in); bond, cobalt; abrasive, 40–50 grit size tough diamond, 30 concentration. Conditions: $V = 32.5$ m s^{-1} (6500 f.p.m.), $v = 1$ m min^{-1} (3.28 f.p.m.), $d = 10$ mm (0.39 in); fluid, water-based at 33 l min^{-1} (8.72 g.p.m.). Forces F_H and F_V are dynamometer values and P represents the spindle power.

when sawing air. The rise in power within each set of six passes is due to an increase in rubbing on the sides of the wheel as the cumulative depth of penetration increases with each pass. The gradual rise in power with each successive slice is due to a change in the wheel surface. This is seen to lead to a steady state maximum value of grinding power of about 5.2 kW (7 h.p.) for both up (U) and down (D) grinding.

Figures 14.7(a) and (b) give corresponding force values obtained using a three component piezoelectric dynamometer. These figures show a gradual approach to equilibrium values as the wheel condition reaches a steady state.

Comparing Figs 14.6 and 14.7, it is seen that whereas directly measured spindle power is the same for up and down grinding, up (U) and down (D) grinding power values derived from dynamometer readings are not the same:

$$P_U = F_{HU} V = 146 \times (6500/33\,000) = 28.7\,\text{h.p.} \quad (21.4\,\text{kW}),$$

$$P_D = F_{HD} V = 45 \times (6500/33\,000) = 8.86\,\text{h.p.} \quad (6.6\,\text{kW}).$$

When this very large difference between dynamometer and spindle power was discovered, it was decided to look further into the matter.

Power Analysis for Large d/D

Figure 14.8(a) shows up grinding with a large value of d/D, while Figure 14.8(b) shows the corresponding down grinding situation. F_H and F_V are the components of force on the wheel as measured by the dynamometer and

$$R = [F_H^2 + F_V^2]^{0.5}, \qquad \beta = \tan^{-1}(F_H/F_V), \qquad (14.2, 14.3)$$

$$\theta = \cos^{-1}(D/2 - d)/D/2 = \cos^{-1}(1 - 2d/D). \qquad (14.4)$$

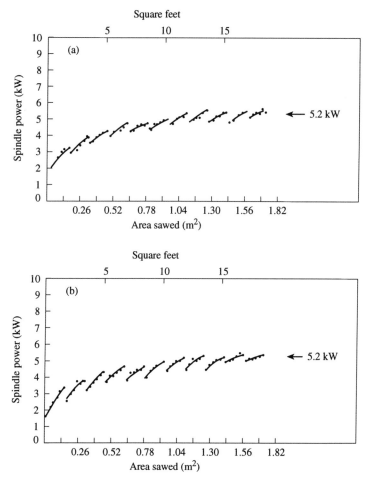

FIG. 14.6. The spindle power versus the area sawed under conditions of Fig. 14.5: (a) up grinding; (b) down grinding (after Brach *et al.* 1988).

The tangential and radial components of the resultant force are:

$$F_T = R \sin \delta, \qquad F_R = R \cos \delta, \qquad (14.5, 14.6)$$

where

$$\delta_U = \beta - K_U \theta. \qquad (14.7)$$

For down grinding (Fig. 14.8(b)) there are two possibilities:

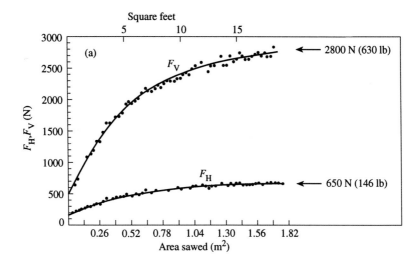

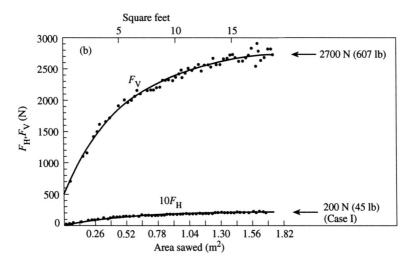

FIG. 14.7. Dynamometer forces F_H and F_V for the tests of Fig. 14.5: (a) up grinding; (b) down grinding (after Brach *et al.* 1988).

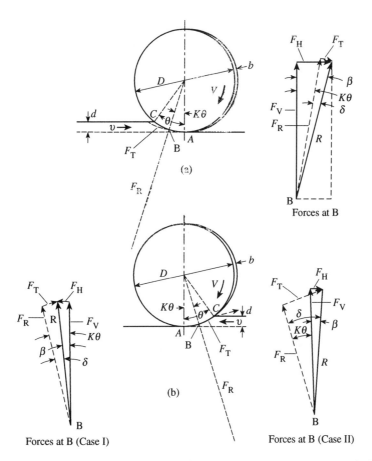

FIG. 14.8. Grinding with a large d/D ratio: (a) up grinding; (b) down grinding.

- Case I, when F_H on the wheel is in the direction of v (work is forced into wheel by feed motor)
- Case II, when F_H on the wheel has a direction opposite to v (work is dragged into wheel by rotating wheel)

These two velocity diagrams are shown in the corresponding inserts in Fig. 14.8(b).

For Case I

$$\delta_D = K_D \theta - \beta, \qquad (14.8a)$$

while for Case II,

$$\delta_D = K_D \theta + \beta. \qquad (14.8b)$$

The values of K in eqns (14.6) and (14.7) depend on the location of the effective point of application of the resultant force R on the arc of wheel–work contact (AC), where

$$K = \text{AB}/\text{AC}. \tag{14.9}$$

Before forces F_T and F_R may be obtained from the workpiece dynamometer values F_H and F_V, some way of estimating the value of K must be found.

Figures 14.7(a) and (b) give corresponding force values obtained using a three-component piezoelectric dynamometer. These figures show a gradual approach to the equilibrium values indicated as the wheel condition reaches a steady state. When these values are substituted into eqns (14.2)–(14.6), the following results are obtained:

$$R_U = 2874 \text{ N } (646 \text{ lb}), \quad R_D = 2707 \text{ N } (609 \text{ lb}),$$
$$\beta_U = 13.07°, \quad \beta_D = 4.24°, \quad \theta_U = \theta_D = 14.25°.$$

From Fig. 14.6(a) and (b) it is evident that the spindle power for up and down grinding is about the same, and is equal to 5.2 kW. From eqns (14.5) and (14.6), the values of δ for up and down grinding may be found as follows:

$$\delta_U = \sin^{-1}(F_T/R)_U = \sin^{-1}(5200/32.5 \times 2874) = 3.19°$$
$$\delta_D = \sin^{-1}(F_T/R)_D = \sin^{-1}(5200/32.5 \times 2707) = 3.39°.$$

From eqns (14.5) and (14.6), the force ratio

$$F_T/F_R = \tan \delta. \tag{14.10}$$

Thus,

$$(F_T/F_R)_U = \tan 3.19° = 0.056, \quad (F_T/F_R)_D = \tan 3.39° = 0.059.$$

These values are an order of magnitude smaller than the corresponding values for metal. This is due to the fact that more fracture and less plastic flow is involved in the grinding of rock and ceramics than for less brittle metals.

From eqs (14.7) and (14.8):

$$K_U = [(\beta - \delta)/\theta]_U = (1307 - 3.19)/14.25 = 0.69,$$
$$K_U = [(\delta + \beta)/\theta]_D = (3.39 + 4.24)/14.25 = 0.54.$$

The value of K is greater for up grinding than for down grinding. This is because in up grinding, chip formation does not begin immediately with contact but only after t, and hence F_R, have exceeded a certain value sufficient to cause a change from rubbing to chip formation. In down grinding, the transition is from chip formation to rubbing, and this will occur at a smaller value of t than for up grinding. Thus, for down grinding the minimum undeformed chip thickness t should be less than for up grinding.

Since the specific grinding energy (u) increases rapidly with a decrease in undeformed chip thickness t, the specific energy in the vicinity of point A

(Fig. 14.8) will be greater in down grinding than in up grinding. This, in turn, will cause the point of application of the resultant force (B) to be closer to A for down grinding than for up grinding and hence $K_D < K_U$.

It is well established (Hahn 1962) that the forces on a grinding wheel drop as soon as the in-feed is sufficient to cause chip formation. For a very small undeformed chip thickness t, rubbing takes less energy than cutting and hence rubbing occurs. This is because for very low t, the effective rake angle is largely negative, due to a finite tool-tip radius. However, when t becomes large enough, the effective rake angle increases and it then takes less energy for cutting than for rubbing and hence cutting occurs.

As previously mentioned, if the power P is taken to be $(F_H V)$, as is usual for conventional fine grinding, the following values are obtained:

$$P_U = (F_H V)_U = 21.4 \text{ kW}, \quad P_D = (F_H V)_D = 21.4 \text{ kW}.$$

These values are seen to be far different from the directly measured value of 5.2 kW and illustrate the fact that F_H may not be substituted for F_T in determining grinding power when F_V/F_H is large. For this example,

$$u = \frac{P}{vbd} = \frac{5200}{(1/60)(0.0045)(0.01)} = 6.93 \text{ N/m}^2 = 6.93 \text{ GPa} (\sim 10^6 \text{ p.s.i.})$$

This is a relatively low value compared with the value in fine grinding of steel (~ 170 GPa, or 25×10^6 p.s.i.), but it is about equal to the value expected for the very coarse grinding of steel ($\simeq 7$ GPa, or 10^6 p.s.i.).

Further data for the diamond sawing of granite are given in Table 14.3 for another material (French Tarn a class II granite) sawed at different wheel depths of cut (d). In all these examples, F_H and F_T are very different, and hence it is not permissible to take the total power consumed as $F_H V$ when grinding a very brittle material when F_V/F_H is relatively large. Values of K_U and K_D are also seen to be very different for this granite than for the Orienta granite.

Abrasive Cut-off

A typical abrasive cut-off process for metal is shown in Fig. 14.9, where representative values are given for an Al_2O_3 wheel on a 25 h.p. machine (19 kW) cutting dry. In this case $u = 1.92 \times 10^6$ p.s.i. (13.2 GPa), the undeformed chip thickness (t) is constant at 300 μin (7.5 μm), and there is no difference between F_H and F_T, regardless of the value of d/D.

Creep Feed Grinding

In creep feed grinding all the metal is removed in a single pass instead of by a large number of shallow passes (Chapter 13). For the same removal rate (vdb) the product vd may be about the same, but d may be 500 times as great for creep feed grinding as for the equivalent 'pendulum' grinding operation.

TABLE 14.3 Grinding of French Tarn granite with a 600 mm Segmental diamond wheel under the following conditions (after Brach et al. 1988)

Machine: 100 kW heavy duty bridges saw with d.c. motor
Wheel dia., D: 608 mm
Wheel width, b: 3.5 mm
Wheel speed, V: 32 m s^{-1}
Water-based coolant flow: 3.5 l/min^{-1}
Removal rate: 540 cm^2 min^{-1}
Abrasive: M8S-760, 30–40 mesh (U.S.)

d (mm)	v (m min^{-1})	Spindle power (kW)	F_H (N)	F_V (N)	R (N)	β (degree)	θ (degree)	δ (degree)	F_T (N)	F_R (N)	F_T/F_R	k	u (GPa)
Up grinding													
10	5.4	10.1	435.0	1200.0	1276.0	19.9	14.7	14.3	316.0	1236.0	0.26	0.38	3.21
15	3.6	11.5	600.0	1500.0	1615.0	21.8	18.0	12.8	359.0	1575.0	0.23	0.50	3.65
20	2.7	11.8	750.0	1600.0	1767.0	25.1	28.8	12.0	369.0	1728.0	0.21	0.63	3.75
30	1.8	14.0	950.0	1800.0	2035.0	27.8	25.5	12.5	438.0	1987.0	0.22	0.60	4.44
Down grinding													
10	5.4	10.1	25 (case II)	1300.0	1300.0	1.1	14.7	14.0	316.0	1261.0	0.25	0.88	3.21
15	3.6	11.5	100 (I)	1600.0	1603.0	3.6	18.1	13.0	359.0	1562.0	0.23	0.92	3.65
20	2.7	11.8	200 (I)	1900.0	1910.0	6.0	20.9	11.1	369.0	1875.0	0.20	0.82	3.75
30	1.8	14.0	350(I)	2400.0	2425.0	8.3	25.7	10.3	438.0	2361.0	0.19	0.73	4.44

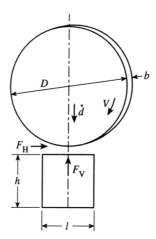

FIG. 14.9. A typical abrasive cut-off operation on AISI 1018 hot rolled steel bars (after Shaw *et al.* 1967). Work, $l = 1$ in (25 mm), $h = 1$ in (25 mm); wheel, A46R6B, $D = 20$ in (508 mm), $b = 0.187$ in (4.75 mm); conditions, $V = 12\,500$ f.p.m. (63.5 m s^{-1}); $d = 15.1$ i.p.m. (384 mm min^{-1}); fluid, air.

Hence, if $D/d = 5000$ for pendulum surface grinding (P), the corresponding value of D/d for creep feed grinding (C) might be 10.

From eqn (14.4):

$$\theta_C = \cos^{-1}(1 - 0.2) = 36.87°, \quad \theta_P = \cos^{-1}(1 - 0.0004) = 1.62°.$$

Assuming that $F_H/F_V = 0.5$ in both cases (see Fig. 14.10):

$$\beta_C = \beta_P = \tan^{-1} 0.5 = 26.57°.$$

Assuming that $K = 0.7$ for up grinding in both cases, then, from eqn (14.6),

$$\delta_C = 26.57 - (36.87 \times 0.7) = 0.76°, \quad \delta_P = 26.57 - (1.62 \times 0.7) = 25.44°.$$

Then, for creep feed up grinding,

$$F_T/F_H = (\sin \delta/\sin b)_{CU} = \sin 0.76/\sin 26.57 = 0.03,$$

and for pendulum up grinding,

$$F_T/F_H = (\sin \delta/\sin b)_{PU} = \sin 25.44/\sin 26.57 = 0.96.$$

It is thus evident that while the difference between F_H and F_T is relatively small for ordinary grinding (3 percent), the difference in creep feed grinding may be as high as 96 percent.

An example of the importance of properly interpreting dynamometer readings in creep feed grinding is given by data from Shiozaki *et al.* (1977). Here a workpiece dynamometer is used to compare the power consumed at the wheel surface for up and down creep feed grinding of hard ball bearing

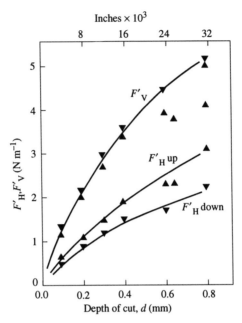

FIG. 14.10. Horizontal and vertical grinding forces per unit width of cut (F'_H and F'_V) for creep feed grinding of $R_c = 61$ bearing steel (after Shiozaki *et al.* 1977). Wheel, WA60G8V, $D = 305$ mm (12 in); conditions, $V = 36 \text{ m s}^{-1}$ (7085 f.p.m.); $v = 0.36 \text{ m s}^{-1}$ (70.9 f.p.m.); fluid, = soluble oil at 25 l min^{-1} (6.6 g.p.m.).

steel using an aluminum oxide wheel. Figure 14.10 gives comparable dynamometer results for different wheel depths of cut d for up and down creep feed plunge surface grinding. The force values given are for unit width of cut. From these results, it was concluded that substantially less power is consumed in down grinding than in up grinding. This conclusion is based on the assumption that power is given by $F_H V$, whereas in reality it is $F_T V$.

The analysis given in Table 14.4 for the case in which $d = 0.8$ mm using $K_U = 0.7$ and $K_D = 0.5$ yields values of F'_{TU} and F'_{HD} that are very much closer than the dynamometer values F'_{HU} and F'_{HD}. In this case, the values assumed for K are relatively unimportant. For example, if K_D and K_U are 0.7 and 0.5 respectively, instead of 0.5 and 0.7 as in Table 14.4, F'_{TU}/F'_{TD} becomes 1.18 instead of 1.16.

Koenig and Schleich (1982) have repeated the experiments of Shiozaki *et al.* (1977) using four different cubic boron nitride (CBN) wheels to grind a hard ($H_{RC} = 63$) high-speed steel, and have similarly misinterpreted their dynamometer results. A workpiece dynamometer was used to measure forces F_H and F_V in creep feed grinding under the conditions indicated in Table 14.5 for the up and down grinding modes. The four CBN wheels, designated as

TABLE 14.4 *A comparison of the power required for up and down creep feed (d = 0.8 mm) grinding based on F_H and on F_T for the grinding conditions of Fig. 14.10 (after Brach et al. 1988)*

Up grinding		
F'_H (N mm^{-1})	3.1	
F'_V (N mm^{-1})	5.1	
R' (N mm^{-1})	6.0	
K_U	0.7	(assumed)
β (degrees)	31.29	
θ (degrees)	5.87	
δ (degrees)	27.18	
F'_T (N mm^{-1})	2.73	
Down grinding		
F'_H (N mm^{-1})	2.1	(case II)
F'_V (N mm^{-1})	5.1	
R' (N mm^{-1})	5.52	
K_D	0.5	(assumed)
β (degrees)	22.38	
θ (degrees)	5.87	
δ (degrees)	25.32	
F'_T (N mm^{-1})	2.36	
F'_{HU}/F'_{HD}	1.48	
F'_{TU}/F'_{TD}	1.16	

A, B, C and D in Table 14.5, had different bond types, grain sizes, and grit concentrations (see Koenig and Schleich 1982 for details).

The equation to be used for δ_D corresponds to Case II (Fig. 14.8), since if Case I were used, δ would be negative, and this is not physically possible since a negative δ means that the system is driving the spindle. The dynamometer values F'_{HU}, F'_{VU}, F'_{HD} and F'_{VD} in Table 14.5 were obtained from Fig. 8 of Koenig and Schleich (1982), while the spindle power (P_U and P_D) values were taken from Fig. 9 of Koenig and Schleich (1982). As in the analysis of the data from Shiozaki *et al.* (1977), the values of K_U and K_D employed are not important and in this case different values of K_U and K_D were used in arriving at the values of Table 14.5.

From the results of Table 14.5 it would appear that on the average it would take 1.92 times the down grinding power for up grinding, assuming that the power is $F_H V$. However, the average difference between up and down creep feed power based on F_T is only 1.09. The mean difference in spindle power between up and down CBN/creep feed grinding was only 1.08, which is in excellent agreement with properly interpreted dynamometer results (1.09).

It thus appears that when creep feed grinding hard steel using either aluminum oxide or CBN wheels, the difference between the power consumed in up and down grinding is relatively small. In up creep feed grinding, the

TABLE 14.5 *A comparison of the power required for up and down creep feed (d = 1 mm) surface grinding based on F_H and on F_T for CBN wheels grinding hard high-speed steel (after Koenig and Schleich 1982)*

Wheel diameter, D: 400 mm
Wheel speed, V: 30 m s^{-1}
Work speed, v: 2 mm s^{-1}
Width of cut, b: 10 mm
$\theta = \cos^{-1}[1 - 2(d/D)] = 5.73°$
$d_U = \beta - K_U \theta$
$\delta_D = \beta + K_D \theta$ (case II)

Wheel	F'_H (N mm^{-1})	F'_V (N mm^{-1})	F'_T (N mm^{-1}) $K_U = 0.5$ $K_D = 0.5$	F'_T (N mm^{-1}) $K_U = 0.4$ $K_D = 0.6$	F'_T (N mm^{-1}) $K_U = 0.6$ $K_D = 0.4$	P (kW)
Up grinding						
A	24.5	105.0	19.22	20.28	18.16	4.40
B	10.0	49.0	7.54	8.03	7.04	4.00
C	9.0	43.0	6.84	7.27	6.40	3.35
D	10.0	47.5	7.61	8.09	7.13	3.90

Down grinding

A	13.0	87.5	17.36	18.22	16.49	4.15
B	5.0	37.0	6.84	7.21	6.48	3.50
C	5.0	32.0	6.59	6.91	6.28	3.06
D	5.0	35.0	6.74	7.09	6.40	3.80

Comparisons

Wheel	F'_{HU}/F'_{HD}	F'_{TU}/F'_{TD} $K_U = 0.5$ $K_D = 0.5$	F'_{TU}/F'_{TD} $K_U = 0.4$ $K_D = 0.6$	F'_{TU}/F'_{TD} $K_U = 0.6$ $K_D = 0.4$	P_U/P_D
A	1.88	1.107	1.113	1.101	1.06
B	2.00	1.102	1.114	1.087	1.14
C	1.80	1.037	1.052	1.020	1.10
D	2.00	1.129	1.141	1.115	1.03
Av.	1.92	1.09	1.11	1.08	1.08

wheel grinds the finished surface when its temperature is minimum; whereas in down grinding, the finished surface is ground when the temperature of the wheel surface is maximum. This, coupled with the small difference in specific grinding energy for up and down grinding, suggests that from the point of view of potential thermal damage to the workpiece, up creep feed grinding is preferable to down creep feed grinding.

It should be noted that on the European Continent, the symbol F_t is used for the dynamometer value that is tangential to the wheel surface at the initial point of contact in grinding. This is what has been designated F_H here, rather than the true tangential component of force, designated F_T. It is F_T and not F_H that should be used to obtain the power consumed in creep feed grinding or in the circular sawing of rock.

Furukawa *et al.* (1979 and 1980) subsequently recognized the importance of using the true tangential force F_T rather than F_H when considering the power consumed under different creep feed grinding conditions. They derived equations based on two assumptions:

(1) that the value of K is 0.5;
(2) that in down creep feed grinding, Case II pertains (the work is forced into the wheel by the wheel, rather than being forced in by the feed motor).

Both of these assumptions are satisfactory for the conditions of their tests but, as we have seen, such do not hold for other situations, as when sawing granite. They purport to prove that K is 0.5 by noting a singular relationship between the extent to which the wheel slows down and the true value of F_T as computed by the use of $K = 0.5$. However, as has been shown, values of F_T in creep feed grinding are relatively insensitive to the value of K assumed and hence the same 'proof' would have been obtained in this case by a variety of other assumed values of K.

In Furakawa *et al.* (1980) creep feed results are extended to a wheel depth of cut of 2.5 mm (0.1 in). At this high wheel depth of cut, It was found that whereas the power based on values of F_H for up and down creep feed grinding of hard ball bearing steel using vitrified aluminum oxide wheels was 2.5 times the up grinding power, the up grinding power based on values of F_T exceeded the down grinding power by a factor of only about 1.3. At lower, more practical values of wheel depth of cut, the difference in the up and down power was considerably less. For example, it had previously been found by Shiozaki *et al.* (1977), that when d was 0.8 mm, the up grinding power exceeded the down grinding power by less than 20 percent. Based on this difference in grinding power, it was concluded that in creep feed grinding there is less danger of thermal damage to the finished surface in the down grinding mode than when up grinding under the same conditions. As previously mentioned, the maximum temperature is located much farther from the finished surface in up grinding than in down grinding and, with the proper application of the coolant, it should be possible to up grind with considerably higher energy consumption without thermal damage to the finished surface. However, the application of the fluid in creep feed grinding is particularly important. When up grinding, the bulk of the fluid should be applied to the region of highest temperature where the wheel is exiting the work rather than to the free surface of the wheel, which has plenty of time to cool before reentry.

Andrew et al. (1985) also mention the importance of using the true tangential force F_T rather than the usual dynamometer value F_H when considering the energy and power consumed in creep feed grinding. However, after identifying the problem, they do not discuss it further and proceed to use F_H instead of F_T throughout their Chapter 7.

Summary: Performance when d/D is Large

It has been shown that sometimes the horizontal component of force on the work may be used directly to obtain the spindle power involved (i.e. $P = F_H V$), while in other cases in which d/D and/or F_V/F_H are large this is not admissible. The conversion of F_H and F_V to the more meaningful values F_T and F_R in general involves three nondimensional quantities, d/D, F_V/F_H, and K. The only safe approach is to convert the measured values of F_H and F_V to the equivalent values of F_T and F_R. In all cases, the spindle power (P) is $F_T V$.

In down grinding conversion of F_H and F_V involves two cases:

- Case I: the feed motor forces the work into the wheel and total power consumed = spindle power + feeding power = $F_T V + F_H v$, the latter term, $F_H v$, being negligible compared with $F_T V$.
- Case II: the grinding wheel forces the work into the wheel. The feed motor then acts as a generator and the total power consumed is the spindle power = $F_T V$.

In up grinding there is only one case – Case I. In down grinding Case I or Case II may possibly pertain and therefore it is important to be cautious in this regard in converting F_H and F_V to F_T and F_R.

In creep feed grinding, circular sawing of rock, and any other situations in which d/D and F_V/F_H are relatively large, it is preferable to measure spinde power directly rather than use a workpiece dynamometer.

Factorial Experimental Design for Power

Pai et al. (1988) have investigated the influence of a number of variables on the power and specific energy associated with the circular sawing of granites. High and low values of six variables in a 2^n factorial experimental design (Chapter 10) are given in Table 14.6, together with other grinding conditions that were held constant.

Values of the specific energy (u) and power required (P) derived from the 64 combinations of the six high and low quantities involved in the study are given

TABLE 14.6 *Operating conditions used in the factorial design study (after Pai et al. 1986)*

Grinding machine	1 h.p. (0.76 kW)
Mode of grinding	Up, Surface Grinding
Grinding fluid	None (air)
Diamond	Tough grade
Binder	Cobalt
Wheel diameter	7.0 in (17.8 mm)
Wheel width	0.125(3.18 mm)

Factor	High value	Low value
Traverse rate (v), f.p.m. (m min^{-1})	$a = 10$ (3)	$1 = 1$ (0.3)
Wheel speed (V), f.p.m. (m s^{-1})	$b = 6100$ (31)	$1 = 9150$ (46.5)
Wheel Depth of cut (d), in (mm)	$c = 0.160$ (4)	$1 = 0.040$ (1)
Grit size (S), U.S. mesh	$d = 50$–60	$1 = 30$–40
Concentration	$e = 35$	$1 = 15$
Type of granite	$f =$ Bright Red (difficult)	$1 =$ Lake Placid Blue (easy)

in Table 14.7. The measured dynamometer values (F_H and F_V) were converted to F_T and F_R values using the method of Brach *et al.* (1988) previously given. Details of the study are available in Pai *et al.* (1988).

Wheel peripheral speed, grit size, and work material were found to be the three dominant variables. Increases in wheel speed (V), wheel depth of cut (d), abrasive concentration, and granite hardness all gave rise to increased specific energy values, albeit in differing degrees. On the other hand, larger grit size and traverse rate (v) values caused the specific energy to decrease. Wheel speed and grit size have the strongest interactive effect. This means that the use of a lower wheel speed accentuates the effect of a higher grit size in diminishing specific energy.

Table speed, depth of cut, and wheel speed are the three major variables relative to power consumption. An increase in these variables and granite hardness all cause an increase in power P. A larger grit size and concentration both cause a decrease in power consumption. Table speed and depth of cut have the strongest interaction, meaning in this case that an increased table speed accentuates the effect of an increased depth of cut in increasing the power required.

In general, the specific energy of chip formation (u) varies inversely with $\bar{t}$. From eqn (4.5), it follows that u should decrease with an increase in v or d, but should increase with an increase in VCr or D. This is found to be consistent with the results of Table 14.7, with the exception of the variation of u with d. In this table it is seen that u actually increases with an increase in d, instead of decreasing. This is so because, for resin- and metal-bonded wheels

TABLE 14.7 *Results of the factorial design study*

	Specific energy (u), k.s.i. (MPa)	Power (P), h.p. (kW)
Mean value	136.19 (939)	0.297 (0.22)
Direct effects		
1. Traverse rate, v	∓ 16.91	± 0.244
2. Wheel speed, V	∓ 36.91	∓ 0.130
3. Wheel depth of cut, d	$\pm\ 9.14$	± 0.223
4. Grit size, S	± 35.05	± 0.072
5. Grit concentration	$\pm\ 7.28$	∓ 0.035
6. Work material	± 31.33	± 0.067

First-order interactions of the three major variables:
$I_{24} = \mp\ 21.58$ $I_{12} = \mp\ 0.123$
$I_{26} = \mp\ \ 8.58$ $I_{13} = \pm\ 0.196$
$I_{46} = \pm\ \ 6.83$ $I_{23} = \mp\ 0.102$

Second-order interactions of the three major variable:
$I_{246} = \mp\ 11.64$ $I_{123} = \mp\ 0.090$

of low porosity, the specific grinding energy also depends upon whether there is a chip storage problem.

Chip Storage

Metal- and resin-bonded wheels have low porosity and there is relatively little space for chip storage between the wheel and the work over the arc of contact. The mean protrusion of active grits beyond the bond (p) and the mean spacing of active grits cutting in the same path (l_r) then play important roles. If the rate of chip formation is greater than the chip storage space can accommodate, then chips will interfere with the grinding process, causing an increase in specific grinding energy (u) due to excessive rubbing between trapped chips and work. The volume of chips to be accommodated per unit time corresponds to the removal rate,

$$\dot{M} = vbd. \qquad (14.11)$$

The chip storage problem may be considered from three points of view:

(1) whether there is sufficient volume to accommodate the chips (volume constraint);
(2) whether the mean deformed chip thickness exceeds the mean protrusion (chip thickness constraint); or
(3) whether the mean deformed chip length exceeds the mean distance between active cutting edges in the wheel surface (length constraint).

From the point of view of the volume constraint, the condition for there to be no chip storage problem is that the chip storage volume swept out per unit time should exceed the removal rate:

$$Vbp > vbd$$

or

$$p > (v/V)d. \qquad (14.12)$$

From the point of view of the chip thickness constraint, the condition for there to be no chip storage problem is that

$$p > t_{max}. \qquad (14.13)$$

From the point of view of the chip length constraint, the condition for there to be no chip storage problem is that the mean distance between consecutive cutting edges in the wheel surface (the repeating distance, l_1) be greater than the mean deformed chip length (l_1):

$$l_r > l_1. \qquad (14.14)$$

Thus, it is evident that two of the chip storage constraints are related to grain protrusion (p) while the third is related to active grit spacing (l_r).

From eqn (14.12) an increase in either v or d has an equal tendency to create a chip storage problem due to a volume constraint. From eqns (14.13) and (4.5), an Increase in v or d has a tendency to create a chip storage problem due to a chip thickness constraint, where t_{max} is proportional to $v^{0.5}$ and $d^{0.25}$ respectively. From Table 14.7 an increase in d has an opposite effect to an increase in v. Thus, an increase in d appears to have a greater influence on u due to chip storage than a change in v. Since this is not in agreement with the above predictions based on volume or chip thickness constraints, it appears that neither of these constraints that depend on the magnitude of p is responsible for the observed unusual influence of d on specific energy. This conclusion is supported by the fact that the mean grit protrusion (p), measured by means of a tracer instrument, was 0.0021 in (53 μm), which is relatively high compared with what is considered to be an optimum protrusion (30 μm, 0.0012 in) according to Yokogawa and Yokogawa (1986a, b).

This suggests that the chip storage problem associated with the present experiments is due to the mean deformed chip length (l_1) being greater than the mean distance between active grits (l_r), which results in the chip being folded back on itself before being released. According to eqns (4.10) and (14.14), this type of chip storage problem should increase with an increase in d, but be independent of v for a given value of l_r. Thus, the reason why an increase in u does not correspond to a decrease in $\bar{t}$ with respect to d, as for all other pertinent variables, appears to be due to the opposing chip storage effect associated with a chip length constraint.

From the foregoing discussion, it is evident that chip storage may be an important consideration under the following conditions:

- for high removal rates (for example, for abrasive cut-off and conditioning billets in the steel industry)
- for wheels of low porosity (metal- or resin-bonded wheels)
- for long arcs of wheel-work contact (vertical spindle surface grinding, creep feed grinding, or sawing of rock)

In this connection, two wheel characteristics are of importance—mean grit protrusion (p) and mean repeating distance (l_r). Mean grit protrusion is controlled by removing the softer bond material by periodically grinding an abrasive stick or by electro-chemical removal of the bond in the case of a metal-bonded wheel. As grit protrusion is increased the holding power of the bond is decreased. Hence, there is an optimum grit protrusion relative to the chip storage and grit retention problems. The active grit spacing (l_r) is dependent on concentration, grit size, and grit-tip geometry (a function of the fracture characteristics of the abrasive and the dressing procedure).

Wheels with a Single Layer of Grits

For wheels where a single layer of grits is anchored to a steel core, greater chip storage space may be provided by anchoring the grits to the core by brazing instead of by nickel plating (Barnard 1990); Chattopadhyay and Hinterman 1992, 1993). Chattopadhyay and Hinterman (1992) found that a thin layer of chromium deposited by chemical vapor deposition (CVD) caused improved wetting of the Ni−P bonding alloy to diamond abrasive particles. This resulted in good grit retention with less bonding material, thus providing more chip storage space. For CBN grits a thin Cr layer gave poor adhesion, but a thin TiC coating did bond well to CBN and was well wetted, by the nickel-based brazing alloy. Figure 14.11 shows (a) Cr-coated D particles and (b) TiC-coated CBN particles anchored to the substrate alloy by a thin layer of brazing alloy. It is evident that in both cases uniform distribution of grits and considerable protrusion is achieved by the brazing technique.

Van der Sande *et al.* (1985) have also studied techniques for improving the adhesion between diamond abrasive particles and the bond in metal-bonded diamond wheels. It was found advantageous to coat the diamonds twice. The first thin coating should be a strong carbide former (Ti, Zr, or Hf) and the second a somewhat thicker metal coating (Al, Si, Ni, or metal alloys). Both coatings may be applied by tumble plating using balls of the material to be deposited. The coated diamonds are then mixed with the matrix material (Ni, Co, or Fe) and hot pressed in a graphite die at about 1000°C for up to 1 h. This causes a strong carbide chemical bond to be established between the diamond and the carbide former (Ti, Zr, or Hf) and between the metal coating, the carbide coating, and the matrix.

Tests used to determine resistance to pull-out were performed on hot pressed cylinders (1 cm diameter × 5 cm = 0.39 × 1.96 in) of coated 20–30 mesh diamonds and a matrix of Ni, Co, or Fe included:

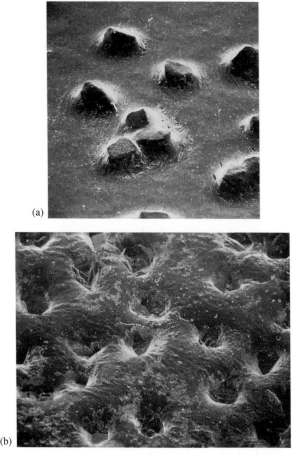

FIG. 14.11. (a) SEM of Cr-coated diamond grits brazed to a substrate by Ni–P brazing alloy. (b) SEM of TiC-coated CBN grits in a Ni–Cr matrix. (after Chattopadhyay and Hinterman 1992).

- a coarse grinding operation performed on the end of the cylinder
- impact fracture of the cylinder

In both cases SEM micrographs at 20 × were used to determine the extent of abrasive pull-out. In the case of both electrodeposited bonding and brazing, the method of sprinkling the abrasive particles on the wheel surface to provide a uniform distribution is important. This has been discussed by Suzuki *et al.* (1993).

Basalt

Basalt is a volcanic rock that has a number of applications. It may be sintered or cast, which gives it a wide range of possibilities. It is more corrosion resistant and has a lower density ($\sim \frac{1}{3}$) and lower thermal expansion ($\sim \frac{2}{3}$) than cast iron. Basalt is used for nonprecision parts requiring corrosion and abrasion resistance in cast form, and for smaller parts with greater dimensional accuracy in sintered form. Grinding characteristics correlate well with the equivalent chip thickness (t_e) discussed in Chapter 4 (eqn 4.15). Figure 4.12 shows values of grinding ration G, specific energy u, wheel life T, and surface roughness R_a, plotted against the equivalent chip thinkness t_e, where

$$t_e = vd/V. \qquad (14.15)$$

Comminution

Comminution is the process by which brittle materials are reduced in size, as in a ball or hammer mill. Because of its industrial importance, comminution has commanded considerable technical attention for over a century. A number of empirical rules have been proposed to enable comminution energy to be predicted. Those bearing the names of Kick and Rittinger have been most widely applied and debated (Haultain 1923). The following discussion of comminution and its relation to grinding with a grinding wheel is from Walker and Shaw (1954).

Kick's law is based on a critical strain energy concept. It states that the energy required to fracture a specimen is directly proportional to its volume and is independent of the number or size of particles into which it is reduced. That is,

$$U = K_1 \, (\text{Vol}), \qquad (14.15)$$

where U is the energy to which a volume (Vol) must be subjected to cause fracture and K_1 is a proportionality constant characteristic of a given material.

Rittinger's law is based on the idea that the energy required to crush a material is that needed to produce the new surface area generated. That is,

$$U = K_2 \, (A_2 - A_1), \qquad (14.16)$$

where K_2 is a proportionality constant characteristic of a given material and $A_2 - A_1$ is the resulting area change.

Investigators using Rittinger's law have found that crushing efficiencies are generally extremely low. Some data of Martin (1926), who crushed quartz in an 18×18 in (46×46 cm) tube ball mill using 1 in (25 mm) diameter balls, are shown in Table 14.8.

The data of Table 14.8 are seen to be in good qualitative agreement with Rittinger's law, but in poor agreement with Kick's law. However, the surface energy of quartz is known to be about 1000 erg cm^{-2} (0.0057 in lb in^{-2}). When this value is compared with the mean value of K_2 in Table 14.8, it is

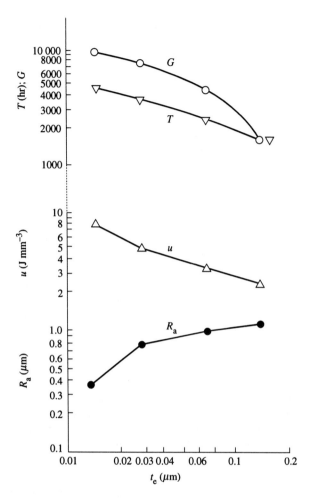

FIG. 14.12. The grinding characteristics of basalt using a metal-bonded wheel (after Marinescu and Bouas 1990). Wheel speed, 25 m s^{-1} (4920 f.p.m.); work speed, 0.27 m s^{-1} (52.5 f.p.m.); width ground, 6 mm (0.24 in).

TABLE 14.8 *Ball mill data for quartz (after Martin 1926)*

Energy consumed, U (ft lb)	New area produced $A_2 - A_1$ (ft^2)	$U/(A_2 - A_1) = K_2$ (ft lb ft^{-1})
243 375	3 971	61.3
470 250	7 852	59.9
699 190	11 170	62.6
892 346	14 941	59.7
1 097 300	17 899	61.3
		Average 60.9

found that only 0.112 percent of the energy associated with the process is being used to supply the necessary surface energy. Thus, from Rittinger's point of view, the efficiency of Martin's ball mill is only 0.112 percent.

Much of the literature in the field of comminution is found to consist of arguments supporting or rejecting one or the other of these two laws. Some investigators have found that their data tend to obey Kick's law, while others have gathered evidence to defend Rittinger's law. Haultain (1923) and Bond (1952) have suggested that the actual law lies halfway between those of Kick and Rittinger.

Experimental work in the field of comminution has been subject to several difficulties. In ball milling experiments the total energy consumed in crushing a charge from one mean particle size to another has been measured. This energy is the energy required to crush particles of a wide variety of sizes. From such data, it is not possible to determine precisely the energy required to crush particles of any one size. The chief variable studied has been the surface area, determined by elaborate screening analyses and by means of gas adsorption and permeability techniques (Kwong *et al.* 1949). In view of the very small percentage of the energy of comminution associated with the development of new surface, it would appear that entirely too much emphasis has been placed on surface area in previous investigations of comminution.

Drop tests have been adopted in some studies to improve the experimental precision. However, interpretation of the data is difficult because of the variety of particle sizes that result in such tests.

If Kick's and Rittinger's laws are expressed in terms of energy per unit volume, we obtain

$$u_K = k_1 \quad \text{(Kick)} \tag{14.17}$$

and

$$u_R = u/(A_2 - A_1)t_R \quad \text{(Rittinger)}, \tag{14.18}$$

where t_R is the mean thickness of the layer removed. Equation (14.18) resembles the variation of specific energy with undeformed chip thickness in grinding, and this suggests that grinding experiments might be used to estimate values K_2 in the comminution regime where Rittinger's law (eqn 14.18) holds.

Grinding tests were performed by Walker and Shaw (1954) on marble, gypsum, and talc using a 46 grit – 8-structure Al_2O_3 wheel for small values of t, and a 24 grit Al_2O_3 belt for large values of t. Figure 14.13 shows plots of specific energy versus undeformed chip thickness (t) for these three minerals. Values of t were obtained from eqn (4.13), taking t, the maximum undeformed chip thickness, to be $2\bar{t}$. It should be mentioned that eqn (4.13) was derived assuming the chips to be long continuous filaments, as produced when metals are ground (see Fig. 14.16, to be discussed later). For brittle materials reduced in size by comminution, the 'chips' will consist of many segments of continuous metal type chips. However, despite this discrepancy it is believed that values of t represent a reasonable approximation to the size of individual particles produced in comminution.

The experimental points in Fig. 14.13 fall on three sets of curves:

- curves (1) for very small values of t
- curves (2) for grinding wheel results for values t above a critical value
- curves (3) for grinding belt results for relatively large values of t

It has been suggested (Walker and Shaw 1954) that below a certain particle size the possibility of encountering a defect becomes so small that the theoretical strength of the material in shear ~ (shear modulus)/2π) is involved, thus giving rise to curve (1) (horizontal). It also appears that for curve (3) the specific energy approaches a constant value above a particle size of about 100 μin (27 μm). Thus, for very small and relatively large particle sizes, Kick's law seems to hold. However, the data appears to vary in accordance with eqn (14.18) at intermediate sizes, thus indicating adherence to Rittinger's law. It is suggested that comminution of very small and very large particles occurs by brittle fracture, whereas ductile behavior is involved in the intermediate size range. Particles that are so small that theoretical strength is involved will exhibit perfectly brittle fracture, while relatively large particles will approach the behavior of the bulk material, which is again brittle for minerals.

Figure 14.14 is a schematic representation of particles being crushed in a ball milling operation. Since the particles will be small compared with the size of the balls, they can readily be forced to shear when compressed and sheared between the balls, as shown in the inset in Fig. 14.14.

Ceramics

Introduction

In the future it is expected that greater quantities of ceramics will require processing in two important areas:

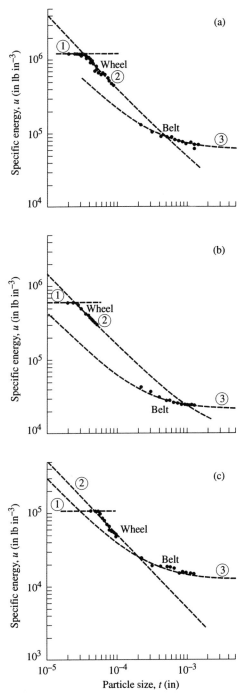

FIG. 14.13. The variation of specific energy with particle size (undeformed chip thickness, t) for (a) marble, (b) talc, and (c) gypsum (after Walker and Shaw 1954).

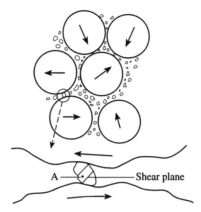

FIG. 14.14. A schematic representation of a particle being crushed in a ball milling operation.

(1) for heat- and wear-resistant machine elements;
(2) for electronic and optical devices.

In the past, ceramics have been primarily oxides. As applications of ceramics have expanded to include structural members, high-strength and temperature-resistant nonoxide ceramics have been introduced, such as silicon carbide and silicon nitride. These materials are usually produced by powder metallurgy techniques and are of course polycrystalline.

Whereas oxide ceramics have a mixture of covalent and ionic bonding, nonoxide ceramics have a much larger percentage of covalent bonds (in the direction of diamond and CBN). This results in a substantial increase in hardness and strength relative to oxide ceramics and is primarily responsible for the difficulty of grinding nonoxide ceramics.

Dislocations have much lower mobility in all ceramics than in metals, making them far less ductile than metals. Fracture toughness values of ceramics are at least an order of magnitude lower than for metals, and for glasses they are an order of magnitude lower than for ceramics. The thermal conductivity of ceramics is generally low (SiC being an exception), giving rise to high surface temperatures. High grinding temperatures coupled with substantial values of the linear expansion coefficient give rise to high thermal stresses that cannot be readily relieved by plastic flow due to the low dislocation density of ceramics.

The behavior of a sharp indenter loaded against a flat brittle surface is useful in visualizing the action of abrasive particles when a ceramic is being ground under conventional FFG conditions. Lawn and Wilshaw (1975) have proposed the sequence of steps shown in Fig. 14.15, where parts (a), (b) and (c) correspond to an increase in the applied load on a sharp indenter; (d) and (e) are for unloading and (f) is for complete removal of the load. At first, a

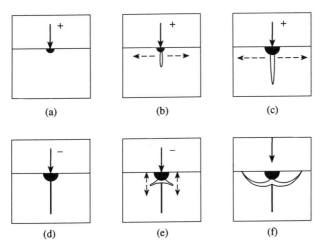

FIG. 14.15. Crack formation in brittle material during loading a sharp indenter into a surface (a-c) and during unloading (e-g) (after Lawn and Wilshaw 1975).

small plastic zone develops, (a), followed by a vertical crack extending downward from the bottom of the plastic zone (b). The dotted lines indicate a transverse tensile stress. As the load is further increased this vertical crack (called a median crack) extends downward and opens up. When the load is relaxed (d), the median crack no longer extends and closes. With a further reduction in load (e), a vertical tensile stress appears, causing lateral vent cracks to develop. When the load is sufficient, these vent cracks extend with further removal of load (f), until they reach the surface and small fragments pop out. Still another type of crack may arise in the surface. This latter type of crack (not shown in Fig. 14.15) is a ring crack that may form in the tensile region of the surface just beyond the contact area when the indenter is blunt. With further loading a ring crack extends downward and outward in a controlled fashion and becomes a cone crack. In ordinary FFG sharp indenters (Fig. 14.15) are involved and the principal defect is the one that appears at the lightest loads, which is the median crack. In ultra-precision diamond grinding, very fine abrasives are involved and the deformation is that of a blunt indenter. This is discussed in Chapter 15.

The situation in grinding is somewhat different from that depicted in Fig. 14.15, in that there is a horizontal as well as a vertical component of force, giving rise to an inclined resultant force. Also, the indenter is moving rapidly, giving rise to a high rate of strain. However, the sequence of events of Fig. 14.15 represents a plausible qualitative explanation of the chip-formation mechanism for conventional grinding of brittle materials.

Somiya (1984) has edited a collection of papers concerning the properties and application of ceramics in the electronics and mechanical industries.

FIG. 14.16. Representative chips produced by conventional pendulum grinding of (a) AISI 1020 steel and (b) Al_2O_3 (after Inasaki 1987).

Included are a chapter devoted to ceramic finishing techniques by Kobayashi and a chronology of ceramic developments.

Pendulum Grinding

Inasaki (1987) has studied the grindability of several structural ceramics using a 3.75 kW (5 h.p.) surface grinding machine in both the pendulum and creep feed grinding modes. Photomicrographs of pendulum grinding chips at high magnification showed a marked difference between metal and Al_2O_3 chips. The metal chips are long and thread-like (Fig. 14.16(a)), while the ceramic chips (Fig. 14.16(b)) are short and blocky, indicating a great deal more fracture in ceramic chip formation than for metals.

Figure 14.17 shows curves of specific grinding force $F_P{'}$ versus wheel depth of cut d for conventional (pendulum) wet diamond grinding of four different structural ceramic materials. Properties for these four materials are given in Table 14.9. It is evident that none of these quantities correlates well with the F_P values (F_P is proportional to specific energy u in this case). From Fig. 14.17 for a wheel depth of cut d of 15 μm (600 μin) the specific energy for Al_2O_3 is 8.8. GPa (1.27×10^6 p.s.i.), while the corresponding value for Si_3N_4 is 44 GPa (6.38×10^6 p.s.i.). Thus, the specific energy for Al_2O_3 is only about 13 percent of the corresponding value for a metal ($\sim 10 \times 10^6$), while that for Si_3N_4 is about 64 percent of that for a metal.

The specific energy associated with chip formation is the product of the stress and strain involved. The reason why the specific energy for Al_2O_3 is only about one-eighth of that for steel is a combination of lower strength and lower strain at fracture for Al_2O_3 than for steels. Similarly, the reason why the specific grinding energy for Si_3N_4 is so much higher than that for Al_2O_3 is due to its greater strength and strain at fracture.

An important difference between metal and ceramic grinding is the ratio of

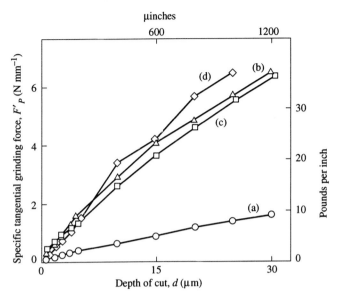

FIG. 14.17. The variation of the specific tangential force F'_P with the wheel depth of cut d for four different ceramics ground with a D170R100B wheel at $V = 22$ m s^{-1} (72.2 f.p.m.) and $v = 8$ m min^{-1} (26.2 f.p.m.): (a) Al$_2$O$_3$, (b) SiC; (c) ZrO$_2$; (d) Si$_3$N$_4$ (after Inasaki 1987).

the normal and tangential components of force (F_Q/F_P). Figure 14.18 shows values of F_Q' plotted against F_P' for comparable tests on Si$_3$N$_4$ and AISI 1020 steel. The ratio of F_Q'/F_P' for the ceramic is seen to be about 2.5 times that for the more ductile steel. This is due primarily to the greater hardness of the ceramic, which requires a greater normal force for the abrasive to penetrate the work. An important consequence of this is that a machine of greater stiffness is required when grinding Si$_3$N$_4$ than when grinding steel under the same conditions. It should be noted that here a diamond wheel is used to grind steel, which is inadvisable on the basis of wear, because of the incompatibility of diamond and hot iron. A more realistic comparison would be to use steel–Al$_2$O$_3$ or steel–CBN in the comparison, where F_Q'/F_P' is about two, and then F_Q'/F_P' for Si$_3$N$_4$ would be over four times that for steel.

Figure 14.19 shows the arithmetic average roughness (R_a) measured transverse to the grinding direction versus the wheel depth of cut (d) for the four ceramics of Fig. 14.17 and Table 14.9. The roughness for Al$_2$O$_3$ is appreciably greater than that for the other three ceramics. This is probably due to the lower value of F_Q' pertaining for Al$_2$O$_3$, which leads to less burnishing action. From this, it appears that high values F_Q' which are detrimental to accuracy are beneficial to surface finish. The finish is seen to

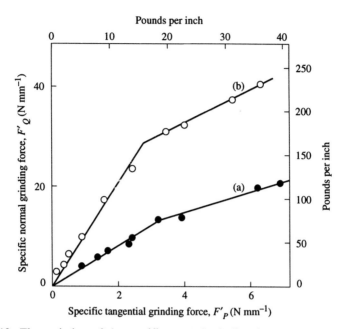

FIG. 14.18. The variation of the specific normal grinding force F'_Q with the specific tangential force F'_P for four different ceramics ground with a D170R100B wheel at $V = 22$ m s^{-1} (72.2 f.p.m.) and $v = 0.8$ m min^{-1} (26.2 f.p.m.): (a) AISI 1045 steel, $F'_Q/F'_P = 3.22$; (b) Si$_3$N$_4$, $F'_Q/F'_P = 8.33$ (after Inasaki 1987).

be independent of wheel depth of cut for all four ceramics, which is essentially the case for metals, where $R_a \sim \bar{t} \sim d^{0.25}$.

Kitajima *et al*. (1992) have studied the grindability of Al$_2$O$_3$, SiC, and Si$_3$N$_4$ ceramics. The grinding forces were least for Al$_2$O$_3$, while those for SiC and Si$_3$N$_4$ were about the same for the same grinding conditions. Grinding temperatures were least for Al$_2$O$_3$ (~ 575 °C) but much higher for Si$_3$N$_4$ (~ 1150 °C) and SiC (~ 1335 °C). The (R_t) surface roughness for Si$_3$N$_4$ was substantially lower than for Al$_2$O$_3$ and SiC, and increased with removal rate in all cases.

The grinding ratio is a strong function of the ceramic composition as well as the grinding conditions. Salje *et al*. (1985) measured the grinding ratio of several ceramics in internal grinding and found the following to hold for all ceramics investigated:

- the grinding ratio improved in going from a diamond grit size of 64 to 125

- the grinding ratio improved in going from a diamond concentration of 75 to 100

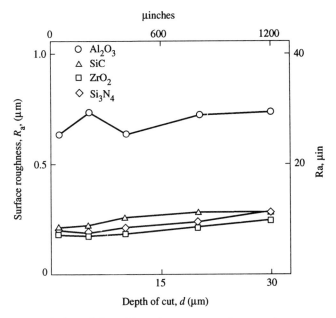

FIG. 14.19. The variation of the arithmetic average surface roughness measured transverse to the grinding direction for the four ceramics of Fig. 14.17 and ground under the same conditions, except for v, which was 4 m min^{-1} (13.1 f.p.m.) instead of 8 m min^{-1} (after Inasaki 1987).

- the grinding ratio improved in going from a resin bond to a metal (bronze) bond

The highest G values obtained when grinding with a diamond wheel of 100 grit size, 100 concentration, and a bronze bond were as follows for $V = 20 \text{ m s}^{-1}$ (3940 f.p.m.), $v = 0.5 \text{ m s}^{-1}$ (1.64 f.p.s.), and removal rate $= 2 \text{ mm}^3 \text{ mm}^{-1} \text{s}^{-1}$ (0.003 $\text{in}^3 \text{ in}^{-1} \text{s}^{-1}$):

Material	G
B_4C	61
SiC	418
Al_2O_3	186
Si_3N_4	78

Due to the brittle nature of ceramics, subsurface cracks produced by grinding may be a problem. Three- or four-point bending tests performed on a specimen with the ground surface on the tensile side provide a convenient test technique for studying this problem. As might be expected, the fracture stress is less when the tensile direction is transverse to the grinding direction than

TABLE 14.9 *Properties of four structural ceramics (after Inasaki 1987)*

Material	Vickers hardness	Bending strength, MPa (k.s.i.)	Young's modulus, GPa (p.s.i.)	Thermal conductivity, W m^{-1} K^{-1}
Al_2O_3	16	304 (44.1)	344 (49.9 × 10^6)	15
ZrO_2	13	980 (142)	206 (29.9 × 10^6)	–
SiC	25	490 (71.1)	392 (56.9 × 10^6)	92
Si_3N_4	15	588 (85.3)	294 (42.6 × 10^6)	14

when it is parallel to the grinding direction. Kayaba and Fijisawa (1986) have determined the depth of subsurface cracks by removing material from the ground surface progressively by lapping before bend testing. It was found in one instance that the depth of the damaged layer was 40 μm when the peak-to-valley roughness was only 2 μm.

Ota and Miyahara (1980) have studied the influence of grinding conditions on the flexural strength of diamond ground Si_3N_4 in four-point bending tests. The flexural strength parallel to the grinding direction was found to increase slightly with a decrease in grit size, as reflected by a small decrease in peak-to-valley roughness (R_t). The strength after grinding was also found to correlate more closely with a reduction in the mean force acting on a single grit (F_Q/lbC) than with the total grinding force F_Q. This important result is discussed in Chapter 15.

Subramanian and Ramanath (1994) have discussed diamond pendulum surface grinding of four ceramic materials (ZrO_2, Si_3N_4, Al_2O_3, and ferrite) using wheels of coarse (91 μm) and fine (6 μm) grit size. It was found that while the total grinding force increased with decreased grit size, the mean extimated force per grit decreased with decrease in grit size. The authors suggest that the lower force per grit was responsible for the observed greater transverse bending strength of surfaces ground with small grit size. While this is a strong possibility, it should be noted that the surface finish was much better with fine grits than with coarse grits and this could contribute to the observed difference in transverse bending strength.

Surface integrity was also found to depend on the fracture toughness of the work materials tested (the fracture toughness of the materials tested decreases in the order ZrO_2, Si_3N_4, and Al_2O_3, ferrite). The higher the fracture toughness of the work, the better the surface integrity was found to be. The authors suggest that the surface integrity of ground ceramic surfaces depends upon:

- grit size and shape
- machine stability (chatter)
- fracture roughness of work
- operating variables (V, v, d, and b')

the grinding temperature will of course also have an important influence.

Creep Feed Grinding

Inasaki (1987) studied the grinding of structural ceramics under creep feed conditions. Figure 14.20 shows the variation of specific normal force (F_Q') versus wheel depth of cut (d) for a constant, very low removal rate of 0.3 mm^3 mm^{-1} s^{-1} (4.7 × 10^{-4} in^3 in^{-1} sec^{-1}) over a wide range of values of d extending from the pendulum griding regime into the creep feed regime. In the pendulum grinding regime P'_Q is essentially constant with an increase in

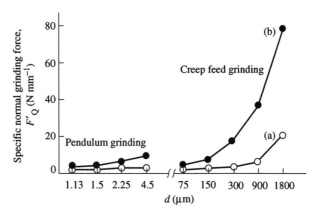

FIG. 14.20. The variation in the specific normal grinding force F'_Q versus the wheel depth of cut d for a constant removal rate of $0.3 \text{ mm}^3 \text{ mm}^{-1} \text{ s}^{-1}$ ($4.7 \times 10^{-4} \text{ in}^3 \text{ in}^{-1} \text{ sec}^{-1}$) (after Inasaki 1987). Wheel, D170-R100 B; $V = 22 \text{ m s}^{-1}$ (4330 f.p.m.). (a) Al_2O_3; (b) Si_3N_4.

d (and a decrease in v), but in the creep feed regime F'_Q rises rapidly with an increase in d for both Al_2O_3 and Si_3N_4.

Inasaki suggests that due to the unusually high values of F_Q' for an advanced structural ceramic such as Si_3N_4 it is questionable whether creep feed grinding is a good choice for such materials.

In Fig. 14.21 Inasaki shows the variation in specific grinding energy (u) with a parameter related to the mean undeformed chip thickness $\bar{t}$ when ceramics are diamond ground under conventional and creep feed conditions. In Chapter 4 the following value is derived for the mean undeformed chip thickness ($\bar{t}$) when long thread-like chips are produced in fine surface grinding of a metal (see eqn 4.13);

$$\bar{t} = \sqrt{\frac{v}{VCr}\sqrt{\frac{d}{D}}}.$$

For a given grinding wheel the quantity Cr should be a constant and, hence,

$$\bar{t}^2 = \frac{v}{V}\sqrt{\frac{d}{D}}. \tag{14.19}$$

Actually, ceramic chips will not be long fibers but fracture particles, as shown in Fig. 14.16. Nevertheless, it appears that, to a first approximation, the volume of individual fracture particles should be related to $\bar{t}^2$, and hence the use of $v/V(d/D)^{0.5}$ as Inasaki has done in Fig. 14.21 is reasonable for 'size effect' purposes.

The log-log plot of Fig. 14.21 clearly shows an unusually rapid increase in specific energy with decrease in chip size parameter $v/V(d/D)^{0.5}$. Since

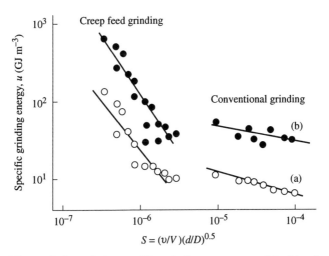

FIG. 14.21. The variation of the specific grinding energy u with chip size parameter $S = v/V(d/D)^{0.5}$ for (a) Al_2O_3 and (b) Si_3N_4 ground with a diamond wheel (D170-R100 B) at a wheel speed of $24\,m\,s^{-1}$ (4723 f.p.m.) (after Inasaki 1987).

grinding temperatures are directly related to specific energy (u) and grinding temperatures play such an important role relative to surface integrity, wheel wear, and accuracy, it appears—as Inasaki (1987) suggests—that unusually high temperatures are to be expected with creep feed grinding of ceramics. The unusually high values of F'_Q when ceramics are subjected to creep feed grinding apparently results in small fracture particles, and this gives rise to very high values of specific grinding energy as a consequence of the size effect.

The AA surface finish (R_a) measured parallel to the grinding direction is only slightly greater than that measured perpendicular to the grinding direction for ceramics. Figure 14.22 shows R_a values versus chip size parameter $v/V(d/D)^{0.5}$ when Al_2O_3 is ground over a very wide range of values extending grom the creep feed regime to the pendulum regime. The surface finish is seen to improve as the chip size parameter decreases. Surface roughness values for SiC, ZrO_2, and Si_3N_4 were found to be nearly the same, and about one-third of the values for Al_2O_3 (Fig. 14.19). This is due to the unusually high values of F'_Q pertaining for SiC, ZrO_2, and Si_3N_4 compared with Al_2O_3, which gives rise to greater burnishing action.

Nakagawa et al. (1986) introduced a cast iron bonded diamond wheel that may stand up to the unusually severe conditions of creep feed grinding of advanced ceramics such as Si_3N_4 when coupled with continuous dressing. This bond is achieved by sintering diamond particles and cast iron fibers compacted at high pressure with a small amount of carbonyl iron ($[Fe(CO)_5]$). Sintering is carried out at 1400 °C in an ammonium atmosphere for 1 hr.

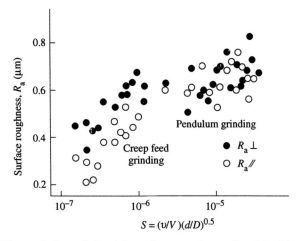

FIG. 14.22. The variation of the AA surface roughness R_a with chip size parameter $S = v/V(d/D)^{0.5}$ for Al_2O_3 ground with a diamond wheel (D170-F100 B) at a wheel speed of 24 m s^{-1} (4723 f.p.m.) (after Inasaki 1987).

Suto *et al.* (1990a) have employed the slotted wheel design shown in Fig. 14.23 to provide better cooling in creep feed grinding of high-temperature alloys and difficult-to-grind ceramics such as Si_3N_4 having a high F_Q/F_P force ratio. A coolant hole is drilled through each slot into a fluid chamber machined into the wheel. Diamond or CBN grits are electolytically attached to the wheel surface between the slots. The coolant is introduced as shown in Fig. 14.23 and forced out through each hole into its slot by centrifugal action. A 205 mm (8 in) diameter wheel was provided with 120 slots and a slot ratio of 30 percent. The specific grinding energy for creep feed grinding ($d = 2$ mm $= 0.08$ in) of four ceramics were as follows, using the wheel design shown in Fig. 14.23:

Ceramic	u	
	$J \text{ mm}^{-3}$	p.s.i.
Si_3N_4	150	21.8×10^6
ZrO_2	180	26.1×10^6
SiC	105	15.2×10^6
Al_2O_3	50	7.3×10^6

The above value for Si_3N_4 was found to be only 64 percent of the value for a conventional wheel–fluid system under the same operating conditions.

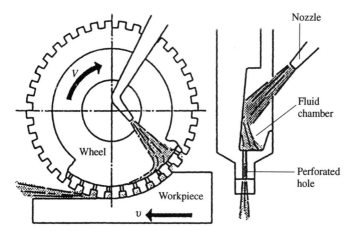

FIG. 14.23. A slotted grinding wheel with a centrifugally induced fluid pressure (after Suto *et al.* 1990*a*).

Suto *et al.* (1990*b*) have also used the wheel design shown in Fig. 14.23 with vitrified diamond segments (D170/200 Q100 V) and a segment ratio of 56 percent to creep feed grind two ceramic matrix composites (aluminum-reinforced Al_2O_3 and carbon-reinforced Si_3N_4). While the ceramic matrix composites required lower specific grinding energy than the monolithic ceramics, the surface roughness was substantially higher.

A centrifugally activated fluid distribution system similar to that of Fig. 14.23 has also been successfully used in the high-speed abrasive cut-off operation employing abrasive segments (Fig. 6.10).

Abrasive Cut-off

Murata *et al.* (1985) have studied the cut-off operation of structural ceramics. Rods of Si_3N_4, Al_2O_3 and SiC, 25 mm (1 in) in diameter, were cut as shown in Fig. 14.24, where a nonrotating workpiece clamped at an angle θ to the horizontal was cut by a metal-bonded 127 mm (5.0 in) diameter wheel that was fed horizontally at a rate v. Force components F_x and F_y were measured and converted to circumferential (F_T) and radial (F_R) values. Wheels containing a tough grade of synthetic diamond with grit sizes ranging from 80 to 400 and a concentration of 100 were used. The width of cut was either 0.5 or 1 mm (0.020 or 0.040 in), the wheel speed was varied from 21 to 36 m s^{-1} (4130–7100 f.p.m.), the table speed was varied from 10 to 30 mm min^{-1} (0.4–1.21 i.p.m.), and a water-based coolant was applied, as shown in Fig. 14.24.

The grinding forces varied as each cut progressed, and Fig. 14.25 shows representative values of peak force for the three work materials and for wheels of different grit sizes. Corresponding values for mild steel are also included

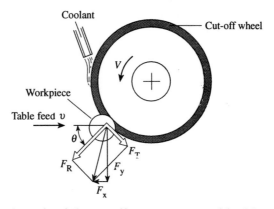

FIG. 14.24. A schematic of the cut-off arrangement used by Murata *et al.* (1985).

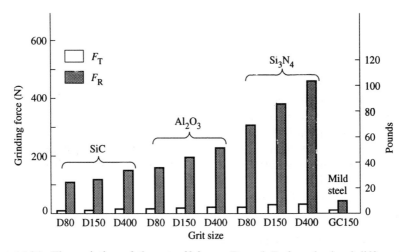

FIG. 14.25. The variation of the cut-off forces F_T and F_R for wheels of different grit sizes when grinding SiC, Al_2O_3, and Si_3N_4 rods (after Murata *et al.* 1985). All wheels had a metal bond and a diamond concentration of 100. Wheel speed, 20 m s^{-1} (3940 f.p.m.); feed rate, 20 mm min^{-1} (0.79 i.p.m.); wheel thickness, 0.5 mm (0.02 in); fluid, water-based coolant.

in Fig. 14.25. Values of F_R for Si_3N_4 were an order of magnitude greater than for mild steel. The force ratio F_R/F_T is also very high for all three ceramics (~10) compared with a value of 2–3 for mild steel. The best results from the standpoint of forces F_R and F_T are seen to occur with the coarse (80) grit size wheel.

The peak power consumed at the maximum removal rate is as follows:

TABLE 14.10 *Representative values of cutting ratio for cut-off of ceramic rods. Wheel, D 150-L 100-M; wheel thickness, 0.5 mm (0.020 in); wheel speed, 20 m s^{-1} (4000 f.p.m.); table feed, 20 mm min^{-1} (0.0008 i.p.m.) (after Murata et al. 1985)*

Si$_3$N$_4$	85
Al$_2$O$_3$	345
SiC	560

$$P = F_T V. \tag{14.20}$$

Substituting values from Fig. 14.25 for Si$_3$N$_4$ and an 80 grit wheel:

$$P = (20\,\text{N})(30\,\text{m s}^{-1}) = 400\,\text{J s}^{-1}(0.4\,\text{kW} = 0.54\,\text{h.p.})$$

This is well within the capacity of the 1.5 kW machine employed.

The specific energy for this operation is

$$u = F_T V/(vbl\sin\theta). \tag{14.21}$$

Substituting values from Fig. 14.25 for Si$_3$N$_4$ and an 80 grit wheel at the peak removal rate ($l = 25$ mm = 1 in):

$$u = 152.4\,\text{N mm}^{-2}(0.02 \times 10^6\,\text{p.s.i.}).$$

Corresponding values of P and u for SiC and Al$_2$O$_3$ are about one-third and one-half of those for Si$_3$N$_4$ respectively.

Even the value of u for Si$_3$N$_4$ is only about one-fiftieth of that for abrasive cut-off of mild steel using coarse Al$_2$O$_3$ grits. This is typical for values of u for cut-off of very brittle materials such as glass and ceramics, since the strain associated with chip formation is greatly limited by fracture when grinding very brittle materials.

The significance of high values of F_R is:

- a tendency for edge chipping
- rapid wear of abrasive particles

Table 14.10 gives representative values of the grinding ratio for the three materials investigated by Murata *et al.* (1985).

The main reason for the high values of the radial force and the relatively low values of the grinding ratio in ceramic cut-off grinding is the rapid rate at which chip storage associated with grit protrusion occurs due to grit tip wear (wheel face glazing). This requires frequent dressing of the wheel face to erode away the bond in order to provide chip storage space associated with increased protrusion.

Electrolytic in-process Dressing

Murata *et al.* (1985) developed the electrolytic-in-process dressing system shown in Fig. 14.26. The coolant (an electrolyte such as an aqueous NaCl

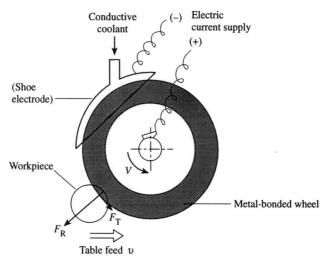

FIG. 14.26. A schematic of the cut-off arrangement with in-process electrolytic dressing (after Murata *et al.* 1985).

solution) is fed into a noncontacting shoe and a low voltage (~4 V) current is passed between the metallic show and the metal-bonded wheel. The polarity is such that electrons flow from the wheel to the shoe. This causes bond material between the grits to be depleted, causing the chip storage space to be maintained as flats develop on the active grits. A control system incorporating a dynamometer adjusts the current to maintain a constant F_R/F_T ratio. Figure 14.27 shows representative results with and without in-process electrolytic dressing. Without in-process dressing, the grinding energy (peak motor current) is found to reach an unacceptable level after relatively few cuts requiring dressing. The downtime associated with controlled in-process dressing is eliminated, resulting in an important improvement in productivity. Details of the in-process control system are to be found in Murata *et al.* (1985).

In addition to elimination of downtime for periodic dressing, the in-process electrolytic dressing system gives greater grinding ratios and a decrease in waviness of the ground surface. There is, however, a small increase in surface roughness due to less burnishing action with a wheel having a more open structure and less glazing. The large reduction in F_R with in-process electrolytic dressing gives less chipping at the edges of the ground surfaces. In order to prevent corrosion of machine elements, a rust inhibitor such as a salt that introduces $[NO_2^{-1}]$ ions into the electrolyte is recommended by Murata *et al.* (1985). It is further mentioned that the effective grade of a wheel may be altered by adjusting the current, thus decreasing the number of wheel grades required to grind a variety of ceramics.

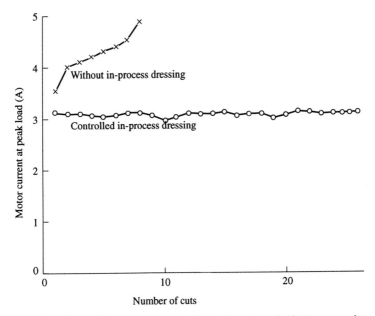

FIG. 14.27. The effect of in-process electrolytic dressing of Si_3N_4 on a change of motor current (A). Grinding wheel, D80 L100 M; wheel thickness, 1 mm (0.040 in); wheel speed, 20 m s^{-1} (3940 f.p.m.); feed, 20 mm min^{-1} (0.79 i.p.m.). Electrolytic current adjusted to 20 A to give $F_R/F_T = 6.67$ (after Murata et al. 1985).

Ferrites

Nickel–zinc and manganese–zinc ferrites used in magnetic storage systems are ceramics that are relatively easy to grind from the standpoint of specific energy and tool life. However, when these materials are sawed, ground, or lapped the surface is found to have a shallow inactive 'dead' layer, which adds to the effective spacing between tape and head that adversely affects the resolution and storage capacity of the system. This 'dead' layer is related to the presence of a positive or negative residual stress.

The specific energy in grinding of ferrite was found to be about one-seventh of that for a comparable metal, due primarily to the low strain associated with chip formation for the more brittle ferrite. While the value of $\beta = (k\rho C)^{0.5}$ for ferrite is only about one-half of that for a comparable metal, the value of u/β which is proportional to the mean surface temperature in grinding will be about $7 \times 0.5 = 3.5$ times as great for metal as for ferrite when grinding under the same conditions. This results in the residual grinding stress for ferrite being of mechanical origin and hence compressive.

To provide a suitable finish, ferrite surfaces must be diamond lapped following grinding (1–3 μm D on bronze lap). Although the specific energy is about 25 times greater for lapping than grinding, the removal rate is only about 1/5000 for lapping, as for grinding. This results in essentially no temperature rise in lapping and hence a net residual stress of mechanical origin. Despite this residual stress being compressive, this gives rise to a detrimental 'dead' zone in a magnetic storage application.

Chandrasekar et al. (1987) have pointed out that the maximum residual stress is greater for lapping of ferrites than for grinding, and hence improvement is more apt to come from better lapping than from better grinding. The following two approaches offer promise:

1. Simultaneous etching (7 v/o solution of phosphoric acid) with lapping will reduce the specific lapping energy by about 30 percent and hence the residual stress.
2. Cooling the workpiece to a low temperature before lapping will make the ferrite more brittle, thus reducing the specific lapping energy and residual stress.

Ueno et al. (1979) studied the grinding characteristics of resin-bonded fine grit diamond wheels on Mn-Zn ferrite using different bonding materials and grinding aids incorporated into the bond. It was found that use of a bond of polyoxybenzoil blended with thermosetting polyimide together with small particles (0.1 μm) of the solid lubricant WS_2 (10 w/o) gave improved grinding performance. With this wheel a peak-to-valley (R_t) surface finish of 0.2 μm (8 μin) and a dead zone of a few μm were obtained. Under the following grinding conditions the specific grinding energy (u) and force ratio (F_Q/F_P) were as follows:

> Wheel: 150 mm (5.9 in) dia., 1500 grit size D, 50 concentration
> Wheel speed, V: 22 m s^{-1} (4330 f.p.m.)
> Work speed, v: 3.33 mm s^{-1} (7.87 i.p.m.)
> Wheel depth of cut, d: 20 μm (800 μin)
> Fluid: water-soluble
> Specific energy, u: 322.7 J mm^{-3} (46.8 × 10^6 p.s.i.)
> Force ratio, F_Q/F_P: 13

The specific energy and force ratio are unusually high for this material, probably due to the very small diamond size employed. The grinding ratio G was 3000 when the WS_2 grinding aid was used but only 500 without the grinding aid.

Farris and Chandrasekar (1989) have measured residual surface stresses by the deflection technique after diamond grinding a Ni-Zn ferrite and typical structural ceramics. Figure 14.28 shows representative values of residual surface stress of ground surfaces versus depth below the surface. The ferrite which is much more brittle (fracture toughness = 1.1 MPa m$^{1/2}$) than either Si_3N_4 (3.5-4.0 MPa m$^{1/2}$) or zirconia (8.5 MPa m$^{1/2}$) is seen to give much lower values of residual stress. All three materials show a compressive residual stress, which is to be expected for very refractory materials for

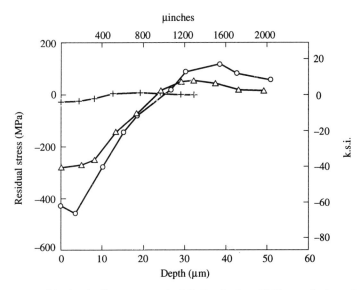

FIG. 14.28. Residual grinding stresses in Ni–Zn ferrite, Si_3N_4, and zirconia. Wheel velocity, 37 m s^{-1} (7282 f.p.m.); table speed, 23 mm s^{-1} (54.3 i.p.m.); wheel depth of cut, 0.02 mm (800 μin) (after Farris and Chandrasekar 1989).

which the residual stress will be of mechanical origin, as discussed in Chapter 10.

Temperatures

Ueda *et al.* (1992) have measured grinding temperatures for horizontal spindle surface grinding of ceramics using an InSb infrared radiation pyrometer with a chalcogenide fiber optic. The fiber optic was mounted in a 0.6 mm blind hole in the workpiece. A chalcogenide fiber was used instead of quartz, since lower temperatures are involved when grinding ceramics than when grinding steel; and a chalcogenide fiber transmits longer wavelengths than quartz, and hence is better for use at lower temperatures. A 200 mm resin-bonded diamond wheel of 60–80 grit size and 80 concentration was used dry in up grinding mode with a wheel depth of cut of 20 μm (800 μin), a wheel speed of 26 m s^{-1} (5150 f.p.m.) and a work speed of 12 m min^{-1} (39.4 f.p.m.). Figure 14.29 shows the output temperatures for Si_3N_4, SiC, and Al_2O_3 workpieces when the fiber was 20 μm (800 μin) below the surface, while Fig. 14.30 shows the variation of temperature with depth below the surface (z) for the maximum temperature, including the peaks. Si_3N_4 gave a much higher maximum temperature (~800 °C at the surface) than SiC or Al_2O_3 (~200 °C). It was suggested that the higher temperature gradient near the surface for Si_3N_4

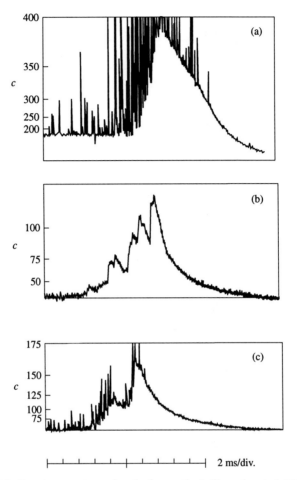

FIG. 14.29. Grinding temperature signals for optical fibers located 20 μm (800 μin) below the surface for ceramic work materials sintered at atmospheric pressure: (a) Si_3N_4; (b) SiC; (c) Al_2O_3, (after Ueda et al. 1992).

than for SiC or Al_2O_3 is due to the greater thermal diffusivity of Si_3N_4 ($\alpha = 11.1 \times 10^{-6} \, m^{-2} s^{-1}$ for Si_3N_4, $31.9 \times 10^{-6} \, m \, s^{-1}$ for SiC, and 6.1×10^{-6} for Al_2O_3).

Single-grit temperature measurements on Ni–Zn ferrite and AISI 1045 steel were made by Chandrasekar et al. (1992) by the infrared radiation technique, both at the surface and at the bottom of a hole 7 μm (280 μin) below the surface. The mean temperature at the surface was found to be 600 °C for the ferrite and 1000 °C for the steel. The corresponding peak temperature measured at the bottom of the small hole 7 μm (280 μin) below the surface was 370 °C for the ferrite.

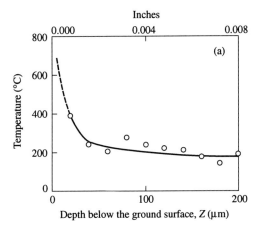
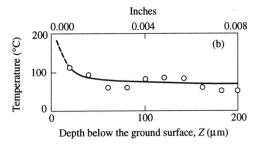
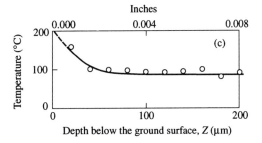

FIG. 14.30. The variation of grinding temperatures with depth below the surface: (a) Si_3N_4; (b) SiC; (c) Al_2O_3 (after Ueda *et al.* 1992). (a) $F_P = 178$ N (40 lb), $F_Q = 25.4$ N (5.7 lb), $P = 665$ W (8.9 h.p.); (b) $F_P = 92.4$ N (20.8 lb), $F_Q = 17.6$ N (3.96 lb), $P = 461$ W (6.2 h.p.); (c) $F_P = 121.7$ N (27.4 lb), $F_Q = 15.7$ N (3.53 lb), $P = 411$ W (5.5 h.p.).

Values of thermal conductivity (k), density (ρ), and specific heat (C) were measured for the ferrite and the following results were obtained:

$k = 8 \text{ W m}^{-1}\,{}^\circ\text{C}^{-1}$,
$\rho = 5300 \text{ kg m}^{-3}$,
$C = 810 \text{ J kg}^{-1}\,{}^\circ\text{C}^{-1}$,
$\beta_F = (k\rho C)^{0.5} = (8 \times 5300 \times 810)^{0.5} = 5860 \text{ N}/(\text{mCs}^{0.5})$.

Comparable values for diamond were as follows:

$k = 1000 \text{ W m}^{-1}\,{}^\circ\text{C}^{-1}$,
$\rho = 3500 \text{ kg m}^{-3}$,
$C = 525 \text{ J kg}^{-1}\,{}^\circ\text{C}^{-1}$,
$\beta_D = (1000 \times 3500 \times 525)^{0.5} = 42\,866 \text{ N}/(\text{mCs}^{0.5})$.

From eqn (9.22), the value of the thermal distribution coefficient for the diamond–ferrite combination is

$$R = 1/[1 = \beta_D/\beta_F] = 0.12.$$

Hence when ferrite is ground with diamond, very little of the energy dissipated goes to the ferrite.

Glasses

Introduction

According to the American Society for Testing Materials (ASTM), glass is an inorganic product of fusion which has cooled to a rigid condition without crystallizing. Figure 14.31(a) shows the structure of the most common glass former (SiO_2) in the glassy (amorphous) state, while Fig. 14.31(b) shows the same material in the crystalline state. The amorphous structure is that of a supercooled liquid that, unlike the crystalline state, has no long-range order. Obsidian is a natural glass of volcanic origin. It is amorphous stone cooled from the molten state under such enormous pressure that it become rigid before crystallizing. This was used for knives, tools, and weapons as long ago as 12 000 BC.

Amorphous glass is very strain rate sensitive and temperatures corresponding to different viscosities are used to characterize the material. As a glass is heated it becomes less viscous and the following four points on the viscosity–temperature curve are usually specified:

1. The working point is the temperature at which the viscosity is suitable for working or forming, and is equal to 10^4 Poise.
2. The softening point is the temperature at which a fiber 1 mm (0.040 in) in diameter will stretch under its own weight at a rate of 1 mm min^{-1} (0.040 i.p.m.). Viscosity = $10^{9.6}$ Poise.
3. The annealing point is the temperature at which residual stress will be removed in 15 min. Viscosity = 10^{13} Poise.

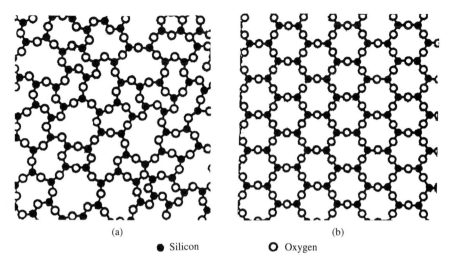

● Silicon   ○ Oxygen

FIG. 14.31. The structure of typical glass former (SiO_2): (a) a glassy (amorphous) structure; (b) crystalline structure (after Adams *et al.* 1984).

4. The strain point is the temperature below which residual stress will not be introduced by rapid cooling. Viscosity = $10^{14.5}$ Poise.

Materials in the glassy (amorphous) state are unusually brittle and are commonly perfectly brittle (fracture stress < yield stress). A completely crystallized glass is a ceramic which has entirely different properties than the glass:

- higher fracture strain
- higher stiffness (Young's modulus)
- more refractory
- lower coefficient of thermal expansion
- higher thermal stress resistance
- opaque as opposed to transparent

Commercial glasses may be broadly classified as follows:

Low-melting glasses
A. Soda-lime = SiO_2 + Na_2O and CaO, softening point ~ 700 °C; window and bottle glass—about 90 percent of all glass made.
B. High lead = up to 80 percent PbO, softening point ~ 440 °C; used for art objects and soldering.

High-melting (hard) glasses
C. Borosilicate = SiO_2 + B_2O_3 + other oxides, softening point ~ 700 °C; good thermal shock and chemical resistance.
D. Aluminosilicate = SiO_2 + Al_2O_3 + other oxides, softening point ~ 900 °C; used for high operating temperatures and good electrical properties.

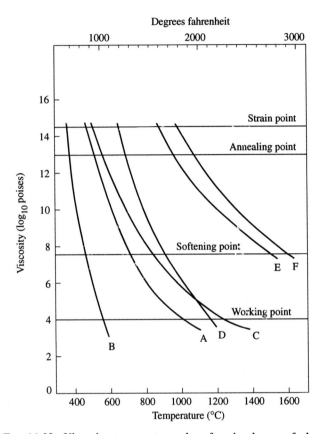

FIG. 14.32. Viscosity–temperature plots for six classes of glass.

Very refractory glasses

E. 96 w/o SiO_2, softening point ~ 1500 °C; excellent thermal shock resistance and high-temperature (900 °C) stability.

F. Fused silica = 99 percent amorphous SiO_2, softening point ~ 1600 °C; refractory and high thermal shock resistance.

Figure 14.32 shows representative viscosity–temperature curves for the six classes of glass.

Indentation hardness may be measured without fracture for all glasses provided that the load on the diamond indenter is low enough. This is because glasses are very much stronger in compression than in tension and exhibit a substantial size effect (increase in fracture stress with decrease in volume subjected to stress). The Knoop hardness numbers for all six classes of glass are essentially the same, ranging from about 450 to 500 kg mm^{-2}. Thus, hardness should not be expected to be a measure of grindability of these glasses, as

TABLE 14.11 *Representative values of Knoop hardness and coefficient of thermal expansion (α) for six classes of glass in order of decreasing grindability*

Glass type	Knoop hardness	$\alpha\ C^{-1}$
Soda-lime	465	94×10^{-7}
High lead	–	84×10^{-7}
Borosilicate	418	34×10^{-7}
Aluminosilicate	514	44×10^{-7}
96 percent SiO_2	487	8×10^{-7}
100 percent SiO_2	489	6×10^{-7}

measured by the grinding specific energy required. The range of hardness for optical glasses is greater (345–750 kg mm^{-2}) and for these glasses specific energy increases with increased hardness.

A more reasonable criterion for grindability would appear to be resistance to thermal shock. Since specific energy is equal to the product of stress and strain, it should be expected to vary as the strain associated with chip formation for different glasses, since the stress ($\sim$ hardness) is about the same for many glasses. When grinding glass, the strain associated with chip formation will be limited by fracture. Since the physical property of glass most clearly related to thermal shock resistance is the coefficient of thermal expansion, it is to be expected that the grindability of glass as measured by specific energy should increase with an increase in the coefficient of thermal expansion (α). Table 14.11 gives approximate values of hardness and α for the six classes of glass previously considered. From this it is to be expected that grinding of soda-lime glass would require the lowest specific energy and fused quartz the highest.

Edge Grinding

Murugesan (1993) has investigated the edge grinding of glass using grooved metal-bonded diamond wheels. When sheet glass is cracked off by scoring with a small sharp edged tungsten carbide wheel, followed by bending, the edges are sharp, wavy, and rough and frequently must be ground. The glass specimens used were 1 mm (0.040 in) thick. Figure 14.33 shows the edge grinding arrangement employed by Murugesan.

Preliminary experiments revealed that the power component of force (F_P) was only slightly less for down grinding than for up grinding, but that the radial force (F_Q) was about twice as high for down grinding as for up grinding. In order to avoid edge chipping that would be greater for high values of F_Q, up grinding was chosen for further study.

Surface finish was not a critical issue and was about the same for up versus down grinding. In fact, surface finish was about the same for all values of the principal operating variables (V, v, and d, shown in Fig. 14.33) but did

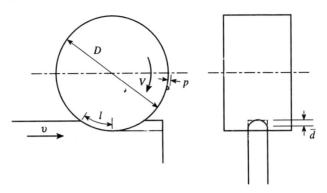

FIG. 14.33. An edge grinding arrangement (after Murugesan 1993).

vary slightly with diamond grit size and concentration. For a 40–50 grit size, 100 concentration wheel, the AA surface finish (R_a) was about 33 μin (0.83 μm) over a wide range of operating conditions, but only about 20 μin (0.50 μm) for a 30–40, 125 concentration wheel. This is consistent with general glass grinding experience, in which it is found that for stock removal grinding a low concentration (~ 50) is used, whereas for optical parts where finish is very important diamond concentrations as high as 200 are used, with fine grit sizes as small as 600 μm (0.024 in) before polishing.

Representative values of specific energy and power are given in Fig. 14.34 versus the material removal rate $\dot{M}$. The specific energy exhibits an increase in u with a decrease of $\dot{M}$, due to the usual size effect in fine grinding. The power consumed is seen to be about the same up to a critical removal rate of about 0.1 in^3 min^{-1} (1639 mm^3 min^{-1})). This was also found to be the case for other wheels having different grit sizes and concentrations.

The ratio of F_Q/F_P was found to increase to values above 5 for removal rates above the critical value of about 0.1 in^3 min^{-1} (1639 mm^3 min^{-1}). This is due to insufficient space between active grits and bond to accommodate the chips over the contact length l, giving rise to an increase in power consumed and hence an increase in temperature and wheel wear.

The ratio of chip storage space available to chip storage space required (R_s) should be expected to be some function ψ of the following variables:

$$R_s = \psi(V, v, \bar{d}, D, p), \qquad (14.22)$$

where p is the mean protrusion of active grits. Since R_s is nondimensional, dimensional analysis gives

$$R_s = \psi_1(V/v, D/\bar{d}, p/D). \qquad (14.23)$$

The nature of the function ψ_1 may be estimated only by a large number of very difficult experiments or by further analysis. Hence, the latter approach will be taken.

HARD WORK MATERIALS

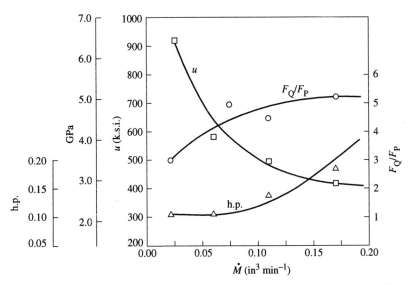

FIG. 14.34. The variation of specific energy (u) power and force ratio (F_Q/F_P) for edge grinding of glass versus the material removal rate (after Murugesan 1993). Wheel, 40-60, 100 conc., metal-bonded diamond; wheel speed, 6000 f.p.m. (30.5 m s^{-1}); work speeds, 13.1 and 39.4 f.p.m. (3.99 and 12.0 m min^{-1}); mean wheel depths of cut, 0.0039 and 0.0078 in (1 and 2 mm); width ground, 1 mm (0.040 in); fluid, 3 v/o water-based glass grinding fluid.

The volume available for storing chips during the wheel–work contact l per unit time will be

$$B_1 = Vpb(l/\pi D) . \qquad (14.24)$$

The volume of chips to be stored per unit time will be

$$B_2 = v\bar{d}b . \qquad (14.25)$$

Hence, after substituting for $l = (D\dot{d})^{0.5}$,

$$R_s = B_1/B_2 = (Vp)/(\pi v\sqrt{D\bar{d}}) . \qquad (14.26)$$

This is in agreement with the dimensional analysis result (eqn 14.23).

In cases such as this, in which chip storage space is a problem, conditions may be improved by:

- increasing V or p, or
- decreasing v, $\bar{d}^{0.5}$, or $D^{0.5}$ (i.e. decreasing v or l)

It is expected that wheel life between dressings will become a problem when the removal rate exceeds that which may be readily accommodated. However, to demonstrate this is practical only in a production environment, since the amount ground between dressing operations to increase grit protrusion (p)

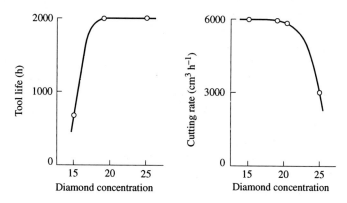

FIG. 14.35. The variation of the wheel life and the cutting ratio with diamond concentration for a 700 mm (27.6 in) diamond wheel (after Seiforth 1985).

and to restore geometry of the groove in the wheel periphery involves many thousands of feet of glass.

Slitting of Glass

Blocks of glass are slit into sections as thin as 2 mm (0.080 in) using diamond saws. Seiforth (1985) gives results using a 700 mm diameter (27.5 in) saw having segments of 3 mm (0.12 in) width, containing 70–80 mesh diamonds with a concentration of 19 in a cobalt–bronze bond. The wheel speed was 25 m s^{-1} (4900 f.p.m.) and a light mineral oil was used as coolant. Kerosene gave somewhat better tool life but was not used due to unsatisfactory odor and fire hazards. Higher wheel speeds gave poorer tool life.

Figure 14.35 shows the variation in wheel life and cutting rate versus diamond concentration. A concentration of about 20 is seen to be optimum. For thinner cuts a finer grit size (120–140 or 170–200 mesh) and a smaller wheel diameter (300 mm, or 13.8 in) is used, resulting in a wheel life about 25 percent of that for Fig. 14.35. It was found that use of a segmented wheel design instead of a monolithic one gave improved wheel life ($\sim 2 \times$) at a greatly increased removal rate (600 percent higher). This was due to improved coolant supply and swarf removal.

The life of a 20 concentration diamond wheel of 700 mm (27.6 in) diameter was found to be about one million cm^3 of glass sawed (61 000 in^3).

Machining of Glass

Giovanola and Finnie (1980) conducted a preliminary study to explore the feasibility of cutting glass just as metals are machined. It was found that soda-lime glass and a lead-doped flint glass gave chips that were continuous and saw-toothed, and similar to those obtained by Nakayama *et al.* (1974) when machining hard metals. A diamond tool with a semicircular face and a rake angle of $-30°$ was used at widths and depths of cut of 100 and 1.6 μm (0.004 in and 64 μin) respectively.

This similarity of chip formation between relatively ductile metals and brittle glass has given rise to the belief that in both cases chip formation involves ductile action. In reality, such serrated or saw-tooth chips result from periodic gross fracture, with a very thin band involving ductile action at the chip-tool interface when the compressive stress is sufficient to arrest these periodic cracks. This occurs only when the undeformed chip thickness is small enough. The matter of so-called ductile mode chip formation of very brittle materials is discussed in more detail in Chapter 15.

Giovanola and Finnie (1980) found that whereas soda-lime and doped flint glass gave saw-tooth chips, fused quartz did not, even at the lowest undeformed chip thicknesses possible with the diamond tool used.

Ultrasonic Grinding

Ultrasonic grinding, which is a nontraditional abrasive operation, is discussed here since it is useful only on very brittle materials such as glasses or ceramics of low fracture toughness.

Materials are brittle only because they contain relatively large imperfections which cause them to rupture before they can develop sufficient stress to cause plastic flow. Furthermore, brittleness is a function of the size of a specimen (Shaw 1954) and is not a fixed property of a material. This becomes clearly evident when one compares the results of making standard Brinell hardness tests on marble with those involving a diamond pyramid and light load in accordance with standard micro-hardness techniques. In the Brinell test the specimen will shatter, while in the micro-hardness test, perfect impressions of the indentor can be obtained without evidence of cracks at even the highest magnification. The Brinell indentation is so large that there is a certainty of encountering a critical imperfection in the plastically deformed region. On the other hand, the indentation in a micro-hardness test is so small that the probability of encountering an imperfection is negligible and the material behaves with perfect plasticity. The grinding action considered here is very similar to that accomplished in a ball mill which, however, has as its objective the reduction of particle size rather than the production of holes and surfaces.

In ultrasonic grinding a tool is given a high-frequency, low-amplitude oscillation along its axis which, in turn, transmits a high velocity to fine abrasive particles that are present between the tool and the workpiece. These abrasive particles cause small pieces to be chipped from the bottom of the hole as they are driven into the workpiece. The shape of the cavities produced corresponds closely to the shape of the tool.

The required oscillatory motion may be imparted to the tool by magnetostriction, which is the property of a metal that causes it to contract in a

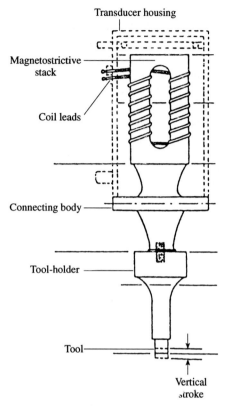

FIG. 14.36. A ultrasonic grinding transducer.

magnetic field. Different metals exhibit different degrees of magnetostriction, and nickel that is properly heat treated exhibits a particularly high degree. High-frequency oscillation is produced by subjecting a stack of nickel plates cemented together by an insulating cement to a high-frequency alternating magnetic field. This stack of nickel plates with its field winding constitutes the transducer which transforms an electrical oscillation into a mechanical one.

The major components of an ultrasonic grinding apparatus consist of the following: an electronic oscillator with an amplifier and a means for adjusting the frequency over a small range, the transducer, a connecting cone (tool-holder), the tool, a machine base and feeding means, and a fluid system for circulating an abrasive slurry. Dual-abrasive systems are frequently used to make the change from roughing to finishing more convenient. A transducer is illustrated in Fig. 14.36.

The amplitude and frequency of the transducer are limited by certain practical considerations. The lower limit of useful frequency lies at the audible threshold (about 15 kHz). Frequencies below this value can be heard and lead to considerable discomfort. The rate of cutting varies as the 1.5 power of the amplitude and hence the amplitude of oscillation should be as high as possible. However, transducer amplitudes in excess of 0.001 inch (25 μm) lead to fatigue failure of the nickel plates. The maximum transducer amplitude is therefore fixed just below 0.001 in (25 μm).

The cutting rate also increases with frequency and hence high frequency is desirable. However, the difficulty of cooling the transducer increases with frequency. It is found that 20–25 kHz represents a convenient upper limit of frequency for which water cooling of the transducer is adequate.

The power required depends upon the size of the transducer. A transducer that has a diameter of 2.5 in requires about 500 W to drive it at 25 kHz with an amplitude of about 0.001 in (25 μm).

An amplitude greater than 0.001 in (25 μm) is desirable at the tool for rapid cutting and can be obtained by use of an element of decreasing cross-section, mounted between the transducer and the tool (Fig. 14.36). The connecting section (usually made from monel metal) will have better fatigue properties than nickel and amplitudes greater than 0.001 in (25 μm) may be used. The increase in amplitude that can be obtained in this way depends upon the diameter of the tool relative to the diameter of the transducer. A gain in amplitude between 2 and 6 can be obtained by proper cone design.

The length of the cone must be carefully adjusted so that the transducer and cone will be in resonance and hence the amplitude will be a maximum. Details of cone design may be found in Neppiras (1953).

Tools are normally made of cold rolled steel and hence are inexpensive. This is important since tool wear is usually high. Tool wear offers little difficulty, since wear is uniform over the face of the tool and hence the shape of the tool changes very slowly.

The abrasive is pumped between the tool and work in a 40 w/o water slurry. Boron carbide (B_4C) is the abrasive most commonly used and grit sizes from 240 to 1000 are used.

Water is the material normally used to transport the abrasive, since it is better able to carry away heat. The ability to get the fluid to all points of the tool is one of the factors limiting the maximum size of tool that can be used. Normally, a thinner slurry gives better results with relatively large tools. When shiny metal spots are evident on the bottom of the ground surfaces, this is an indication that the tool is too large for the unit. In such a case the amplitude is not sufficient and the shiny spots result from continuous metal-to-metal contact. Normally, those materials that machine more readily can accommodate the largest tools.

The average performance shown in Table 14.12 was obtained with the several materials listed on a 700 W ultrasonic grinder. This unit has a 2.5 in (6.35 cm) diameter transducer section, and the maximum tool diameter recommended is about 2 in (5 mm) for the most easily ground material (glass). In this case the rate of tool wear is very low ($G = 100$) and the rate of penetration

TABLE 14.12 *Representative ultrasonic grinding performance*

Material	Ratio of stock removed to tool wear	Maximum practical grinding area (in^2)	Average grinding rate, $\frac{1}{2}$ in dia. tool, $\frac{1}{2}$ in Deep (in min^{-1})
Glass	100:1	4.0	0.150
Ceramic	75:1	3.0	0.060
Germanium	100:1	3.5	0.085
Tungsten carbide	1.5:1	1.2	0.014
Tool steel	1:1	1.2	0.010
Mother of pearl	100:1	4.0	0.150
Synthetic ruby	2:1	0.9	0.020
Carbon–graphite	100:1	3.0	0.080
Ferrite	100:1	3.5	0.125
Quartz	50:1	3.0	0.065
Boron carbide	2.5:1	0.9	0.015
Glass-bonded mica	100:1	3.5	0.125

is relatively rapid (0.15 in min^{-1}, or 3.8 mm min^{-1}). The performance on tool steel is poor (G about 1 and with penetration rate of only about 0.01 i.p.m. (0.25 mm min^{-1}).

In considering possible removal mechanisms in ultrasonic grinding, the following actions are evident:

(1) throwing abrasive particles against the work surface;

(2) pressing or hammering abrasive particles into the work surface;

(3) cavitation erosion;

(4) chemical action associated with the fluid employed.

Items (3) and (4) would be the only ones present if no abrasive particles were suspended in the fluid. Since the grinding rate is very low in the absence of abrasive particles, these last two items must be considered to be secondary factors which tend to weaken the workpiece surface. The action of primary interest is clearly a mechanical one associated with the abrasive particles.

Shaw (1956) has derived expressions for the rate of removal by actions (1) and (2) above and has found action (2) to be predominant. The analysis of action (2) results in a removal rate that varies directly with the first power of the mean abrasive diameter, which is in agreement with the experimental results of Nishimura *et al.* (1955). This suggests that ultrasonic grinding primarily involves a hammering action rather than a throwing action.

It is possible that cavitation may play a secondary role in both the rate of cutting and the rate of tool wear. During the upward swing of the tool, the pressure in the fluid will decrease and vacuous cavities may develop on the surfaces. When the tool reverses its direction these cavities will be subjected to very high pressure and will rapidly collapse. A portion of the liquid at the surface of the cavity frequently becomes detached from the main surface during collapse, and is projected inward against the solid surface with great impact. If the amplitude of the oscillating suface is greater than 0.0005 in (0.013 mm) and its frequency above 7.5 kHz, particles can be removed from the metal surface by impact of the liquid particles. Damage of this sort is called cavitation erosion, and it is likely that this action augments the mechanical abrasive action described above. Cavitation erosion is discussed in more detail in Shaw and Macks (1949).

A detailed discussion of ultrasonic grinding, abrasive life, and economics is to be found in Kaczmarek *et al.* (1966).

One of the expensive items associated with die sinking by electrodischarge machining (EDM) is the cost of tool production. Ultrasonic grinding offers an effective way of producing and redressing EDM tools made of graphite (Gilmore 1990). Ultrasonic grinding is also used at low amplitude for polishing metal dies after engraving.

References

Adams, P. B., Britton, M. G., Lonergan, J. R., and McLellan, G. W. (1984). Company Literature, Corning, Inc. N.Y.
Andrew, C., Howes, T. D., and Pearce, T. R. A. (1985). *Creep feed grinding.* Holt, Rinehart and Winston, New York, p. 139.
Barnard, J. M. (1990). *Proc. 4th Int. Grinding Conf., MR90-508.* SME, Dearborn, Michigan.
Birle, J. D. and Ratterman, E. (1986). *Dimensional Stone* **2**, 12.
Bond, F. (1952). *Trans. ASME* **193**, 484.
Brach, K., Pai, D. ., Ratterman, E., and Shaw, M. C. (1988). *J. Engng Ind. (Tran. ASME)* **110**, 25.
Chandrasekar, S., Shaw, M. C., and Bhushan, B. (1987). *J. Engng Ind.* **109**, 76.
Chandrasekar, S., Farris, T. N., and Bhushan, B. (1992). *J. Tribology* **112**, 535.
Chattopadhyay, A. K. and Hinterman, H. E. (1992). *Ann. CIRP* **41**(1), 381.
Chattopadhyay, A. K. and Hinterman, H. E. (1993). *Ann. CIRP* **42**(1), 413.
Farris, T. N. and Chandrasekar, S. (1989). *J. Mach. Working Technol.* **20**, 69.
Furukawa, Y., Ohishi, S., and Shiozaki, S. (1979). *Ann. CIRP* **28**(1), 213.
Furukawa, Y., Ohishi, S., and Shiozaki, S. (1980). *Bull. Japan Soc. Prec. Engrs.* **14**, 85.
Gilmore, R. (1990), *4th SME Int. Grinding Conf., MR90-536.*
Giovanola, J. H. and Finnie, I. (1980). *J. Mat. Sci.* **15**, 2508.
Hahn, R. S. (1962). *Proc. 3rd Int. MTDR Conf.*, Pergamon Press, Oxford, p. 129.
Haultain, H. E. T. (1923). *Trans. AIME* **69**, 183.
Hawkins, A. C. and Brauninger, G. (1991). *SME Technical Paper MR91-165.*
Hawkins, A. C., Antenen, A. P., and Johnson, G. (1990). *Dimensional Stone* **6**.
Inasaki, I. (1987). *Ann. CIRP* **36**(2), 227.
Ito, S. and Murata, R. (1987). *Japan Mech. Engng Lab.* **41**, 236.
Kaczmarek, J., Kops, L., and Shaw, M. C. (1966). *J. Engng Ind.* **88**, 49.
Kayaba, N. and Fujisawa, M. (1986). *Proc. Japan Soc. Prec. Engrs* **20**, 355.

Kitajima, K., Cai, G. O., Kumagai, N., Tanaka, Y., and Zheng, H. W. (1992). *Ann. CIRP* **41**(1), 361.
Koenig, W. and Schleich, H. (1982). *Ultrahard Mat. Technol.* **1**, 24. DeBeers, Ascot, U.K.
Kwong, J. M. S., Adams, J. T., Johnson, J. F., and Piret, E. L. (1949). *Chem. Engng Prog.* **45**, 508, 655, 708.
Lawn, B. and Wilshaw, R. (1975). *J. Mat. Sci.* **10**, 1049.
Marinescu, I. D. and Bouas, D. (1990). *Proc. 4th SME Int. Grind. Conf., MR 90-545*.
Martin, G. (1926). *J. Soc. Chem. Ind.* **45**, 1601.
Murugesan, S. (1993). M.Sc. thesis, Arizona State University, Tempe.
Murata, R., Okano, K., and Tsutsumi, C. (1985). *ASME PED 16*, 261.
Nakagawa, T., Suzuki, K., and Uematsu, T. (1986). *Ann. CIRP* **35**(1), 205.
Nakayama, K., Takagi, J., and Nakano, T. (1974). *Proc. Int. Conf. Prod. Engng*, Tokyo, p. 572.
Neppiras, E. A. (1953). *J. Sci. Instr.* **30**, 72.
Nishimura, Y., Jimbo, Y. and Shimakawa, K. (1955). *J. Fac. Engng Univ. Tokyo*, **24**(3), 65.
Ota, M. and Miyahara, K. (1980). *4th Int. Grinding Conf., MR90-538*, SME, Dearborn, Michigan.
Pai, D. M., Ratterman, E., and Shaw, M. C. (1988). *Proc. Conf. on Intersoc. Symp. on Machining of Advanced Ceramic Materials and Components*, ASME, p. 99.
Salje, E., Damlos, H., and Moehlen, H. (1985). *Ann. CIRP* **34**(1), 263.
Seiforth, M. (1985). *Superabrasive Conf.* SME, Dearborn, Michigan.
Shaw, M. C. (1954). *Proc. U.S. Natl Acad. Sci.* **40**, 394.
Shaw, M. C., (1956). *Ann. CIRP* **5**(1), 25.
Shaw, M. C., and Macks, E. F. (1949). *Analysis and lubrication of bearing*. McGraw-Hill, New York.
Shaw, M. C., Farmer, D. A. and Nakayama, K. (1967). *J. Engng. Ind* **89**, 495.
Shiozaki, S. Furukawa, Y. and Ohishi, S. (1977). *Bull. Japan Soc. Prec. Engrs* **11**, 97.
Somiya, S. (1984). *Advanced technical ceramics*, Academic Press, New York.
Subramanian, S. and Ramanath, S. (1994). ASME vol PED **58**, 1.
Suto, T., Waida, T., Naguchi, H., and Inoue, H. (1990*a*). *Bull. Japan Soc. Prec. Engrs* **24**, 339.
Suto, T., Waida, T., Naguchi, H., and Inoue, H. (1990*b*). *Bull. Japan Soc. Prec. Engrs* **24**, 227.
Suzuki, K., Sato, K., and Yokoyama, T. (1993). *Int. J. Japan Soc. Prec. Engrs* **27**, 225.
Ueda, K., Yamada, K., and Sugita, T. (1992). *J. Engng Ind.* **114**, 317.
Ueno, Y., Miyake, S., Kuroda, H., and Watanabe, J. (1979). *Ann. CIRP* **28**(1), 219.
Van der Sande, J. V., Uhlmann, D. R., and Akeson, A. (1985). *Chicago Superabrasives Conference, MR85-308*. SME, Dearborn, Michigan.
Walker, D. R. and Shaw, M. C. (1954). *Trans. AIME* **199**, 313.
Yokogawa, M. and Yokogawa, K. (1986*a*). *Grinding performance of CBN wheels*, Kasaku, Gijyutsu Shinko Zaidun, Japan.
Yokogawa, M. and Yokogawa, K. (1986*b*). *Bull. Japan Soc. Prec. Engrs* **20**, 231.

15

PRECISION FINISHING

Introduction

Until about 1960, precision grinding involved a maximum precision of about 10 μm (400 μin). According to Taniguchi (1992) this figure had been reduced to about 1.0 μm (40 μin) by 1970, and is expected to reach 100 nm (0.10 μm = 4 μin) by the year 2000. Taniguchi (1992) also suggests that the ultimate precision obtainable is when chip size approaches atomic size ($\simeq 0.3$ nm = 0.0003 μm, or 0.012 μin).

Nanotechnology is of particular interest in the optical and electronic industries and involves single-point diamond turning (SPDT) as well as ultra-precision diamond grinding (UPDG). These two processes are similar in that chips of unusually small size are involved. Both are capable of producing surfaces with mirror finish without polishing, using specially designed machine tools of high rigidity with air bearing spindles. Yoshioka *et al.* (1985) have discussed the requirements and design philosophy for such machines. Details concerning precision turning machines are reviewed in Aronson (1994).

In this chapter, SPDT will be considered first, followed by UPDG of very brittle materials such as glass.

The major problem with both SPDT and UPDG is the appearance of subsurface defects usually in the form of microcracks. UPDG is usually performed on very hard brittle materials, such as glasses and ceramics, while SPDT is performed on very soft ductile metals, such as pure copper. However, this latter difference in the two processes is only apparent.

Consequences of Small Chip Size

A major consequence of small chip size is that the normally ductile metal in SPDT behaves as though it were as brittle as glass when the volume deformed is limited to a very small size. Herring and Galt (1952) found that very small whiskers produced when a normally ductile metal was deposited electrolytically at a very small rate were essentially defect free. When these whiskers were tested in bending they were found to be completely brittle (yield stress > fracture stress), to have unusually high elastic fracture strain (>0.1), and to have a fracture stress approaching the theoretical strength of the material (shear modulus/n, where n is in the range 2π to 10, depending on the approximations involved in the estimate). This clearly demonstrates that when the size of the deformation zone approaches the mean defect structure spacing, normally ductile materials behave as though they were very brittle.

Another consequence of the small undeformed chip thicknesses involved in

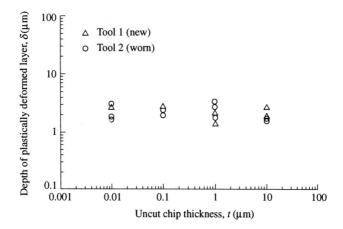

FIG. 15.1. The measured depth of plastically deformed layer (δ) versus the uncut chip thickness (t) after machining a Te–Cu alloy with a sharp and a worn diamond tool (after Lucca and Seo 1994).

SPDT and UPDG is that the chip-forming model shifts from one involving concentrated shear (Fig. 1.1) to microextrusion (Fig. 1.3). When the undeformed chip thickness (t) becomes less than the radius (r) at the tool or grit tip, the effective rake angle of the tool has such a large negative value that Fig. 1.3 replaces Fig. 1.1 (Shaw 1972). In the mechanism of Fig. 1.3, a relatively large volume of material has to be brought to the fully plastic state in order for a relatively small amount of material to escape as a chip. This is one of the reasons for the exponential increase in specific energy with a decrease in undeformed chip thickness (the size effect) in UPDG and SPDT.

A further increase in specific energy with a decrease in the undeformed chip thickness is due to a decrease in the probability of encountering a stress-reducing dislocation or a defect capable of inducing dislocations as the deformation zone decreases in volume. Lucca and Seo (1994) have measured the depth of subsurface deformation (δ) versus undeformed chip thickness (t) and have obtained the results shown in Fig. 15.1. This is surprising in that it indicates that δ is essentially independent of t. The ratio δ/t is about 200 at $t = 0.01\ \mu$m, but two orders of magnitude smaller at $t = 1.0\ \mu$m. The mean defect spacing for this material is about 1 μm. Thus, in order that a dislocation or dislocation-generating defect be encountered for plastic flow to occur at $t = 0.01\ \mu$m would require the elastic stress field to penetrate very deeply and to have a much higher mean value than when t is 1 μm or greater. This further accounts for the very high value of specific energy and deformation depth for $t = 0.01\ \mu$m in Fig. 15.1.

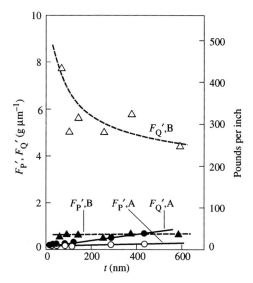

FIG. 15.2. The variation of the cutting forces per width cut, F_P' and F_Q', with the undeformed chip thickness t for tool A, which is sharper (has a smaller radius of curvature at the cutting edge) than tool B (after Ikawa *et al.* 1987).

Single-point Diamond Turning (SPDT)

The important role of tool tip sharpness in SPDT is illustrated in Fig. 15.2. Both tools A and B are relatively sharp but A is sharper than B. Tool sharpness is seen to be relatively unimportant relative to cutting force per unit width cut (F_P') but extremely important relative to feed force per unit width (F_Q'). The less sharp tool B has values of F_Q' that are more than one order of magnitude greater than those for the very sharp tool A and exhibits a pronounced size effect. The difference in behavior of these two tools is due to a difference in the radius at the tip of the tool. The tip radius for tool B is large enough relative to the undeformed chip thickness t so that the removal model for FFG (Fig. 1.3) holds, while that for tool A is small enough so that the metal cutting model (Fig. 1.1) is approached. Thus, when cuts are made having values of undeformed chip thickness that are smaller than the tool tip radius, the removal process in SPDT approaches that of fine FFG.

Ikawa *et al.* (1987) suggest that with care, it is possible to polish diamond tools to a roughness of about 1 nm and a sharpness (tip radius) of about 10 nm. It should be possible therefore to remove material by SPDT with an undeformed chip thickness of about 50 nm and still have the concentrated shear model as the one pertaining.

Figure 15.3 shows chips produced by sharp tool A in (a) and chips produced by less sharp tool B (b). The chips shown in Fig. 15.3(a) resemble those in

FIG. 15.3. SEM micrographs of chips produced at $t = 100$ nm (4 μin), (a) for sharp tool A and (b) for less sharp tool B of Fig. 15.2 (after Ikawa *et al.* 1987).

ordinary metal cutting, with one side smooth and the other side rough. These clearly conform to the concentrated shear metal cutting model of Fig. 1.1. The chips in Fig. 15.3(b) have an entirely different appearance and suggest formation by the extrusion-like mechanism of FFG shown in Fig. 1.3.

Figure 15.4 shows a long continuous chip obtained when diamond face turning electrodeposited Cu with an undeformed chip thickness of 1 nm. This was apparently produced by a very sharp tool, and resembles the chips of Fig. 15.3(a) except that there are many more microcracks in the surface, having a pitch of 50–100 nm. There is obviously a great deal of fracture involved for the chip of Fig. 15.4. The continuity of the chip is due to one or a combination of the following:

- individual microcracks do not extend far across the width of the chip

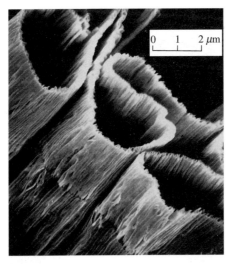

FIG. 15.4. SEM micrograph of an electrodeposited Cu chip produced at $t = 1$ nm (0.04 μin) in an ultra-precision turning operation (after Donaldson et al. 1987).

- individual microcracks do not extend all the way across the thickness of the chip
- microcracks reweld as the chip proceeds up the tool face

The greater amount of fracture involved in Fig. 15.4 compared with Fig. 15.3(a) is undoubtedly due to the presence of more stress-concentrating defects in electrodeposited copper than in wrought copper, and also the smaller undeformed chip thickness for Fig. 15.4. Both Fig. 15.3(a) and 15.4 clearly indicate that microfracture and the defect structure of the metal cut play important roles in ultra-precision machining. It would therefore appear that any attempt mathematically to model this process that does not include normal or hydrostatic stress in the constitutive relationship to take fracture into account will be inadequate. It also appears that any attempt to model ultra-precision machining by molecular dynamic simulation—discussed in Ikawa et al. (1991)—that does not include a defect structure will be of limited value.

Lucca and Seo (1994) have studied the effect of undeformed chip thickness (t) and tool tip radius (ρ) on the cutting forces and the depth of the deformed layer in SPDT of elecrodeposited copper and a 0.5 w/o Te–Cu alloy. Tool tip radii were measured with an atomic force microscope (AFM) using standard carboxylate spheres of 519 ± 10 nm diameter to estimate the radius of the cantilever tip of the microscope, which was found to be about 30 nm (see Lucca and Seo 1994 for details of the basic method). A piezoelectric dynamometer was used to measure cutting forces F_P and F_Q and the taper section technique was used to estimate the depth of deformation (δ) after cutting. The nominal

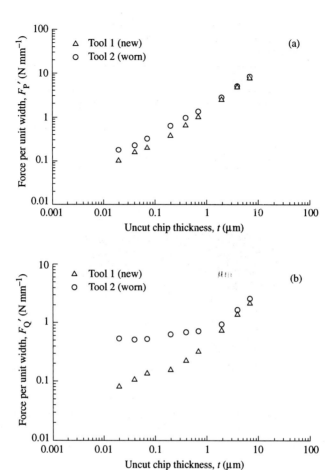

FIG. 15.5. The variation of the forces per unit width, (a) F'_P and (b) F'_Q, with undeformed chip thickness (t) for a new diamond tool (1) and a used tool (2) when cutting a 0.5 w/o Te–Cu alloy (after Lucca and Seo 1994).

rake angle of the tool was 0° and the clearance angle was 5°. The undeformed chip thickness values used were from 10 to 10^4 nm in orthogonal cutting. The surfaces of the cut specimens were carefully protected to prevent edge rounding when being taper sectioned at an inclination of 5°. The depth of the deformed layer was made visible by etching with an aqueous solution of $Fe(NO_3)_3 \cdot 9H_2O$ + HCl for several seconds.

The cutting tool tip radius was found to be 75 ± 15 nm for the new tool and 105 nm for a used tool. The variation of unit force component in the cutting direction (F'_P) versus undeformed chip thickness (t) is shown in Fig. 15.5(a) and for the orthogonal direction (F'_Q) in Fig. 15.5(b). Approximate values of forces F'_P, F'_Q, and the ratio F'_P/F'_Q are given in Table 15.1 for

TABLE 15.1 *Approximate values of force components F_P' and F_Q' and the ratio F_P'/F_Q' for SPDT of a Te-Cu alloy with new and used tools (after Lucca and Seo 1994)*

Item	New tool at t of			Used tool at t of		
	10 nm	10^3 nm	10^4 nm	10 nm	10^3 nm	10^4 nm
F'_P (N mm^{-1})	0.1	1.0	10.0	0.1	1.0	10.0
F'_Q (N mm^{-1})	0.06	0.6	3.0	0.6	0.6	3.0
F'_P/F'_Q	1.66	1.66	3.33	6	0.6	0.3

FIG. 15.6. Top and side views of an *in-situ* SEM chip of amorphous metal cut as follows: V 370 μm min^{-1} (0.015 i.p.m.); rake angle = 20°; t = 5 μm (200 μin) (after Ueda and Manabe 1992).

the new and used tools for a wide range of undeformed chip thicknesses (t). It is seen that the ratio F'_P/F'_Q is greater than one for the sharp tool, as it should be when cutting is predominantly in the concentrated shear mode. However, this ratio is less than one (except for the very light cut at 10 nm) for the used tool. A value less than one is to be expected when cutting is predominantly in the microextrusion mode. The unusually high value of F'_P/F'_Q for the dull tool operating at small t is undoubtedly due to most of the energy being associated with friction without removal for such a light cut with a dull tool.

Ueda and Manabe (1992) have observed microcutting chip formation of an amorphous metal (Fe$_{78}$B$_{13}$Si$_9$) by *in-situ* cutting in an SEM at extremely low cutting speeds (15-850 μm/min, or 600-34 000 μin min^{-1}). A typical chip is shown in Fig. 15.6, where a card-like structure is clearly seen. The authors report that the concentrated shear bands which were less than 0.03 μm (12 μin) in width showed no evidence of fracture at very high magnification. The

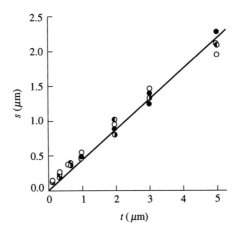

FIG. 15.7. The relation of lamella spacing (S) and undeformd chip thickness (t) for cutting speeds ranging from 15 μm min^{-1} to 850 μm min^{-1} (600–34 000 μin min^{-1}) and a rake angle of 20° (after Ueda and Manabe 1992).

spacing of the 'cards' (S) was found to scale with the undeformed chip thickness t, as shown in Fig. 15.7.

The spacing of the shear bands (S) in Fig. 15.6 is approximately half the undeformed chip thickness t and the cutting ratio is approximately one-third of t. This chip is unusual in that the side of the chip contacting the tool face (A in Fig. 15.6) is not burnished. Because of the extremely low cutting speed in these tests (370 μm/min, or 0.15 i.p.m.), temperature effects will be nil.

Since the geometry of the tool tip plays such an important role, the wear of diamond tools is extremely important in SPDT. Diamond tool wear depends on structural anisotropy, imperfections present (influencing strength and microchipping), the affinity for the metal being cut (influencing adhesive wear), and the role of temperature and atmosphere (influencing weight loss of diamond due to oxidation and graphitization). The questions of optimum crystal orientation and identification of defects in diamonds are complex issues that are still being pursued in the literature (Ikawa et al. 1987). Metals that show little affinity for diamond are Cu, Al, Pb, electroless Ni, and brass, while Be, Ta, Fe, Ni, Mo, Ti, U, and W have relatively high affinity for diamond and hence cause relatively high rates of wear (Ikawa et al. 1987).

A number of workers have observed a pronounced increase in specific energy with a decrease in undeformed chip thickness t, as shown in Fig. 15.8, when diamond turning copper under ultra-precision conditions. The approximate specific energy for machining copper in the conventional finish machining regime ($t \approx 0.005$ in, or 0.13 mm) is shown by the dashed line in Fig. 15.8.

Moronuki et al. (1994) have performed slow-speed ultra-precision cutting tests on, several materials, using the cutting arrangement shown in Fig. 15.9(a).

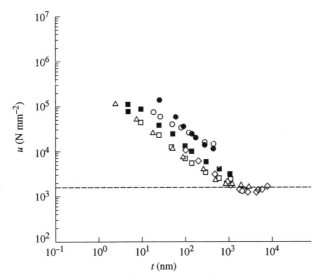

FIG. 15.8. The variation in the specific energy u with the undeformed chip thickness t in diamond micromachining of copper, from several independent investigators (after Ikawa *et al.* 1991).

The cutting forces shown in Fig. 15.9(b) were obtained when machining pure polycrystalline aluminum under the conditions indicated in the figure caption. The cutting force F_P varied from a minimum value of about 0.1 N (0.022 lb) to a maximum value of about 0.3 N (0.067 lb), with a peak of varying magnitude occurring at each mm of travel. Force F_Q was only about one-quarter of F_P and showed a similar threefold fluctuation for each 1 mm (0.040 in) of tool travel. The authors attribute the rather large fluctuations in the cutting forces to grain to grain anisotropy, and suggest that crystalline anisotropy plays an important role when the undeformed chip thickness approaches mean crystal size. Liang *et al.* (1994) have used finite element analysis to verify the influence of grain boundaries and crystal orientation on microcutting mechanics.

High-precision cutting and grinding of optical components of glass offers the possibility of obtaining surfaces of excellent finish and accuracy without the need for time-consuming polishing. However, the problem with this approach is that unless great care is taken, the finished surface will not be crack free.

Fujita and Shibata (1993) have performed single-groove cutting tests on a glass cylinder using 90° diamond cones having a negative rake angle of 2°. In order to obtain a groove free of cracks, it was necessary for the normal force on the diamond to be <0.03 N (0.007 lb), for the radius at the tip of the diamond to be <5 μm (200 μin) and for the depth of cut to be <0.2 μm

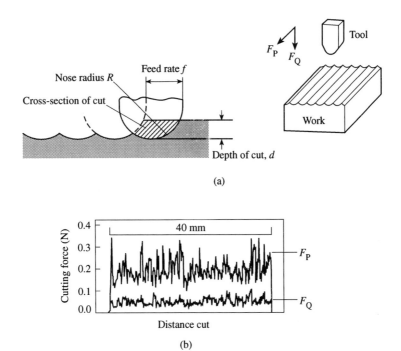

FIG. 15.9. (a) An ultra-precision slow speed ($V = 6 \text{ m s}^{-1}$, or 1180 f.p.m.) cutting test arrangement. (b) Fluctuations in cutting forces when cutting polycrystalline aluminum (mean grain size, 640 μm (0.026 in) and Vickers hardness = 56) with $V = 6 \text{ m s}^{-1}$ (1180 f.p.m.), $f = 40$ μm (0.0016 in per stroke), $d = 6$ μm (240 μin), $R = 0.5$ mm (0.020 in). (after Moronuki *et al.* 1994).

(8 μin). The critical depth of cut to avoid subsurface cracking was found to increase with cutting speed to 500 m min^{-1} (1640 f.p.m.) and then to remain constant. This could be explained in terms of an increase in temperature with speed, causing the glass to be less brittle.

Takahashi *et al.* (1993) have also performed single-point diamond cutting tests on glass in the fly cutting mode, as shown in Fig. 15.10. The tools were Vickers indenters with values of tip radii of 2 μm and 0.5 μm (80 and 20 μin). At a cutting speed of 250 m s^{-1} (49 200 f.p.m.), wavy marks were found on each side of the groove when examined with a scanning tunneling microscope. These marks had a pitch of 2 μm (80 μin). This corresponds to a time between peaks of 8 ns and a frequency of 124×10^6 Hz, thus ruling out the possibility of a vibrational origin. Similar grooves were found with both values of tip radius, but were more pronounced with the 2 μm (80 μin) radius than with the 0.5 μm (20 μin) radius. Etched grooves had a molten appearance, suggesting that they were of thermal origin. The critical maximum depth of cut at which

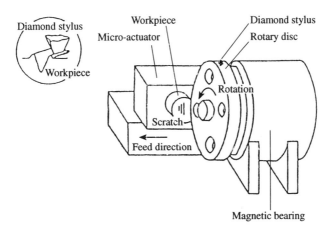

FIG. 15.10. An apparatus for making single diamond grit micro-fly milling cuts on a glass workpiece at speeds up to 250 m s^{-1} (50 000 f.p.m.) (after Takahashi *et al.* 1993).

subsurface cracks appeared (t_c) is shown plotted in Fig. 15.11 together with the method of deriving t_c from measured values of groove width l and tip radius ρ. The critical depth of cut decreases with increase in speed to 10 000 m min^{-1} (32 800 f.p.m.) and then remains constant at about 0.2 μm (8 μin) with further increase in speed.

While the critical depths of cut (t_c) found by Fujita and Shibata (1993) and Takahashi *et al.* (1993) were approximately the same, the former found t_c to increase with cutting speed, while the latter found t_c to decrease with cutting speed. This could be due to Fujita and Shibata employing continuous cutting at constant depth, while Takahashi *et al.* employed intermittent cutting at a varying depth of cut.

Although SPDT is best adapted to relatively ductile materials (Cu, Al, Pb, etc.), because of the rapid loss of cutting edge sharpness with harder materials, Blake (1988) has applied this process to diamond turning of two very brittle materials — single-crystal germanium and silicon. It was found that surfaces of excellent finish that were free of surface pits could be produced by use of a gem grade diamond tool having a large (1/8 in, 3.18 mm) nose radius in the feed direction, as shown in Fig. 15.12. The tool tip radius at the cutting edge was estimated to be 10 nm. The actual operation was face turning with a tool having a large nose radius, that was fed slowly toward the center of the work, which had a circumferential velocity V at the tool. The kinematics were similar to a phonograph (record = work and needle = tool) with the tool fed inward at a fixed rate. The width of cut b was about 50 times the feed per revolution v. The undeformed chip thickness was not constant as in orthogonal cutting but varied from 0 at A to a maximum (t_m) at B (Fig. 15.12(a)).

The tool was quickly raised out of the cut after equilibrium was established

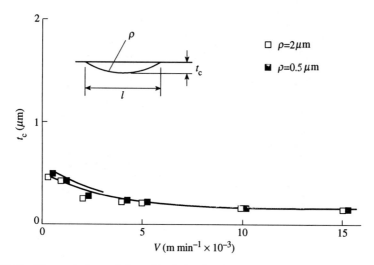

FIG. 15.11. The variation of the critical depth of cut t_c with the cutting speed V for Vickers diamond tools with two different tip radii R when cutting BK7 glass (after Takahashi et al. 1993). $t_c = \rho - (\rho^2 - l^2/4)^{1/2}$.

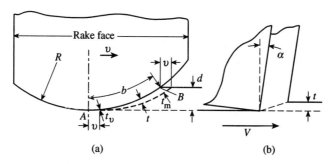

FIG. 15.12. The cutting arrangement used by Blake (1988) to determine the maximum critical undeformed chip thickness in an SPDT operation (face turning) of polycrystaline Ge. a) Front view in V direction b) side view. Typical values employed were as follows: V, up to 14.3 m s^{-1} (2814 f.p.m.); $v = 2.5$ μm (100 μin); $d = 2.5$ μm (100 μin); $R = 3125$ μm (0.125 in); $\alpha = -10°$; $b = 125$ μm (0.005 in); $t_m = 0.1$ μm (0.4 μin); clearance angle, $-6°$, fluid, distilled water; estimated tool tip radius, 10 nm (0.4 μin). Under these conditions, the R_a surface roughness was found to be 80 nm (3.2 μin) and the critical undeformed chip thickness to avoid subsurface damage was about 45 nm (1.8 μin).

FIG. 15.13. SEM micrograph of chips produced under the conditions of Fig. 15.12 (after Blake 1988).

so that the cut shoulder could be examined for finish and defects. The defects were in the form of pits that increased in depth and density as t increased up the shoulder. It was found that there was a value of undeformed chip thickness (t_c) below which no pits were found. In addition, the value of t_c was so small over the feed per revolution distance that there was appreciable burnishing, leading to very good finish ($R_a \simeq 80$ nm, or $3.2\,\mu$in). This was called ductile regime machining, even though practically all of the material removed involved very brittle behavior. Figure 15.13 is an SEM of germanium chips produced under 'ductile regime' cutting, which clearly indicates appreciable fracture associated with their formation.

The most important results of Blake's study are as follows:

- By use of a large nose radius and a small feed, the bulk of the material may be removed without regard for fracture as long as t at the feed distance (t_v in Fig. 15.12) is less than t_c.
- There is a critical value of undeformed chip thickness below which no sub-surface cracks are obtained. Thus, t_c under a standard set of conditions (V, v, R, α, and fluid) may be used to characterize a material in SPDT.
- The depth of cut (d) is of negligible importance but, instead, undeformed chip thickness (t) is all important.
- Values of t_c increase with a decrease (more negative) in the rake angle and with a decrease in the clearance angle.
- The critical undeformed chip thickness is greater for Si than for Ge, but Ge gives the better finish.
- Crystal orientation is important, t_c being lowest for the $\langle 110 \rangle$, cutting direction.

The poorer finish obtained with silicon was attributed to microchipping of the cutting edge of the diamond. It appears that SPDT is best suited to ultraprecision machining of relatively soft materials (Cu, Al, etc.), because it requires a tool having a very small tip radius which will relatively quickly wear when cutting hard materials, even with a diamond tool. It appears that UPDG discussed in the next section offers greater promise than SPDT for precision finishing of hard, brittle materials such as glass and ceramics.

Ultraprecision Diamond Grinding (UPDG)

Yoshioka et al.(1985) have analyzed the requirements for UPDG machines of high accuracy and have given examples of two prototypes—a vertical spindle surface grinding machine and a centerless grinding machine. In their discussion they identify two copying principles:

(1) pressure copying, which includes the classical method of lens production by lapping and polishing and employs a machine of low stiffness and a high-precision tool;
(2) motion copying, which is the essence of UPDG, requiring a machine of high stiffness, precision, and control.

The move from pressure copying to motion copying has been made possible by the application of machine design and control technology that has been rapidly evolving in recent years.

UPDG is capable of producing surfaces requiring no polishing, and is of particular interest for aspheric lens production where ordinary polishing kinematics renders polishing unsuitable. The main problem of using UPDG in lens production is in producing surfaces free of microcracks with sufficient smoothness and accuracy. Special machines of suitable stiffness, stability, and control are being developed to enable glass surfaces that require no polishing to be produced by UPDG.

Namba and Abe (1993) have studied the UPDG of 11 optical glass compositions having Knoop hardness values ranging from 345 to $750 \, \text{kg mm}^{-2}$. A variety of resin-bonded diamond wheels were used with grit screen sizes ranging from 200 to 3000. Figure 15.14(a) shows a Nomarski interference micrograph of an optical glass surface ground with an SD200 75 B wheel, while Fig. 15.14(b) shows the same glass ground with an SD1500 75 B wheel at the same wheel speed V and wheel depth of cut d, but a work speed only one-eight as large as that for Fig. 15.14(a). The surface finish of Fig. 15.14(a) was $R_a = 0.43 \, \mu\text{m}$ (17 μin), $R_t = 2.6 \, \mu\text{m}$ (104 μin) and $R_t/R_a = 6$. The corresponding values of Fig. 15.14(b) were $R_a = 1.35$ nm (0.054 μin), $R_t = 11.7$ nm (0.47 μin), and $R_t/R_a = 8.7$. The surface finish of Fig. 15.14(b) is actually better than that produced by conventional diamond grinding followed by polishing. The sample of Fig. 15.14(a) was opaque until polished. Grinding performance was found to be independent of wheel depth of cut d for values below 10 μm (400 μin), but to depend upon feed per revolution and grit size as follows (Namba and Abe 1993).

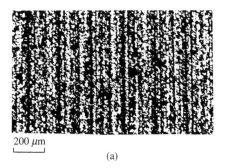 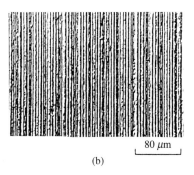

200 μm		80 μm
(a)		(b)

FIG. 15.14. Nomarski interference micrographs of optical glass surfaces produced by UPDG in a vertical spindle surface grinder with cup wheels at a wheel speed V of 20 m s^{-1} (3940 f.p.m.) and a wheel depth of cut d of 1 μm (40 μin): (a) wheel SD200 75 B and work speed $v = 0.61 \text{ m min}^{-1}$ (2.0 f.p.m.); (b) wheel SD1500 75 B and work speed $v = 0.01 \text{ m min}^{-1}$ (0.033 f.p.m.) (after Namba and Abe 1993).

The force ratio F_Q/F_P:

- increases markedly with a decrease in grit size
- increases slightly with a increase in the feed per revolution
- increases by a factor of 30 as the Knoop hardness of the glass goes from 345 to 750 kg mm^{-2}

The surface roughness:

- decreases markedly with a decrease in grit size
- increases slightly with the increase in feed per revolution
- decreases by a factor of 10 as the Knoop hardness of the glass goes from 345 to 750 kg mm^{-2}

Yoshido and Ito (1990) have also reported an excellent finish ($R_t = 0.05$ μm $= 2$ μin) and shape accuracy (0.5 μm $= 20$ μin) when grinding glass with a metal-bonded diamond wheel of very fine grit size (3000) on a very rigid machine. The specific grinding energy (u) also increased substantially with a decrease in the mean undeformed chip thickness $\bar{t}$, but the wheel depth of cut had little influence on u.

Hashimoto *et al.* (1993) have developed the machine shown in Fig. 15.15, that enables roughing and finishing operations to be performed in production without moving the workpiece from one machine to another. Two air bearing spindles are used. Grinding wheel A (D400M) is for roughing, while B (D3000M) is for finishing. A flywheel is attached to the rough grinding spindle to suppress fluctuations in r.p.m. The wheel speed reported appears to be abnormally low at 471 m min^{-1} (1545 f.p.m.).

The normal grinding force (F_Q) in roughing was only 0.8 N (3.6 lb) for a

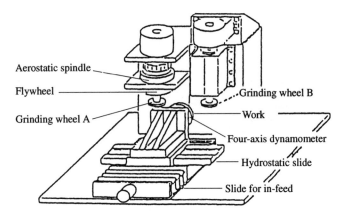

FIG. 15.15. A two-spindle UPDG machine for rough and finish grinding of optical parts without moving the work from one machine to another (after Hashimoto *et al.* 1993).

slide velocity $v = 150 \text{ m min}^{-1}$ (5.9 i.p.m.) and wheel depth of cut $d = 2 \, \mu\text{m}$ (80 μin). The corresponding normal force (F_Q) in finishing was about 24 N (5.4 lb), which is 30 times that for roughing. Electrolytic dressing, discussed in the next section, was applied to the roughing wheel at 10 V, 1 A for 8 min. Surface finish in roughing was higher with electrolytic dressing than without and was about $R_a = 0.5 \, \mu\text{m}$ (20 μin) after equilibrium of the wheel surface was established after dressing. The surface roughness in a finishing operation with the 3000 grit size wheel with a bronze bond was $R_a = 3.4 \text{ nm}$ (0.14 μin).

Hashimoto *et al.* (1993) used etching with a dilute hydrofluoric acid solution (0.2 percent HF and 0.1 percent HCl) to detect subsurface microcracks. Small pits were found at the bottom of deep scratches when finish grinding under the above conditions with a 3000 M diamond wheel, but none were found when a 5000 M wheel was used. The pits obtained at the bottom of deep scratches were more or less regularly spaced, with a mean spacing of about 0.25 μm (10 μin). They did not appear until an etching depth of about 0.5 μm (20 μin) was reached and were clearly evident to a depth of 1 μm (40 μin). It was also found that the surface ground with the 3000 M wheel showed an increased etching rate for the first 0.2 μm (8 μin) but a lower etching rate below this level. When a comparable polished surface was etched under the same conditions, the rate was the same as the lower etching rate for the ground surface from the start. These etching tests suggest that there is no subsurface change in polishing or when grinding with the 5000 M wheel, but a small amount of subsurface damage when grinding with the 3000 M wheel.

The method used by Hashimoto *et al.* (1993) in their studies involved suspending a thin ground specimen in the etching solution and recording the change in weight. Etching appears to be a very useful test to establish the presence of subsurface change following UPDG, even though it does not indicate the

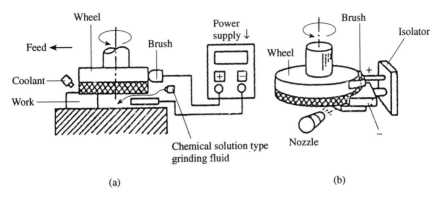

FIG. 15.16. An arrangement for the continuous electrolytic dressing of a metal-bonded diamond wheel: (a) general arrangement; (b) brush and fluid nozzle detail (after Ohmori 1992).

nature of the change (microcrack, localized residual stress, rewelded microcrack, etc.). In this sense it is like acoustic emission, which yields monitoring results that are not readily interpreted relative to cause.

A problem that develops with very fine grit abrasive wheels is maintaining the protrusion of the grits as they wear.

Continuous Electrolytic Dressing

Electrolytic in-process dressing (ELID; Ohmori 1992) has been used with metal (cast iron) bonded, very fine grit superabrasive wheels to produce mirror finish surfaces and with coarser grit size abrasive particles for rough grinding. Figure 15.16 shows the process applied to a vertical spindle surface grinding operation. This represents an extension of the method of reestablishing protrusion in electrochemical grinding (ECG, Chapter 13) from an intermittent to a continuous process. In the conventional ECG process, grinding is interrupted periodically, the potential is reversed to make the metal bonded wheel the cathode, and a highly conductive metal (Cu) is ground as the metal bond is deplated back from the grits to reestablish protrusion. In the ELID process the bond is deplated essentially continuously by a pulsed d.c. power supply. This enables optimum protrusion to be had at all times. The ELID process is applicable to grinding of either electrical conducting or nonconducting work materials, but only with metal-bonded wheels. This is is essentially the same as the process described in Chapter 14 (Murata et al. 1985), used in the abrasive cut-off of ceramics.

When applied to the grinding of conducting work materials, ELID could be combined with conventional ECG by adding a second continuous power supply with reversed potential (wheel+, work −) to the pulsed ELID power

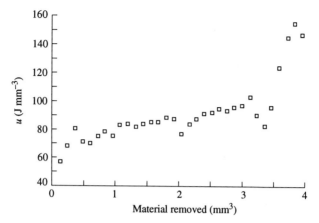

FIG. 15.17. The variation in the specific energy u with the volume removed in UPDG of CVD-deposited SiC (after Bifano et al. 1994).

supply (wheel $-$, brush $+$). The importance of ELID to grinding of optical glass to a mirror finish apparently lies in the possibility of using finer grit sizes (as small as 4 μm, or 160 μin).

Bifano et al. (1994) have studied UPDG of chemically vapor deposited SiC for use in reflective optics and lens molds. This single-phase polycrystalline material is less likely to give subsurface defects than fused silica when ground at the same undeformed chip thickness. When using a 100 mm (3.94 in) diameter diamond cup wheel (4–8 μm natural diamonds, 75 concentration, and resin bond) at a wheel speed of 28.3 m s^{-1} (5564 f.p.m.) and distilled water as a coolant, the specific energy results shown in Fig. 15.17 were obtained. This shows a very rapid initial increase in u with volume ground and a relatively short wheel life ($\sim$3 mm^3, or 83 $\times$ 10^{-4} in^3). The initial value of u was about 50 J mm^{-3} (7.25 $\times$ 10^6 p.s.i.) and the value at the point at which u increases very rapidly (wheel life) was about 100 J mm^{-3} (14.5 $\times$ 10^6 p.s.i.). It would appear that continuous dressing would be necessary in any practical application of this technology.

'Ductile' Regime Removal

It has been observed by many that the tendency for subsurface cracks to develop decreases with a decrease in undeformed chip thickness and to disappear at a critical value. This has been thought to be due to the material being less brittle below a critical value of t and has therefore been termed ductile mode cutting or grinding. In reality, materials behave in a more brittle fashion with a decrease in loaded specimen size, as the previously discussed whisker experiments of Herring and Galt (1952) clearly indicate. The answer to this

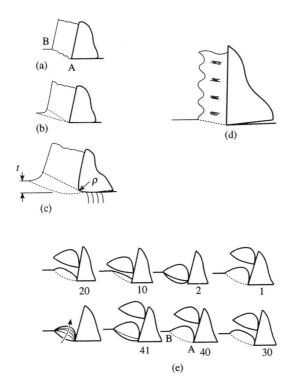

FIG. 15.18. Types of chips obtained in metal cutting: (a) concentrated shear model; (b) pie-shaped shear zone; (c) cutting with a blunt tool; (d) wavy chip; (e) disconinuous chip formation with a crack running periodically from A to B. Types (a), (b), and (c) are steady state chips, while (d) and (e) are cyclic chips.

paradox lies in the nature of the chip-formation mechanism that pertains when very brittle materials are cut under conditions of normal friction and compressive stress at the tool tip (Shaw and Vyas 1993). This is called saw-tooth chip formation.

Chip formation in SPDT is really an extension of ordinary cutting mechanics, and therefore it is useful to review the types of chips that are found in cutting.

The classical model is that due to Merchant (1945). shown in Fig. 15.18(a). This is a steady state model that is sometimes called the concentrated shear model because most of the energy ($\simeq 75$ percent) is associated with a very high shear strain that occurs abruptly as the material crosses a narrow shear plane (A–B in Fig. 15.18(a)). Deviations from this steady state model involve:

- the formation of a built-up edge when the temperature on the tool face is relatively low (low cutting speed, V) and sticking friction occurs over part of the tool–chip contact area

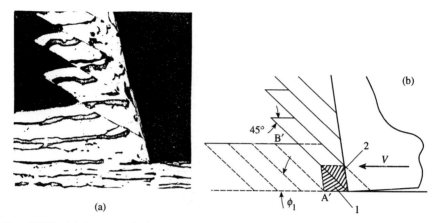

FIG. 15.19. (a) An optical photomicrograph of a partially formed chip of 60–40 cold worked brass (rolled to 60 percent reduction in area before cutting). Rake angle, $-15°$; undeformed chip thicknesss, 0.16 mm (0.0063 in); cutting speed, 0.075 m min^{-1} (2.95 i.p.m.) (courtesy of Nakayama, Toyama University). (b) A diagrammatic represenation of (a), where dashed lines show the locations of future shear fracture planes running from A′ to B′. The cross-hatched area 1 is displaced to 2 as the tool face side of the chip is burnished.

- the formation of a pie-shaped shear zone when the material cut has a tendency to strain harden during chip formation and friction on the tool face is relatively high (Fig. 15.18(b))
- when cutting with a blunt tool (Fig. 15.18(c))

In addition to these steady state models there are several cyclic ones. The friction on the tool face may vary cyclically, giving rise to a wavy chip (Fig. 15.18(d)) as the shear angle cycles over a range of values. Materials that are brittle and give low friction on the tool face (gray cast iron and unleaded brass) will produce discontinuous chips. Figure 15.18(e) shows a series of high-speed motion picture frames illustrating this type of chip formation. A tensile crack will form periodically *at the tool tip* and run outward and upward to the free surface, shifting from a tensile crack to a shear crack as it goes. The material left behind is then removed by an extrusion-like operation as the cutting forces rise. Subsequently, a new crack forms at the tool tip and the extruded material flies off with the sudden release of energy.

If the material cut is relatively hard and brittle due to strain hardening, heat treatment, or some other reason it may fracture periodically in shear, with a crack running along the shear plane *downward from the free surface*. Figures 15.19(a) and (b) show this type of chip formation, where the shear fracture plane extends all the way from the uncut surface to the level of the new surface, and possibly even below.

As the shear fracture plane approaches point A′ in Fig. 15.19(b), it will be

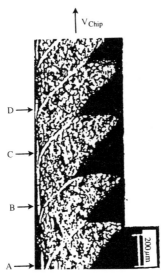

FIG. 15.20. An optical photomicrogaph of a continuous saw-tooth chip when turning AISI 4340 steel ($H_B = 325$ kg mm^{-2}) with an Al$_2$O$_3$-TiC ceramic tool at 250 m min^{-1} (0.018 i.p.r.); depth of cut, 3.75 mm (0.150 in); rake angle, $-5°$; fluid, none (after Komanduri et al. 1982).

subjected to even higher compressive stress and the crack may be quenched before it reaches the level of the new surface, as shown in the chip of Fig. 15.20. The material between the end of the crack and the new surface must then be removed by a process involving very high shear strain along the tool face.

The types of chips in Fig. 15.19 and 15.20 are called saw-tooth chips, and Nakayama (1974) has presented some important observations concerning saw-tooth chip formation. Saw-tooth chips are continuous even when the shear fracture planes extend all the way to the level of the new surface (to A' in Fig. 15.19) due to rewelding along the tool face side of the chip. Saw-tooth chips of the type shown in Fig. 15.20 have a tendency to form at high cutting speeds. At very high cutting speeds the temperature developed due to block-wise sliding along the fracture planes and along the tool face may exceed the phase transformation temperature of iron, and nonetching white layers of untempered martensite appear in these regions. Figure 15.20 is an example of this. The variation in thickness of the white layer along the tool face side of the chip suggests stick-slip frictional behavior, the chip moving over the tool face at A, B, C, and D but being stationary at points between.

An important characteristic of saw-tooth chip formation is the unusually high values of cutting ratio (chip length ratio) that pertain. For the case of Fig. 15.19, where the periodic shear planes are inclined at about 45° to the cutting direction, the cutting ratio will approach 1 (compared with a value of about 1/3 for the continuous concentrated shear model for a metal, shown

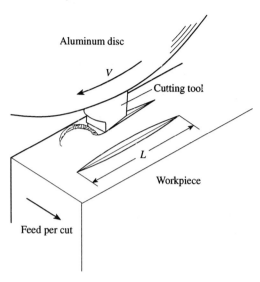

FIG. 15.21. A test with a 120° single-point tool to determine the cutting ratio for different cutting conditions (after Nakayama et al. 1974).

in Fig. 15.18(a)). For very high-speed cutting of the type shown in Fig. 15.20, cutting ratios well above 1 have been observed. This is possible because the material between the end of the shear cracks and the tool face is being upset under hot working conditions. The value of the resulting cutting ratio will depend upon the thickness of the uncracked layer adjacent to the tool face, the frictional resistance as one block of material slides over its neighbor in the cracked region, and the flow stress of the material, σ_f, adjacent to the tool face. Further discussion of saw-tooth chip formation may be found in Shaw and Vyas (1993).

Nakayama et al. (1974) performed microcutting tests under a wide range of conditions using a 200° V-shaped tool as shown in Fig. 15.21. Chips were collected and their lengths measured. The mean values of cutting ratio obtained are shown in Fig. 15.22. The cutting ratio is seen to vary significantly with cutting speed and workpiece hardness, but only slightly with rake angle α. For the hardest material ($H_v = 500$) and the highest cutting speed (1570 m min^{-1}, or 5150 f.p.m.), the cutting ratio approached 1 indicating saw-tooth chip formation. Figure 15.23 shows complete chips for soft steel ($H_v 150$) *in (a) and for hard steel* ($H_v = 500$) at (b). The steady state soft steel chips are seen to be much shorter than the saw-tooth hard steel chips.

When similar cutting ratio tests were performed by grinding a thin workpiece, as shown in Fig. 15.24, the soft steel gave shorter chips than the hard steel and both were of the saw-tooth type (Fig. 15.25). This suggests there was more fracture, sliding, and rewelding action for the hard steel chips than for the soft steel chips.

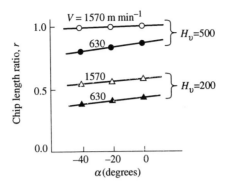

FIG. 15.22. The variation in the cutting ratio r for different values of V, workpiece Vickers hardness (H_v), and rake angle α for the test arrangement of Fig. 5.21.

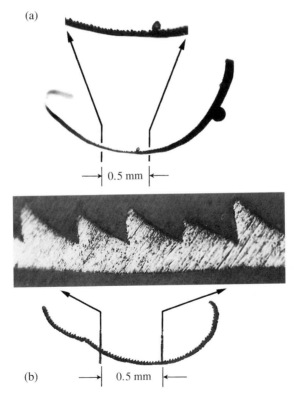

FIG. 15.23. Sample chips obtained in test of Fig. 5.21 for (a) soft steel ($H_v = 200$) and (b) hard steel ($H_v = 500$) (after Nakayama et al. 1974).

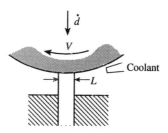

FIG. 15.24. The test arrangement used to obtain the cutting ratio of chips produced when a thin specimen is subjected to plunge grinding (after Nakayama *et al.* 1974).

FIG. 15.25. Sample chips obtained in test of Fig. 15.24 for (a) soft steel ($H_v = 200$) and (b) hard steel ($H_v = 500500$) (after Nakayama *et al.* 1974). Wheel WA 46J V; $V = 1450 \text{ m min}^{-1}$ (4756 f.p.m.); downfeed, 30 mm s^{-1} (1.18 i.p.s.); fluid, water-soluble coolant.

Figures 15.3(a), 15.4, and 15.13 show SPDT chips produced under conditions in which there are no subsurface cracks (very small undeformed chip thickness). It is evident that fracture plays a dominant role over most of the chip volume. Only on the tool face side is the material being plastically deformed without fracture.

It is a misnomer to refer to such chips as being produced by ductile regime cutting. The convention for current flow from plus to minus instead of from minus to plus is a similar case of inappropriate definition, which occurred because experiment preceded understanding. Since experiment often precedes understanding, this is probably not the last time that unfortunate nomenclature will appear as technology develops.

Subsurface Crack Detection

Convenient ways of detecting subsurface defects following SPDT or UPDG are as follows:

- a decrease of tensile strength in bending of a finished surface
- the etching of a finished surface

Mayer and Fang (1994) studied the effect of undeformed chip thickness on the tensile strength of resin-bonded diamond ground surfaces in a standard four-point bending test. Hot pressed Si_3N_4 was surface ground under a range of conditions, including table feed rate, wheel depth of cut, and grit size. Values of the maximum undeformed chip thickness were estimated by the following two equations (4.13 and 4.15, respectively):

$$\bar{t} = \sqrt{\frac{v}{VCr}\sqrt{\frac{d}{D}}}, \quad t_e = \frac{vd}{V}.$$

Equation (4.13), from Reichenbach et al. (1956) takes both the operating variables (v, V, d, and D) and the wheel face geometry (characterized by C and r) into account. Equation (4.15) considers the material to be removed as an undivided sheet and considers the operating variables (v, V, and d) but ignores the influence of the wheel diameter (D) and the wheel face geometry variables (C and r). The more approximate measure of chip size (t_e) is called the chip equivalent.

Mayer and Fang (1994) found $\bar{t}$ to correlate quite well with flexural fracture strength for bending in the grinding direction. A critical value of $\bar{t}$ above which strength was found to decrease was 80 μm (2 μm). No decrease in strength for $\bar{t}$ greater than this critical value was found when the bending arc was normal to the grinding direction. Correlation with t_e was not nearly as good.

Mean Radial Force per Grit

It was found by Ota and Mayahara (1990) that the bending strength of a structural ceramic correlates more closely with the mean radial force acting on a single grit F_Q/lbC in fine surface grinding than with the total grinding force F_Q. Since subsurface microcracks are particularly troublesome when grinding very hard brittle material, for which F_Q/F_P is unusually large, it appears reasonable that indentation as characterized by a hardness indenter should play an important role.

It has been customary to consider the behavior of a sharp indenter in connection with the grinding of glasses and ceramics. Fig. 15.26(a) shows a sequence of indentation steps with accompanying cracks for a sharp indenter. Theory for such an indenter assumes the material to be plastic–rigid, and two-dimensional plasticity theory (Prandl 1920; Hill 1950; etc.) leads to a mean stress on the contact area that is three times the uniaxial flow stress. This is in excellent agreement with experiment. The plastic–rigid approximation for a sharp indenter leads to a plastic zone that is unsupported from above (Fig. 15.26(a)). However, when a spherical (blunt) indenter is employed (Fig. 15.26(b)), there is no upward flow as long as the depth of material beneath the indentation is several times the depth of the plastic zone.

For this type of indentation an plastic–elastic assumption is more appropriate, and Shaw and DeSalvo (1970) have presented such an analysis. This leads to the result shown in Fig. 15.26(c) where an elastic–plastic

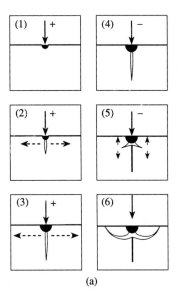

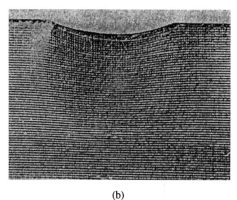

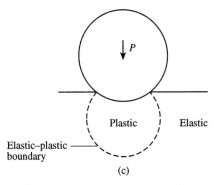

FIG. 15.26. (a) Crack formation in a brittle material during *loading* of a sharp indenter [(1), (2), (3)] and during unloading [(4), (5), (6)] (after Lawn and Wilshaw 1975). (b) Deformation beneath a blunt indenter (after Shaw and DeSalvo 1970). (c) An interpretation of (b).

boundary extends below the original surface, and has about 1.8 times the diameter of the permanent dent left in the plane of the original surface.

When brittle hard materials are ground with very small grit size to produce good finish, indentation of a *blunt* indenter and the radial component of force (F_Q) appears to play a major role. The cutting force component (F_P) should then be relatively small, since it merely involves extruding a chip from a plastic zone that has been fully developed by the radial force component.

The mean force per grit that Ota and Mayahara (1990) found to correlate well with tensile strength in bending may be found as follows:

$$F_Q'' = \left(\frac{F_Q}{F_P}\right)\frac{F_P}{lbC}, \qquad (15.1)$$

where l is the wheel-work contact length and b is the grinding width.

The following equations also pertain to surface grinding with a small undeformed chip thickness:

$$\text{power } F_P V = u(vbd), \qquad \text{size effect } u \simeq 1/t, \qquad (15.2, 15.3)$$

where u is the grinding energy per unit volume (specific energy) and b is the width ground.

$$F_Q' = \left(\frac{F_Q}{F_P}\right) rt \qquad (15.4)$$

To a good approximation,

$$r \simeq \frac{2\sqrt{gt}}{t} \simeq \sqrt{\frac{g}{t}} \qquad (15.5)$$

or, from eqn (15.4), combining eqns (15.1)–(15.5),

$$F_Q'' \simeq \left(\frac{F_Q}{F_P}\right)\sqrt{gt}. \qquad (15.6)$$

From the standpoint of a critical mean force per grit to avoid subsurface median cracks, the following should pertain:

- small undeformed chip thickness t
- small grit size g
- small F_Q/F_P

The first of these is in agreement with the experimental results of Mayer and Fang (1994), and the above example explains why this is so.

The second is in agreement with the previously described experimental results of Hashimoto *et al.* (1993), who used etching with dilute hydrofluoric acid to highlight subsurface damage. Small etch pits were found at the bottom of the deepest grinding scratches with a 3000 grit diamond wheel. However, no pits were found under similar conditions with a 5000 grit size.

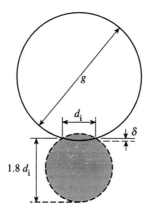

FIG. 15.27. A reworking of Fig. 15.26(c) pertaining to fine grinding when indentation plays a dominant role. The volume deformed per unit time is appoximately proportional to the cross-hatched area, while the volume removed per unit time is proportional to the solid black area.

The third item, which is most important (F_Q/F_P to the first power), explains why subsurface damage is particularly troublesome for very hard work materials and for a very small undeformed chip size where the force ratio F_Q/F_P is unusually large.

Volume Deformed in Fine Grinding

It was previously mentioned that in fine grinding, where the removal mechanism of Fig. 1.3 is predominant, much more material is plastically deformed than escapes as chips. The ratio of these two volumes may be estimated as follows with the aid of Fig. 15.27, which is a reworking of Fig. 15.26(c) pertaining to fine grinding when indentation plays a dominant role:

$$\frac{\text{volume deformed per unit time}}{\text{volume removed per unit time}} = \frac{(\pi/4)(1.8d_i)^2 v}{\frac{2}{3}d_i \delta v} = 7.63\sqrt{g/\delta}, \quad (15.7)$$

where

$$d_i \simeq 2\sqrt{g\delta}. \quad (15.8)$$

In fine grinding g/δ will be of the order of 100, and hence (volume deformed/(volume removed) will be of the order of 75.

The Possibility of Indentation Involving Pulverization

Zang and Howes (1994) have suggested that the material undergoing permanent deformation beneath a very small indenter is due to pulverization,

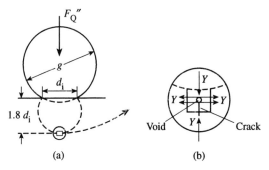

FIG. 15.28. (a) An abrasive grit of diameter g with an elastic–plastic boundary (dashed line). (b) A circular void at the bottom of the elastic–plastic boundary, showing tensile stress of magnitude Y at the top and bottom of void, causing a median crack to develop.

displacement of pulverized particles, and rewelding. They speak of a powder regime instead of a plastic regime for indentations that are very shallow. For indentations that are small compared with the spacing of structural defects, ordinary plastic flow involving dislocations is less probable than for large indentations.

Microfracture followed by gross slip and rewelding is involved in metal cutting chip formation (Shaw, 1980, 1993). If fracture were not involved in chip formation the shear stress on the shear plane would be independent of normal stress, in accordance with the Von Mises flow critierion. However, shear stress on the shear plane in cutting is a function of normal stress, as the experimental results of Merchant (1945) clearly show. Fracture in metal cutting chip formation and in micro-indentation are of different origins. In micro-indentation, fracture is due to lack of dislocations or defects, giving rise to dislocations based on the small volume involved. In metal cutting, fracture is due to the excessive strains involved in chip formation.

While an indentation theory based on pulverization has not been developed, it appears that the elastic–plastic approach should provide a reasonable approximation relative to the size of the deformation zone involved. A perfectly brittle material will remain elastic until fracture occurs and hence the extent of the elastic zone should correspond to the extent of the pulverized zone. However, the magnitude of the indentation force is not expected to be well approximated by the elastic-plastic theory of hardness, since frictional resistance of one particle sliding over its neighbor is not included in the elastic–plastic theory of indentation.

The Specal Situation for Glass

The median cracks that are formed in UPDG of glass may be explained by Fig. 15.28. Here a spherical indenter is shown, with its elastic-plastic boundary. It is convenient to characterize the stress concentration responsible for brittle crack initiation as a circular void. In the case of glass this might correspond to one of the 'holes' involved in the classical molecular theory of viscosity of an amorphous liquid. The intensified stress at the top and bottom of a circular void in a uniaxial compressive field will have the value of the nominal compressive stress with reversed sign, as shown in Fig. 15.28(b). Since glass is a perfectly brittle material (fracture stress < yield stress), then a brittle median crack may initiate at the bottom of a circular void located at the plastic–elastic boundary when the plastic zone has become sufficiently large to encounter what approximates a circular void in the more or less random amorphous structure of glass. The mean hydostatic stress at the bottom of the elastic–plastic boundary will be a minimum, inducing brittle fracture at this point.

References

Aronson, R. B. (1994). *Manufactoring Engng* **113**, 63.
Bifano, T., Yuan, Y., and Kahl, W. K. (1994). *Prec. Engng* **16**, 106.
Blake, P. N. (1988). Ph.D. dissertation, North Cardina State University, Raleigh, North Cardina.
Donaldson, R. R., Syn, C. K., Taylor, J. S., Ikawa, N., and Shimada, S. (1987). *UCRL 97606*.
Fujita, S. and Shibata, J. (1993). *J. Japan Soc. Prec. Engrs.* **27**, 138.
Hashimoto, H., Takeda, J., Imai, K., and Blaedell, K. (1993). *Int. J. Japan Soc. Prec. Engrs* **27**, 95.
Herring, C. and Galt, J. (1952). *Phys. Rev.* **85**, 1060.
Hill, R. (1950). *The mathematical theory of plasticity*. Clarendon Press, Oxford.
Ikawa, N., Shimada, S., and Morooka, H. (1987). *Bull. Japan Soc. Prec. Engrs* **21**, 233.
Ikawa, N., Donaldson, R. R., Komanduri, R., Koenig, W., McKeon, P. A., and Moriwaki, T. (1991). *Ann. CIRP* **40**(2), 587.
Komanduri, R., Schroeder, T. H., Hayra, J., von Turkovich, B. F., and Flom, D. G. (1982). *J. Engng Ind.*, **104**, 121.
Lawn, B. and Wilshaw, R. (1975). *J. Mat. Sci.* **10**, 1049.
Liang, Y., Moronuki, N., and Furukawa, Y. (1994). *Prec. Engng* **16**, 132.
Lucca, D. A. and Seo, Y. W. (1994). *Ann. CIRP* **43**(1), 43.
Lucca, D. A., Seo, Y. W., and Rhorer, R. L. (1993). *Proc. NSF Design and Mfg. Conf.* **1**, 225. (Pub. by SME.)
Mayer, J. E. and Fang, G. F. (1994). *Ann. CIRP* **43**(1), 309.
Merchant, M. E. (1945). *J. appl. Phys.* **16**, 267, 318.
Moronuki, N., Liang, Y., and Furukawa, Y. (1994). *Prec. Engng* **16**, 124.
Murata, R., Okano, K., and Tsutsumi, C. (1985). *ASME PED* **16**, 261.
Namba, Y. and Abe, M. (1993). *Ann. CIRP* **42**(1), 417.
Nakayama, K. (1974). *Proc. Int. Conf. Prod. Engng, Tokyo*, p. 572.
Nakayama, K., Takagi, J. and Nakano, T. (1974). *Ann. CIRP* **23**(1), 89.
Ohmori, H. (1992). *Int. J. Japan Soc. Prec. Engrs* **26**, 273.
Ota, M. and Mayahara, K. (1990). *4th SME Grinding Conf., MR 90 538*.

Prandl, L. (1920). *Nach der Akad-Wissenschaften in Gottingen*, Math.-Phys. K., 74.
Reichenbach, G. S., Mayer, J. E., Kalpakjian, S., and Shaw, M. C. (1956). *Trans. ASME* **78**, 847.
Shaw, M. C. and DeSalvo, G. J. (1970). *Trans. ASME* **92**, 480.
Shaw, M. C. (1972). *Mech. Chem. Engng Trans., Inst. Engrs (Australia)* **MC8**, 73.
Shaw, M. C. (1980). *Int. J. Mech. Sci.* **22**, 673.
Shaw, M. C. (1993). *Appl. Mech. Rev.* **46**, 74.
Shaw, M. C. and Vyas, A. (1993). *Ann. CIRP* **42**(1), 29.
Takahasahi, M., Ueda, S., and Kurobe, T. (1993). *Japan Soc. Prec. Engng* **27**, 140.
Taniguchi, N. (1992). *Int. J. Japan Soc. Prec. Engrs* **27**, 14.
Ueda, K. and Manabe, K. (1992). *Ann. CIRP* **41**(1), 129.
Yoshido, S. and Ito, H. (1990). *Bull. Japan Soc. Prec. Engrs* **24**, 239.
Yoshioka, J., Hashimoto, F., Miyashita, M., Kanai, A., Abo, T., and Daito, M. (1985). *ASME PED 16*, 209.
Zang, B. and Howes, T. (1994). *Ann. CIRP* **43**(1), 305.

AUTHOR INDEX

Abbay, P. A. 376
Abe, M. 548–9
Abe, T. 153
Abo, T. 535, 548
Abrahamson, G. R. 359, 375
Acheson, E. G. 15
Adams, J. T. 499
Adams, P. B. 523
Aghan, R. L. 461
Agrawal, K. 457
Ahmed, O. I. 325–8
Akeson, A. 495
Alden, G. 91
Alman, J. O. 298
Althaus, P. 309
Alvi, S. S. 239–40
Amer. Nat. Stds. Inst. (ANSI) 13
Andrew, C. 151, 259, 392, 491
Antenen, A. P. 471–2
Anzai, M. 458–60
Appell, B. 187
Appun, J. 5, 147, 343
Archer, J. F. 461
Aronson, R. B. 535
Atkins, N. 6, 178
Avery, J. P. 178

Babu, N. R. 149
Backer, W. R. 82–93, 101, 104, 109, 112
Baird, E. D. 467–8
Banerjee, J. K. 161
Barash, M. 373–4, 416–17
Barber, G. C. 359
Barnard, J. M. 495
Barin, I. 330
Basuray, P. K. 280
Batiashvili, B. 454–5
Bauer, R. 17
Becker, K. 432–5
Beilby, G. T. 461
Bhushan, B. 244, 518–20
Bielawski, E. J. 56
Bifano, T. 552
Birle, J. D. 159, 474–7
Blaidell, K. 549–50
Blake, P. N. 545–8
Blok, H. 224
Bokuchava, G. V. 28, 55, 73

Bond, F. 499
Borkowski, J. 41
Bose, A. 360
Bouas, D. 498
Bovenkirk, H. P. 24
Brach, K. 478–4
Brauninger, G. 471–2
Brecker, J. N. 32, 44–7, 53, 67–8, 82–3, 98–100, 127–33, 153, 192–9, 264–9
Bridgman, P. W. 22, 184–5
Brinksmeier, E. 286, 301, 395
Britton, M. G. 523
Brown, D. 10, 286
Brown, R. H. 65, 126–7, 134–7, 251
Brueckner, D. K. 404
Bundy, F. P. 22, 25
Burnham, C. 445
Butskhrididze, D. S. 454–5

Cadwell, D. E. 358
Cadwell, E. J. 358–9
Cai, G. D. 506
Cammet, E. 208, 286, 308
Chalmers, B. 219
Chapman, W. H. 91
Chandrasekar, S. 185–7, 244–6, 309, 366–73, 518–20
Chandhri, M. M. 309
Chasmar, R. P. 244
Chattopadhyay, A. B. 360
Chattopadhyay, A. K. 360, 495–6
Cherapanov, G. P. 185
Chvorinov, N. 219
Cichy, P. 17
Coes, L. 17, 21, 55, 58, 232–3
Coffin, L. F. 17
Colding, B. 104
Colwell, L. V. 113–15, 303–9
Cook, L. M. 462
Cook, N. H. 95
Cosmano, R. J. 359, 375
Cottringer, T. 17
Courtel, R. 338
Cozminca, M. 318
Crecraft, D. I. 300
Crisp, J. 82, 98
Cullity, B. D. 303
Cuntze, E. O. 284

Daimon, M. 146
Daito, M. 535, 548
Dall, A. H. 415
Danlos, H. H. 418, 506
Dawihl, G. J. 55
Decneut, A. 104, 120, 401–3
Degnan, W. 300
DeSalvo, G. I. 84, 89, 559–60
Des Ruisseaux, N. R. 227, 259
DiLullo, T. D. 286
Dimberg, A. P. 453
Dodd, H. D. 308
Doeblin, E. O. 244
Donaldson, R. R. 539
Dugdale, D. S. 325–8
Dulis, E. J. 331
Dunkley, J. J. 321
Duwell, E. 57–8, 339–40, 358–9, 375, 429–31

Eda, H. 293–4
Evans, J. C. 463

Fang, G. F. 558
Farber, E. 393
Farmer, D. 167, 170–1, 191–5, 201–12, 264–9, 485
Farris, T. N. 244–6, 518–20
Federation of European Producers of Abrasive Products (FEPA) 13
Field, M. 261, 286, 290–4, 303
Finnie, I. 528–9
Fix, R. M. 300
Flom, D. G. 555
Fox, M. 455–7
Frost, M. 416
Fujisawa, M. 509
Fujita, S. 543
Fuller, E. R. 309
Funck, A. 455
Fursdon, P. M. T. 416
Furukawa, Y. 390, 416, 485–90, 542–4

Gagliardi, J. J. 359, 429–31
Gall, D. A. 178
Galileo, G. 221
Galt, J. 464–5, 535
Gatto, L. R. 286
Geopfert, G. J. 56, 68, 328
Gill, R. 418
Gilmore, R. 533
Giovanola, J. H. 528–9
Glucklilck, J. 41
Grdzelshivili, G. L. 454–5
Greenwood, J. A. 127

Griffith, A. 185–6
Grisbrook, H. 65, 95, 404
Grof, H. E. 155
Gruehring, K. 344
Gruen, F. J. 153
Gu, D. Y. 127
Guest, J. J. 91
Gurney, J. P. 416

Hahn, R. S. 6, 12, 126, 151, 216, 224, 281, 286, 381–5, 408, 483
Hall, H. T. 22, 25
Hashimoto, F. 390, 418, 535, 543, 548–50
Hashish, M. 445–8
Hauk, V. 303
Haultain, H. E. T. 497–9
Hawkins, A. C. 471–2
Hayra, J. 555
Hebbar, R. 245–7
Heidenfelder, H. 409
Heinz, W. B. 91
Herring, C. 464–5, 535
Hill, R. 559
Hillier, M. J. 161
Hinterman, H. E. 495–6
Hitchcock, A. L. 444–5
Honcia, G. 127
Hong, I. S. 339–40, 358–9, 429–31
Hosokawa, A. 241, 444
Howes, T. D. 151, 259–60, 491, 562
Hutchinson, R. V. 79, 80–2, 91
Hyler, S. 301–2

Ikawa, N. 73, 79, 80–2, 463–5, 537–42
Imai, K. 549–50
Imanaka, O. 458–60
Inasaki, I. 235, 504–11
Inoue, H. 512–13
Ishakawa, T. 146
Ito, H. 549
Ito, S. 473

Jacobs, C. F. 16
Jacobs, U. 127
Jaeger, J. C. 217–18
Jen, T. C. 259
Jessops, R. S. 22
Jimbo, Y. 532
Johnson, G. 300, 471–2
Johnson, J. F. 499
Jones, P. E. 244

AUTHOR INDEX

Kaczmarek, J. 533
Kahl, W. K. 552
Kahles, J. F. 261, 286, 290-4, 303
Kalpakjian, S. 91, 100, 104
Kanai, A. 390, 418, 535, 548
Kasak, A. 331
Kawamura, S. 134
Kawashima, E. 458
Kayaba, N. 509
King, R. I. 12, 151, 382, 385, 408
Kingery, W. D. 131
Kino, G. S. 300
Kitajima, K. 506
Kitano, M. 376
Kiyoshi, K. 360-1
Knocke, O. 330
Knowles, J. E. 301
Koenig, W. 84, 286, 308, 377, 486, 539
Koenigsberger, F. 416-17
Koistenen, D. P. 302
Komanduri, R. 19, 27, 32, 71-3, 79, 96, 134, 155, 178, 330-6, 419, 457, 467-8, 539, 555
Kopalinski, E. M. 259
Kops, L. 244-5, 533
Koster, W. P. 304, 306
Krauss, G. 287
Krug, C. 91, 127
Kumagai, N. 360-1, 506
Kumar, K. V. 27, 214, 287, 308, 318, 362-5
Kurobe, T. 458, 544-5
Kuroda, H. 518
Kwong, J. M. S. 499

Lal, G. K. 62, 64, 75-8, 93, 280, 444
Lane, R. O. 113-18, 151
Lardner, D. L. 182
Lauer-Schmaltz, H. 377
Lauterline, T. 300
Lavine, A. S. 259
Lawn, B. 309, 502-3, 560
Lee, O. G. 227
Leithauser, M. C. 17
Leskovar, P. 286, 308
Letner, H. R. 301-2
Liang, Y. 542-4
Lindbeck, B. 452
Lindenbeck, D. A. 308
Lindsay, R. 216, 224, 381
Littman, W. E. 236, 286
Liverton, J. W. 395
Loladze, T. N. 28, 55, 73, 454-5
Lonergan, J. R. 523
Lucca, D. A. 536-41
Ludewig, T. 84

Luikov, L. V. 219
Lundberg, C. O. 286

McClellan, G. W. 523
McClure, R. 465
McDonald, W. J. 57-8, 339-40, 358
Macherauch, E. 303
McKeon, P. A. 539
Macks, E. F. 464-5
Maksond, T. M. A. 259
Malkin, S. 12, 95, 155, 216, 224, 226, 230-1, 286, 359, 404
Mamulashvili, G. L. 454-5
Manabe, K. 541-2
Marburger, R. E. 301
Marinescu, I. D. 498
Maris, M. 5, 147, 259
Markevich, J. I. 454-5
Marshall, E. R. 82, 87, 91, 93, 101, 104, 109, 112
Martin, G. 497-9
Matsui, S. 426-8
Matsuo, T. 64-5, 195-8
Matsushima, K. 458
Mayahara, K. 559-61
Mayer, J. E. 91, 100, 104, 239-41, 558
Merchant, M. E. 2, 89, 553, 563
Mereness, E. 358
Metcut Research Associates 449-50
Minke, H. 395
Mittal, B. 359
Miyahara, K. 509
Miyake, S. 518
Miyashita, M. 393, 416, 418, 535, 548
Mochida, M. 390
Moehlen, H. 506
Montgomery, D. C. 264
Moran, H. 65
Morgan, M. W. 127
Mori, Y. 463-5
Moriwaki, T. 539
Moronuki, N. 542-4
Morooka, H. 537
Moyan, S. 298
Murata, R. 6, 180, 473, 513-16, 551
Murugesan, S. 525-7

Naguchi, H. 512-13
Nakagawa, T. 458-60, 511
Nakamoto, T. 376
Nakano, T. 529, 556-8
Nakayama, K. 82, 97, 127-33, 153, 170-1, 178, 201, 272-85, 485, 529, 555-6
Namba, Y. 427-9, 465, 548-9
Naoyuki, S. 185

Navarro, N. P. 33
Neppiras, E. A. 531
Neumeyer, T. A. 331
Nishimura, Y. 532
Nishiyama, Z. 287–9

Oertenblad, B. 187
Ohishi, S. 390, 485–90
Ohmori, H. 551
Ohmura, E. 293–4
Okano, K. 513–16, 551
Olsen, J. H. 443
Ono, K. 390
Opitz, H. 344, 407
Orioka, T. 96, 270–3
Oshima, E. 195–8
Ota, M. 509, 559–61
Otaki, H. 458–60
Outwater, J. 55, 233, 235–6, 337–8

Pahlitzsch, G. 147, 281–4, 343
Pai, D. M. 5, 157, 478–84, 491
Parks, R. E. 452
Patterson, H. B. 157, 159, 341
Paul, T. 235
Pauling, L. 348
Paulmann, R. 127
Pearce, T. R. A. 151, 491
Peets, W. J. 414
Pekelharing, A. J. 281–2
Peklenik, J. 11, 91, 96–7, 113–18, 126, 142, 237–43
Peters, J. 104, 120, 147, 235, 286, 308, 403
Phillips, R. E. 439
Piispanen, V. 2
Piret, E. L. 499
Pollock, C. 212
Poritsky, H. 139
Powell, J. W. 259, 392
Prandl, L. 559
Purcell, J. 161

Qi, H. S. 127

Rabnowicz, E. 461
Radhakrishnan, V. 147, 377, 417–18
Rahman, J. F. 147
Ramanath, S. 156–7, 223, 226, 299, 301, 310, 509
Ramaraj, T. C. 156–7
Ratterman, E. 157, 159, 474–7, 478–84, 491
Read, W. T. 301
Reeka, D. 416

Reichenbach, G. S. 91, 100–4, 232–3
Rhoades, L. J. 451
Rhorer, R. L. 536–41
Ridgway, R. R. 16, 35
Riefenstall, J. 343
Rossini, F. P. 22
Rowe, W. B. 127, 416–18

Sachs, G. 300
Safto, G. R. 392
Sahay, B. V. 280
Saini, D. P. 131, 251
Saito, K. 126–7, 134–7
Salje, E. 127, 343, 409, 413, 506
Salmon, S. C. 12
Samuel, R. 310–11
Samuels, L. E. 461
Sata, T. 44
Sato, K. 44, 89, 226, 268, 282, 496
Sato, Y. 185
Sauer, W. 139–42, 155, 178, 203–4, 259, 426
Saunders, L. E. 16
Schey, J. 111
Schleich, H. 486
Schmitt, R. 148
Schonert, K. 244
Schroeder, T. H. 555
Schwartz, K. E. 127
Scorah, R. L. 393
Scrutton, R. F. 64
Seidel, J. R. 82, 98
Seiforth, M. 528
Seo, Y. W. 536–41
Shafto, G. R. 259
Shen, C. H. 282
Shephard, J. H. 65
Shewmon, P. G. 289
Shibata, J. 543
Shimada, S. 537
Shimamune, T. 390
Shimeno, K. 460
Shinmura, T. 457
Shiozaki, S. 390, 416, 485–90
Shunmugam, M. S. 417–18
Sidhwa, A. P. 131
Smith, R. A. 244
Snoeys, R. 104, 119, 120, 127, 147, 259, 286, 401–3
Soderlung, K. N. 113, 115
Somiya, S. 503
Sonoda, S. 195–8
Sowman, H. G. 17
Spreggett, S. 418
Spur, G. 432–5
Steffens, K. 84
Steijn, R. P. 57

Stelson, T. S.   178, 419
Steven, J.   331
Stokey, W. F.   82, 98
Stowers, I. F.   467–8
Strong, H. M.   25
Subramanian, S.   509
Sugito, T.   519–20
Sugiyama, K.   463–5
Suits, C. G.   22
Suto, T.   458–60, 512–13
Suzuki, K.   376, 496, 511
Syn, C. K.   539
Syoji, K.   426–8
Szymanski, A.   41

Tabor, D.   89
Takagi, J.   44, 153, 186, 529, 556–8
Takahashi, M.   544–5
Takazawa, K.   236–9
Takeda, J.   549–50
Takenaka, N.   65
Tanaka, T.   73, 79, 80–2
Tanaka, Y.   134, 318, 339, 506
Taniguchi, N.   535
Tarasov, L. P.   32, 155, 203–6, 286, 325–8, 346
Teiwes, H.   409
Toenshoff, H. K.   235, 286, 308
Tone, F.   16
Treuting, R. G.   301
Tripp, J. H.   127
Tsutsumi, C.   513–16, 551
Tsuwa, H.   134, 148–50, 427–9, 463–5

Udupa, N. G.   417–18
Ueda, K.   519–20, 541–2
Ueda, S.   544–5
Ueda, T.   241–4, 374, 386
Ueguchi, T.   339
Ueltz, H. F. G.   17
Uematsu, T.   511
Ueno, Y.   518
Uhlmann, D. R.   495
Umehara, N.   458–9

van der Merwe, R. H.   17
Van der Sande, J. V.   495
van Lutterwelt, C. A.   142–4
Van Vlack, J.   292

Verkerk, J.   127, 146–8, 153, 281–2
Verma, A. P.   444
Visser, R. G.   453
von Turkovich, B. F.   555
Vyas, A.   363, 553

Wager, J. G.   127, 251
Waida, T.   512–13
Walker, D. R.   497–591
Wang, I. C.   127
Warnecke, G.   153
Watanabe, J.   518
Watson, D.   127
Waugh, A.   131
Weichert, R.   244
Weisbecker, H. L.   358
Wells, A. F.   348
Wentdorf, R. H.   25–6
Westbrook, J. M.   308
Wetton, A. G.   156
White, R. H.   16
Williams, J. L.   56, 68, 328
Wilshaw, R.   502–3, 560
Woxen, R.   104–5
Wulff, J.   236

Yamada, K.   519–20
Yamaguchi, K.   376
Yamamoto, A.   386
Yamauchi, S.   293–4
Yamoto, A.   241–4, 374
Yang, C. T.   32–6, 269, 325–6, 346–50
Yasui, H.   148–50
Yokogawa, K.   494
Yokowaga, M.   494
Yokoyama, T.   496
Yoshido, S.   549
Yoshikawa, H.   44
Yoshioka, J.   535, 548
Yuan, Y.   552
Yule, A. J.   321

Zang, B.   562
Zerkle, R. D.   227
Zheng, H. W.   127, 506
Zhou, Q. E.   397
Zhou, Z. X.   142–4
Zorn, F. J.   456

SUBJECT INDEX

abrasive cut-off 164–88
  economics 166–9
  machine considerations 177–80
  mechanics 169–74
  optimization 178–9
  oscillation and rotation 179–81
  other methods 181–7
  performance 171–7
  power 171, 180–1
  specific energy 172–3
  undeformed chip thickness 170–1
  wheel speed 174–5
Abrasive Grain Association 6
abrasives
  principal types 15–17, 22–27, 34
  surface morphology 17–22, 27–34
atomized tool steel 331–6
Aujer electron spectroscopy (AES) 329
axial feed 160–1, 413–2

basalt 497
belt grinding 424–36
  contact wheel 425–7
  finish 430–1
  future possibilities 435–6
  life 427–30
  special operations 431–5
bench tests 80
brazed grits 495
buffing 461

calorimetric tests (SRG) 25
centerless grinding 414–24
  applications 415
  machines 418
  mechanics 415–18
  optimization 419–24
ceramics 500–22
chatter 283–6
chip formation 154–60
  cutting 1–3
  ductile vs brittle work 157–60
  FFG 4
  spherical chips 155–6
  SRG 5
chip storage *see* protrusion
coated products 40–1

comminution 497–500
concrete *see* rock
conditioning *see* dressing
conditioning of steel 189–200
  abrasive type 195–7
  economics 192–5
  high speed 198–9
  mechanics 190–1
  size effect 198
  wear 191–2
conformity 102
constant force grinding 382–5
contact length, wheel/work 102
continuous electrolytic dressing 661–2
coolant 226–44
cracking off 184–8
creep feed grinding 386
  cooling 391–3
  cubic boron nitride 394
  cylindrical grinding 395–9
  high efficiency deep grinding (HEDG) 395
  large depth of cut 483–93
  low/wheel speed 390
  machines 393
  surface finish 393–4
  temperature 388–90
  up versus down grinding 393
cutting
  chip types 3, 552–8
  overview 1–3
  specific energy 3
cylindrical grinding 400–14
  axial force 413–14
  chatter 408–9
  CIRP study 401–3
  rake angle effect 90
  speed ratio 409–13
  wear 403–7
  wheel life 408–9

dalmation wheel 151
deburring 449–52
  abrasive flow machining (AFM) 451–2
  brushing 450
  grinding 450
deflection 125–44, 324–5
depth of cut, large 478–91